CARL C. COWEN
Purdue University

BARBARA D. MACCLUER
University of Richmond and University of Virginia

Composition Operators on Spaces of Analytic Functions

CRC PRESS

Boca Raton New York London Tokyo

Library of Congress Cataloging-in-Publication Data

Catalog record is available from the Library of Congress

No claim to original U.S. Government works
International Standard Book Number 0-8493-8492-3
Printed in the United States of America 1 2 3 4 5 6 7 8 9 0
Printed on acid-free paper

Studies in Advanced Mathematics

Contents

Preface

The study of composition operators lies at the interface of analytic function theory and operator theory. As a part of operator theory, research on composition with a fixed function acting on a space of analytic functions is of fairly recent origin, dating back to work of E. Nordgren in the mid 1960's. The first explicit reference to composition operators in the Mathematics Subject Classification Index appeared in 1990. As a glance at the bibliography will show, over the intervening years the literature has grown to a point where it would be difficult for a novice to read all of the papers in the subject. At the same time, there are themes developing so that it is possible to see important groups of papers as exploring the same theme. This book is an attempt to synthesize the achievements in the area so that those who wish to learn about it can get an overview of the field as it exists today. At the same time, we hope to bring into clearer focus the themes from the literature so it is easier to see the broad outlines of the developing theory. We have taken this opportunity to present, in addition to material that is well known to experts, some results that are appearing here for the first time. Many interesting and seemingly basic problems remain open and it is our hope that this book may point out areas in which further exploration is desirable and serve to entice others into thinking about some of these problems.

One of the attractive features of this subject is that the prerequisites are minimal. This book should be suitable for second or third year graduate students who have had basic one semester courses in real analysis, complex analysis, and functional analysis. We have included a large number of exercises with the student in mind. While these exercises vary in difficulty, they are all intended to be accessible; we have *not* used the exercises as a place to collect major results from the literature that space did not permit a discussion of in the text. Since the exercises both illustrate and extend the theory, we urge all readers, students and non-students alike, to consider the exercises as an integral part of the book. Rather than seeking the utmost generality, the theory is developed in a context that is comfortable and illustrates the nature of the general results.

In several places we consider composition operators acting on function spaces in the unit ball in C^N. Typically, our study of composition operators in the several variable setting is done in separate sections (or separate chapters) from the one variable theory. Exceptions to this only occur in places where the several variable reasoning is identical to that in one variable, and where the reader interested only in the latter situation would not find it burdensome to read the arguments with N set equal to 1. However, only very little from the extensive field of complex analysis in several variables is ever needed here, and we give a complete discussion of much of the several complex variable background that is central to our subject — the automorphisms of the ball and their fixed point properties, angular derivatives and the Julia–Carathéodory theory in B_N, and iteration properties of self-maps of the ball. Most of the theory of several complex variables that we use without proof can be found in the first 45 pages of W. Rudin's *Function Theory in the Unit Ball of C^n* [Ru80]. Statements of results used and relevant definitions are all included here, so that the reader unfamiliar with these results but willing to accept some of them without proof should find the several variable sections of the book readable. Indeed, one of our goals is to convince the reader that the unit ball of C^N is an interesting place to do function-theoretic operator theory because one can quickly get to phenomena that are not seen in the disk but which can nevertheless be handled with a minimum of technical machinery. In short, we believe this is an ideal place for a first excursion into several variable function theory.

This book is written from a philosophy that mathematics develops best from a base of well chosen examples and that its theorems describe and generalize what is true about the characteristic objects in a subject. This is a book about the concrete operator theory that arises when we study the operation of composition of analytic functions in the context of the classical spaces. In particular, we study the relationship between properties of C_φ and properties of the symbol map φ: the goal is to see the norm, the spectrum, normality, etc., of C_φ as consequences of particular geometric and analytic features of the function φ. The theory of multiplication operators, arising from the spectral theorem for normal operators, has developed and branched into the study of Toeplitz operators, subnormal operators and so on. We believe composition operators can similarly inform the development of operator theory because they are very diverse and occur naturally in a variety of problems. Composition operators have arisen in the study of commutants of multiplication operators and more general operators and play a role in the theory of dynamical systems. De Branges' original proof of the Bieberbach conjecture depended on composition operators (he called them substitution operators) on a space of analytic functions.

Ergodic transformations are sometimes thought of as inducing composition operators on L^p spaces, for example, but not on analytic spaces; except in the introductory material, we do not discuss the theory of composition operators on non-analytic spaces because that theory seems to be developing in rather different ways. In recent years there has been broad interest in complex dynamics, in the theory of iteration of rational functions in the plane, and so on. In general, this book will not touch on these studies; their emphasis is on the sets where the iteration is chaotic, whereas our study of composition operators will emphasize the regions in which the iteration is regular.

The introductory chapter, as its title implies, sets the stage for the remainder of the book by giving the basic definitions, proving a few theorems that hold in very great generality, and posing the basic questions that will be addressed. The second chapter defines the Hardy and Bergman spaces and their generalizations that we will be working in and develops the analytic tools that are not usually covered in basic graduate courses but are needed in the study of composition operators. Readers familiar with this material may wish to skim the chapter to pick up our notation and see what we consider the basic background for our study. The third chapter contains the core material on boundedness and compactness of composition operators and estimates for their norms. Many of these computations are based on estimates arising from Carleson type measure considerations. In general, the emphasis will be on the standard spaces of analytic functions, but in chapters four and five we discuss smaller and larger spaces of analytic functions and illustrate the differences between composition operators on these spaces and the standard spaces. While the majority of the theory develops in parallel ways in one and several variables, some more subtle phenomena specific to the study of compactness and boundedness questions in several variables are investigated in chapter six. In chapter seven, computation of the spectra of composition operators is described. This description is most complete in the case of compact operators. In the case of non-compact operators, the theory is more complete in the cases in which a weighted shift analogy can be used, and less complete when the weighted shift analogy fails. For compact and invertible operators, the spectral theory is developed in one and several variables, but for the more difficult cases, we consider the theory only on one variable Hardy space and its close relatives. It turns out that composition operators are rarely normal, subnormal, or hyponormal; some results concerning such phenomena are described in chapter eight. Chapter nine consists of several sections devoted to less developed parts of the theory, such as results on equivalences, the topological structure of the space of composition operators, and an application of composition operators to a problem in polynomial approximation. After chapter three, the chapters are largely independent of each other, although chapters four and five on small and large spaces are best appreciated as a package. The results of the last three chapters are largely restricted to Hilbert spaces and especially $H^2(D)$, while the earlier chapters include many results on Banach spaces.

While we have tried to summarize the existing literature on composition operators on spaces of analytic functions, constraints of time, space, and probably the reader's patience have prevented us from including every interesting topic. If your favorite topic has been left out, please accept our sincere regrets. To assist in the reader's further study, the bibliography attempts to be a comprehensive list of works on composition operators on analytic spaces. We apologize in advance to the authors of those papers that we have inadvertently omitted — it was not intentional! At the end of each section, we include historical notes on the origins of the results and exercises, relationships between papers, and references for further reading.

Both authors gratefully acknowledge their appreciation to the National Science Foundation for its support of their work on composition operators over many years. Additionally, the completion of this project was greatly facilitated by support from the NSF-VPW program and Purdue University which permitted the second named author to be in residence at Purdue during the 1994–95 year. We also wish to express our sincere appreciation to friends and colleagues who read portions of the manuscript and offered valuable advice for its improvement, including Randy Campbell–Wright, David Cruz–Uribe, Mirjana Jovovic, Peter Mercer, Sivaram Narayan, and Doug Szajda. We would also like to thank two of our students, Paul Hurst, who helped with some of the exercises, and Rebecca Wahl, who assisted in the preparation of the Bibliography. Special thanks go to Paul Bourdon for his insightful critique of early versions of many parts of the manuscript and to Thomas Kriete for his enthusiastic encouragement of this project from its inception, and for valuable suggestions for improvement throughout its development. While we are grateful to these colleagues for lending their expertise to improve this book, we acknowledge that any errors or shortcomings remaining are our responsibility.

We hope that others will find the reading of this book to be as stimulating as we have found the writing to be.

Carl C. Cowen
Barbara D. MacCluer
West Lafayette, Indiana
December 1994

1

Introduction

The classical Banach spaces are all realized by norming a collection of real or complex valued functions on a set X. If φ maps X into itself, it is natural to consider the ***composition operator*** C_φ defined by

$$\left(C_\varphi f\right)(x) = f(\varphi(x))$$

for x in X and functions f in the Banach space. Such operators are clearly linear. We will study other properties of these operators for various Banach spaces and φ's, principally Banach spaces of analytic functions and analytic maps φ, but always for spaces of complex valued functions. If, in addition, ψ is a complex valued function defined on X, the ***weighted composition operator*** $W_{\varphi,\psi}$ is defined by $\left(W_{\varphi,\psi} f\right)(x) = \psi(x) f(\varphi(x))$.

Although the substitution operation is basic to mathematics and studies involving composition of functions have been pursued for a very long time (for example, iteration of rational functions or ergodic theory), the study, as a part of operator theory, of linear operators induced by composition with fixed functions has a relatively short history. This is an interesting contrast to the much older study of linear operators induced by multiplication by fixed functions, which has its roots in the spectral theorem and is being pursued today in such guises as the theory of subnormal operators and the theory of Toeplitz operators.

Especially in view of the relative neglect of the study of composition operators, it is reasonable to point out some of the reasons for their study now. The pat answer to this question is that they are "natural objects" that are "interesting in themselves". While true, this answer satisfies no one not already convinced of their interest. A better answer, one that illuminates their naturality, is that they are surprisingly general and occur in settings other than the obvious ones. For example, a backward shift of any multiplicity can be represented as a composition operator. The Hilbert space $\ell^2(N)$ can be regarded as a space of complex valued functions on the set of non-negative integers N, and for φ defined on N by $\varphi(n) = n + 1$ the composition operator C_φ on $\ell^2(N)$ is

$$(f(0), f(1), \ldots) \mapsto (f(1), f(2), \ldots)$$

which is a backward shift of multiplicity one. Forelli [Fo64] proved that all isometries of the Hardy spaces $H^p(D)$ for $p \neq 2$ are weighted composition operators. They arise naturally in studying other questions from operator theory. For a multiplication operator, $(M_h f)(x) = h(x)f(x)$, consider the question "Which operators A satisfy $AM_h = M_h A$?" Clearly, other multiplication operators do. Less obvious is that if there is φ so that $h \circ \varphi = h$ then C_φ commutes with M_h:

$$C_\varphi M_h f = (h \circ \varphi)(f \circ \varphi) = h(f \circ \varphi) = M_h C_\varphi f$$

Although there need not always be non-trivial φ that satisfy this identity, if h is a reasonably nice bounded analytic function, then the commutant of M_h, at least on $H^2(D)$, is generated by the multiplication operators and the composition operators that arise in this way [Co78, Co80a, Co80b]. The operator point of view is beginning to draw interest from other areas of mathematics, for example, dynamical systems.

Research on composition operators is part of what can be called "concrete" operator theory. Mathematics advances principally when there is a well understood class of examples on which to base a general theory. In operator theory, the normal operators are the only class of operators on infinite dimensional spaces that can be said to be well understood at this time, although great strides have been made in the past decade in the understanding of subnormal operators and operators whose spectrum is a spectral set. Composition operators generally do not fit into these classes. They form a collection of examples of operators that, we hope, can be understood and in so doing can lead to a better understanding of properties of general operators.

The goal of our study of composition operators will be to answer (or at least try to answer) the questions operator theory demands we ask, questions of invertibility, boundedness, spectrum, membership in special classes, and so forth. We raise several such questions now, but in raising them, we will see that we need to develop technical tools in the theory of Banach spaces of analytic functions before we can answer them. This development will constitute the second chapter of the book.

Although composition operators have been studied on a variety of spaces, the majority of the literature concerns spaces whose functions are analytic on some set and in which the vector and norm structures are closely connected to the analytic structure. Moreover, such spaces have played an important role in the development of analysis, and composition operators on these spaces can be studied without additional special hypotheses.

DEFINITION 1.1 *A Banach space of complex valued functions on a set X is called a **functional Banach space on** X if the vector operations are the pointwise operations, $f(x) = g(x)$ for each x in X implies $f = g$, $f(x) = f(y)$ for each function in the space implies $x = y$, and for each x in X, the linear functional $f \mapsto f(x)$ is continuous.*

A functional Banach space whose functions are analytic on the underlying set will usually be called a ***Banach space of analytic functions***. The principal subject of this book is composition operators on Hilbert spaces of analytic functions. The following result and its corollary illustrate the connection between the structures in a functional Banach space. Let K_x be the linear functional for evaluation at x, that is, $K_x(f) = f(x)$. For functional Hilbert spaces, the Riesz representation theorem implies that there is a function (which we will usually call K_x) in the Hilbert space that induces this linear functional: $f(x) = \langle f, K_x \rangle$. In this case, the functions K_x are called the ***reproducing kernels*** and the functional Hilbert space is also called a ***reproducing kernel Hilbert space***. Theorem 2.10 and its corollaries illustrate these kernels in several typical cases.

PROPOSITION 1.2
If a sequence in a functional Banach space converges weakly to a function f, then the sequence is bounded and converges pointwise to f. Conversely, if the span of the point evaluation linear functionals is dense in the dual of a functional Banach space, then a bounded sequence that converges pointwise to f also converges weakly to f.

PROOF The sufficiency of the condition follows from the uniform boundedness principle and the fact that point evaluation is a continuous linear functional.

To prove the converse, suppose f_n is a sequence such that $\|f_n\| \leq M$ and $f_n(x) \to f(x)$ for all x in X. If ℓ is a continuous linear functional, then there is a finite linear combination of point evaluation linear functionals close to ℓ and since

$$|\ell(f_n) - \ell(f)| \leq$$

$$\left| \left(\ell - \sum_j \alpha_j K_{x_j} \right)(f_n) \right| + \sum_j |\alpha_j| |f_n(x_j) - f(x_j)| + \left| \left(\ell - \sum_j \alpha_j K_{x_j} \right)(f) \right|$$

$$\leq \left\| \ell - \sum_j \alpha_j K_{x_j} \right\| M + \sum_j |\alpha_j| |f_n(x_j) - f(x_j)| + \left\| \ell - \sum_j \alpha_j K_{x_j} \right\| \|f\|$$

the sequence converges weakly. ∎

COROLLARY 1.3
A sequence in a reflexive functional Banach space converges weakly if and only if it is bounded and converges pointwise.

PROOF We only need prove the density condition of Proposition 1.2 holds for a reflexive functional Banach space $\mathcal{Y}$. If the span of the point evaluation linear functionals were not dense in the dual space $\mathcal{Y}'$, we could find a non-zero linear functional in $\mathcal{Y}''$ that has value 0 at each K_x. Since $\mathcal{Y}$ is reflexive, this means that there is g in $\mathcal{Y}$ so that $0 = K_x(g) = g(x)$ for each x in X but $g \neq 0$, which is impossible. ∎

Composition operators on functional Banach spaces, like multiplication operators, can be characterized by considering the set of point evaluation linear functionals. The formula for the action of the adjoint of a composition operator on the point evaluation linear functionals that appears in this theorem will be at the heart of much of our study.

THEOREM 1.4
If A is a bounded linear operator mapping a functional Banach space into itself, then A is a composition operator if and only if the set $\{K_x : x \in X\}$ is invariant under A^*. In this case, $A = C_\varphi$ where φ and A are related by $A^*(K_x) = K_{\varphi(x)}$.

PROOF   If $A = C_\varphi$, then for each function f

$$\left(A^*(K_x)\right)(f) = K_x(Af) = K_x(f \circ \varphi) = f(\varphi(x)) = K_{\varphi(x)}(f)$$

so $A^*(K_x) = K_{\varphi(x)}$, and the set of point evaluation linear functionals is invariant under A^*.

Conversely, if the set of point evaluation linear functionals is invariant under A^*, then define the map φ on X by $A^*(K_x) = K_{\varphi(x)}$. (This defines φ since the vectors in a functional Banach space separate the points of X.) Then

$$(Af)(x) = K_x(Af) = \left(A^*(K_x)\right)(f) = K_{\varphi(x)}(f) = f(\varphi(x))$$

so $A = C_\varphi$. ∎

This characterization can be used to study the invertibility of composition operators; indeed, it is apparent that the inverse of C_φ ought to be the composition operator whose symbol is φ^{-1}. However, as we shall see, to draw this conclusion, we need further assumptions on the functional Banach space. The following definition isolates a property that can be used for the characterization, namely, that partially multiplicative linear functionals all come from point evaluations. Thus, we are making an assumption about a connection between the algebraic structure of the Banach space and its description as a space of functions. In the classical functional Banach spaces, this consistency is implicit in the definition of the space and is not usually consciously noted (see Theorem 2.15 and Exercise 2.1.15).

DEFINITION 1.5   *Suppose $\mathcal{Y}$ is a functional Banach space on X. We call $\mathcal{Y}$* ***algebraically consistent*** *if for each non-zero, bounded linear functional k on $\mathcal{Y}$ such that $k(fg) = k(f)k(g)$ whenever f, g, and fg are in $\mathcal{Y}$, there is a point y in X so that $k = K_y$.*

As the proof below shows, what is needed for the application of Theorem 1.4 to the invertiblility problem is a way to show that all the point evaluation linear functionals that ought to be included in the space are there, that is, that the underlying set includes all the points it should. The multiplicativity of the point evaluations is a natural way to pick them out. The majority of the spaces we will consider in this book are Banach spaces of functions analytic on the open unit disk or the open unit ball and the hypotheses of the following theorem are satisfied.

THEOREM 1.6
Let $\mathcal{Y}$ be an algebraically consistent functional Banach space on the set X. Suppose that the interior of X is connected and dense in X and that the functions in $\mathcal{Y}$ are analytic on the interior of X and continuous on X. Suppose φ is a map of X into itself that is analytic on the interior of X and C_φ is bounded on $\mathcal{Y}$. If C_φ is invertible on $\mathcal{Y}$, then φ is a one-to-one map of X onto itself and $C_\varphi{}^{-1} = C_{\varphi^{-1}}$. Conversely, if φ is a one-to-one map of X onto itself and $C_{\varphi^{-1}}$ is a bounded composition operator, then C_φ is invertible and $C_\varphi{}^{-1} = C_{\varphi^{-1}}$.

PROOF Let C_φ be invertible. If x and y are points of X so that $\varphi(x) = \varphi(y)$, then $C_\varphi^* K_x = K_{\varphi(x)} = K_{\varphi(y)} = C_\varphi^* K_y$. Since C_φ^* is invertible, $K_x = K_y$ and $x = y$. That is, if C_φ is invertible, φ is one-to-one.

To show that φ is an onto mapping, suppose x is a point of X. Since C_φ^* is invertible, there is a linear functional k so that $C_\varphi^*(k) = K_x$. If $K_x = 0$, then the invertibility of C_φ^* implies $k = 0 = K_x$. This means $K_{\varphi(x)} = C_\varphi^*(K_x) = 0 = K_x$, so $\varphi(x) = x$ and x is in the range of φ. If $K_x \neq 0$, suppose f, g, and fg are in $\mathcal{Y}$. Then, since C_φ is invertible, there are F, G and H in $\mathcal{Y}$ so that $C_\varphi(F) = f$, $C_\varphi(G) = g$, and $C_\varphi(H) = fg$. Now

$$F(\varphi(y))G(\varphi(y)) = f(y)g(y) = (fg)(y) = H(\varphi(y))$$

Since φ is univalent, φ is an open map and the image of the interior of X is an open set on which FG and H agree which means that FG and H agree on the interior of X. Since the interior of X is dense in X and since F, G, and H are continuous on X, we see $FG = H$. Therefore,

$$k(fg) = k(C_\varphi(H)) = C_\varphi^*(k)(H) = H(x) = F(x)G(x)$$
$$= C_\varphi^*(k)(F)C_\varphi^*(k)(G) = k(C_\varphi(F))k(C_\varphi(G)) = k(f)k(g)$$

That is, k is a non-zero linear functional on $\mathcal{Y}$ that is partially multiplicative, so there is y so that $k = K_y$ for some y in X. In particular, $K_x = C_\varphi^*(K_y) = K_{\varphi(y)}$ which means $x = \varphi(y)$. Since x was arbitrary, φ maps X onto itself! Finally, since φ maps X one-to-one, onto itself, the equality $K_x = C_\varphi^*(K_y)$ can be rewritten as

$$(C_\varphi{}^{-1})^*(K_x) = (C_\varphi^*)^{-1}(K_x) = K_{\varphi^{-1}(x)}$$

which means, by Theorem 1.4, that $C_\varphi{}^{-1} = C_{\varphi^{-1}}$.
The converse is clear. ∎

Exercises 2.1.14 and 3.1.6 give an example of a Hilbert space of analytic functions that is not algebraically consistent and an invertible "composition operator" whose inverse is not a "composition operator". The example illustrates the importance of the algebraic consistency that is built into classical spaces.

The most elementary issue concerns the boundedness of C_φ. By the closed graph theorem, for most spaces of interest (see Exercise 1.1.1), this is equivalent to the question "For which φ does f in the Banach space $\mathcal{Y}$ imply $f \circ \varphi$ is in $\mathcal{Y}$?" This question clearly depends on the Banach space and does not always have an easy answer; as we study each class of Banach spaces, we will be forced to carefully consider this issue. When X is a subset of the complex plane and the identity function $I(x) = x$ is in $\mathcal{Y}$, an affirmative answer implies that φ must be in $\mathcal{Y}$.

One of the difficulties in the study of composition operators is their lack of algebraic properties. Only rarely is a linear combination of composition operators another composition operator (see Exercise 1.1.10). However, if φ and ψ are both maps of X, then $C_\psi C_\varphi = C_{\varphi \circ \psi}$. To find the spectral radius of C_φ, we need to find $\lim_{n \to \infty} \|C_\varphi^n\|^{1/n}$. That is, from the preceding calculation, we see that we need to study the iterates of φ:

$$\varphi_1 = \varphi, \quad \varphi_2 = \varphi \circ \varphi, \quad \varphi_n = \varphi \circ \varphi_{n-1}$$

and find $\lim_{n \to \infty} \|C_{\varphi_n}\|^{1/n}$.

The easiest part of the spectrum to find should be the eigenvalues. If λ is an eigenvalue of C_φ, then there is a non-zero function f so that $\lambda f = C_\varphi f = f \circ \varphi$, which is Schroeder's functional equation. We see that we will need to consider solutions of this functional equation to find the point spectrum.

With this motivation, we begin our study of Banach spaces of analytic functions. Chapter 2 includes the basic facts from analysis that will be exploited throughout the remainder of the book.

Exercises

1.1.1 Suppose $\mathcal{Y}$ is a functional Banach space on the set X. Show that if φ is a map of X into X then $(C_\varphi f)(x) = f(\varphi(x))$ defines a closed operator on the domain

$$\{f \in \mathcal{Y} : f \circ \varphi \in \mathcal{Y}\}$$

1.1.2 An operator S on a Hilbert space $\mathcal{H}$ is called a **unilateral shift** [Hal82, pp. 44, 77] if it is an isometry and $\cap_{k=1}^\infty (S^k \mathcal{H}) = (0)$. The **multiplicity** of the shift is the dimension of $(S\mathcal{H})^\perp$. A **backward shift** is the adjoint of a unilateral shift.

  (a) Justify the assertion that a backward shift of any multiplicity can be represented as a composition operator. That is, for each integer k and for infinity, find a map φ defined on N so that the composition operator C_φ on $\ell^2(N)$ is a backward shift of the given multiplicity.

(b) Show that if $\varphi(x) = x + 1$, the operator C_φ^* on $L^2(0, \infty)$ is a unilateral shift of infinite multiplicity.

1.1.3 A multiplication operator on a functional Banach space is an operator defined by $(M_h f)(x) = h(x)f(x)$ for some complex valued function h on the set X.

    (a) Show that a bounded operator A is a multiplication operator if and only if the non-zero evaluation functionals K_x are eigenvectors for A^*.

    (b) Show that if M_h is a bounded operator then h is a bounded function on X.

1.1.4 Let X be a subset of the complex plane and let $\mathcal{Y}$ be a functional Banach space on X. Suppose the multiplication operator defined by $(Mf)(x) = xf(x)$ is bounded and that the polynomials are dense in $\mathcal{Y}$ in such a way that if f, g, and fg are in $\mathcal{Y}$, there are polynomials p_n and q_n so that $p_n \to f$, $q_n \to g$, and $p_n q_n \to fg$. Show that $\mathcal{Y}$ is algebraically consistent if and only if the only eigenvectors for M^* are multiples of the evaluation functionals K_x.

1.1.5 Suppose A is a bounded operator on the algebraically consistent functional Banach space $\mathcal{Y}$. Show that A is a composition operator if and only if $A(pq) = A(p)A(q)$ whenever p, q, and pq are in $\mathcal{Y}$.

1.1.6 Let $X = [0, 1]$, or more generally let X be a compact Hausdorff space, and let $\mathcal{Y} = C(X)$, the Banach space of continuous, complex valued functions on X with the supremum norm. Show that every continuous function φ of X into itself induces a bounded composition operator on $\mathcal{Y}$. For which φ is C_φ invertible?

1.1.7 Let φ be a measurable transformation of $[0, 1]$ into itself, that is, φ maps $[0, 1]$ into itself and for each measurable subset E, we suppose the set $\varphi^{-1}(E)$ is measurable. Let m be Lebesgue measure on $[0, 1]$ and let $m\varphi^{-1}$ be the measure defined by $m\varphi^{-1}(E) = m(\varphi^{-1}(E))$. Show that C_φ on $L^p(0, 1)$ is a bounded operator if and only if $m\varphi^{-1}$ is absolutely continuous with respect to m and the Radon–Nikodym derivative

$$\frac{d(m\varphi^{-1})}{dm}$$

is essentially bounded. Show that in this case,

$$\|C_\varphi\| = \left\| \frac{d(m\varphi^{-1})}{dm} \right\|_\infty^{1/p}$$

1.1.8 Let φ be a monotone function mapping $(0, \infty)$ into itself. Find sufficient conditions on φ so that C_φ is bounded on $L^2(0, \infty)$ and in this case, find $\|C_\varphi\|$.

1.1.9 (a) Let $\{b_j\}$ be a sequence of positive numbers, let $\mathcal{H}$ be the Hilbert space of complex sequences with norm

$$\|(x_0, x_1, \dots)\|^2 = \sum_0^\infty |x_j|^2 b_j$$

and let T be the backward shift

$$T(x_0, x_1, \dots) = (x_1, x_2, \dots)$$

Show that T is unitarily equivalent to a weighted composition operator on $\ell^2(N)$ by finding the mapping φ and the weight function ψ.

    (b) Let $\{w_j\}$ be a sequence of positive numbers and let S be the backward weighted shift on $\ell^2(N)$ (with orthonormal basis $\{e_j\}$) given by $Se_0 = 0$ and $S(e_{j+1}) = w_j e_j$ for $j > 0$. Renorm the complex sequences to get a Hilbert space $\mathcal{H}$ so that S is the composition operator C_φ for $\varphi(j) = j + 1$ on $\mathcal{H}$.

1.1.10 Let $\mathcal{Y}$ be a Banach space of functions analytic on the open unit disk that contains the polynomials and let K_x be the linear functional in $\mathcal{Y}'$ for evaluation of the functions in $\mathcal{Y}$ at the point x.

(a) Suppose

$$K_x = \sum_{j=1}^{n} \lambda_j K_{y_j}$$

where $\lambda_j \neq 0$ for each j and the y_j's are distinct points. Show that $n = 1$, $\lambda_1 = 1$, and $y_1 = x$.

(b) Suppose C_φ and C_{ψ_j} are bounded on $\mathcal{Y}$ such that

$$C_\varphi = \sum_{j=1}^{n} \lambda_j C_{\psi_j}$$

where the ψ_j's are distinct analytic maps of the disk into itself and $\lambda_j \neq 0$ for each j. Show that $n = 1$, $\lambda_1 = 1$, and $\psi_1 = \varphi$.

Notes

The notion of functional Banach space we use in this chapter seems to have originated with P. R. Halmos [Hal82, ShW70]. Variations of many of the results and exercises in this section can be found in E. A. Nordgren's survey [No78]. Theorem 1.6 and Exercise 1.1.5 in the case of the Hardy space are results in H. J. Schwartz's thesis [Scz69] but the treatment in the generality here appears to be new. Exercises 1.1.4 and 1.1.5 arise from studying the sorts of algebraic hypotheses that should be included in an appropriate setting for the study of composition operators.

2

Analysis Background

In this chapter we collect information from geometric function theory as well as material on spaces of analytic functions that provides the necessary background for our study of composition operators. We develop this material in the form we need later, so our presentation does not constitute complete or comprehensive coverage for some topics. You may prefer to skip or skim this material on the first reading and return to it later as needed.

2.1   A menagerie of spaces

The classical Banach spaces of analytic functions are derived from various L^p spaces. We begin by defining the Hardy spaces of the unit disk $D = \{z : |z| < 1\}$ in the complex plane. More complete discussions of the Hardy spaces may be found in books of Hoffman [Hof62], Duren [Dur70], Garnett [Ga81], or Koosis [Koo80]. For a function f analytic in the unit disk and $0 < r < 1$, define the dilate f_r by $f_r(e^{i\theta}) = f(re^{i\theta})$. The functions f_r are continuous for each r, so they are in $L^p(\partial D, d\theta/2\pi)$.

DEFINITION 2.1   *For $0 < p < \infty$ the **Hardy space** $H^p(D)$ is the set of functions analytic on the unit disk for which*

$$\sup_{0<r<1} \int_0^{2\pi} |f_r(e^{i\theta})|^p \frac{d\theta}{2\pi} < \infty$$

Denote the pth root of this supremum by $\|f\|_p$. The Hardy space $H^\infty(D)$ is the set of analytic functions that are bounded in D, with supremum norm $\|f\|_\infty$.

Application of Corollary 2.23 to the subharmonic function $|f|^p$ shows that the L^p- norms $\|f_r\|_p$ increase as r tends to 1 so the supremum in the definition of $H^p(D)$ above is actually a limit. For $p \geq 1$, $H^p(D)$ is a Banach space with norm $\|f\|_p$

and we will see shortly that it is a functional Banach space on D. For $0 < p < 1$, $H^p(D)$ is a non-locally-convex topological vector space and $d(f, g) = \|f - g\|_p^p$ is a complete metric for it. Many of the results in this book extend to the Hardy spaces with $p < 1$, although we will usually consider only the Banach spaces.

As temporary notation, for $p \geq 1$, let H_*^p be the subspace of $L^p(\partial D)$ consisting of those functions whose negative Fourier coefficients are all zero.

THEOREM 2.2

Suppose f is in $H^p(D)$ for some $p > 0$. Then for almost all θ

$$f^*(e^{i\theta}) = \lim_{r \to 1} f(re^{i\theta})$$

exists and the mapping $f \to f^*$ is an isometry of $H^p(D)$ to a closed subspace of $L^p(d\theta/2\pi)$. For $p \geq 1$ the boundary function f^* is in H_*^p. Moreover, the mapping $f \mapsto f^*$ is an isometric isomorphism of $H^p(D)$ onto H_*^p.

We will use the isomorphism from this theorem, which we will not prove, to abuse notation: we will write $f(e^{i\theta})$ for $f^*(e^{i\theta})$ except when confusion could arise and write $H^p(D)$ for either $H^p(D)$ or H_*^p when $p \geq 1$. For $p < \infty$ an equivalent way to define H_*^p would be to say that it is the closed linear span in $L^p(\partial D)$ of the set $1, e^{i\theta}, e^{2i\theta}, \ldots$. Since these functions are the boundary functions of the analytic functions $1, z, z^2, \ldots$, we frequently say $H^p(D)$ is the closure of the analytic polynomials in $L^p(\partial D)$. This holds even when $p < 1$.

As another temporary notation, let $\mathcal{H}^p$ ($0 < p < \infty$) denote the set of analytic functions in the disk such that $|f|^p$ has a harmonic majorant. For f in $\mathcal{H}^p$, let h_f denote the least harmonic majorant of $|f|^p$, that is, h_f is harmonic on the disk, $|f(z)|^p \leq h_f(z)$ for all z in D, and $h_f(z) \leq u(z)$ for all harmonic majorants u of $|f|^p$ (see, for example, Duren [Dur70, p. 28]). Fix a point a in D, and for each f in $\mathcal{H}^p$, let $\|f\|_a' = h_f(a)^{1/p}$. When $p \geq 1$, $\|\cdot\|_a'$ is a norm that makes $\mathcal{H}^p$ into a Banach space. By Harnack's inequality, the norms for $\mathcal{H}^p$ using different points are equivalent, that is, for a and b in the disk, there are positive constants c_{ab} and C_{ab} so that

$$c_{ab}\|f\|_a' \leq \|f\|_b' \leq C_{ab}\|f\|_a'$$

THEOREM 2.3

For $0 < p < \infty$, the spaces $H^p(D)$ and $\mathcal{H}^p$ are the same. Moreover, if f is in $H^p(D)$, then

$$\|f\|_p = h_f(0)^{1/p}$$

This characterization of $H^p(D)$ will be useful in obtaining norm inequalities for composition operators but aside from this, we will not use this point of view.

Now $H^2(D)$ is a Hilbert space with the inner product

$$\langle f, g \rangle = \int_0^{2\pi} f(e^{i\theta})\overline{g(e^{i\theta})} \frac{d\theta}{2\pi}$$

and an easy calculation shows that the monomials $1, z, z^2, \ldots$ form an orthonormal set. Since $H^2(D)$ is spanned by the monomials, $\{1, z, z^2, \ldots\}$ is an orthonormal basis for $H^2(D)$, and we regard this as the standard basis. Since every function analytic in the open disk has a Maclaurin expansion that converges absolutely and uniformly on compact subsets of D, we have for $f(z) = \sum_{j=0}^{\infty} a_j z^j$

$$\|f\|_2^2 = \sup_{0<r<1} \int_0^{2\pi} |f(re^{i\theta})|^2 \frac{d\theta}{2\pi} = \sup_{0<r<1} \int_0^{2\pi} \sum_{j=0}^{\infty} \sum_{k=0}^{\infty} a_j \overline{a_k} r^{j+k} e^{i(j-k)\theta} \frac{d\theta}{2\pi}$$

$$= \sup_{0<r<1} \sum_{j=0}^{\infty} \sum_{k=0}^{\infty} \int_0^{2\pi} a_j \overline{a_k} r^{j+k} e^{i(j-k)\theta} \frac{d\theta}{2\pi}$$

$$= \sup_{0<r<1} \sum_{j=0}^{\infty} |a_j|^2 r^{2j} = \sum_{j=0}^{\infty} |a_j|^2$$

This calculation makes it clear that

$$H^2(D) = \{f = \sum_{j=0}^{\infty} a_j z^j : \sum_{j=0}^{\infty} |a_j|^2 < \infty\}$$

and it is easy to use the calculation to show that the (forward) unilateral shift on $\ell^2(N)$ is unitarily equivalent to the operator of multiplication by z on $H^2(D)$.

We will use the Cauchy formula to show that $H^p(D)$, $p \geq 1$, is a functional Banach space. This approach applies to many analytic spaces on many domains; later we will give an approach tailored specifically to the disk. For w in D, Cauchy's formula shows that if $|w| < r < 1$ and Γ_r is the circle of radius r with center at the origin, then for any f in $H^p(D)$,

$$f(w) = \frac{1}{2\pi i} \int_{\Gamma_r} \frac{f(\zeta)}{\zeta - w} d\zeta = \frac{1}{2\pi i} \int_0^{2\pi} \frac{f(re^{i\theta})}{re^{i\theta} - w} rie^{i\theta} d\theta$$

$$= \int_0^{2\pi} f(re^{i\theta}) \frac{r}{r - we^{-i\theta}} \frac{d\theta}{2\pi}$$

Applying Hölder's inequality, we see

$$|f(w)| \leq \|f_r\|_p \left\| \frac{r}{r - we^{-i\theta}} \right\|_q$$

where $1/p + 1/q = 1$. Since $|r/(r - we^{-i\theta})|$ converges uniformly to the bounded function $(1 - we^{-i\theta})^{-1}$ as r tends to 1 and $\|f_r\|_p \leq \|f\|_p$, we see that evaluation at w is a continuous linear functional.

If Ω is a domain (i.e. open connected set) in the plane, we might wish to define Hardy spaces $H^p(\Omega)$ in some analogous way. The definition in terms of harmonic majorants can be extended immediately. To find an analogue of the first definition, we need an exhaustion of Ω by rectifiable curves Γ_r so that Γ_r converges to the boundary as r tends to 1. For the second definition, we need a natural measure on the boundary of Ω. It turns out that the analogues of the three definitions given above for $H^p(D)$ are, in general, different from each other. Each of the definitions leads to an interesting theory of Hardy spaces.

We next define the Bergman spaces of the unit disk. More complete discussions of the Bergman spaces may be found in the paper of S. Axler [Ax88] or the book of K. Zhu [Zh90b].

DEFINITION 2.4   *For $0 < p < \infty$ the **Bergman space** $A^p(D)$ is the set of functions analytic on the unit disk for which*

$$\int_D |f(z)|^p \frac{dA(z)}{\pi} < \infty$$

where $dA(z)$ is Lebesgue area measure on the unit disk and $\|f\|_p$ is the pth root of this integral.

For $p \geq 1$, $A^p(D)$ is a Banach space with norm $\|f\|_p$. For $0 < p < 1$, $A^p(D)$ is a non-locally convex topological vector space and $d(f, g) = \|f - g\|_p^p$ is a complete metric for it.

Moreover, $A^2(D)$ is a Hilbert space with the inner product

$$\langle f, g \rangle = \int_D f(z)\overline{g(z)} \frac{dA(z)}{\pi}$$

An easy calculation using polar coordinates shows that the monomials $1, z, z^2, \ldots$ are orthogonal in $A^2(D)$ and $\|z^j\|^2 = 1/(j+1)$. Using the Maclaurin expansion, we have for $f(z) = \sum_{j=0}^{\infty} a_j z^j$ and $\rho < 1$

$$\int_{\rho D} |f(z)|^2 \frac{dA(z)}{\pi} = \int_0^\rho \int_0^{2\pi} \sum_{j=0}^{\infty} \sum_{k=0}^{\infty} a_j \overline{a_k} r^{j+k+1} e^{i(j-k)\theta} \frac{dr\, d\theta}{\pi}$$

$$= \sum_{j=0}^{\infty} \sum_{k=0}^{\infty} \int_0^\rho \int_0^{2\pi} a_j \overline{a_k} r^{j+k+1} e^{i(j-k)\theta} \frac{dr\, d\theta}{\pi}$$

$$= \sum_{j=0}^{\infty} |a_j|^2 \int_0^\rho 2r^{2j+1}\, dr = \sum_{j=0}^{\infty} \frac{|a_j|^2}{j+1} \rho^{2j+2}$$

It follows that

$$\|f\|^2 = \lim_{\rho \to 1} \int_{\rho D} |f(z)|^2 \frac{dA(z)}{\pi} = \lim_{\rho \to 1} \sum_{j=0}^{\infty} \frac{|a_j|^2}{j+1} \rho^{2j+2} = \sum_{j=0}^{\infty} \frac{|a_j|^2}{j+1}$$

We may therefore characterize the Bergman space by

$$A^2(D) = \{f = \sum_{j=0}^{\infty} a_j z^j : \sum_{j=0}^{\infty} \frac{|a_j|^2}{j+1} < \infty\}$$

and multiplication by z on $A^2(D)$ is a weighted unilateral shift.

To see that the Bergman spaces are functional Banach spaces when $p \geq 1$, we again use the Cauchy formula. For w in D, let $\rho = (1 + |w|)/2$. As in the Hardy space case, Cauchy's formula shows that if $\rho < r < 1$ and Γ_r is the circle of radius r with center the origin, then for any f in $A^p(D)$,

$$f(w) = \frac{1}{2\pi i} \int_{\Gamma_r} \frac{f(\zeta)}{\zeta - w} \, d\zeta = \frac{1}{2\pi} \int_0^{2\pi} \frac{f(re^{i\theta})}{re^{i\theta} - w} re^{i\theta} \, d\theta$$

Integrating this formula for $\rho < r < 1$, we get

$$(1 - \rho)f(w) = \int_\rho^1 f(w) \, dr = \int_\rho^1 \int_0^{2\pi} f(re^{i\theta}) \frac{e^{i\theta}}{2(re^{i\theta} - w)} \frac{r \, d\theta \, dr}{\pi}$$

Writing K for the function

$$K(z) = \begin{cases} 0 & |z| \leq \rho \\ \dfrac{z}{2(1 - \rho)|z|(z - w)} & \rho < |z| < 1 \end{cases}$$

we get

$$f(w) = \int_D f(z)K(z) \frac{dA(z)}{\pi}$$

Since K is bounded, we see that evaluation at w is a continuous linear functional.

Bergman spaces for domains other than the unit disk are defined analogously by replacing the integral over the disk by the area integral over the domain.

DEFINITION 2.5   *The **Dirichlet space** $\mathcal{D}$ is the set of functions analytic on the unit disk for which*

$$\int_D |f'(z)|^2 \frac{dA(z)}{\pi} < \infty$$

with norm given by

$$\|f\|_{\mathcal{D}}^2 = |f(0)|^2 + \int_D |f'(z)|^2 \frac{dA(z)}{\pi}$$

and inner product

$$\langle f, g \rangle_{\mathcal{D}} = f(0)\overline{g(0)} + \int_D f'(z)\overline{g'(z)} \frac{dA(z)}{\pi}$$

We have included the term $|f(0)|^2$ in the expression for the norm so that $\|1\| \neq 0$. Clearly, there are other ways to accomplish this, for example, adding $\|f\|_{H^2}^2$ in place of $|f(0)|^2$ is one such. Another approach is to consider equivalence classes of functions that differ by a constant. Although there is no general agreement of which should be chosen, for most purposes, there is little difference in the resulting theory. Analogously to the calculations for Hardy and Bergman spaces, one can show that $\mathcal{D}$ is a functional Hilbert space, that the monomials $1, z, z^2, \ldots$ form an orthogonal basis for $\mathcal{D}$, and that

$$\mathcal{D} = \{f = \sum_{j=0}^{\infty} a_j z^j : \sum_{j=0}^{\infty} |a_j|^2 (j+1) < \infty\}$$

The Dirichlet space is equivalently normed by letting $\|f\|_{\mathcal{D}}^2$ be the latter sum.

Recently, various authors have introduced weighted versions of the Bergman and Dirichlet spaces by replacing the measure $dA(z)$ in the classical Bergman and Dirichlet spaces by a weighted measure $G(|z|)\, dA(z)$ for a positive, continuous weight function G satisfying $\int_0^1 G(r) r\, dr < \infty$. The monomials are an orthogonal basis for the classical Hardy space, Bergman space, and Dirichlet space and many of the weighted spaces; these examples motivate the following generalization.

DEFINITION 2.6   *A Hilbert space $\mathcal{H}$ whose vectors are functions analytic on the unit disk will be called a **weighted Hardy space** if the monomials $1, z, z^2, \ldots$ constitute a complete orthogonal set of non-zero vectors in $\mathcal{H}$.*

The assumption that each function in $\mathcal{H}$ is analytic in D and not some smaller set is a pertinent part of the definition. The term "complete" is used here in its inner product space sense and the completeness is equivalent to the density of the polynomials in $\mathcal{H}$. We will usually assume that the norm satisfies the normalization $\|1\| = 1$. Writing $\beta(j) = \|z^j\|$, the orthogonality is easily seen to imply that the norm on $\mathcal{H}$ is given by

$$\left\| \sum_{j=0}^{\infty} a_j z^j \right\|^2 = \sum_{j=0}^{\infty} |a_j|^2 \beta(j)^2$$

and the inner product by

$$\langle \sum_{j=0}^{\infty} a_j z^j, \sum_{j=0}^{\infty} c_j z^j \rangle = \sum_{j=0}^{\infty} a_j \overline{c_j} \beta(j)^2$$

The weighted Hardy space with weight sequence $\beta(j)$ will be denoted $H^2(\beta)$ or $H^2(\beta, D)$ if needed for clarity. The emphasis in the definition is on revealing the weighted Hardy space structure in a functional Hilbert space; Exercise 2.1.10 gives conditions for constructing a weighted Hardy space from a given weight sequence.

The classical Hardy space, the classical Bergman space, and the classical Dirichlet space are weighted Hardy spaces with $\beta(j) \equiv 1$, $\beta(j) = (j+1)^{-1/2}$, and (up to an equivalent norm) $\beta(j) = (j+1)^{1/2}$ respectively. Weighted Bergman and Dirichlet spaces are also weighted Hardy spaces if the normalization $\|1\| = 1$ is made (see Exercise 2.1.5). Moreover, the relationship of the Cauchy integral formulas for f and f' shows that a weighted Bergman space is also a weighted Dirichlet space with a different weight. There is no standard terminology in the literature for distinguishing between these weighted spaces. We adopt the convention that a space will be described as a weighted Hardy space if we wish to emphasize that the norm is given in terms of the power series coefficients, a weighted Bergman space to emphasize the norm is obtained by integrating $|f|^2$, and a weighted Dirichlet space to emphasize the norm is obtained by integrating $|f'|^2$. Although the norm may change to an equivalent norm in this transition, most of the topics of interest to us depend only on the size of the norm and not the actual numerical value so there will be few differences in the resulting theories.

The relationship between the weighted Hardy spaces and unilateral weighted shifts, which can be viewed as multiplication by z on the weighted Hardy spaces, was explored by Shields [Shi74, §3] who introduced the notation $H^2(\beta)$ for them. We record for reference the following facts about the operator of multiplication by z on $H^2(\beta)$ (see [Shi74, p. 59]). Since this operator is one-to-one, it is bounded below if and only if it has closed range. In items (3) and (4) of the proposition, $\|\cdot\|_e$ refers to the essential norm of an operator, that is, the distance of the operator from the compact operators.

PROPOSITION 2.7
Let T be the operator of multiplication by z on $H^2(\beta)$.

(1) T is bounded if and only if $B = \sup \beta(j+1)/\beta(j)$ is finite.

(2) T is bounded below if and only if $b = \inf \beta(j+1)/\beta(j)$ is positive.

(3)   If T is bounded, then

$$\|T^*T\|_e = \limsup_{j\to\infty} \left(\frac{\beta(j+1)}{\beta(j)} \right)^2$$

(4)   If T is bounded below, then

$$\|(T^*T)^{-1}\|_e^{-1} = \liminf_{j\to\infty} \left(\frac{\beta(j+1)}{\beta(j)} \right)^2$$

PROOF   Since T is bounded (bounded below) if and only if T^*T is bounded (bounded below), each of the statements is an easy consequence of the observation that T^*T is a diagonal operator with diagonal entries $(\beta(j+1)/\beta(j))^2$.   ∎

The properties of the weighted Hardy space $H^2(\beta)$ clearly depend on the weights $\beta(j)$. These weights can be encoded in a generating function and many properties of $H^2(\beta)$ can be described in terms of its generating function.

DEFINITION 2.8   *The **generating function** for the weighted Hardy space $H^2(\beta)$ is the function*

$$k(z) = \sum_{j=0}^{\infty} \frac{z^j}{\beta(j)^2}$$

We begin by showing that the generating function is analytic on the unit disk. The analyticity is a consequence of our assumption in the definition of weighted Hardy space that all the functions of $H^2(\beta)$ are analytic in D.

LEMMA 2.9
If k is the generating function for a weighted Hardy space, then k is analytic on the open unit disk.

PROOF   Let k be the generating function for the weighted Hardy space $H^2(\beta)$.

Since $H^2(\beta)$ is complete, the Taylor series $\sum a_j z^j$ represents an analytic function in $H^2(\beta)$ if and only if $\sum |a_j|^2 \beta(j)^2 < \infty$. Let $a_0 = 0$, let $a_j = 1/(j\beta(j))$ for $j > 0$, and let $f(z) = \sum a_j z^j$. Now

$$\sum_{j=0}^{\infty} |a_j|^2 \beta(j)^2 = \sum_{j=1}^{\infty} \frac{1}{(j\beta(j))^2} \beta(j)^2 = \sum_{j=1}^{\infty} \frac{1}{j^2} < \infty$$

so f is in $H^2(\beta)$.

Every function in a weighted Hardy space is analytic in the open unit disk, by definition. Since f is in $H^2(\beta)$, the radius of convergence of its power series is at least 1, so

$$\limsup_{j \to \infty} \left(\frac{1}{\beta(j)^2} \right)^{1/j} = \left(\limsup_{j \to \infty} \left(\frac{1}{j\beta(j)} \right)^{1/j} \right)^2 \leq 1$$

which implies k is analytic in the open unit disk.   ∎

Up to now, we have shown that the classical Hardy, Bergman and Dirichlet spaces are functional Banach spaces but we have not found the norms of the evaluation functionals. The next theorem shows the principal role of k: it generates the reproducing kernels for $H^2(\beta)$. It follows that $H^2(\beta)$ is a functional Hilbert space on the disk and we calculate the norm of the evaluation functionals.

THEOREM 2.10
Let $H^2(\beta)$ be a weighted Hardy space. For each point w in the open unit disk, evaluation of functions in $H^2(\beta)$ at w is a bounded linear functional and, for all f in $H^2(\beta)$, $f(w) = \langle f, K_w \rangle$ where $K_w(z) = k(\overline{w}z)$. Moreover, $\|K_w\|^2 = k(|w|^2)$.

PROOF For $|w| < 1$, the analyticity of k on the open unit disk implies that K_w is in $H^2(\beta)$. Indeed,

$$\|K_w\|^2 = \left\| \sum_{j=0}^{\infty} \frac{\overline{w}^j}{\beta(j)^2} z^j \right\|^2 = \sum_{j=0}^{\infty} \left| \frac{\overline{w}^j}{\beta(j)^2} \right|^2 \beta(j)^2 = \sum_{j=0}^{\infty} \frac{|w|^{2j}}{\beta(j)^2} = k(|w|^2) < \infty$$

Now for $f(z) = \sum a_j z^j$ in $H^2(\beta)$,

$$\langle f, K_w \rangle = \sum_{j=0}^{\infty} a_j \frac{\overline{\overline{w}^j}}{\beta(j)^2} \beta(j)^2 = \sum_{j=0}^{\infty} a_j w^j = f(w)$$

∎

COROLLARY 2.11
In the Hardy space $H^2(D)$, evaluation at w in the disk is given by $f(w) = \langle f, K_w \rangle$ where

$$K_w(z) = \frac{1}{1 - \overline{w}z} \quad \text{and} \quad \|K_w\| = \frac{1}{\sqrt{1 - |w|^2}}$$

COROLLARY 2.12
In the Bergman space $A^2(D)$, evaluation at w in the disk is given by $f(w) = \langle f, K_w \rangle$ where

$$K_w(z) = \frac{1}{(1 - \overline{w}z)^2} \quad \text{and} \quad \|K_w\| = \frac{1}{1 - |w|^2}$$

COROLLARY 2.13
In the weighted Hardy space $H^2(\beta)$ defined by $\beta(j) = (j + 1)^{1/2}$, which is equivalent to the Dirichlet space $\mathcal{D}$, evaluation at w in the disk is given by $f(w) = \langle f, K_w \rangle$ where

$$K_w(z) = \frac{1}{\overline{w}z} \log \left(\frac{1}{1 - \overline{w}z} \right) \quad \text{and} \quad \|K_w\|^2 = \frac{1}{|w|^2} \log \left(\frac{1}{1 - |w|^2} \right)$$

Since the polynomials are dense in the Hardy and Bergman Banach spaces for $1 \leq p < \infty$, we can use the Hilbert space results to get a nicer formula for the evaluation functionals.

COROLLARY 2.14

Suppose $1 \le p < \infty$ and $|w| < 1$. If f is in $H^p(D)$ then

$$f(w) = \int_0^{2\pi} \frac{f(e^{i\theta})}{1 - we^{-i\theta}} \frac{d\theta}{2\pi}$$

If f is in $A^p(D)$ then

$$f(w) = \int_D \frac{f(z)}{(1 - w\bar{z})^2} \frac{dA(z)}{\pi}$$

Moreover, on $H^p(D)$, the norm of evaluation at w is

$$\left(\frac{1}{1 - |w|^2} \right)^{1/p}$$

PROOF   Evaluation at w is a continuous linear functional on $H^p(D)$ and $A^p(D)$. Since $1/(1 - we^{-i\theta})$ and $1/(1 - w\bar{z})^2$ are bounded functions, the integrals above give continuous linear functionals on $H^p(D)$ and $A^p(D)$ respectively. The polynomials are dense in these spaces and the integral formulas are correct for the polynomials because they are in the Hardy and Bergman Hilbert spaces. Thus, the left and right sides of the asserted equalities are continuous linear functionals that agree on a dense set and the equalities follow.

To compute the norm of evaluation at w in $H^p(D)$, we note first that if F is in $H^p(D)$ with inner–outer factorization $F = uf$ (see Theorem 2.27 or [Hof62, Chap. 5]), then f is non-zero, $\|F\|_p = \|f\|_p$ and $|F(w)| \le |f(w)|$. So it is enough to check the norm for non-zero functions f. Now if a non-zero function f is in $H^p(D)$, then $f^{p/2}$ is single-valued, is in $H^2(D)$, and

$$\|f\|_p^p = \int |f|^p = \int |f^{p/2}|^2 = \|f^{p/2}\|_2^2$$

Thus,

$$|f(w)|^p = |f^{p/2}(w)|^2 \le \frac{1}{1 - |w|^2} \|f^{p/2}\|_2^2 = \frac{1}{1 - |w|^2} \|f\|_p^p$$

Equality is attained for the function $f(z) = (1 - \bar{w}z)^{-2/p}$.   ∎

The generating function provides an easy way to see that a variety of functional Hilbert spaces, including the classical spaces, are algebraically consistent Hilbert spaces of analytic functions on the open unit disk.

THEOREM 2.15

If k, the generating function for the weighted Hardy space $H^2(\beta)$, is analytic on the open unit disk and $k(1) = \infty$, then $H^2(\beta)$ is an algebraically consistent Hilbert space of analytic functions on the open unit disk.

PROOF Suppose h gives a bounded linear functional on $H^2(\beta)$ that satisfies $\langle fg, h \rangle = \langle f, h \rangle \langle g, h \rangle$ whenever f, g, and fg are in $H^2(\beta)$. Since the polynomials are in $H^2(\beta)$, this holds when f and g are polynomials. Letting $w = \langle z, h \rangle$, it follows that $\langle z^2, h \rangle = \langle z, h \rangle \langle z, h \rangle = w^2$, and, more generally, $\langle z^j, h \rangle = w^j$ for $j = 1, 2, \ldots$. Now

$$w = \langle z, h \rangle = \langle 1z, h \rangle = \langle 1, h \rangle \langle z, h \rangle = \langle 1, h \rangle w$$

so that either $w = 0$ or $\langle 1, h \rangle = 1$. Moreover

$$\langle 1, h \rangle = \langle 1^2, h \rangle = \langle 1, h \rangle \langle 1, h \rangle = \langle 1, h \rangle^2$$

so either $\langle 1, h \rangle = 0$ or $\langle 1, h \rangle = 1$.

If $\langle 1, h \rangle = 0$ and $w = 0$ then $\langle z^j, h \rangle = w^j = 0$ for $j = 0, 1, 2, \ldots$ and hence $\langle p, h \rangle = 0$ for all polynomials p. Since the polynomials are dense in $H^2(\beta)$, this shows h gives the zero linear functional which is excluded from consideration in the definition of algebraic consistency.

Thus $\langle 1, h \rangle = 1$ and since 1, $z/\beta(1)$, $z^2/\beta(2)$, $\ldots$, forms an orthonormal basis for $H^2(\beta)$ we have

$$h(z) = \langle 1, h \rangle 1 + \sum_{j=1}^{\infty} \langle h, \frac{z^j}{\beta(j)} \rangle \frac{z^j}{\beta(j)} = 1 + \sum_{j=1}^{\infty} \frac{\overline{w}^j}{\beta(j)^2} z^j$$

For $|w| < 1$ this is $h(z) = k(\overline{w}z) = K_w(z)$ and h gives the linear functional of evaluation at w. If $|w| \geq 1$, then

$$\|h\|^2 = \sum_{j=0}^{\infty} \frac{|w|^{2j}}{\beta(j)^4} \beta(j)^2 = \sum_{j=0}^{\infty} \frac{|w|^{2j}}{\beta(j)^2} \geq k(1) = \infty$$

which contradicts the boundedness of the linear functional determined by h.

Thus every bounded, partially multiplicative linear functional is either zero or a point evaluation for a point of the open unit disk and $H^2(\beta)$ is an algebraically consistent Hilbert space of analytic functions on the open unit disk. ∎

Since the boundedness of point evaluations in the classical spaces can be proved using the Cauchy integral formula and since the derivatives may be computed from a similar integral formula, it is not surprising that evaluation of any derivative at a point of the disk is a bounded linear functional. The proof of this theorem is left to the exercises.

THEOREM 2.16

Let $H^2(\beta)$ be a weighted Hardy space. For each point w in the open unit disk and positive integer m, evaluation of the mth derivative of functions in $H^2(\beta)$ at w is *a bounded linear functional and* $f^{(m)}(w) = \langle f, K_w^{(m)} \rangle$ *where*

$$K_w^{(m)}(z) = \sum_{j=m}^{\infty} \frac{j!}{(j-m)!} \overline{w}^{j-m} \frac{z^j}{\beta(j)^2}$$

This expression for the kernel can be easily remembered and understood as

$$K_w^{(m)}(z) = \frac{d^m}{d\overline{w}^m} k(\overline{w}z)$$

We can see why this should be true from the following calculation involving the first derivative. It can be made into a proof in many cases by worrying about convergence.

$$f'(w) = \lim_{\zeta \to w} \frac{1}{\zeta - w}(f(\zeta) - f(w)) = \langle f, \lim_{\zeta \to w} \frac{1}{\overline{\zeta} - \overline{w}}(K_\zeta - K_w) \rangle$$

The behavior of the reproducing kernels for points near the unit circle has important consequences for composition operators. The conclusion of Theorem 2.17 should be contrasted with the behavior for spaces in which the generating function is smooth on the closed disk (see Exercise 2.1.6).

THEOREM 2.17
In a weighted Hardy space for which the series $\sum \beta(n)^{-2}$ diverges, the normalized reproducing kernels

$$\frac{K_w}{\|K_w\|}$$

tend to 0 weakly as $|w|$ tends to 1.

PROOF   Let w_j be a sequence in D such that $|w_j|$ tends to 1. The values of the reproducing kernels are given by the generating function k and $\|K_{w_j}\|^2 = k(|w_j|^2)$. Since $\lim_{j\to\infty} k(|w_j|^2) = \sum 1/\beta(n)^2 = \infty$, it follows that $\|K_{w_j}\|$ tends to infinity and for every polynomial p,

$$\lim_{j\to\infty} |\langle p, \frac{K_{w_j}}{\|K_{w_j}\|} \rangle| = \lim_{j\to\infty} \frac{|p(w_j)|}{\|K_{w_j}\|} = 0$$

The closed unit ball of a Hilbert space is weakly compact, so every sequence tending to the circle has a subsequence for which the normalized reproducing kernels $K_{w_j}/\|K_{w_j}\|$ converge weakly. Since the polynomials are dense in $H^2(\beta)$, the estimate above shows the only possible limit is zero. Since this is true for every subsequence, the conclusion follows.   ∎

Spaces of functions of several variables

If Ω is an open set in C^N, a complex valued function f on Ω is called analytic if it is an analytic function (of one variable) in each variable separately. Though not obvious, Hartogs' theorem guarantees that such functions are continuous in Ω. If φ is a map of Ω into C^m such that each of the component functions $\varphi_1, \ldots, \varphi_m$ is analytic, then φ will be called an ***analytic (or holomorphic) map***. It follows from Hartogs' theorem and standard results of elementary analysis that φ is differentiable on Ω. If φ is analytic and has an inverse that is also analytic, then φ will be called ***biholomorphic***.

DEFINITION 2.18 *The **unit polydisk** in C^N is the set*

$$D^N = \{z = (z_1, \ldots, z_N) : |z_j| < 1, \text{ for } j = 1, \ldots, N\}$$

*The **torus** or **distinguished boundary of** D^N, is the set*

$$T^N = \{z = (z_1, \ldots, z_N) : |z_j| = 1, \text{ for } j = 1, \ldots, N\}$$

*The **unit ball** in C^N is the set*

$$B_N = \{z = (z_1, \ldots, z_N) : \sum_{j=1}^{N} |z_j|^2 < 1\}$$

*The **unit sphere** in C^N is the set*

$$S_N = \{z = (z_1, \ldots, z_N) : \sum_{j=1}^{N} |z_j|^2 = 1\}$$

For $z = (z_1, \ldots, z_N)$ and $w = (w_1, \ldots, w_N)$ in C^N, we write $\langle z, w \rangle$ for the Euclidean inner product $\sum_1^N z_j \overline{w_j}$ and $|z| = \sqrt{\langle z, z \rangle}$. With this notation, $B_N = \{z \in C^N : |z| < 1\}$ and $S_N = \{z \in C^N : |z| = 1\}$, analogously to the unit disk and circle for $N = 1$.

If f is analytic in the polydisk and continuous on the closed polydisk, then the Cauchy formula generalizes to

$$f(z) = \int_{T^N} f(w) K(z, w) \, d\tau_N(w)$$

where τ_N is the measure on T^N that is the product of normalized Lebesgue measure on the circles $|z_j| = 1$ and for z in D^N and w in T^N

$$K(z, w) = \prod_{j=1}^{N} (1 - \overline{w_j} z_j)^{-1} = \sum_{\alpha} \overline{w}^{\alpha} z^{\alpha}$$

where $\alpha = (\alpha_1, \ldots, \alpha_N)$ is a multi-index and $z^{\alpha} = z_1^{\alpha_1} z_2^{\alpha_2} \cdots z_N^{\alpha_N}$. Moreover, f has a power series $f(z) = \sum c(\alpha) z^{\alpha}$ that converges absolutely and uniformly in every compact subset of the polydisk D^N, where the coefficients are given by

$$c(\alpha) = \int_{T^N} f(w) \overline{w}^{\alpha} \, d\tau_N(w)$$

Since the ball B_N is a union of polydisks centered at the origin, every analytic function f in the ball has a *global* power series representation of the form $\sum_\alpha c(\alpha)z^\alpha$ converging to f at every point of B_N. If we set

$$f_s(z) = \sum_{|\alpha|=s} c(\alpha)z^\alpha$$

where $|\alpha| = \alpha_1 + \cdots + \alpha_N$ then f_s is a homogeneous polynomial of degree s (meaning $f_s(\lambda z) = \lambda^s f_s(z)$ for all λ in C and z in C^N) and $f = \sum f_s$ is referred to as the homogeneous expansion of f. As in one variable, we also write f_r for the function $f_r(z) = f(rz)$; it should be clear from the context whether we are speaking of this dilate of f or the rth term in the homogeneous expansion.

It is clear that the monomials z^α are orthogonal in $L^2(\tau_N)$. Perhaps less obvious, but of fundamental importance is the fact that the monomials z^α are orthogonal in $L^2(\sigma_N)$, where σ_N is the positive Borel measure on S_N that is rotation invariant (that is, invariant under the unitary group) and normalized so that $\sigma_N(S_N) = 1$. (In the notation σ_N the subscript N will occasionally be supressed when the dimension is clear from the context). The proof of this orthogonality can be found in [Ru80, § 1.4] which also includes a computation of the $L^2(\sigma_N)$ norms of the monomials:

$$\|z^\alpha\|^2_{L^2(\sigma_N)} = \frac{(N-1)!\alpha!}{(N-1+|\alpha|)!} \tag{2.1.1}$$

where $\alpha! = \alpha_1! \cdots \alpha_N!$.

DEFINITION 2.19   *For $0 < p < \infty$ the **Hardy space** $H^p(D^N)$ is the set of functions analytic in the polydisk D^N such that*

$$\sup_{0<r<1} \int_{T^N} |f_r|^p \, d\tau_N < \infty$$

Define the norm $\|f\|_p$ to be the pth root of this supremum. The Hardy space $H^\infty(D^N)$ is the set of bounded analytic functions on D^N with the supremum norm.

As in one variable, these Hardy spaces are functional Banach spaces for $p \geq 1$ and for $f(z) = \sum c(\alpha)z^\alpha$, the orthonormality of the monomials z^α in $L^2(\tau_N)$ shows f is in $H^2(D^N)$ if and only if $\sum |c(\alpha)|^2 < \infty$.

We next define the Hardy and Bergman spaces of the unit ball.

DEFINITION 2.20   *For $0 < p < \infty$ the **Hardy space** $H^p(B_N)$ is the set of functions analytic in the ball B_N such that*

$$\|f\|^p_p \equiv \sup_{0<r<1} \int_{S_N} |f_r|^p \, d\sigma_N < \infty$$

*The Hardy space $H^\infty(B_N)$ is the set of bounded analytic functions on B_N with the supremum norm. For $0 < p < \infty$ the **Bergman space** $A^p(B_N)$ is the set of* *functions analytic in B_N such that*

$$\|f\|_{A^p}^p \equiv \int_{B_N} |f|^p \, d\nu_N < \infty$$

Here, ν_N denotes Lebesgue volume measure in C^N normalized by $\nu_N(B_N) = 1$.

As expected, for $p \geq 1$, the Hardy and Bergman spaces are functional Banach spaces. For f in the Hardy space $H^p(B_N)$, $p \geq 1$, and $|w| < 1$, we have

$$f(w) = \int_{S_N} \frac{f(\zeta)}{(1 - \langle w, \zeta \rangle)^N} \, d\sigma_N(\zeta)$$

where $f(\zeta) = \lim_{r \to 1} f(r\zeta)$ (which exists almost everywhere $[\sigma_N]$), while for f in the Bergman space $A^p(B_N)$, $p \geq 1$, and $|w| < 1$,

$$f(w) = \int_{B_N} \frac{f(z)}{(1 - \langle w, z \rangle)^{N+1}} \, d\nu_N(z)$$

It follows that $K_w(z) = (1 - \langle z, w \rangle)^{-N}$ and $K_w(z) = (1 - \langle z, w \rangle)^{-(N+1)}$ are the reproducing kernels for the Hardy and the Bergman spaces on the ball.

Analogous to the definition in the disk, a weighted Hardy space is a Hilbert space of functions analytic in the unit ball in which the monomials z^α, $|\alpha| \geq 0$ form a complete orthogonal set of non-zero vectors with

$$\frac{\|z^{\alpha_1}\|}{\|z^{\alpha_1}\|_2} = \frac{\|z^{\alpha_2}\|}{\|z^{\alpha_2}\|_2}$$

whenever $|\alpha_1| = |\alpha_2|$, where $\| \cdot \|_2$ denotes the norm in $L^2(\sigma_N)$. Then we may define numbers $\beta(s)$ by $\beta(|\alpha|) = \|z^\alpha\| / \|z^\alpha\|_2$. If f_s is a homogeneous polynomial of degree s, we have $\|f_s\| = \|f_s\|_2 \beta(s)$ and we denote the Hilbert space by $H^2(\beta, B_N)$. It follows that if $f = \sum_0^\infty f_s$ is the homogeneous expansion of a function analytic in B_N then f is in $H^2(\beta, B_N)$ if and only if

$$\sum_0^\infty \|f_s\|_2^2 \beta(s)^2 \tag{2.1.2}$$

is finite and this sum is $\|f\|_{H^2(\beta)}^2$. (Note that for $N = 1$, the homogeneous expansion of f is just its Maclaurin expansion and this definition agrees with the earlier definition.) The Hardy space $H^2(B_N)$ is the weighted Hardy space with $\beta(s) = 1$ and the Bergman space $A^2(B_N)$, with an equivalent norm, is the weighted Hardy space with $\beta(s) = (s + 1)^{-1}$ (see Exercise 2.1.13). As in the single variable case, we will assume $H^2(\beta, B_N)$ has been normalized to have $\|1\| = 1$.

By using the polarization identity relating the norm in a Hilbert space to its inner product, the sum for the norm given by Expression (2.1.2) implies the inner product on $H^2(\beta, B_N)$ is given by

$$\langle f, g \rangle = \sum_0^\infty \langle f_s, g_s \rangle_2 \beta(s)^2$$

where $\langle f_s, g_s \rangle_2$ is the inner product of the homogeneous polynomials f_s and g_s in $L^2(\sigma_N)$. The reproducing kernels for the space $H^2(\beta, B_N)$ can be computed in terms of the β's. For w in the ball, the kernel $K_w(z) = \sum_\alpha c(\alpha) z^\alpha$ for evaluation at w satisfies $\langle f, K_w \rangle = f(w)$ for all f in $H^2(\beta, B_N)$, so for each monomial, z^α, we must have

$$w^\alpha = \langle z^\alpha, K_w \rangle = \langle z^\alpha, c(\alpha) z^\alpha \rangle_2 \beta(|\alpha|)^2 = \overline{c(\alpha)} \frac{(N-1)! \alpha!}{(N-1+|\alpha|)!} \beta(|\alpha|)^2$$

from which it follows that

$$
\begin{aligned}
K_w(z) &= \sum_\alpha \frac{(N-1+|\alpha|)!}{(N-1)! \alpha!} \frac{z^\alpha \overline{w}^\alpha}{\beta(|\alpha|)^2} \\
&= \sum_{s=0}^\infty \frac{(N-1+s)!}{(N-1)! s!} \frac{1}{\beta(s)^2} \sum_{|\alpha|=s} \frac{s!}{\alpha!} z^\alpha \overline{w}^\alpha \\
&= \sum_{s=0}^\infty \frac{(N-1+s)!}{(N-1)! s!} \frac{\langle z, w \rangle^s}{\beta(s)^2}
\end{aligned}
$$

This gives an expression for the norms of the kernel functions as well

$$\|K_w\|^2 = \langle K_w, K_w \rangle = K_w(w) = \sum_{s=0}^\infty \frac{(N-1+s)!}{(N-1)! s!} \frac{|w|^{2s}}{\beta(s)^2} \qquad (2.1.3)$$

Parallel with the one variable case, we define the ***generating function*** for the weighted Hardy space $H^2(\beta, B_N)$ to be the function

$$k(u) = \sum_{s=0}^\infty \frac{(N-1+s)!}{(N-1)! s!} \frac{u^s}{\beta(s)^2} \qquad (2.1.4)$$

for u in the unit disk so that $K_w(z) = k(\langle z, w \rangle)$ for weighted Hardy spaces in several variables also.

There is a useful connection between each weighted Hardy space in B_N and a related weighted Hardy space in the disk. For a sequence $\{\beta(j)\}$ of positive numbers and an integer $N \geq 2$, let $\{\widetilde{\beta}(j)\}$ be defined by

$$\widetilde{\beta}(j)^2 = \frac{(N-1)! j!}{(N-1+j)!} \beta(j)^2 \qquad (2.1.5)$$

for $j = 1, 2, \ldots$. We then define extension and restriction operators, acting respectively on analytic functions on the disk D and ball B_N, by

$$Ef(z_1, z') = f(z_1)$$

for $(z_1, z') = (z_1, z_2, \ldots, z_N)$ in B_N and

$$RF(\lambda) = F(\lambda, 0')$$

for λ in D. Thus Ef is analytic on B_N whenever f is analytic on D and RF is analytic on D whenever F is analytic on B_N. The following proposition is more explicit.

PROPOSITION 2.21
(1) E is an isometry of $H^2(\widetilde{\beta}, D)$ into $H^2(\beta, B_N)$.
(2) R is a norm-decreasing map of $H^2(\beta, B_N)$ onto $H^2(\widetilde{\beta}, D)$.
    Moreover, the generating functions for $H^2(\widetilde{\beta}, D)$ and $H^2(\beta, B_N)$ are the same.

PROOF   Let $f = \sum c(j) z^j$ be in $H^2(\widetilde{\beta}, D)$. Then $Ef(z_1, z') = \sum c(j) z_1^j$ so

$$\|Ef\|_\beta^2 = \sum \|c(j) z_1^j\|_2^2 \beta(j)^2 = \sum |c(j)|^2 \frac{(N-1)!j!}{(N-1+j)!} \beta(j)^2$$

$$= \sum |c(j)|^2 \widetilde{\beta}(j)^2 = \|f\|_{\widetilde{\beta}}^2$$

which gives (1).
    For the second part, let $F = \sum f_s$ be in $H^2(\beta, B_N)$ where

$$f_s = c_s^1 z_1^s + \sum_{|\alpha|=s, \alpha \neq (s,0')} c(\alpha) z^\alpha$$

By definition $RF(\lambda) = F(\lambda, 0') = \sum c_s^1 \lambda^s$ and $\|RF\|_{\widetilde{\beta}}^2 = \sum |c_s^1|^2 \widetilde{\beta}(s)^2$. But

$$\|f_s\|_2^2 = \sum |c(\alpha)|^2 \|z^\alpha\|_2^2 \geq |c_s^1|^2 \frac{(N-1)!s!}{(N-1+s)!}$$

from which it follows that

$$\|RF\|_{\widetilde{\beta}}^2 \leq \sum \|f_s\|_2^2 \frac{(N-1+s)!}{(N-1)!s!} \widetilde{\beta}(s)^2 = \sum \|f_s\|_2^2 \beta(s)^2 = \|F\|_\beta^2$$

To see that R maps onto $H^2(\widetilde{\beta}, D)$, simply note that if $f = \sum c_s^1 \lambda^s$ is in $H^2(\widetilde{\beta}, D)$ then $f = RF$ where $F(z_1, z') = \sum c_s^1 z_1^s$. An easy computation shows that F is in $H^2(\beta, B_N)$.
    From Equation (2.1.5), the generating function for $H^2(\widetilde{\beta}, D)$ is

$$\widetilde{k}(u) = \sum_{j=0}^\infty \frac{u^j}{\widetilde{\beta}(j)^2} = \sum_{j=0}^\infty \frac{(N-1+j)!}{(N-1)!j!} \frac{u^j}{\beta(j)^2}$$

Comparing this with Equation (2.1.4) gives the last part of the result.   ∎

Chapter 2: Background

In weighted Hardy spaces in one variable, the monomials $1, z, z^2, \ldots$ form an orthogonal basis and they have a natural order. In more than one variable, the monomials also form an orthogonal basis, but in this case there is not an obvious order. It will simplify some arguments for us to use a standard order for this basis in the several variable case as well. We will order the monomials z^α by ordering the multi-indices. When $|\alpha| < |\beta|$, we say $\alpha < \beta$. When $|\alpha| = |\beta|$, we say $\alpha < \beta$ if there is j_0 so that $\alpha_j = \beta_j$ for $j < j_0$ and $\alpha_{j_0} > \beta_{j_0}$. For example, for $N = 3$, the ordering begins

$$1, \ z_1, \ z_2, \ z_3, \ z_1^2, \ z_1 z_2, \ z_1 z_3, \ z_2^2, \ z_2 z_3, \ z_3^2, \ \ldots$$

This ordering has the convenient property that if z^α precedes $z^{\alpha'}$ and z^β precedes $z^{\beta'}$, then $z^\alpha z^\beta$ precedes $z^{\alpha'} z^{\beta'}$. We will refer to the orthonormal basis $z^\alpha / \|z^\alpha\|$, with this ordering, as the **standard basis** for the weighted Hardy space.

For $\zeta \neq 0$ in C^N we denote the complex line through ζ and 0 by $[\zeta]$; that is

$$[\zeta] = \{\lambda\zeta : \lambda \in C\}$$

If ζ is in ∂B_N and f is holomorphic in B_N then the function f_ζ defined on D by $f_\zeta(\lambda) = f(\lambda\zeta)$ is called a slice function. Integrals over ∂B_N can be computed by first averaging over slices:

$$\int_{\partial B_N} f \, d\sigma_N = \int_{\partial B_N} \int_0^{2\pi} f(e^{i\theta}\zeta) \, \frac{d\theta}{2\pi} \, d\sigma_N(\zeta) \tag{2.1.6}$$

Formula (2.1.6) is often referred to as the slice integration formula; it holds for arbitrary Lebesgue integrable f. A proof can be found in [Ru80, p. 15].

Exercises

2.1.1 Show that
$$f(z) = e^{(1+z)/(1-z)}$$
is analytic in the unit disk and for $0 < \theta < 2\pi$
$$\lim_{r \to 1} |f(re^{i\theta})| = 1$$
but that f is not in $H^p(D)$ for any p. Thus, it is *not* the case that $H^p(D)$ consists of those analytic functions in D whose boundary values are in L^p.

2.1.2 Suppose f is in $L^2(D, dA)$ and is analytic in the unit disk except possibly at 0. Show that actually f is analytic at 0 which means f is in the Bergman space $A^2(D)$.

2.1.3 Verify the assertions made in the text about the Dirichlet space:
  (a) Show that for $f = \sum_{j=0}^\infty a_j z^j$
$$\|f\|^2_{H^2(D)} + \int_D |f'(z)|^2 \, \frac{dA(z)}{\pi} = \sum_{j=0}^\infty |a_j|^2 (j+1)$$

so
$$D = \{f = \sum_{j=0}^{\infty} a_j z^j : \sum_{j=0}^{\infty} |a_j|^2 (j+1) < \infty\}$$
and the series gives an equivalent norm on the Dirichlet space.

(b) Show that D is a functional Hilbert space.

(c) Find an expression for K_w in D analogous to that given in Corollary 2.13.

2.1.4 (a) Let w be in D and $c > 0$. Show that
$$\int_0^{2\pi} \frac{1}{|1 - \overline{w}e^{i\theta}|^{1+c}} \frac{d\theta}{2\pi} \sim (1 - |w|^2)^{-c}$$
where $\sim$ indicates that the ratio of the two expressions has positive finite limit as $|w| \to 1$.

Hints: Start with the binomial series formula
$$\left(\frac{1}{1 - \overline{w}e^{i\theta}}\right)^{\lambda} = \sum_{k=0}^{\infty} \frac{\Gamma(k+\lambda)}{k!\Gamma(\lambda)} (\overline{w}e^{i\theta})^k$$
if λ is not a negative integer. Apply this with $2\lambda = c + 1$ to show
$$\int_0^{2\pi} \frac{1}{|1 - \overline{w}e^{i\theta}|^{1+c}} \frac{d\theta}{2\pi} = \sum_{k=0}^{\infty} \frac{\Gamma^2(k+\lambda)}{k!^2\Gamma^2(\lambda)} |w|^{2k}$$
Use Stirling's formula
$$\Gamma(n+1) \sim \sqrt{2\pi n} \left(\frac{n}{e}\right)^n$$
to show that the coefficients in this series are asymptotic to k^{c-1} as $k \to \infty$.

(b) If w is in D and $c > 0$ show
$$\int_D \frac{1}{|1 - \overline{w}z|^{2+\alpha+c}} (1 - |z|^2)^{\alpha} \, dA(z) \sim (1 - |w|^2)^{-c}$$
Hint: Convert to polar coordinates and use (a).

(c) Generalize (a) and (b) to B_N, $N > 1$: if w is in B_N and $c > 0$ then
$$\int_{\partial B_N} \frac{1}{|1 - \langle w, \zeta \rangle|^{N+c}} \, d\sigma_N(\zeta) \sim (1 - |w|^2)^{-c}$$
and
$$\int_{B_N} \frac{1}{|1 - \langle w, \zeta \rangle|^{N+1+\alpha+c}} (1 - |z|^2)^{\alpha} \, d\nu_N(z) \sim (1 - |w|^2)^{-c}$$

2.1.5 For $\alpha > -1$, define a weighted Bergman space by
$$A_\alpha^2(D) = \{f \text{ analytic in } D : \|f\|^2 = \int_D |f(z)|^2 (1 - |z|^2)^{\alpha} \frac{dA(z)}{\pi} < \infty\}$$
and a weighted Dirichlet space by
$$D_\alpha = \{f \text{ analytic in } D : \|f\|^2 = |f(0)|^2 + \int_D |f'(z)|^2 (1 - |z|^2)^{\alpha} \frac{dA(z)}{\pi} < \infty\}$$

(a) Show that the reproducing kernel functions in $A_\alpha^2(D)$ are
$$K_w(z) = \frac{\alpha + 1}{(1 - \overline{w}z)^{\alpha+2}}$$

Hint: Recall the formula
$$\int_0^1 s^{x-1}(1 - s)^{y-1} ds = B(x, y) = \frac{\Gamma(x)\Gamma(y)}{\Gamma(x+y)}$$
where $B(x, y)$ is the Beta function and Γ is the Gamma function.

(b) Show that $A_\alpha^2(D)$ and $\mathcal{D}_\alpha$ are weighted Hardy spaces and find estimates for the weights $\beta(j)$.

(c) Show that $A_\alpha^2(D) = \mathcal{D}_{\alpha+2}$ and that the norms on these spaces are equivalent.

(d) Show that $A_\alpha^2(D)$ and $\mathcal{D}_\alpha$ are functional Hilbert spaces.

2.1.6 (a) Suppose $H^2(\beta)$ is a weighted Hardy space for which the generating function k is continuous on the closed disk. Show that all functions in $H^2(\beta)$ can be extended continuously to the closed disk.

(b) Use part (a) to find an example of a Hilbert space of analytic functions on the disk consisting of functions continuous on the closed disk but including at least one function that is not differentiable at 1.

2.1.7 Prove Theorem 2.16.

2.1.8 Suppose $\mathcal{H}$ is a functional Hilbert space on a set X and suppose $\mathcal{K}$ is a closed subspace of $\mathcal{H}$.

(a) Given that K_x is the function in $\mathcal{H}$ such that $f(x) = \langle f, K_x \rangle$ for all f in $\mathcal{H}$, find the reproducing kernels for $\mathcal{K}$ (note that they should be vectors in $\mathcal{K}$).

(b) Is $\mathcal{K}$ also a functional Hilbert space?

2.1.9 Let $A(D)$ be the Banach space of functions that are continuous on the closed unit disk and analytic on the open unit disk, with the supremum norm.

(a) Show that $A(D)$ is a functional Banach space on D.

(b) Show that $A(D)$ is the closure in $C(\overline{D})$ of the analytic polynomials.

(c) Let $\Omega = \{z : .5 < |z| < 1\}$. Describe the closure in $C(\overline{\Omega})$ of the analytic polynomials and describe the space of functions continuous on $\overline{\Omega}$ and analytic in Ω.

2.1.10 Show that if $\{\beta(n)\}$ is a sequence with $\beta(0) = 1$, $\beta(j) > 0$ for all j, and $\liminf \beta(j)^{1/j} \geq 1$, then

$$\mathcal{H} = \left\{ f(z) = \sum_{j=0}^{\infty} a_j z^j : \sum_{j=0}^{\infty} |a_j|^2 \beta(j)^2 < \infty \right\}$$

with $\langle f, g \rangle = \sum_j a_j \overline{b_j} \beta(j)^2$ defines a weighted Hardy space. Thus, a weighted Hardy space can be given as a Hilbert space of functions and the weight sequence constructed or can be given as a weight sequence and the Hilbert space constructed.

2.1.11 Show that if $\{\beta(n)\}$ is a positive, bounded sequence such that 0 is not a limit point of $\{\beta(n)\}$ then f is in $H^2(\beta)$ if and only if f is in $H^2(D)$.

2.1.12 Show that a weighted Hardy space $H^2(\beta)$ contains $H^\infty(D)$ if and only if the weight sequence $\{\beta(n)\}$ defining the space is bounded.

2.1.13 (a) Use the polar coordinate formula

$$\int_{C^N} f \, d\nu_N = 2N \int_0^\infty \int_{S_N} f(r\zeta) r^{2N-1} \, d\sigma_N(\zeta) \, dr$$

to calculate

$$\int_{B_N} |z^\alpha|^2 \, d\nu_N(z)$$

(b) Show that the Bergman space $A^2(B_N)$ is equivalent to the weighted Hardy space arising from the choice $\beta(s) = (s+1)^{-1}$.

2.1.14 Let D be the open unit disk and let $\mathcal{H}$ denote the set of analytic functions in the unit disk such that

$$f(z) = \sum_{n=0}^{\infty} a_n (z+1)^n \text{ and } \|f\|^2 = \sum_{n=0}^{\infty} 4^n |a_n|^2 < \infty$$

(a) Show that $\mathcal{H}$ is a Hilbert space of analytic functions on D.

(b) Find the kernels K_z for $|z| < 1$.

(c) Find a linear functional k so that $\langle pq, k \rangle = \langle p, k \rangle \langle q, k \rangle$ for all polynomials p and q but $k \neq K_z$ for $|z| < 1$. This shows that $\mathcal{H}$ is not algebraically consistent.

2.1.15 Prove that the Hardy spaces $H^p(D)$ and the Bergman spaces $A^p(D)$ are algebraically consistent Banach spaces of analytic functions on the open unit disk D.

2.1.16 Prove the assertion in the text that if z^α precedes $z^{\alpha'}$ and z^β precedes $z^{\beta'}$ in the ordering of monomials in the several variable case, then $z^\alpha z^\beta$ precedes $z^{\alpha'} z^{\beta'}$.

2.1.17 Generalize Lemma 2.9 by showing that if k is the generating function for a weighted Hardy space $H^2(\beta, B_N)$ on the ball, then k is analytic on the open unit disk.

2.1.18 Prove that if the weighted Hardy space $H^2(\beta, B_N)$ contains H^∞, then $\|K_w\|$ tends to infinity as w approaches the boundary and generalize Theorem 2.17 for this case. (Hint: Generalize Exercise 2.1.12 and use Equation (2.1.3).)

Notes

Theorem 2.3, due to W. Rudin [Ru55], provides motivation for a particularly useful definition of the Hardy spaces on domains with nonsmooth boundaries.

The property of norms of kernel functions in Exercise 2.1.18, from C. Cowen and B. MacCluer's paper [CoM94], was suggested by S. Axler; the proof outlined here is due to P. Bourdon. The fact that the kernel functions for the weighted Hardy spaces, even in several variables, are the composition of a complicated single variable function (the generating function) and a simple function of several variables ($\langle z, w \rangle$) seems to persist in a variety of situations; see S. Bell's recent paper [Bel94].

The computation in Exercise 2.1.4 can be found in [Ru80, p. 17]. The results of Exercises 2.1.11 and 2.1.12 appear in the work of N. Zorboska [Zo89a].

2.2 Some theorems on integration

The Littlewood Subordination Theorem [Lit25] can be used to show that composition operators whose symbol fixes 0 are bounded in a variety of spaces. It is frequently quite a different problem to investigate the boundedness of composition operators whose symbols do not fix 0. The subordination theorem is an integral inequality for subharmonic functions. Recall that a continuous real valued function G on a plane domain Ω is called **subharmonic** if for every domain U with $\overline{U}$ in Ω and every function u harmonic on U and continuous on $\overline{U}$ for which $G(z) \leq u(z)$ on the boundary of U, then $G(z) \leq u(z)$ on U also. Subharmonic functions are those that satisfy the local sub-mean-value property, that is, the value of $G(z)$ is no more than the average value of G on small circles centered at z. The most important subharmonic functions for our work are the functions $|f(z)|^p$ for f analytic and $p > 0$.

THEOREM 2.22   (*Littlewood Subordination Theorem*)
Let φ be an analytic map of the unit disk into itself such that $\varphi(0) = 0$. If G is a subharmonic function in D, then for $0 < r < 1$

$$\int_0^{2\pi} G(\varphi(re^{i\theta}))\, d\theta \leq \int_0^{2\pi} G(re^{i\theta})\, d\theta$$

PROOF   Let H be the function harmonic in rD and continuous on $\overline{rD}$ that agrees with G on the circle of radius r. By Schwarz's lemma, $|\varphi(z)| \leq r$ for $|z| \leq r$, so $H(\varphi(z))$ is well defined and harmonic in rD also. Since $G(z) \leq H(z)$ in rD and $H(\varphi(0)) = H(0)$, we have

$$\int_0^{2\pi} G(\varphi(re^{i\theta})) \frac{d\theta}{2\pi} \leq \int_0^{2\pi} H(\varphi(re^{i\theta})) \frac{d\theta}{2\pi} = H(\varphi(0))$$

$$= H(0) = \int_0^{2\pi} H(re^{i\theta}) \frac{d\theta}{2\pi} = \int_0^{2\pi} G(re^{i\theta}) \frac{d\theta}{2\pi}$$

∎

Exercise 2.2.1 gives conditions for equality to hold in this inequality. As immediate corollaries, we show that means of subharmonic functions increase as the radii of the circles increase and derive the inequality that implies a composition operator on $H^p(D)$ whose symbol fixes 0 is a contraction.

COROLLARY 2.23
If G is subharmonic function in D, then for $0 < r_1 < r_2 < 1$

$$\int_0^{2\pi} G(r_1 e^{i\theta})\, d\theta \leq \int_0^{2\pi} G(r_2 e^{i\theta})\, d\theta$$

PROOF   Apply the theorem to the analytic function $\varphi(z) = r_1 z / r_2$ with $r = r_2$. ∎

COROLLARY 2.24
If φ is an analytic map of the unit disk into itself such that $\varphi(0) = 0$ and f is in $H^p(D)$, then $\|f \circ \varphi\|_p \leq \|f\|_p$.

PROOF   If f is analytic, then $G = |f|^p$ is subharmonic and $\|f \circ \varphi\|_p$ and $\|f\|_p$ are the pth roots of the integrals in the theorem. ∎

In Section 2.1 we announced our intention to identify the space $H^p(D)$ with the closed subspace of $L^p(\partial D)$ spanned by $e^{in\theta}$ for $n \geq 0$. The following result of J. Ryff, showing that for f in $H^p(D)$ the radial limit of $f \circ \varphi$ is almost everywhere equal to the composition of the radial limits of f and φ, justifies ignoring the distinction in connection with our study composition operators. (In the expression $f^* \circ \varphi^*$, we mean by f^* the function on the closed disk that takes the value of f in D and the value of the radial limit of f on ∂D.)

PROPOSITION 2.25
If f is in $H^p(D)$ where $p > 0$ and φ is an analytic map of the unit disk into itself, then $(f \circ \varphi)^* = f^* \circ \varphi^*$ almost everywhere.

PROOF   Since every function in $H^p(D)$ is a quotient of two $H^\infty(D)$ functions ([Dur70, p. 16]) it is enough to prove the theorem for f in $H^\infty(D)$. Given such an f there is a set $E \subset \partial D$ of full measure so that $(f \circ \varphi)$ and φ have radial limits at every point of E. Write $E = E_1 \cup E_2$ where φ has radial limits of modulus one on E_1 and radial limits of modulus less that one on E_2. For ζ in E_2 we clearly have $(f \circ \varphi)^* = f \circ \varphi^* = f^* \circ \varphi^*$ by the continuity of f in D. For ζ in E_1, the definition of E guarantees that f has a limit along the arc $\varphi(r\zeta)$ for $0 < r < 1$. Lindelöf's Theorem (for example, see [Ru87, p. 259]) then applies to show that f has radial limit at $\varphi^*(\zeta)$ equal to $(f \circ \varphi)^*(\zeta)$, in other words, $f^*(\varphi^*(\zeta)) = (f \circ \varphi)^*(\zeta)$ as desired.   ∎

The inner–outer factorization theory of $H^p(D)$ functions, which plays a major role in the study of multiplication operators and their extensions, is less important in the study of composition operators. We present only a small part of what is known and interesting from this theory.

DEFINITION 2.26   *The function g is called an **inner function** if it is a bounded analytic function on the unit disk such that $\lim_{r \to 1} |g(re^{i\theta})| = 1$ almost everywhere. A nonzero function F in $H^1(D)$ is called an **outer function** if*

$$\log|F(0)| = \int_0^{2\pi} \log|F(e^{i\theta})| \frac{d\theta}{2\pi}$$

THEOREM 2.27   (Inner–Outer Factorization)
If f is a nonzero function in $H^1(D)$ there is an inner function g and an outer function F in $H^1(D)$ such that $f = gF$. Moreover, g and F are unique up to multiplication by a constant of modulus 1.

A proof of this theorem can be found in [Hof62] or [Dur70]. The outer function F is not zero in the disk, but there are also inner functions that do not vanish in the disk. We will content ourselves with proving only a more modest factorization into the product of an inner function carrying the information about the zeros and a nonzero function that is not necessarily outer (Theorem 2.28).

The simplest inner functions are the *finite Blaschke products*. The function B is a finite Blaschke product if there are non-zero points of the disk $\alpha_1, \alpha_2, \ldots, \alpha_n$, not necessarily distinct, $|\lambda| = 1$, and k a non-negative integer, so that

$$B(z) = \lambda z^k \prod_{j=1}^{n} \frac{|\alpha_j|}{\alpha_j} \frac{\alpha_j - z}{1 - \overline{\alpha_j} z}$$

In this case, B vanishes at the points α_j (and 0 if $k > 0$). Geometrically, a finite Blaschke product is a branched cover of the unit disk, and the finite Blaschke products are precisely the proper analytic maps of D to itself.

An infinite Blaschke product is a non-zero function of the same form but with infinitely many points α_j prescribed. Since the partial products are all bounded by 1 in the disk, the infinite product converges to a non-trivial analytic function if and only if the product without the factor z^k converges to a non-zero value for $z = 0$, that is, if and only if $\prod |\alpha_j|$ converges, which is equivalent to $\sum(1 - |\alpha_j|) < \infty$. A sequence satisfying this condition is called a *Blaschke sequence*.

If B is an infinite Blaschke product, then we can see that it is an inner function. Indeed, if B_p and B_q are partial products of B with $p < q$,

$$B_q(e^{i\theta})\overline{B_p(e^{i\theta})} = \frac{B_q(e^{i\theta})}{B_p(e^{i\theta})} = \prod_{p+1}^{q} \frac{|\alpha_j|}{\alpha_j} \frac{\alpha_j - e^{i\theta}}{1 - \overline{\alpha_j} e^{i\theta}}$$

since these functions have modulus 1 on the unit circle. Thus, the boundary values of $B_q \overline{B_p}$ agree with the boundary values of the function B_q/B_p which is analytic in D. Therefore,

$$\|B_p - B_q\|^2 = \int_0^{2\pi} |B_p|^2 - 2\mathrm{Re}(B_q \overline{B_p}) + |B_q|^2 \, \frac{d\theta}{2\pi} = 2\mathrm{Re} \int_0^{2\pi} 1 - \frac{B_q}{B_p} \frac{d\theta}{2\pi}$$

$$= 2\mathrm{Re}\left(1 - \frac{B_q(0)}{B_p(0)}\right) = 2\left(1 - \prod_{p+1}^{q} |\alpha_j|\right)$$

Since the infinite product $\prod |\alpha_j|$ converges, this shows that the partial products are a Cauchy sequence in $H^2(D)$. The limit of the sequence of partial products must be B because $H^2(D)$ is a functional Hilbert space and the analytic functions converge to B in the open disk. Now a norm convergent sequence in $H^2 \subset L^2$ has a subsequence that converges pointwise almost everywhere. The fact that each of the partial products has modulus 1 on the circle, implies B does also, that is, B is an inner function. This observation leads to an easy factorization theorem for bounded analytic functions.

THEOREM 2.28
If φ is an analytic map of D into itself such that $\varphi(0) \neq 0$, then there is a Blaschke product B and a non-vanishing analytic function ψ with $|\psi(z)| \leq 1$ in the disk so that

$$\varphi(z) = B(z)\psi(z)$$

In particular, $|\varphi(0)| \leq \prod |\alpha_j|$ where the α_j are the zeros of φ.

PROOF Let $\alpha_1, \alpha_2, \ldots$ be the zeros of φ and let B_p be the finite Blaschke product with the first p of the zeros. Since B_p has modulus 1 on the unit circle, $|\varphi/B_p| \leq 1$ on the circle and therefore in D. Thus, for every p,

$$|\varphi(0)| \leq |B_p(0)| = \prod_1^p |\alpha_j|$$

It follows that the infinite product $\prod |\alpha_j|$ converges and, from the above discussion, the finite Blaschke products B_p converge to an infinite Blaschke product B. Clearly, $\psi = \varphi/B$ satisfies the conditions of the theorem. ∎

It is an immediate consequence of Proposition 2.25 and the definition of inner function that the composition of two inner functions is again an inner function. Not all inner functions are Blaschke products and it is *not* the case that the composition of two Blaschke products is again a Blaschke product.

Littlewood's inequality, which we will see is a restatement of the inequality of Theorem 2.28, extends the Schwarz lemma: the Schwarz lemma says that if φ maps the disk to itself and $\varphi(0) = 0$ then $|z_j| \geq |w|$ for each z_j with $\varphi(z_j) = w$, whereas Littlewood's inequality implies the product of all the $|z_j|$'s is greater than or equal to $|w|$. The statement of the inequality involves the ***Nevanlinna counting function***, which is a tool from value distribution theory. If φ is an analytic mapping on the disk and $w \neq \varphi(0)$ is a point of the plane, let z_j be the points of the disk for which $\varphi(z_j) = w$, with multiplicities. The Nevanlinna counting function is $N_\varphi(w) = \sum_j \log(1/|z_j|)$, where we understand $N_\varphi(w) = 0$ for w not in $\varphi(D)$. The case in which equality holds in Littlewood's Inequality is explored somewhat in Lemma 3.27.

THEOREM 2.29   (*Littlewood's Inequality*)
Let φ be an analytic map of the unit disk into itself.
For each w in $D \setminus \{\varphi(0)\}$,

$$N_\varphi(w) \leq \log \left| \frac{1 - \overline{\varphi(0)}w}{\varphi(0) - w} \right|$$

PROOF Suppose ψ is an analytic map of D into itself and $\psi(0) \neq 0$. If $\{z_j\}$ is the set of zeros of ψ, Theorem 2.28 shows that $|\psi(0)| \leq \prod |z_j|$. Since the logarithm is an increasing function, we get

$$\log |\psi(0)| \le \log \left(\prod |z_j| \right) = \sum \log |z_j|$$

Multiplying by -1, we have

$$N_\psi(0) = \sum \log \frac{1}{|z_j|} \le \log \frac{1}{|\psi(0)|}$$

This is the case $w = 0$ of the desired inequality.

To get the general case of the inequality, let

$$\psi(z) = \frac{\varphi(z) - w}{1 - \overline{w}\varphi(z)}$$

If φ is an analytic mapping of the disk into itself and w is in the disk, but $\varphi(0) \ne w$, then ψ satisfies $|\psi(z)| < 1$ and $\psi(0) \ne 0$ so the inequality holds. Since $\psi(z) = 0$ if and only if $\varphi(z) = w$, this is the inequality of the theorem. ∎

The following is a version of the Littlewood–Paley identity which gives the $H^2(D)$ norm of an analytic function (possibly infinity) in terms of a weighted area integral. A version for f in $L^1(\partial D)$, with a proof using Green's theorem, can be found in Garnett [Ga81, p. 236].

THEOREM 2.30 *(Littlewood–Paley Identity)*
If f is analytic in D, then

$$\|f\|_2^2 - |f(0)|^2 = \int_D |f'(z)|^2 \log \frac{1}{|z|^2} \frac{dA(z)}{\pi}$$

where $\| \sum a_j z^j \|_2^2 = \sum |a_j|^2$.

PROOF Since $\int_0^1 |\log(t)| \, dt$ is finite, the singularity of the logarithm at 0 causes no difficulty. For $0 < \rho < 1$ and $f(z) = \sum a_j z^j$, we have

$$\int_{\rho D} |f'(z)|^2 \log \frac{1}{|z|^2} \frac{dA}{\pi}$$

$$= -\frac{2}{\pi} \sum_{j=1}^\infty \sum_{k=1}^\infty jk a_j \overline{a_k} \int_0^\rho \int_0^{2\pi} r^{j-1} e^{i(j-1)\theta} r^{k-1} e^{-i(k-1)\theta} (\log r) r \, d\theta \, dr$$

$$= -4 \sum_{j=1}^\infty j^2 |a_j|^2 \int_0^\rho r^{2j-1} \log r \, dr = \sum_{j=1}^\infty (\rho^{2j} - 2j\rho^{2j} \log \rho)|a_j|^2$$

Now $\rho^{2j} - 2j\rho^{2j} \log \rho$ is an increasing function of ρ for each j, so we may take the limit as ρ tends to 1 term by term in the series to obtain the desired result. ∎

C. S. Stanton [Es85, Sta86] has proved a change of variables formula that extends the Littlewood–Paley identity. We present a specialization of his formula due to J. H. Shapiro [Sho87a]

THEOREM 2.31
If f is analytic in the unit disk and φ is a non-constant analytic mapping of D into itself, then

$$\|f \circ \varphi\|_2^2 - |f(\varphi(0))|^2 = 2 \int_D |f'(w)|^2 N_\varphi(w) \frac{dA(w)}{\pi}$$

where $\| \sum a_j z^j \|_2^2 = \sum |a_j|^2$.

PROOF Applying the Littlewood–Paley identity to the function $f \circ \varphi$ and using the chain rule, we obtain

$$\|f \circ \varphi\|_2^2 - |f(\varphi(0))|^2 = \int_D |(f \circ \varphi)'(z)|^2 \log \frac{1}{|z|^2} \frac{dA(z)}{\pi}$$

$$= 2 \int_D |f'(\varphi(z))|^2 |\varphi'(z)|^2 \log \frac{1}{|z|} \frac{dA(z)}{\pi}$$

We wish to change variables in the final integral. The function φ is locally univalent on D except at the (at most countably many) points at which the derivative vanishes, so we can find a countable collection of disjoint open sets Ω_j so that the area of $D \setminus \cup \Omega_j$ is zero and the restriction of φ to each Ω_j is univalent with $\psi_j(w)$ the inverse mapping of φ that takes $\varphi(\Omega_j)$ onto Ω_j. Then for each j, by the usual change of variables formula, if $w = \varphi(z)$ then $dA(w) = |\varphi'(z)|^2 dA(z)$ and

$$\int_{\Omega_j} |f'(\varphi(z))|^2 |\varphi'(z)|^2 \log \frac{1}{|z|} \frac{dA(z)}{\pi} = \int_{\varphi(\Omega_j)} |f'(w)|^2 \log \frac{1}{|\psi_j(w)|} \frac{dA(w)}{\pi}$$

Writing χ_j for the characteristic function of $\varphi(\Omega_j)$, we obtain

$$\int_D |f'(\varphi(z))|^2 |\varphi'(z)|^2 \log \frac{1}{|z|} \frac{dA(z)}{\pi}$$

$$= \sum_j \int_{\Omega_j} |f'(\varphi(z))|^2 |\varphi'(z)|^2 \log \frac{1}{|z|} \frac{dA(z)}{\pi}$$

$$= \sum_j \int_{\varphi(\Omega_j)} |f'(w)|^2 \log \frac{1}{|\psi_j(w)|} \frac{dA(w)}{\pi}$$

$$= \int_D |f'(w)|^2 \sum_j \chi_j(w) \log \frac{1}{|\psi_j(w)|} \frac{dA(w)}{\pi}$$

$$= \int_D |f'(w)|^2 N_\varphi(w) \frac{dA(w)}{\pi}$$

∎

We will also need a generalization of the change of variable argument of Theorem 2.31. In the following formula, $\{z_j(w)\}$ denotes the sequence of zeros of $\varphi(z) - w$, repeated according to multiplicity.

THEOREM 2.32   *(Area Formula)*
If g and W are non-negative measurable functions on D, then

$$\int_D g(\varphi(z))|\varphi'(z)|^2 W(z)\,dA(z) = \int_{\varphi(D)} g(w)\left(\sum_{j\geq 1} W(z_j(w))\right) dA(w)$$

PROOF   As in the proof of Theorem 2.31 we may find a countable collection of disjoint open sets Ω_j with the area of $D \setminus \cup\Omega_j$ equal to zero and the restriction of φ to each Ω_j univalent. Write $\psi_j(w)$ for the inverse of φ taking $\varphi(\Omega_j)$ onto Ω_j. Then the standard change of variable formula gives

$$\int_{\Omega_j} g(\varphi(z))W(z)|\varphi'(z)|^2\,dA(z) = \int_{\varphi(\Omega_j)} g(w)W(\psi_j(w))\,dA(w)$$

where $z = \psi_j(w)$. Now denoting the characteristic function of $\varphi(\Omega_j)$ by χ_j we have

$$\int_D g(\varphi(z))W(z)|\varphi'(z)|^2\,dA(z)$$

$$= \int_{\varphi(D)} g(w)\left(\sum_{j\geq 1}\chi_j(w)W(\psi_j(w))\right) dA(w)$$

$$= \int_{\varphi(D)} g(w)\left(\sum_{j\geq 1}W(z_j(w))\right) dA(w)$$

∎

Carleson measure theorems

In 1962, in connection with his work on the corona problem, L. Carleson [Cal62] developed inequalities relating behavior of an $H^p(D)$ function in the disk with its behavior on the unit circle. The inequalities assert the continuity of inclusion maps from $H^p(D)$ into certain measure spaces. These inequalities and their generalizations to other inclusion maps are frequently referred to as Carleson inequalities and the underlying measures as Carleson measures.

For b on the unit circle and $0 < h < 1$, let

$$S(b, h) = \{z \in D : |z - b| < h\}$$

that is, the unit disk intersected with the disk of radius h centered at b (see Figure 2.1). The next result [Cal62, p. 548] characterizes measures μ on the disk for which $H^p(D)$ is contained in $L^p(\mu)$. By the closed graph theorem (which holds even when $p < 1$ [DuS88, p. 57]), this is equivalent to the continuity of the inclusion map from $H^p(D)$ into $L^p(\mu)$.

THEOREM 2.33 (*Carleson's Theorem*)
For μ a finite, positive Borel measure on D and $0 < p < \infty$, the following are equivalent:

(1) *There is a constant $K < \infty$ so that $\mu(S(b,h)) < Kh$ for $|b| = 1$ and $0 < h < 1$.*

(2) *There is a constant C so that*

$$\int_D |f|^p \, d\mu \le C\|f\|_p^p$$

for all f in $H^p(D)$.

(3) *There is a constant C' so that*

$$\int_D \left|\frac{1}{1 - \overline{w}z}\right|^2 d\mu(z) \le C' \frac{1}{1 - |w|^2}$$

for every w in D.

DEFINITION 2.34 *Measures that satisfy these equivalent conditions are called* **Carleson measures** *for the Hardy spaces in D.*

The sets $S(b, h)$ can be replaced by equivalent choices which, in specific applications, may be easier to use. In particular, one sometimes uses the **Carleson windows** (see Figure 2.1) defined for b in the unit circle and $0 < h < 1$:

$$W(b, h) = \{z \in D : 1 - h < |z| < 1 \text{ and } z/|z| \in S(b, h)\}$$

Since there is a constant $c > 1$ so that $W(b, h/c) \subset S(b, h) \subset W(b, h)$, condition (1) holds for the sets $S(b, h)$ if and only if it holds for the sets $W(b, h)$ (where the value of K may change).

We will base the proof of the hard part of Carleson's theorem on a result about nontangential maximal functions for $H^p(D)$ functions (which we will not prove). As in Figure 2.2, for $0 \le \theta \le 2\pi$, let

$$G_\theta = \{z : |e^{i\theta} - z| < 3(1 - |z|)\}$$

(See Section 2.3 for a more complete description of nontangential approach regions.) For f defined on the disk, let M_f be the function

$$M_f(e^{i\theta}) = \sup\{|f(z)| : z \in G_\theta\}$$

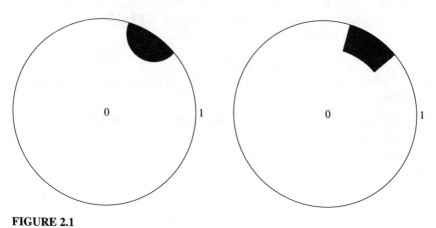

FIGURE 2.1

The sets $S(b, .3)$ and $W(b, .3)$

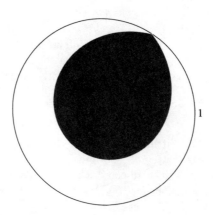

FIGURE 2.2

A typical G_θ

which is a nontangential maximal function. There is a constant c such that for all f in $H^p(D)$, the nontangential maximal function satisfies

$$\int_0^{2\pi} M_f(e^{i\theta})^p \, d\theta \leq c \int_0^{2\pi} |f(e^{i\theta})|^p \, d\theta$$

The constant c does not depend on p. A proof of this theorem can be found in [Ga81, p. 57] or [Koo80, p. 247].

PROOF (of Theorem 2.33) If (2) holds for some $0 < p < \infty$, it holds for $p = 2$. To see this, factor f in $H^2(D)$ as $f = BF$ where B is Blaschke and F is non-zero with $\|F\|_2 = \|f\|_2$ and consider $F^{2/p}$. Condition (3) of the theorem is then the special case of condition (2) where $p = 2$ and f is the reproducing kernel K_w, so (2)$\Rightarrow$(3) is trivial.

The implication (3)$\Rightarrow$(1) follows from the estimate on the kernel functions

$$|K_w(z)|^2 \geq \frac{1}{4h^2}$$

for b on the circle, $0 < h < 1$, $w = (1 - h)b$, and z in $S(b, h)$. Thus,

$$\int_D |K_w|^2 \, d\mu \geq \int_{S(b,h)} |K_w|^2 \, d\mu \geq \frac{\mu(S(b,h))}{4h^2}$$

which together with (3) gives

$$\mu(S(b,h)) \leq 4h^2 \int_D |K_w|^2 \, d\mu \leq C\frac{4h^2}{1 - |w|^2} = C\frac{4h^2}{2h - h^2} \leq 4Ch$$

We are ready to prove (1) implies (2). Let f in $H^p(D)$ be given. For $\lambda > 0$, note that, by lower semicontinuity, the set $\{e^{i\theta} : M_f(e^{i\theta}) > \lambda\}$ is open, so it consists of countably many open arcs I_j on the unit circle. For λ large enough that this set is not the whole circle, let S_j be the set $S(b, h)$ for which b is the midpoint of I_j and the radius h is such that the circle $|b - z| = h$ passes through the end points of I_j. If $z = re^{i\theta_0}$ and $|f(z)| > \lambda$, let J_z be the arc $\{e^{i\theta} : z \in G_\theta\}$. The point $e^{i\theta_0}$ is the midpoint of J_z and if ζ is an end point of J_z, by the definition of G_θ, $|z - \zeta| = 3(1 - |z|) = 3|e^{i\theta_0} - z|$. Thus,

$$|e^{i\theta_0} - \zeta| \geq |z - \zeta| - |e^{i\theta_0} - z| = 2|e^{i\theta_0} - z|$$

which means that z is in $S(e^{i\theta_0}, |e^{i\theta_0} - \zeta|)$. Now z in G_θ implies $M_f(e^{i\theta}) > \lambda$ so $M_f(e^{i\theta}) > \lambda$ for $e^{i\theta}$ in J_z. This means J_z is contained in I_j for some j so $S(e^{i\theta_0}, |e^{i\theta_0} - \zeta|)$ is a subset of S_j and z is in S_j. Now (1) implies

$$\mu(\{z : |f(z)| > \lambda\}) \leq \sum \mu(S_j)$$

$$\leq K \sum |I_j| = K \left|\{e^{i\theta} : M_f(e^{i\theta}) > \lambda\}\right|$$

Therefore, using the change of variables formula for distribution functions [Ru87, p. 172],

$$\int_D |f(z)|^p \, d\mu = \int_0^\infty p\lambda^{p-1}\mu(\{z : |f(z)| > \lambda\}) \, d\lambda$$

$$\leq K \int_0^\infty p\lambda^{p-1} \left|\{e^{i\theta} : M_f(e^{i\theta}) > \lambda\}\right| \, d\lambda = K \int_0^{2\pi} M_f(e^{i\theta})^p \, d\theta$$

Using the maximal function result, we obtain the conclusion

$$\int_D |f(z)|^p \, d\mu \leq Kc\|f\|_p^p$$

∎

Note that the proof shows that the constant C of (2) can be chosen to be a constant multiple of the constant K of (1), so that when K is small, so is C.

We will have occasion to use a variation of Theorem 2.33 for measures on the closed disk $\overline{D}$. Let $\mathcal{S}(b, h)$ denote the set

$$\mathcal{S}(b, h) = \{z : |z| \leq 1 \text{ and } |z - b| < h\}$$

where $|b| = 1$ and $0 < h < 1$.

THEOREM 2.35
For μ a finite, positive Borel measure on $\overline{D}$ and $0 < p < \infty$, the following are equivalent:

(1)   There is a constant K so that $\mu(\mathcal{S}(b, h)) < Kh$ for $|b| = 1$ and $0 < h < 1$.

(2)   There is a constant C so that

$$\int_{\overline{D}} |f|^p \, d\mu \leq C\|f\|_p^p$$

for all f in $H^p(D)$.

The implication (2)⇒(1) follows exactly as before, setting $p = 2$ and using the test function K_w for $w = (1 - h)b$. For the other direction, let $Q(b, h) = \mathcal{S}(b, h) \cap \partial D$. Of course, $Q(b, h)$ is an arc of the unit circle centered at b and its Lebesgue measure is comparable to h. If λ is a positive measure on the circle with $\lambda(Q(b, h)) \leq Ch$ for all $|b| = 1$ and $h > 0$, then the Radon–Nikodym Theorem implies λ is absolutely continuous with respect to Lebesgue measure and its Radon–Nikodym derivative g satisfies $\|g\|_\infty \leq C$. Now if μ is a measure on $\overline{D}$, we can write it as $\mu = \mu_0 + \lambda$ where μ_0 is supported on D and λ is supported on ∂D. Thus, (1)⇒(2) follows from these observations concerning λ and Theorem 2.33.

Extensions of the Carleson Theorem can be proved for the weighted Bergman spaces defined for $\alpha > -1$ by

$$A_\alpha^p(D) = \left\{ f \text{ analytic in } D : \int_D |f(z)|^p (1 - |z|^2)^\alpha \frac{dA(z)}{\pi} < \infty \right\}$$

While we could use the sets $S(b,h)$ as the "Carleson sets" in the theorem (see Exercise 2.2.7), we make the more convenient choice of pseudohyperbolic disks. Fix a value for r with $0 < r < 1$. For w in the disk, denote by $D(w,r)$, or simply $D(w)$ when no confusion can result, the disk whose pseudohyperbolic center is w and whose pseudohyperbolic radius is r:

$$D(w,r) = \left\{ z : \left| \frac{z - w}{1 - \overline{w}z} \right| < r \right\}$$

Since $D(w,r)$ is the image of the (Euclidean) disk of radius r centered at 0 under the linear fractional map $(w - z)/(1 - \overline{w}z)$, it is also a Euclidean disk but its Euclidean center and its pseudohyperbolic center do not coincide unless $w = 0$. Some basic facts about pseudohyperbolic disks are outlined in the exercises.

THEOREM 2.36
Let $0 < p < \infty$ and let $\alpha > -1$. Fix r with $0 < r < 1$. For μ a finite, positive Borel measure on D the following are equivalent:

(1) There is a constant $K < \infty$ so that for all w in the disk,

$$\mu(D(w,r)) \le K(1 - |w|^2)^{\alpha+2}$$

(2) There is a constant $C_1 < \infty$

$$\int_D |f(z)|^p \, d\mu \le C_1 \int_D |f(z)|^p (1 - |z|^2)^\alpha \, dA(z)/\pi$$

for every f in $A_\alpha^p(D)$.
(3) There is a constant $C_2 < \infty$ so that

$$\int_D \left| \frac{1}{1 - \overline{w}z} \right|^{2\alpha+4} d\mu(z) \le C_2 \frac{1}{(1 - |w|^2)^{\alpha+2}}$$

for every w in D.

The right hand side of (1) is best interpreted (see Exercise 2.2.7) as being comparable to $\int_{D(w,r)} (1 - |z|^2)^\alpha \, dA(z)$.

PROOF Condition (3) of the theorem is the special case of Condition (2) obtained (see Exercise 2.1.5) by choosing f to be

$$\left(\frac{1}{1 - \overline{w}z} \right)^{(2\alpha+4)/p}$$

The implication (3)⇒(1) again follows from an estimate for the kernel functions on the Carleson sets. Indeed, since

$$1 - \left| \frac{z-w}{1-\overline{w}z} \right|^2 = \frac{(1-|w|^2)(1-|z|^2)}{|1-\overline{w}z|^2}$$

we have, for z in $D(w,r) \equiv D(w)$,

$$\frac{1}{|1-\overline{w}z|^2} \geq \frac{1-r^2}{(1-|w|^2)(1-|z|^2)}$$

Squaring both sides and multiplying by $|1-\overline{w}z|^2$ we get

$$\frac{1}{|1-\overline{w}z|^2} \geq \frac{(1-r^2)^2|1-\overline{w}z|^2}{(1-|w|^2)^2(1-|z|^2)^2} \geq \frac{(1-r^2)^2(1-|z|)^2}{(1-|w|^2)^2(1-|z|^2)^2}$$

$$= \frac{(1-r^2)^2}{(1-|w|^2)^2(1+|z|)^2} \geq \frac{(1-r^2)^2}{4} \frac{1}{(1-|w|^2)^2}$$

Thus, raising this inequality to the $\alpha+2$ power and using (3), we obtain

$$C_2 \frac{1}{(1-|w|^2)^{\alpha+2}} \geq \int_D \left| \frac{1}{1-\overline{w}z} \right|^{2\alpha+4} d\mu(z)$$

$$\geq \int_{D(w)} \left| \frac{1}{1-\overline{w}z} \right|^{2\alpha+4} d\mu(z)$$

$$\geq \left(\frac{(1-r^2)^2}{4} \right)^{\alpha+2} \frac{1}{(1-|w|^2)^{2(\alpha+2)}} \mu(D(w))$$

Now, multiplying by $(1-|w|^2)^{2\alpha+4}$ gives the desired result.

As in Theorem 2.33, the difficult implication is (1)⇒(2). We will not give the proof here but instead refer the interested reader to [Ax88]. ∎

We emphasize that in (1) of Theorem 2.36, the sets $D(w,r)$ (for fixed r) can be equivalently replaced by the sets $S(b,h)$ (as b ranges over the unit circle and $0 < h < 1$) or the windows $W(b,h)$ and in future applications we will pass between these choices as suits our purposes.

There are versions of Carleson's Theorem that apply to the analogous spaces in the unit ball B_N of C^N for $N > 1$. For the Hardy spaces, the Carleson sets are certain "non-isotropic balls". For b in C^N with $|b| = 1$, and $0 < h < 1$, let

$$S(b,h) = \{z \in B_N : |1 - \langle z, b \rangle| < h\}$$

and

$$\mathcal{S}(b,h) = \{z \in \overline{B_N} : |1 - \langle z, b \rangle| < h\}$$

(Note that these sets specialize to those previously defined when $N = 1$.) To visualize these sets, consider $b = e_1 = (1, 0, \ldots, 0)$ so that $S(e_1, h) = \{z \in B_N : |1 - z_1| < h\}$. Its intersection with the complex line through 0 and e_1 is the set $S(1, h)$ in one variable which is the part of a disk of radius h. On the other hand, if z' lies in the $N - 1$ dimensional ball of radius $\sqrt{2h - h^2}$, which is much larger than h when h is small, then $(1 - h, z')$ will be in the closure of $S(e_1, h)$. Since $U(S(b, h)) = S(Ub, h)$ for any unitary map U of C^N, the same comments apply for an arbitrary b. If we set $Q(b, h) = S(b, h) \cap \partial B_N$, then the above remarks should make it plausible that the surface measure of $Q(b, h)$ is on the order of $h(\sqrt{h})^{2N-2} = h^N$. A precise verification of this fact can be found in Rudin's book [Ru80, p. 67]. Carleson measure theorems for Hardy spaces on the ball take the following form.

THEOREM 2.37
For μ a finite, positive Borel measure on $\overline{B_N}$ and $0 < p < \infty$, the following are equivalent:

(1) *There is a constant $K < \infty$ so that $\mu(S(b, h)) < Kh^N$ for $|b| = 1$ and $0 < h < 1$.*

(2) *There is a constant $C < \infty$ so that*

$$\int_{\overline{B_N}} |f|^p \, d\mu \le C \|f\|_p^p$$

for all f in $H^p(B_N)$.

When K is small, so is C. We will not prove this result here except to note that a proof of the easier direction (2)⇒(1), which uses appropriate "test" functions, is essentially contained in the proof of Theorem 3.35 in the next chapter. A complete proof can be obtained by combining the result in [Pow85] with the remarks in [Mc85, Theorem 1.5].

Similarly we have a Carleson measure theorem for weighted Bergman spaces in B_N, $N > 1$. For $\alpha > -1$ and $0 < p < \infty$ define the standard weighted Bergman spaces $A^p_\alpha(B_N)$ to be the analytic functions f on B_N for which

$$\int_{B_N} |f(z)|^p (1 - |z|^2)^\alpha \, d\nu_N(z) < \infty$$

Write $\|f\|_\alpha^p$ for this integral.

THEOREM 2.38
Let $0 < p < \infty$ and let $\alpha > -1$. For μ a finite, positive Borel measure on B_N the following are equivalent:

(1) *There is a constant $K < \infty$ so that for all w in the disk,*

$$\mu(S(b, h)) \le Kh^{\alpha+N+1}$$

(2)   *There is a constant $C < \infty$*

$$\int_{B_N} |f(z)|^p \, d\mu \leq C \int_{B_N} |f(z)|^p (1 - |z|^2)^\alpha \, d\nu_N(z)$$

for all f in $A_\alpha^p(B_N)$.

Exercises

2.2.1   In Theorem 2.22, show that there is r, $0 < r < 1$, such that

$$\int_0^{2\pi} G(\varphi(re^{i\theta})) \, d\theta = \int_0^{2\pi} G(re^{i\theta}) \, d\theta$$

if and only if φ is a rotation or G is harmonic.

2.2.2   Suppose φ is a non-constant analytic map of D into D.

    (a) Use Littlewood's inequality, the Littlewood–Paley Identity and Theorem 2.31 to prove that if $\varphi(0) = 0$ then on $H^2(D)$, $\|f \circ \varphi\| \leq \|f\|$. This gives an alternative approach to Corollary 2.24 for the case $p = 2$.

    (b) Can the method of part (a) be used to show that C_φ is bounded on $H^2(D)$ even if $\varphi(0) \neq 0$? Hint: $-\log x \sim 1 - x$ for x near 1.

2.2.3   Let $0 \leq r_1 \leq r_2 \leq r_3 \leq \cdots \leq 1$ be a sequence and let δ_j be point mass at r_j, that is, δ_j is the Borel measure on D given by $\delta_j(E) = 1$ if r_j is in E and $\delta_j(E) = 0$ if r_j is not in E. Find sufficient conditions on $\{r_j\}$ so that $\mu = \sum(1 - r_j)\delta_j$ is a Carleson measure for the Hardy space.

2.2.4   Let $0 \leq r_1 \leq r_2 \leq r_3 \leq \cdots \leq 1$ be a sequence so that $\mu = \sum(1 - r_j)\delta_j$ is a Carleson measure, as in the exercise above. Show that if f is any function in $H^2(D)$, then the sequence $\{f(r_j)(1 - r_j)^{1/2}\}$ is in ℓ^2.

2.2.5   Show that Lebesgue measure on the unit interval is a Carleson measure for the Hardy space and use this fact to prove Hilbert's inequality: There is a constant A so that

$$\sum_{j,k=0}^{\infty} \frac{c_j \overline{c_k}}{1 + j + k} \leq A \sum_{j=0}^{\infty} |c_j|^2$$

(This inequality shows that the Hilbert matrix defines a bounded operator ℓ^2.)

2.2.6   Fix $0 < r < 1$ and define the pseudohyperbolic disk

$$D(w) = \{z : \left| \frac{z - w}{1 - \overline{w}z} \right| < r\}$$

    (a) Show that $D(w)$ is a Euclidean disk with Euclidean center

$$\frac{1 - r^2}{1 - r^2|w|^2} w$$

and Euclidean radius

$$r \frac{1 - |w|^2}{1 - r^2|w|^2}$$

In particular,

$$\frac{|D(w)|}{(1 - |w|^2)^2}$$

is bounded above and below by a constant depending only on r, where $|D(w)|$ denotes the area of $D(w)$.

(b) Show that for each $D(w)$ there is a b in ∂D so that $D(w) \subset S(b,t)$ for $t \sim 1 - |w|$; where $\sim$ means that the ratio of the two quantities is bounded above and below by positive constants independent of w.

2.2.7 (a) Noting that $\varphi_w(z) = (w - z)/(1 - \overline{w}z)$ has derivative $\varphi'_w(z) = (|w|^2 - 1)/(1 - \overline{w}z)^2$ use a change of variables to show that for fixed r, $0 < r < 1$

$$\int_{D(w)} (1 - |z|^2)^\alpha dA(z)$$

is bounded above and below by a constant multiple (depending on r) of $(1 - |w|^2)^{\alpha+2}$.

(b) Show that condition (1) of Theorem 2.36 can be replaced by

(1′) There is a constant K' so that $\mu(S(b,h)) \le K' h^{\alpha+2}$

2.2.8 Show that

$$\int_{S(b,h)} (1 - |z|^2)^\alpha \, d\nu_n(z) \sim h^{\alpha+N+1}$$

and thus the right hand side of (1) in Theorem 2.38 can be replaced by $\nu_\alpha S(b,h)$ where $d\nu_\alpha = (1 - |z|^2)^\alpha \, d\nu_N$.

Notes

The presentation of Theorem 2.22 is due to P. Duren [Dur70]. Proposition 2.25 appears in the work of J. V. Ryff [Ry66] on subordination of $H^p(D)$ functions. The treatment of inner–outer factorization given here is very incomplete; the books of K. Hoffman [Hof62] and Duren [Dur70] devote a great deal of attention to this theory. The formulation of Theorem 2.27 on factorization and the proof that an infinite Blaschke product is an inner function follow Hoffman [Hof62, pp. 63,64]. J. H. Shapiro [Sho87a] was apparently the first to use C. Stanton's change of variables theorem (Theorem 2.31) in the study of composition operators and the statement and proof of that theorem follows Shapiro.

Carleson measure theorems in various settings have an extensive history, beginning with the $H^p(D)$ version due to L. Carleson [Cal62]. L. Hormander gave a Carleson measure theorem for Hardy spaces on strictly pseudoconvex domains in [Hor67] and S. Power [Pow85] gave a simplified proof in the case of the unit ball. The unweighted Bergman space version for the disk and polydisk appears in W. Hastings' paper [Hast75]. A proof which emphasizes the naturalness of pseudohyperbolic disks as the Carleson sets is due to S. Axler [Ax88]. Versions for weighted Bergman spaces in the disk can be found in work of D. Stegenga [Stg80] and for weighted Bergman spaces in the ball in work of J. Cima and W. Wogen [CiW82]. D. Luecking [Lue83] describes a general technique for characterizing Carleson measure on a variety of Bergman spaces. Luecking's result is used by J. A. Cima and P. R. Mercer [CiMe94] who prove a Carleson measure theorem for Bergman spaces on strictly pseudoconvex domains in C^N.

The Exercises 2.2.3, 2.2.4 and 2.2.5 stem from Carleson's original paper [Cal62] and H. S. Shapiro and A. L. Shields' paper [ShS61]. Hilbert's inequality had several proofs in the early part of this century with that of Hardy [Hay20] among the most famous.

2.3   Geometric function theory in the disk

It is elementary to prove that the **automorphisms** of the unit disk, that is, the one-to-one analytic maps of the disk onto itself, are just the functions

$$\varphi(z) = \lambda \frac{a - z}{1 - \overline{a}z}$$

where $|\lambda| = 1$ and $|a| < 1$. In terms of the mapping, $\lambda = -\varphi'(0)/|\varphi'(0)|$ and $a = \varphi^{-1}(0)$. When $\lambda = 1$, this automorphism is an involution, that is, $\varphi^{-1} = \varphi$, that exchanges 0 and a. It is easy to check that for any automorphism ψ of the disk

$$1 - |\psi(z)|^2 = \frac{(1 - |z|^2)(1 - |a|^2)}{|1 - \overline{a}z|^2} \tag{2.3.1}$$

where $a = \psi^{-1}(0)$ and $|z| \leq 1$. The group of automorphisms of the disk will be denoted $\text{Aut}(D)$.

Every disk automorphism is an automorphism of the Riemann sphere and has two fixed points on the sphere, counting multiplicity. The automorphisms are classified according to the location of their fixed points: elliptic if one fixed point is in the disk and the other is in the complement of the closed disk, for example $\varphi(z) = iz$ which has fixed points 0 and ∞, hyperbolic if both fixed points are on the unit circle, for example $\varphi(z) = (z + .5)/(1 + .5z)$ which has fixed points ± 1, and parabolic if there is one fixed point on the unit circle (of multiplicity 2), for example, $\varphi(z) = [(1 + i)z - i]/[iz + 1 - i]$ which has fixed point 1. (The other combinations of locations cannot occur as fixed points of a disk automorphism.) Dynamically, the elliptic automorphisms are "rotations" around the fixed point, hyperbolic automorphisms are flows from one fixed point to the other, and the

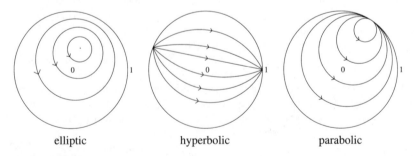

elliptic   hyperbolic   parabolic

FIGURE 2.3

Flow lines of automorphisms.

parabolic automorphisms are flows around the disk from the fixed point and back to it from the other side. A hyperbolic automorphism is conformally equivalent to translation (parallel to the edges) in a strip or dilation in the right halfplane. A parabolic automorphism is conformally equivalent to translation (parallel to the edge) in a halfplane.

From the point of view of analytic functions, the usual Euclidean metric on the disk is inappropriate. The automorphisms of the disk are isometries with respect to the Poincaré metric in which the length of a curve γ is

$$\int_\gamma \frac{|dz|}{1 - |z|^2}$$

or with respect to the (equivalent) pseudohyperbolic metric in which the distance between two points ζ_1, ζ_2 is

$$\left| \frac{\zeta_1 - \zeta_2}{1 - \overline{\zeta_1}\zeta_2} \right|$$

so these metrics are more useful than the Euclidean metric.

One view of automorphisms is that they represent a change of variables. In the study of a problem, we frequently wish to change variables to highlight a particular aspect of the problem. That the automorphisms preserve pseudohyperbolic distances gives a restriction on how much can be accomplished in changing variables. Exercise 2.3.1 illustrates how much flexibility we have; in particular, it illustrates the transitivity of $\text{Aut}(D)$.

It is frequently helpful to give theorems in a form which is invariant under the application of automorphisms. The invariant form of the Schwarz Lemma is the following. Many of the results of this section are consequences of this basic inequality.

THEOREM 2.39   *(Schwarz–Pick Theorem)*
If φ is an analytic map of the disk into itself, then

$$\left| \frac{\varphi(w) - \varphi(z)}{1 - \overline{\varphi(w)}\varphi(z)} \right| \leq \left| \frac{w - z}{1 - \overline{w}z} \right|$$

and if equality holds for any $z \neq w$, then φ is an automorphism of the disk.

PROOF   Let $u = \varphi(w)$. Since φ maps the disk into itself so does

$$\left(\frac{u - \varphi(z)}{1 - \overline{u}\varphi(z)} \right)$$

Since $(w - z)/(1 - \overline{w}z)$ is zero at $z = w$ and has modulus 1 on the unit circle, if we define ψ by

$$\left(\frac{u - \varphi(z)}{1 - \overline{u}\varphi(z)} \right) = \left(\frac{w - z}{1 - \overline{w}z} \right) \psi(z)$$

then ψ is analytic in the disk and by the maximum modulus theorem, satisfies $|\psi(z)| \leq 1$ for all z in D, which is the inequality we were to prove. If equality holds for some z in D with $z \neq w$, then $|\psi(z)| = 1$, so ψ is constant and

$$\left(\frac{u - \varphi(z)}{1 - \overline{u}\varphi(z)} \right) = \lambda \left(\frac{w - z}{1 - \overline{w}z} \right)$$

from which the equality condition follows. ∎

Geometrically, the Schwarz–Pick Theorem says that the image of the pseudohy-perbolic disk $D(w, r)$ under φ is contained in $D(\varphi(w), r)$. A helpful interpretation of the Schwarz–Pick Theorem is that in the Poincaré or pseudohyperbolic met-rics, a map of the disk into itself is a contraction (although not necessarily a strict contraction). As a corollary, we get an upper bound on the modulus of $\varphi(z)$.

COROLLARY 2.40
If φ is an analytic map of the disk into itself, then

$$|\varphi(z)| \leq \frac{|z| + |\varphi(0)|}{1 + |z||\varphi(0)|}$$

PROOF By Equation (2.3.1) we have

$$1 - \left| \frac{\varphi(0) - \varphi(z)}{1 - \overline{\varphi(0)}\varphi(z)} \right|^2 = \frac{(1 - |\varphi(0)|^2)(1 - |\varphi(z)|^2)}{|1 - \overline{\varphi(0)}\varphi(z)|^2}$$

so that

$$\left| \frac{\varphi(0) - \varphi(z)}{1 - \overline{\varphi(0)}\varphi(z)} \right|^2 \geq 1 - \frac{(1 - |\varphi(0)|^2)(1 - |\varphi(z)|^2)}{(1 - |\varphi(0)||\varphi(z)|)^2}$$

$$= \frac{(|\varphi(z)| - |\varphi(0)|)^2}{(1 - |\varphi(0)||\varphi(z)|)^2}$$

The Schwarz–Pick Theorem (Theorem 2.39) gives

$$\left| \frac{\varphi(0) - \varphi(z)}{1 - \overline{\varphi(0)}\varphi(z)} \right| \leq \left| \frac{0 - z}{1 - \overline{0}z} \right| = |z|$$

Thus

$$|z| \geq \frac{||\varphi(z)| - |\varphi(0)||}{1 - |\varphi(0)||\varphi(z)|} \geq \frac{|\varphi(z)| - |\varphi(0)|}{1 - |\varphi(0)||\varphi(z)|}$$

from which the result follows. ∎

In particular, notice that the estimate of Corollary 2.40 shows that for any analytic map φ of the disk into itself we have

$$\frac{1 - |\varphi(z)|}{1 - |z|} \geq \frac{1 - |\varphi(0)|}{1 + |\varphi(0)|}$$

an observation which is relevant to Julia's lemma below.

For $k > 0$ and ζ in the unit circle, let

$$E(k, \zeta) = \{z \in D : |\zeta - z|^2 \leq k(1 - |z|^2)\}$$

A computation shows that $E(k, \zeta)$ is a closed disk internally tangent to the circle at ζ with center $\frac{1}{1+k}\zeta$ and radius $\frac{k}{1+k}$. The boundary circle of this disk is called an **oricycle**. The parabolic automorphisms that fix ζ map these oricycles onto themselves and are distinguished from all other non-identity maps of the disk by this property in the sense that is made precise in Exercises 2.3.7 and 2.3.8. Moreover, these oricyles play an important role for arbitrary self-maps of the disk, as described by the following geometric lemma.

LEMMA 2.41 *(Julia's Lemma)*
Suppose ζ is in the unit circle and

$$d(\zeta) = \liminf_{z \to \zeta} \frac{1 - |\varphi(z)|}{1 - |z|}$$

is finite where the lower limit is taken as z approaches ζ unrestrictedly in D. Suppose $\{a_n\}$ is a sequence along which this lower limit is achieved and for which $\varphi(a_n)$ converges to η. Then $|\eta| = 1$ and for every z in D

$$\frac{|\eta - \varphi(z)|^2}{1 - |\varphi(z)|^2} \leq d(\zeta) \frac{|\zeta - z|^2}{1 - |z|^2}$$

Moreover, if equality holds for some z in D, then φ is an automorphism of the disk.

The geometric interpretation of this result is that φ maps each disk $E(k, \zeta)$ into the corresponding disk $E(kd(\zeta), \eta)$.

PROOF   We are assuming $a_n \to \zeta$ and $\varphi(a_n) \to \eta$ with

$$d(\zeta) = \lim_{n \to \infty} \frac{1 - |\varphi(a_n)|}{1 - |a_n|} < \infty$$

Clearly $|\eta| = 1$. The Schwarz–Pick Theorem (2.39) gives, for all z in D,

$$1 - \left|\frac{\varphi(z) - \varphi(a_n)}{1 - \varphi(z)\overline{\varphi(a_n)}}\right|^2 \geq 1 - \left|\frac{z - a_n}{1 - \overline{a_n}z}\right|^2$$

which by the fundamental identity for automorphisms (Equation (2.3.1)) is equivalent to

$$\frac{(1 - |\varphi(z)|^2)(1 - |\varphi(a_n)|^2)}{|1 - \varphi(z)\overline{\varphi(a_n)}|^2} \geq \frac{(1 - |a_n|^2)(1 - |z|^2)}{|1 - \overline{a_n}z|^2}$$

or

$$\frac{|1 - \varphi(z)\overline{\varphi(a_n)}|^2}{1 - |\varphi(z)|^2} \leq \frac{(1 - |\varphi(a_n)|^2)|1 - \overline{a_n}z|^2}{(1 - |a_n|^2)(1 - |z|^2)}$$

Letting n go to ∞ we obtain

$$\frac{|\eta - \varphi(z)|^2}{1 - |\varphi(z)|^2} = \frac{|1 - \overline{\eta}\varphi(z)|^2}{1 - |\varphi(z)|^2} \leq d(\zeta)\frac{|1 - \overline{\zeta}z|^2}{1 - |z|^2} = d(\zeta)\frac{|\zeta - z|^2}{1 - |z|^2}$$

as desired.

The proof of the equality condition is left as an exercise (Exercise 2.3.9). ∎

The quantity $d(\zeta)$ plays an important role in the study of the geometry of self-maps of the disk. While $d(\zeta)$ may be $+\infty$, as a consequence of Julia's Lemma (or by the estimate following the proof of Corollary 2.40), it must always be strictly greater than 0.

The geometric interpretation of Julia's Lemma is particularly satisfying in the case $\zeta = \eta$. Clearly, in this case, the point ζ deserves to be called a fixed point, but since we have not assumed (and do not want to) that φ is continuous on the boundary, we need to extend our notion of fixed point to include the case of a fixed point on the unit circle.

DEFINITION 2.42 *If φ is an analytic mapping of the disk into itself and b is a point of the closed disk, we will call b a **fixed point of** φ if*

$$\lim_{r \to 1} \varphi(rb) = b$$

The Schwarz–Pick Theorem implies that an analytic function on the disk has at most one fixed point inside the disk, but analytic functions can have many fixed points on the circle. The fixed point set must have measure zero and for a univalent function, the fixed point set has capacity zero, but, at least for compact sets, these are the only requirements (see [CoP82] and Exercise 2.3.13). The Schwarz–Pick Theorem tells us about the behavior of an analytic function φ near an interior fixed point: φ maps pseudohyperbolic disks centered at the fixed point into other (smaller) pseudohyperbolic disks centered at the fixed point. Julia's Lemma gives a similar statement for a fixed point ζ on the boundary when $d(\zeta)$ is finite: φ maps internally tangent disks at ζ into (other) internally tangent disks at ζ.

For ζ on the unit circle and $\alpha > 1$ we define a **nontangential approach region at** ζ by

$$\Gamma(\zeta, \alpha) = \{z \in D : |z - \zeta| < \alpha(1 - |z|)\}$$

Of course, the term "nontangential" refers to the fact that the boundary curves of $\Gamma(\zeta, \alpha)$ have a corner at ζ, with angle less than π (see Exercise 2.3.11). A function f is said to have a **nontangential limit at** ζ if $\lim_{z \to \zeta} f(z)$ exists in each nontangential region $\Gamma(\zeta, \alpha)$.

DEFINITION 2.43 *We say φ has a **finite angular derivative** at ζ on the unit circle if there is η on the circle so that $(\varphi(z) - \eta)/(z - \zeta)$ has a finite nontangential limit as $z \to \zeta$. When it exists (as a finite complex number), this limit is denoted $\varphi'(\zeta)$.*

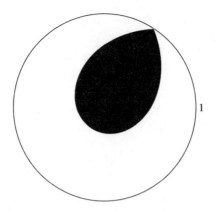

FIGURE 2.4

A typical nontangential approach region.

Our next result, the Julia–Carathéodory Theorem, is a circle of ideas which makes precise the relationship between the angular derivative $\varphi'(\zeta)$, the limit of $\varphi'(z)$ at ζ, and the quantity $d(\zeta)$ from Julia's Lemma.

THEOREM 2.44 *(Julia–Carathéodory Theorem)*
For $\varphi : D \to D$ analytic and ζ in ∂D, the following are equivalent:

(1) $d(\zeta) = \liminf_{z \to \zeta}(1 - |\varphi(z)|)/(1 - |z|) < \infty$, *where the limit is taken as z approaches ζ unrestrictedly in D.*

(2) φ *has finite angular derivative $\varphi'(\zeta)$ at ζ.*

(3) *Both φ and φ' have (finite) nontangential limits at ζ, with $|\eta| = 1$ for $\eta = \lim_{r \to 1} \varphi(r\zeta)$.*

Moreover, when these conditions hold, we have $\lim_{r \to 1} \varphi'(r\zeta) = \varphi'(\zeta) = d(\zeta)\bar{\zeta}\eta$ and $d(\zeta)$ is the nontangential limit $\lim_{z \to \zeta}(1 - |\varphi(z)|)/(1 - |z|)$.

The proof uses the following simple lemma about nontangential approach regions in D.

LEMMA 2.45
Given $1 < \alpha < \beta$, let $\delta = (\beta - \alpha)/(\alpha + \alpha\beta)$. If z is in $\Gamma(\zeta, \alpha)$ and $|\lambda| \le \delta|\zeta - z|$, then $z + \lambda$ is in $\Gamma(\zeta, \beta)$.

PROOF We have

$$
\begin{aligned}
|z + \lambda - \zeta| &\le |z - \zeta| + |\lambda| \\
&< \alpha(1 - |z|) + \delta|\zeta - z| \\
&\le \alpha(1 - |z|) + \delta\alpha(1 - |z|)
\end{aligned}
$$

$$= (\alpha + \delta\alpha)(1 - |z|)$$

But since $|\lambda| \leq \delta|\zeta - z|$ and $|\zeta - z| < \alpha(1 - |z|)$ we get

$$1 - |z + \lambda| \geq 1 - |z| - |\lambda| \geq (1 - |z|)(1 - \delta\alpha)$$

Thus,

$$|z + \lambda - \zeta| \leq (\alpha + \delta\alpha)(1 - |z|) \leq \frac{\alpha + \delta\alpha}{1 - \delta\alpha}(1 - |z + \lambda|)$$

Since $\beta = (\alpha + \delta\alpha)/(1 - \delta\alpha)$, the conclusion follows. ∎

PROOF (of Theorem 2.44)   We will show (1) ⇒ (2) ⇒ (3) ⇒ (1).

For (1) ⇒ (2) recall that by Lemma 2.41 there exists η on the unit circle so that for all z in D

$$\frac{|\eta - \varphi(z)|^2}{1 - |\varphi(z)|^2} \leq d(\zeta)\frac{|\zeta - z|^2}{1 - |z|^2}$$

We first consider the radial limit of $(\varphi(z) - \eta)/(z - \zeta)$ at ζ. Now

$$\frac{1 - |\varphi(r\zeta)|}{1 - r}\frac{1 + r}{1 + |\varphi(r\zeta)|} \leq \frac{|\eta - \varphi(r\zeta)|^2}{1 - |\varphi(r\zeta)|^2}\frac{1 - r^2}{(1 - r)^2}$$

$$\leq d(\zeta)\frac{|\zeta - r\zeta|^2}{1 - r^2}\frac{1 - r^2}{(1 - r)^2} = d(\zeta)$$

Since $d(\zeta)$ is the lower limit of $(1 - |\varphi(z)|)/(1 - |z|)$ at $z = \zeta$ we must have

$$\lim_{r \to 1}\frac{1 - |\varphi(r\zeta)|}{1 - r} = d(\zeta) \tag{2.3.2}$$

and $\lim_{r \to 1}|\varphi(r\zeta)| = 1$. Furthermore, since

$$\frac{(1 - |\varphi(r\zeta)|)^2}{(1 - r)^2} \leq \frac{|\eta - \varphi(r\zeta)|^2}{(1 - r)^2} \leq d(\zeta)\frac{1 - |\varphi(r\zeta)|^2}{1 - r^2}$$

we have

$$\lim_{r \to 1}\frac{|\eta - \varphi(r\zeta)|}{1 - r} = d(\zeta) \tag{2.3.3}$$

Comparing Equations (2.3.2) and (2.3.3) yields

$$\lim_{r \to 1}\frac{1 - |\varphi(r\zeta)|}{|\eta - \varphi(r\zeta)|} = 1$$

and a computation with this shows that $\arg(1 - \bar{\eta}\varphi(r\zeta))$ tends to 0 as r goes to 1. Using Equation (2.3.3) we see that

$$\lim_{r \to 1}\frac{\eta - \varphi(r\zeta)}{\zeta - r\zeta} = d(\zeta)\bar{\zeta}\eta$$

To finish, we must extend this from radial convergence to nontangential convergence. To this end, fix an arbitrary nontangential approach region $\Gamma(\zeta, \alpha)$. For z in $\Gamma(\zeta, \alpha)$, we have $|\zeta - z| < \alpha(1 - |z|) \leq \alpha(1 - |z|^2)$ so by Julia's Lemma

$$\frac{|\eta - \varphi(z)|^2}{1 - |\varphi(z)|^2} \leq d(\zeta)\frac{|\zeta - z|^2}{1 - |z|^2} \leq \alpha|\zeta - z|d(\zeta)$$

This implies

$$\frac{|\eta - \varphi(z)|}{|\zeta - z|} \leq \alpha d(\zeta)(1 + |\varphi(z)|)\frac{1 - |\varphi(z)|}{|\eta - \varphi(z)|} \leq 2\alpha d(\zeta)$$

Thus $(\eta - \varphi(z))/(\zeta - z)$ is bounded in $\Gamma(\zeta, \alpha)$, and since it has radial limit $d(\zeta)\eta\bar{\zeta}$ at ζ, Lindelöf's Theorem ([Dur70, p. 6]) shows that it tends to the same limit in $\Gamma(\zeta, \beta)$ for any $\beta < \alpha$. Since α, and hence β, is arbitrary, we are done.

Next we show (2) $\Rightarrow$ (3). Suppose φ has finite angular derivative at ζ and $\eta = \lim_{r \to 1} \varphi(r\zeta)$. Fix a nontangential approach region $\Gamma(\zeta, \alpha)$ and fix w in this region. If r is small enough that $\{w + re^{i\theta} : 0 \leq \theta \leq 2\pi\}$ lies in D then by the Cauchy Integral Formula for $(\varphi - \eta)'(w)$ we have

$$\varphi'(w) = (\varphi - \eta)'(w) = \frac{1}{2\pi}\int_0^{2\pi} \frac{\varphi(w + re^{i\theta}) - \eta}{re^{i\theta}}\, d\theta$$

$$= \frac{1}{2\pi}\int_0^{2\pi} \left(\frac{\varphi(w + re^{i\theta}) - \eta}{w + re^{i\theta} - \zeta}\right)\left(\frac{w + re^{i\theta} - \zeta}{re^{i\theta}}\right) d\theta$$

If we choose $r = \delta|w - \zeta|$ where $\delta = (1 + 2\alpha)^{-1}$ then Lemma 2.45 guarantees that the circle $w + re^{i\theta}$ is contained in $\Gamma(\zeta, \beta)$ where $\beta = 2\alpha$. It follows from Condition (2) that

$$\frac{\varphi(w + re^{i\theta}) - \eta}{w + re^{i\theta} - \zeta}$$

is bounded, independent of θ and w in $\Gamma(\zeta, \alpha)$. Since

$$|(w + re^{i\theta} - \zeta)/(re^{i\theta})| = |1 + (w - \zeta)/re^{i\theta}| \leq 1 + 1/\delta$$

we have φ' bounded in $\Gamma(\zeta, \alpha)$. Moreover, setting $w = t\zeta$ for $0 < t < 1$ and letting t tend to 1, we can use the bounded convergence theorem together with the fact that $(\varphi(z) - \eta)/(z - \zeta)$ approaches $\varphi'(\zeta)$ nontangentially to conclude that $\lim_{t \to 1} \varphi'(t\zeta) = \varphi'(\zeta)$. The boundedness of φ' in $\Gamma(\zeta, \alpha)$ and this radial convergence imply nontangential convergence in any smaller approach region. Since α is arbitrary we are done.

Finally, we show (3) $\Rightarrow$ (1). Let $M < \infty$ be such that $|\varphi'(r\zeta)| \leq M$ for $r > 0$. Then

$$|\eta - \varphi(r\zeta)| = \left|\int_r^1 \varphi'(t\zeta)\, dt\right| \leq M(1 - r)$$

Hence

$$\frac{1 - |\varphi(r\zeta)|}{1 - |r\zeta|} \le \frac{|\eta - \varphi(r\zeta)|}{1 - r} \le M$$

and $d(\zeta)$, being a lower limit, is finite.

In the proof of (1) $\Rightarrow$ (2), we saw that $(\eta - \varphi(z))/(\zeta - z)$ converges to $d(\zeta)\overline{\zeta}\eta$ as z tends to ζ nontangentially. This is the same as saying that $(1 - \overline{\eta}\varphi(z))/(1 - \overline{\zeta}z)$ converges to $d(\zeta)$ as z tends to ζ nontangentially. In particular, since $d(\zeta)$ is positive, $|1 - \overline{\eta}\varphi(z)|/|1 - \overline{\zeta}z|$ also converges to $d(\zeta)$ and

$$\lim_{z \to \zeta} \frac{1 - \overline{\eta}\varphi(z)}{|1 - \overline{\eta}\varphi(z)|} \bigg/ \frac{1 - \overline{\zeta}z}{|1 - \overline{\zeta}z|} = 1$$

As a consequence, we see that when z approaches ζ nontangentially, $\varphi(z)$ approaches η nontangentially also. Nontangential convergence of z to ζ implies

$$\left| \text{Im} \frac{1 - \overline{\zeta}z}{|1 - \overline{\zeta}z|} \right| \le C\text{Re} \frac{1 - \overline{\zeta}z}{|1 - \overline{\zeta}z|}$$

for some constant C, so

$$\lim_{z \to \zeta} \frac{\text{Re}\,(1 - \overline{\eta}\varphi(z))}{|1 - \overline{\eta}\varphi(z)|} \bigg/ \frac{\text{Re}\,(1 - \overline{\zeta}z)}{|1 - \overline{\zeta}z|} = 1$$

or

$$\lim_{z \to \zeta} \frac{\text{Re}\,(1 - \overline{\eta}\varphi(z))}{\text{Re}\,(1 - \overline{\zeta}z)} = \lim_{z \to \zeta} \frac{|1 - \overline{\eta}\varphi(z)|}{|1 - \overline{\zeta}z|} = d(\zeta)$$

Finally, the nontangential convergence implies

$$\lim_{z \to \zeta} \frac{\text{Re}\,(1 - \overline{\zeta}z)}{1 - |z|} = 1 \quad \text{and} \quad \lim_{z \to \zeta} \frac{\text{Re}\,(1 - \overline{\eta}\varphi(z))}{1 - |\varphi(z)|} = 1$$

so

$$\lim_{z \to \zeta} \frac{1 - |\varphi(z)|}{1 - |z|} = \lim_{z \to \zeta} \frac{\text{Re}\,(1 - \overline{\eta}\varphi(z))}{\text{Re}\,(1 - \overline{\zeta}z)} = d(\zeta)$$

as z approaches ζ nontangentially, as we wished to prove. ∎

The final conclusion shows that if φ has finite angular derivative at ζ then $|\varphi'(\zeta)| = d(\zeta)$. Moreover, since $d(\zeta) = \infty$ if and only if φ does not have finite angular derivative at ζ we are justified in writing $|\varphi'(\zeta)| = d(\zeta)$ even when $d(\zeta) = \infty$ and we will do so whenever convenient. Because of this identification the number $d(\zeta)$, whether finite (and positive) or infinite, is itself sometimes referred to as the angular derivative at ζ. It is useful to notice that the infimum of the angular derivative is always attained:

PROPOSITION 2.46
If φ is an analytic map of the disk into itself, then

$$\inf_{\zeta \in \partial D} |\varphi'(\zeta)| = \liminf_{|z| \to 1} \frac{1 - |\varphi(z)|}{1 - |z|} = |\varphi'(\zeta_0)|$$

for some ζ_0 in the unit circle.

PROOF Set $\beta = \inf_{\zeta \in \partial D} |\varphi'(\zeta)|$. If we assume $|\varphi'(\zeta)| > \beta$ for all ζ in the circle, we may find for each ζ an open disk $\Delta(\zeta)$ with center ζ and radius $\epsilon(\zeta) > 0$ such that

$$\frac{1 - |\varphi(z)|}{1 - |z|} > \beta + \epsilon(\zeta)$$

for all z in $\Delta(\zeta) \cap D$. Cover ∂D by a finite subcollection $\Delta(\zeta_1), \Delta(\zeta_2), \ldots, \Delta(\zeta_m)$ and find $r_0 > 0$ so that the annulus $\{r_0 < |z| < 1\}$ is contained in $\cup_{j=1}^{m} \Delta(\zeta_j)$. Set $\epsilon = \min\{\epsilon(\zeta_j) : j = 1, 2, \ldots, m\}$. Then on this annulus

$$\frac{1 - |\varphi(z)|}{1 - |z|} > \beta + \epsilon$$

which contradicts the definition of β. ∎

From the proof of the Julia–Carathéodory Theorem, we see that $d(\zeta) < \infty$ implies that $\varphi'(\zeta) = d(\zeta)\overline{\zeta}\eta$ where η is the radial limit of φ at ζ. This says $\arg \varphi'(\zeta) = \arg \eta - \arg \zeta$ which leads, by the standard argument, to the conclusion that φ is conformal at ζ whenever $d(\zeta) < \infty$. In particular, note that at a fixed point of φ on the unit circle, then $\varphi'(\zeta) = d(\zeta) > 0$. Another geometric consequence of the Julia–Carathéodory Theorem is contained in the following corollary.

COROLLARY 2.47
If $|\zeta| = 1$ and ζ is a fixed point of φ with $\varphi'(\zeta) < \infty$, then for any nontangential approach region $\Gamma(\zeta, \alpha)$ there is $r < 1$ such that φ is univalent on $\Gamma(\zeta, \alpha) \cap \{z : r < |z| < 1\}$.

PROOF We know that $\varphi'(z)$ has nontangential limit equal to $\varphi'(\zeta)$ at ζ. Fix a nontangential approach region $\Gamma(\zeta, \alpha)$ and choose r sufficiently close to 1 so that $|\varphi'(z) - \varphi'(\zeta)| < \frac{1}{2}|\varphi'(\zeta)|$ for z in $\Gamma(\zeta, \alpha) \cap \{r < |z| < 1\}$. If both z_1 and z_2 lie in $\Gamma(\zeta, \alpha) \cap \{r < |z| < 1\}$ then so does the line segment L joining them and

$$|\varphi(z_2) - \varphi(z_1) - \varphi'(\zeta)(z_2 - z_1)| = \left| \int_L (\varphi'(z) - \varphi'(\zeta))\, dz \right| \leq \frac{1}{2}|\varphi'(\zeta)||z_2 - z_1|$$

so that $\varphi(z_2) \neq \varphi(z_1)$ unless $z_1 = z_2$. ∎

Our next result identifies a distinguished fixed point on the unit circle in the case φ has no fixed points in D and shows, for all fixed points ζ of φ on the unit circle except this one, that $d(\zeta) > 1$. In other words, except for this distinguished fixed point, $\varphi' > 1$ at fixed points on the circle; this inequality is improved in Exercise 2.3.15.

THEOREM 2.48   (*Wolff's Lemma*)

If φ is an analytic map of the disk into itself that has no fixed points in D, then there is a unique fixed point a of φ on the unit circle with $d(a) \leq 1$. If φ, not the identity, has a fixed point in D, then $d(\zeta) > 1$ for all fixed points ζ of φ on the unit circle.

PROOF   First assume φ has no fixed point in D. Choose any sequence r_n increasing to 1 and consider the maps $f_n \equiv r_n\varphi : D \to r_nD$. Since f_n is continuous as a map of $r_n\overline{D}$ into itself, the Brouwer Fixed Point Theorem guarantees that f_n has a fixed point a_n in D. By passing to a subsequence, we may assume the sequence $\{a_n\}$ converges to a point a in $\overline{D}$. If $|a| < 1$, then the continuity of φ near a shows $\varphi(a_n)$ converges to $\varphi(a)$ and, since r_n increases to 1, this means $r_n\varphi(a_n)$ tends to $\varphi(a)$ also. But $r_n\varphi(a_n) = a_n$ so we must have $\varphi(a) = a$, contradicting the assumption that φ has no fixed points in the open disk. Thus $|a| = 1$.

We claim that a is a fixed point of φ and that $d(a) \leq 1$. This second fact is immediate:

$$\liminf_{z \to a} \frac{1 - |\varphi(z)|}{1 - |z|} \leq \liminf_{n \to \infty} \frac{1 - |\varphi(a_n)|}{1 - |a_n|} = \liminf_{n \to \infty} \frac{1 - \frac{|a_n|}{r_n}}{1 - |a_n|} \leq 1$$

and therefore $d(a) \leq 1$. Since $f_n(a_n) = a_n$, the Schwarz–Pick Theorem (Theorem 2.39) shows that for arbitrary w in D,

$$1 - \left|\frac{a_n - f_n(w)}{1 - \overline{a_n}f_n(w)}\right|^2 \geq 1 - \left|\frac{a_n - w}{1 - \overline{a_n}w}\right|^2$$

Equation (2.3.1) and some algebra show this is equivalent to

$$\frac{|1 - \overline{a_n}f_n(w)|^2}{1 - |f_n(w)|^2} \leq \frac{|1 - \overline{a_n}w|^2}{1 - |w|^2} \tag{2.3.4}$$

On recalling that $f_n = r_n\varphi$ and letting n tend to infinity in Inequality (2.3.4), we obtain

$$\frac{|a - \varphi(w)|^2}{1 - |\varphi(w)|^2} \leq \frac{|a - w|^2}{1 - |w|^2}$$

for all w in D. Thus $\varphi(a) = a$ and the claim is verified.

Uniqueness follows from Julia's Lemma (2.41): if ζ_1 and ζ_2 are fixed points of φ on the unit circle with $d(\zeta_1) \leq 1$ and $d(\zeta_2) \leq 1$, choose k_1 and k_2 so that the oricycles $\partial E(k_1, \zeta_1)$ and $\partial E(k_2, \zeta_2)$ are tangent to each other at w in D (see Figure 2.5). But then $\varphi(w)$ is in $\overline{E(k_1, \zeta_1)} \cap \overline{E(k_2, \zeta_2)} = \{w\}$, contradicting the hypothesis that φ is fixed point free in D.

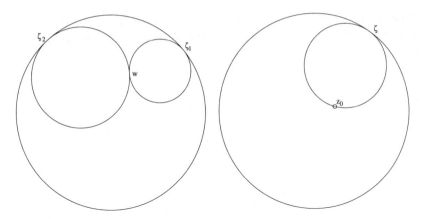

FIGURE 2.5

Oricycles at contractive fixed points.

Similarly, suppose φ has a fixed point z_0 in D and ζ is a fixed point of φ on the unit circle with $d(\zeta) \leq 1$. If ρ is chosen so that the oricycle $\partial E(\rho, \zeta)$ passes through z_0 (see Figure 2.5), then in the inequality in Julia's Lemma (2.41), we have $\eta = \zeta$, $d(\zeta) = 1$ and equality holds for $z = z_0$. This means that φ is an automorphism of the disk that fixes z_0 and ζ which is impossible since φ is not the identity. ∎

We will use this theorem to investigate the iterates φ_n of a fixed point free map φ, where $\varphi_1 = \varphi$ and $\varphi_{n+1} = \varphi \circ \varphi_n, n = 1, 2, \dots$. Since the analytic maps of D into D form a normal family, every sequence of iterates of φ contains a subsequence which converges uniformly on compact subsets of D. By the maximum modulus theorem any such limit is either an analytic map of D into D or a constant function of modulus 1. Our next lemma shows that only one constant function is possible among the subsequential limits of $\{\varphi_n\}$.

LEMMA 2.49
Suppose φ is an analytic map of the disk into itself that has no fixed points in D and let a be the unique fixed point of the circle with $d(a) \leq 1$. The only constant function that can appear as a limit of iterates of φ is $f(z) \equiv a$.

PROOF Suppose φ_{n_j} converges uniformly on compact subsets of D to the constant w in the closed disk and assume $w \neq a$. We may find a neighborhood V of w in $\overline{D}$ and ρ sufficiently small so that $E(\rho, a)$ and V are disjoint. By Julia's Lemma, $\varphi_n(E(\rho, a)) \subset E(\rho, a)$ for all n since $d(a) \leq 1$. Hence for any z in $E(\rho, a), \varphi_n(z)$ is not in V for all n, which contradicts the assumption that φ_{n_j} converges to w. ∎

We also note for future reference that if φ (not the identity map) has a fixed point a in D, then $\varphi_n(a) = a$ for every n and the only constant function among the subsequential limits of φ_n is $f(z) \equiv a$.

LEMMA 2.50

Let φ be an analytic map of the disk into itself and suppose $\{\varphi_n\}$ has a subsequence which converges to a non-constant function. Then φ is an automorphism of the disk.

PROOF We assume φ_{n_j} converges to g uniformly on compact subsets of the disk, where g is non-constant. Of course, $g(D) \subset D$. Set $m_j = n_{j+1} - n_j$ and choose a subsequence $\varphi_{m_{j_k}}$ which converges to, say, h. On the one hand $\varphi_{m_{j_k}} \circ \varphi_{n_{j_k}}$ converges to $h \circ g$, but also $\varphi_{m_j} \circ \varphi_{n_j} = \varphi_{n_{j+1}}$ which converges to g. Thus h is the identity on the range of g, a set with more than one point so, by the Schwarz Lemma, h is the identity on all of D. Passing to a further subsequence if necessary, we may assume

$$\varphi_{m_{j_k}-1} \to f$$

and hence

$$\varphi_{m_{j_k}} = \varphi_{m_{j_k}-1} \circ \varphi \to f \circ \varphi$$

so that $f \circ \varphi$ is the identity. Since $f(D) \subset D$ so that also $\varphi \circ \varphi_{m_{j_k}-1} \to \varphi \circ f$, we have $\varphi \circ f$ is the identity. This implies φ is a one-to-one map of the disk onto itself. ∎

We use these lemmas to prove the Denjoy–Wolff theorem.

THEOREM 2.51 (Denjoy–Wolff Theorem)

If φ, not the identity and not an elliptic automorphism of D, is an analytic map of the disk into itself, then there is a point a in $\overline{D}$ so that the iterates φ_n of φ converge to a uniformly on compact subsets of D.

PROOF We consider first the case that φ is not an automorphism of D. By Lemmas 2.49 and 2.50 the only function that can appear as a limit of any sequence of iterates of φ is a constant function, where the constant is either the interior fixed point of φ, if there is one, or the unique unimodular a with $d(a) \leq 1$. Since $\{\varphi_n\}$ is a normal family, the entire sequence of iterates must converge to this constant, uniformly on compact subsets of the disk.

If φ is a hyperbolic or parabolic automorphism of D we still know that there is a unique fixed point a of the unit circle with $\varphi'(a) \leq 1$. It is not difficult to show directly (Exercise 2.3.2 and 2.3.3) that in this case, φ_n converges uniformly on compact subsets to a. ∎

DEFINITION 2.52 *The limit point a of Theorem 2.51 will be referred to as the Denjoy–Wolff point of φ.*

The preceding few results show that if φ is a map with no fixed point in D, then there is unique fixed point a on the unit circle with $d(a) \leq 1$ and that the iterates of φ converge to this fixed point. On the other hand, if φ, not the identity or an elliptic automorphism, has a fixed point a in D, then the Schwarz Lemma implies a is unique and $|\varphi'(a)| < 1$ and Theorem 2.48 shows that φ has no fixed points on the boundary with $d(\zeta) \leq 1$. In this case, the iterates of φ also converge to the fixed point a. Thus, we see that the Denjoy–Wolff point of φ can be described as the unique fixed point of φ in $\overline{D}$ with $|\varphi'(a)| \leq 1$. In the next section, we will study the iteration of functions in the disk. We can use the geometric ideas of this section to better understand the nature of the iteration that occurs in the Denjoy–Wolff theorem. We will pay particular attention to the behavior of the mapping in the neighborhood of the Denjoy–Wolff point.

Exercises

2.3.1   (a) Show that if $z_1 \neq z_2$ and $w_1 \neq w_2$ are points of D for which
$$\left| \frac{z_1 - z_2}{1 - \overline{z_1} z_2} \right| = \left| \frac{w_1 - w_2}{1 - \overline{w_1} w_2} \right|$$
then there is a unique automorphism φ of the disk so that $\varphi(z_1) = w_1$ and $\varphi(z_2) = w_2$.

     (b) Show that if ζ_1, ζ_2, and ζ_3 are distinct points arranged counterclockwise on the unit circle and η_1, η_2, and η_3 are also distinct points arranged counterclockwise on the unit circle, then there is a unique automorphism φ of the disk so that $\varphi(\zeta_j) = \eta_j$ for $j = 1, 2, 3$.

2.3.2   Suppose ψ is an automorphism of the upper halfplane Π^+ which fixes 0 and ∞ only. Show that $\psi(w) = tw$ for some $t > 0$, $t \neq 1$. (Hint: ψ is a linear fractional transformation, so $\psi(w) = (aw + b)/(cw + d)$ where we may assume a, b, c, and d are real since ψ maps the extended real axis to itself.) Show that, consequently, any hyperbolic automorphism of the disk is conformally equivalent to a dilation of Π^+ and that its iterates converge to its Denjoy–Wolff point as asserted in Theorem 2.51.

2.3.3   Suppose ψ is an automorphism of Π^+ which fixes ∞ only. Show that $\psi(w) = w + b$ for some real $b \neq 0$. Conclude that a parabolic automorphism of the disk is conformally equivalent to a translation of Π^+ and that its iterates converge to its Denjoy–Wolff point as asserted in Theorem 2.51.

2.3.4   Suppose φ is a map of the disk into itself that is analytic in a neighborhood of the closed disk and has Denjoy–Wolff point a. Suppose ψ is an automorphism of the disk with $\psi(b) = a$. Show that $\widetilde{\varphi}$ defined by $\widetilde{\varphi}(z) = \psi^{-1}(\varphi(\psi(z)))$ has Denjoy–Wolff point b and $\widetilde{\varphi}'(b) = \varphi'(a)$. Find a formula relating $\widetilde{\varphi}''(b)$ to $\varphi''(a)$.

2.3.5   (a) Show that an elliptic disk automorphism φ is conformally equivalent to λz where $\lambda = \varphi'(a)$ and a is the fixed point of φ in D.

(b) Show that a hyperbolic disk automorphism φ is conformally equivalent to $(z+r)/(1+rz)$ where $r = (1 - \varphi'(a))/(1 + \varphi'(a))$ and a is the Denjoy–Wolff point of φ.

(c) Show that a parabolic disk automorphism φ is conformally equivalent to $((1+i)z - 1)/(z + i - 1)$ or $((1 - i)z - 1)/(z - i - 1)$.

2.3.6 Verify the halfplane version of the Schwarz–Pick Theorem:

If ψ is an analytic map of the halfplane $\{w : \operatorname{Re} w > 0\}$ into itself, then

$$\left| \frac{\psi(w) - \psi(w_0)}{\psi(w) + \overline{\psi(w_0)}} \right| \leq \left| \frac{w - w_0}{w + \overline{w_0}} \right|$$

2.3.7 Suppose that φ is a parabolic automorphism of the disk with fixed point ζ on the unit circle. Show that φ maps each oricycle $|\zeta - z|^2 = k(1 - |z|^2)$ onto itself.

2.3.8 Suppose φ is an analytic map of the disk into itself that satisfies

$$\frac{|\zeta - \varphi(z)|^2}{1 - |\varphi(z)|^2} \leq \frac{|\zeta - z|^2}{1 - |z|^2}$$

for some ζ on the unit circle and all z in D. Show that if equality holds for some z_0 in D and φ is not the identity map, then φ is a parabolic automorphism with fixed point ζ.

2.3.9 Prove the equality condition in Lemma 2.41.

2.3.10 (a) Verify the halfplane version of Julia's Lemma:

Let $\psi = u + iv$ be an analytic map of the right halfplane into itself, that is, $u(x + iy) > 0$ for $x > 0$, and let

$$\kappa = \inf_{0 < x} \frac{u(x + iy)}{x}$$

Prove that $u(x + iy) > \kappa x$.

(b) Verify the halfplane version of the Julia–Carathéodory Theorem:

Let $\psi = u + iv$ be an analytic map of the right halfplane into itself, that is, $u(x + iy) > 0$ for $x > 0$, and let

$$\kappa = \inf_{0 < x} \frac{u(x + iy)}{x}$$

Prove that for $\alpha > 0$,

$$\lim_{\substack{z \to \infty \\ |y| < \alpha x}} \frac{\psi(z)}{z} = \lim_{\substack{z \to \infty \\ |y| < \alpha x}} \psi'(z) = \kappa$$

2.3.11 Find the relationship between α and the angle at ζ in the nontangential approach region $\Gamma(\zeta, \alpha)$.

2.3.12 Find all the fixed points of the listed functions and their derivatives there. Then find the Denjoy–Wolff point. ($\sqrt{}$ means the branch of the square root that is positive on the positive axis.)

(a) $\varphi(z) = \exp((z + 1)/(z - 1))$

(b)

$$\varphi(z) = \left(\frac{z + 1/3}{1 + z/3} \right)^2$$

(c) $\varphi(z) = w^{-1}(\psi(w(z))$ where $w(z) = (1 + z)/(1 - z)$ maps the disk to the right halfplane and

$$\psi(w) = \sqrt{4w^2 + 3}$$

(d)
$$\varphi(z) = \frac{1 + z + 2\sqrt{1 - z^2}}{3 - z + 2\sqrt{1 - z^2}}$$

2.3.13 Use Rudin's theorem [Ru56] "Given a compact set K of measure zero in the unit circle and a continuous complex valued function f on K, there is a function ψ that is analytic on D and continuous on $\overline{D}$ such that $\psi|_K = f$ and $\|\psi\|_\infty = \|f\|_\infty$" to prove that for every compact set K of measure zero on the unit circle, there is an analytic map φ of D into itself whose fixed point set is $K \cup \{0\}$.

2.3.14 Prove: If Ω is a convex region and ψ is an analytic function on Ω such that ψ' has positive real part, then ψ is univalent on Ω.

2.3.15 Improve the angular derivative inequalities implicit in Theorem 2.48 to

(a) *If φ is an analytic map of the disk into itself with Denjoy–Wolff point a, where $|a| = 1$, and b is another fixed point of φ, then*
$$\varphi'(b) \geq \frac{1}{\varphi'(a)}$$

(b) *If φ is an analytic map of the disk into itself with Denjoy–Wolff point a, where $|a| < 1$, and b is another fixed point of φ, then*
$$\varphi'(b) \geq 2\frac{1 - \operatorname{Re}\varphi'(a)}{1 - |\varphi'(a)|^2} \geq \frac{2}{1 + |\varphi'(a)|}$$

(c) Show that these inequalities are best possible.

2.3.16 Suppose the angular derivative $|\varphi'(\zeta)|$ of φ at ζ in ∂D satisfies $|\varphi'(\zeta)| \leq B_0$ where $0 < B_0 < \infty$. Let $B_0 < B < \infty$. Show that there exists $r_0, 0 < r_0 < 1$, with $|\varphi(r\zeta)| \geq r^B$ for all r in $(r_0, 1)$.

2.3.17 For φ a non-constant analytic map of D into D let $A_0 = \inf\{|\varphi'(\zeta)| : \zeta \in \partial D\}$ and $M(r) = \max_\theta |\varphi(re^{i\theta})|$ show that
(a)
$$\lim_{r \to 1} \frac{1 - M(r)}{1 - r}$$
and
$$\lim_{r \to 1} \frac{\log M(r)}{\log r}$$

exist and equal A_0.

(b) If $A < A_0$ then there exists $r_0 < 1$ so that for $|z| > r_0$ we have $|\varphi(z)| \leq |z|^A$.

2.3.18 For $M(r)$ as defined in the last problem show

(a) Show that there exists $r_0 < 1$ and $c > 0$ such that
$$M(r) - M(s) \geq c(r - s)$$
if $r_0 \leq s < r < 1$.

(b) For $r_0 < 1$ define A by $M(r_0) = r_0^A$. Show that for $r_0 < r < 1$
$$M(r) \leq r^A$$

Hint: Use Hadamard's Three Circles Theorem.

Notes

Julia's Lemma can be found in [Ju20] and Wolff's Lemma in [Wol26]. There are various versions and proofs of the Julia–Carathéodory Theorem (sometimes also called the Julia–Wolff–Carathéodory Theorem); a standard reference is [Cac60]. The proof we have given borrows heavily from the exposition in W. Rudin's book [Ru80]. A Hilbert space proof of the Julia–Carathéodory Theorem can be found in a paper by D. Sarason [Sar88]. In a series of papers in 1926, J. Wolff [Wol26], and A. Denjoy [Den26] gave essentially simultaneous proofs of Theorem 2.51.

The criterion for univalence given in Exercise 2.3.14 is attributed by various authors to S. E. Warschawski [Wa35], J. Wolff [Wol34], or K. Noshiro [Nos34], but it surely has been discovered by many people and will continue to be.

A paper of D. Behan [Beh73] includes the proof of the improved inequality of Exercise 2.3.15, part (a). C. Cowen and Ch. Pommerenke [CoP82] did a more thorough study of fixed point sets and associated derivative inequalities. The inequalities show that compared to the derivative at the Denjoy–Wolff point, the derivatives at the other fixed points must be large collectively, not just individually. These inequalities were given shorter proofs, and in one case improved, by K. Y. Li [Lik89].

2.4   Iteration of functions in the disk

In this section, we present a model for iteration of analytic functions mapping the open unit disk D into itself and apply it to the solution (Theorem 2.63) of Schroeder's functional equation $f \circ \varphi = \lambda f$ that arises as the eigenvalue equation for composition operators. In the main theorem (Theorem 2.53), we show that, under very general conditions, an analytic map of the disk, φ, into itself can be intertwined with a linear fractional transformation, that is, there is a linear fractional map Φ and an analytic map σ so that

$$\Phi \circ \sigma = \sigma \circ \varphi$$

Since many problems involving iteration or solution of functional equations can be explicitly solved for linear fractional transformations and the intertwining relates these answers to solutions of the problems for the given function φ, we obtain the information we seek. In order to emphasize the more important conceptual issues, we have deferred the proofs of some of the technical geometric function theory to end of the section.

Statement and solution of some of these problems go back more than a century to Schroeder [Scr71] and Koenigs [Koe84]. The setting in that work was the analysis of a contractive map analytic in a neighborhood of a fixed point, and the solution (σ in the theorem below) of the basic equation was obtained as a concrete limit, now frequently called Koenigs' function in this case. The reader is invited to reproduce Koenigs' original solution in Exercise 2.4.1. Just as the vanishing of a derivative creates difficulties in many problems of elementary complex analysis, a derivative being 1 at a fixed point causes difficulty for iteration. A fixed point of φ is a zero of the function $\varphi(z) - z$; a is a multiple zero of this function if and only if $\varphi'(a) = 1$. (More generally, difficulties arise when $|\varphi'(a)| = 1$ because $\varphi'(a)^n = 1$ means a is a multiple zero of $\varphi_n(z) - z$, but, except when φ is an automorphism, $|\varphi'(a)| = 1$, $\varphi'(a) \neq 1$ cannot occur for a function mapping D into itself.) Progress came much later in the study of iteration problems where the derivative at the fixed point is 1 (see [Sie42]) or where the fixed point is on the boundary of the domain of analyticity (see [Pom79] and [BaP79]). However, each of these attacks on the problem was by way of calculating an appropriate concrete limit. The model presented here involves an abstract limiting process. By so doing, one gets a unified point of view in the various cases that occur and gets insight into questions of uniqueness. In addition, with the exception of the (critical) application of the Riemann Mapping Theorem, most of what is done abstractly can be generalized to several variables, although the interpretation of the outcome has, so far, not been accomplished. The hypothesis $\varphi'(a) \neq 0$ cannot be avoided; it is not difficult to show that no intertwining by a non-constant map is possible if $\varphi'(a) = 0$ (see Exercise 2.4.6).

The main result of this section is the following theorem; the intertwining is illustrated in Figure 2.6. The concept of a fundamental set that appears in the statement of the theorem is defined below.

THEOREM 2.53 *(The Model for Iteration)*
Let φ be an analytic mapping of D into itself, φ non-constant and not an automorphism of D, and let a be the Denjoy–Wolff point of φ. If $\varphi'(a) \neq 0$, then there is a fundamental set V for φ on D, a domain Ω, either a halfplane or the plane, an automorphism Φ mapping Ω onto Ω, and a mapping σ of D into Ω such that φ and σ are univalent on V, $\sigma(V)$ is a fundamental set for Φ on Ω and

$$\Phi \circ \sigma = \sigma \circ \varphi$$

Moreover, Φ is unique up to conjugation by an automorphism of Ω onto Ω, and Φ and σ depend only on φ, not on the particular fundamental set V.

From the intertwining $\Phi \circ \sigma = \sigma \circ \varphi$, it follows that the iterates are intertwined $\Phi_n \circ \sigma = \sigma \circ \varphi_n$. Since Φ_n is univalent, we see that if $\varphi_n(w) = \varphi_n(z)$ then $\sigma(w) = \sigma(z)$. This leads to the basic equivalence relation used in the construction. The proof of the theorem begins with the geometric observation that φ is univalent on a fundamental set V. For each z in V, $\varphi_n(z)$ is a point of V for $n = 0, 1, 2, 3, \ldots$. Using the equivalence relation, we construct a Riemann surface by adjoining to V abstract points $\varphi_n(z)$, $n = -1, -2, -3, \ldots$, and extending the action of φ to a univalent map of the surface onto itself. This surface is simply connected and non-compact, so is conformally equivalent to either a halfplane or the whole plane. The extension of φ from V to the surface is the map Φ which is a univalent map

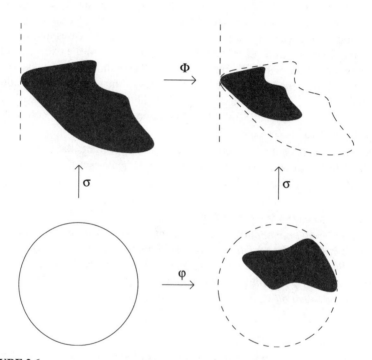

FIGURE 2.6

The intertwining $\Phi \circ \sigma = \sigma \circ \varphi$.

of the surface onto itself. After the identification of the surface as the plane or the halfplane, we see this means Φ is a linear fractional transformation.

The "fundamental set" in the theorem is a technical concept introduced to describe a set of points near the Denjoy–Wolff point small enough that φ is well behaved, but large enough that $\varphi_n(z)$ is eventually in this set. That $\sigma(V)$ is a fundamental set for Φ gives the uniqueness of Φ: Exercises 2.4.13 and 2.4.14 illustrate that without this property, there may be a variety of essentially different models. We give the definition.

DEFINITION 2.54 *If ψ maps a domain Δ into itself, we say V* **is a fundamental set for** *ψ **on** Δ if V is an open, connected, simply connected subset of Δ such that $\psi(V) \subset V$ and for every compact set K in Δ, there is a positive integer n so that $\psi_n(K) \subset V$.*

Clearly, the Denjoy–Wolff point must be in the closure of any fundamental set for φ on D since the iterates of φ converge to it. We need to show that φ is well-behaved near its Denjoy–Wolff point. If the fixed point is in the disk, the Schwarz Lemma guarantees this good behavior. When there is no fixed point in the disk, we will show (Lemmas 2.66 and 2.68) that the iterates of a point converge to the Denjoy–Wolff point through a set on which the function is univalent. Corollary 2.47 of the previous section shows that φ is univalent on a part of a nontangential approach region sufficiently near the Denjoy–Wolff point. For some mappings, that is all we need, but for other mappings, the convergence is like that of parabolic automorphisms and this corollary does not help.

Before beginning any proofs, we present an example that is simple enough that the construction can be carried out explicitly and concretely so that the process can be easily understood. This example motivates the principal steps in the development of the model.

EXAMPLE 2.55
If $\varphi(z) = sz + 1 - s$ for $0 < s < 1$, we see that φ maps the disk D onto the subdisk $\{z : |z - (1 - s)| < s\}$ and has Denjoy–Wolff point $a = 1$ with $\varphi'(a) = s$. Figure 2.7 shows the unit circle ($n = 0$) and its images under φ_n as well as the preimages ($n = -1, -2, \ldots$) for $s = 1/\sqrt{2}$. Since φ is univalent on the whole plane, the equivalence relation to be used in the proof is equality and the fundamental set V will be D. Since Φ is to be an automorphism, $\sigma(-.5)$ must have a preimage, even though $-.5$ may not have a preimage under φ as a map of the disk to itself. Clearly, though, the crescent between the circle of radius s^{-1} and center $1 - s^{-1}$ and the unit circle may be thought of as the preimage of the crescent between the unit circle and $\overline{\varphi(D)}$. We must also add the preimage of this crescent (which should be the crescent between the circle with radius s^{-2} and center $1 - s^{-2}$ and the circle of radius s^{-1} and center $1 - s^{-1}$) and so on. If we put all the preimages of the crescents together to get a Riemann surface, the

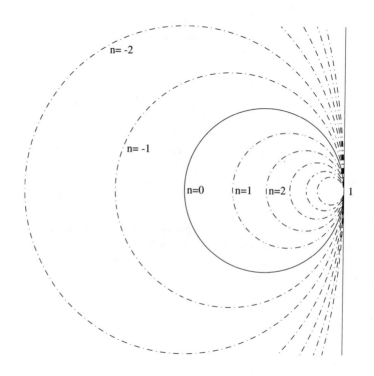

FIGURE 2.7

Constructing Φ.

map Φ will act as an automorphism and we will have accomplished our goal: in this case the union of all the preimages is the halfplane $\{z : \text{Re}(z) < 1\}$. (To conform with the notation to follow, we would choose $\Omega = \{z : \text{Re}(z) > 0\}$, with $\sigma(z) = 1 - z$, and $\Phi(z) = sz$. It is easily checked that $\Phi \circ \sigma = \sigma \circ \varphi$.)

Because φ is univalent, the fundamental set V in the disk can be taken to be the whole disk. The situation for Φ illustrates the need for the concept of fundamental set: $\Phi(z) = sz$ maps the plane onto itself in addition to the halfplane $\{z : \text{Re}(z) > 0\}$. The uniqueness statement in Theorem 2.53 is possible because of the requirement that $\sigma(V)$ be a fundamental set for Φ on Ω. In this case, $\sigma(V) = \{z : |z - 1| < 1\}$ and $\sigma(V)$ is fundamental for Φ on the halfplane $\Omega = \{z : \text{Re}(z) > 0\}$ but $\sigma(V)$ is not fundamental for Φ on the whole plane. ☐

The uniqueness statement of the theorem means that the only way in which Ω, σ and Φ of the model can be varied is in the obvious way: changing variables by conjugation by an automorphism. Indeed, suppose φ is a map of the disk into itself, Ω is a domain with Φ an automorphism of Ω, and σ is a map of the disk into Ω such that $\Phi \circ \sigma = \sigma \circ \varphi$. If ψ is a univalent map of Ω onto Ω_1, then let $\sigma_1 = \psi \circ \sigma$ and $\Phi_1 = \psi \circ \Phi \circ \psi^{-1}$. It is easily checked that Φ_1 is an automorphism of Ω_1 and σ_1 is a map of the disk into Ω_1 such that $\Phi_1 \circ \sigma_1 = \sigma_1 \circ \varphi$.

The construction of the model is based on an understanding of the iteration of the function φ near its Denjoy–Wolff point. This iteration is elementary at an interior fixed point because the function is analytic in the neighborhood of the fixed point. When the Denjoy–Wolff point is on the unit circle, however, because we are not assuming any smoothness on the boundary, studying the iteration is more difficult. One of the more difficult aspects is that the iterates of a point need not converge to the boundary nontangentially. For clarity of the exposition, we will postpone the geometric details until the end of the section and content ourselves now with the statement of the property necessary for the construction.

PROPOSITION 2.56

If φ is an analytic mapping of the disk into itself, not an automorphism of the disk, with Denjoy–Wolff point a and $\varphi'(a) \neq 0$, then there is a fundamental set for φ on D on which φ is univalent.

We are now ready to prove Theorem 2.53. The proof of the theorem is broken into three parts. In the first step, the Riemann surface S is constructed as a set of equivalence classes of pairs of points of V and integers; the pair (z, j) should be thought of as representing $\varphi_j(z)$. The second step is to produce the abstract version (Ψ and π) of the maps Φ and σ. Finally, Ω is identified and the maps Φ and σ are defined and their properties verified.

PROOF (of Theorem 2.53) Let a be the Denjoy–Wolff point of φ. Since $\varphi'(a) \neq 0$, Proposition 2.56 shows that there is a fundamental set V for φ on D on which φ

is univalent. We will use V to construct an abstract Riemann surface S and a map Ψ of S into itself which correspond to Ω and Φ.

I. *The construction of the Riemann surface S*. We introduce S as a point set, topologize it, and give it an analytic structure. For this purpose we will restrict our attention to the action of φ on V. Some points of V are not in the image of φ, some are not in the image of $\varphi \circ \varphi$, etc. As in the discussion accompanying Figure 2.7, we may think of S as being made from V by gluing on (abstract) preimages of points in V so that every point of V is in the image of φ_j acting on S for all non-negative integers j.

If j and k are integers and z and w are in V we say (z, j) is equivalent to (w, k) and write $(z, j) \sim (w, k)$ if there is an integer m, $m \geq \max\{-j, -k\}$, so that $\varphi_{j+m}(z) = \varphi_{k+m}(w)$. Since φ is one-to-one on V, $\varphi_{j+m}(z) = \varphi_{k+m}(w)$ if and only if $\varphi_{j+m'}(z) = \varphi_{k+m'}(w)$ for any m and m' not smaller than $-j$ and $-k$. The relation $\sim$ is an equivalence relation between pairs; we use the symbol $[(z, j)]$ to denote the equivalence class containing (z, j). Let

$$S = \{[(z, j)] : z \in V,\ j \text{ an integer}\}$$

The pair (z, j) should be thought of as representing $\varphi_j(z)$.

If U is open in V and j is an integer, let $\mathcal{N}_j^U = \{[(z, j)] : z \in U\}$. We claim that the sets $\{\mathcal{N}_j^U\}$ form a basis for a Hausdorff topology on S (see, for example, [Dug66, p. 67]). We need to show that we can find disjoint basic neighborhoods of any two points and that the intersection of two basic neighborhoods of a point contains a basic neighborhood of that point. To check the separation condition, suppose $z_1^* = [(z_1, j_1)] \neq [(z_2, j_2)] = z_2^*$. Let $m = \max\{-j_1, -j_2\}$. For $k = 1$ and $k = 2$, we have $z_k^* = [(\varphi_{m+j_k}(z_k), -m)]$ since $(z_k, j_k) \sim (\varphi_{m+j_k}(z_k), -m)$ and $\varphi_{m+j_1}(z_1) \neq \varphi_{m+j_2}(z_2)$, since $z_1^* \neq z_2^*$. Choose U_1 and U_2 disjoint open sets in V containing $\varphi_{m+j_1}(z_1)$ and $\varphi_{m+j_2}(z_2)$ respectively. Since φ is univalent on V and U_1 and U_2 are subsets of V, it follows that $\mathcal{N}_{-m}^{U_1}$ and $\mathcal{N}_{-m}^{U_2}$ are disjoint neighborhoods of z_1^* and z_2^* respectively. To check the intersection condition, suppose w^* is in $\mathcal{N}_{j_1}^{W_1} \cap \mathcal{N}_{j_2}^{W_2}$, that is suppose $w^* = [(w_1, j_2)] = [(w_2, j_2)]$ where w_k is in W_k for $k = 1, 2$. Thus we have $\varphi_{m+j_1}(w_1) = \varphi_{m+j_2}(w_2)$ where $m = \max\{-j_1, -j_2\}$. Since $\mathcal{N}_k^U = \mathcal{N}_{-m}^{\varphi_{k+m}(U)}$, if $Y = \varphi_{m+j_1}(W_1) \cap \varphi_{m+j_2}(W_2)$ then $\mathcal{N}_{-m}^Y \subset \mathcal{N}_{j_1}^{W_1} \cap \mathcal{N}_{j_2}^{W_2}$ and w^* is in $\mathcal{N}_{-m}^Y$. Thus $\{\mathcal{N}_j^U\}$ is the basis for a Hausdorff topology on S.

We introduce an analytic structure on S by defining the coordinate maps $c_j : V \to S$ given by $c_j(z) = [(z, j)]$. The map c_j is easily seen to be one-to-one (since φ is) and continuous. Since V is locally compact and S is Hausdorff, c_j is a homeomorphism of V onto $c_j(V)$ (see, for example, [Dug66, pp. 226, 237]). If $j = k + m$ where $m \geq 0$ then $c_k^{-1} c_j$ is defined on V and $c_k^{-1} c_j(z) = c_k^{-1}([(z, j)]) = c_k^{-1}[(z, k+m)] = c_k^{-1}[(\varphi_m(z), k)] = \varphi_m(z)$, so $c_k^{-1} c_j$ is analytic on V. Since φ_m is one-to-one on V, $c_j^{-1} c_k = \varphi_m^{-1}$ which is analytic on $\varphi_m(V)$, where it is defined. Thus S is a Riemann surface.

The surface S is simply connected, for suppose $\gamma : [0, 1] \to S$ is a loop. Since

$\gamma([0,1])$ is compact, $\mathcal{S} = \cup_j c_j(V)$, and the sets $c_j(V)$ are open and nested, we can find an integer j so that $\gamma([0,1])$ is in $c_j(V)$. But $c_j(V)$ is homeomorphic to V, which is simply connected, so γ is null homotopic in $c_j(V)$.

The surface $\mathcal{S}$ is not compact since if z is any point in V different from a then the sequence $z_j^* = [(z, -j)], j = 1, 2, 3, \ldots$, does not have a convergent subsequence. Indeed, suppose U is an open subset of V with compact closure and suppose m is a fixed integer. Then $z_j^* \in \mathcal{N}_m^U$ means $z \in \varphi_{m+j}(U)$, for $j > -m$, but φ_k converges uniformly on U to $a \neq z$. Thus, the sequence z_j^* can have at most finitely many terms in any neighborhood with compact closure, and cannot have a convergent subsequence.

By the Riemann Mapping Theorem, the surface $\mathcal{S}$ is analytically equivalent to either a halfplane or the whole complex plane. (We will see below that both cases can occur.)

II. *The maps Ψ and π.* Define the map $\pi : V \to \mathcal{S}$ by $\pi(z) = [(z, 0)]$ and the map $\Psi : \mathcal{S} \to \mathcal{S}$ by $\Psi([(z, j)]) = [(\varphi(z), j)]$. One easily verifies that Ψ is well defined, that π and Ψ are analytic and that π and Ψ are one-to-one. Since $[(z, j)] = [(\varphi(z), j - 1)] = \Psi([(z, j - 1)])$, we see that Ψ is surjective. Clearly $\Psi \circ \pi = \pi \circ \varphi$.

We note that $\pi(V)$ is a fundamental set for Ψ on $\mathcal{S}$. Indeed if K is a compact subset of $\mathcal{S}$, then we can find an integer j so that $K \subset c_j(V)$. If $j \geq 0$, then $c_j(V) \subset \pi(V)$ and if $j < 0$, then $\Psi_{-j}(K) \subset \pi(V)$.

Since Ψ is a map of $\mathcal{S}$ onto itself and V is a fundamental set for φ on D, we extend π to all of D by letting $\pi(z) = \Psi_{-j}(\pi(\varphi_j(z)))$ where j is a positive integer large enough that $\varphi_j(z)$ is in V, and Ψ_{-j} denotes the jth iterate of Ψ^{-1}. The equivalence relation on pairs shows that the extension is well defined. The extension is single valued, analytic, and maps D into $\mathcal{S}$ because Ψ_{-n}, π, and φ_n are single valued and analytic and Ψ^{-1} maps $\mathcal{S}$ onto $\mathcal{S}$.

III. *The maps Φ and σ, and the domain Ω.*

We have already noted that the Riemann surface $\mathcal{S}$ is analytically equivalent to either a halfplane or the whole plane: let Ω denote this domain, and let ρ be a Riemann map of $\mathcal{S}$ onto Ω. We define σ on D by $\sigma = \rho \circ \pi$ and define Φ on Ω by $\Phi = \rho \circ \Psi \circ \rho^{-1}$. One easily verifies that Φ is a one-to-one map of Ω onto itself (hence a linear fractional transformation), that σ is a univalent map of V into Ω, that $\sigma(V)$ is a fundamental set for Φ, and that $\Phi \circ \sigma = \sigma \circ \varphi$.

To prove uniqueness, we suppose that $\widetilde{V}$ is a fundamental set for φ on D such that φ is univalent on $\widetilde{V}$ and that $\widetilde{\Omega}$, $\widetilde{\Phi}$ and $\widetilde{\sigma}$ are the corresponding objects as above. Let $K = \{t\varphi(0) : 0 \leq t \leq 1\}$, the line segment in D connecting 0 and $\varphi(0)$. Now K is compact, connected, and has the property that for each n, the set $\cup_{j=n}^{\infty} \varphi_j(K)$ is connected; it is a curve starting at $\varphi_n(0)$ and passing through the iterates $\varphi_k(0)$ for $k > n$. Since V and $\widetilde{V}$ are each fundamental sets for φ, there is an integer n so that $\varphi_n(K) \subset V \cap \widetilde{V}$. Since $\varphi(V) \subset V$ and $\varphi(\widetilde{V}) \subset \widetilde{V}$, we see that the whole curve $\cup_{j=n}^{\infty} \varphi_j(K)$ is a subset of $V \cap \widetilde{V}$.

Let W be the connected component of $V \cap \widetilde{V}$ that contains $\cup_{j=n}^{\infty} \varphi_j(K)$. We claim that W is a fundamental set for φ on D on which φ is univalent. It is clear that W is open, connected, and simply connected. Since $\varphi(V \cap \widetilde{V}) \subset V \cap \widetilde{V}$, by Exercise 2.4.4, we have $\varphi(W) \subset W$ and since φ is univalent on $V \cap \widetilde{V}$, we see that φ is univalent on W. If L_0 is a compact subset of the disk, let $r_0 = \max\{|z| : z \in L_0\}$ and let $L = \{z : |z| \leq r_0\}$. Since V and $\widetilde{V}$ are each fundamental sets for φ on D, there is an integer m so that $\varphi_m(L) \subset V$ and $\varphi_m(L) \subset \widetilde{V}$. Since 0 is in L and L is connected, this means that if j is greater than m and n, then $\varphi_j(L)$ is contained in the component of $V \cap \widetilde{V}$ that contains $\cup_{j=n}^{\infty} \varphi_j(K)$. Putting this together, we see that for each compact set L_0 in D, there is an integer k so that $\varphi_k(L_0) \subset W$.

It follows that $\sigma(W)$ and $\widetilde{\sigma}(W)$ are fundamental sets for Φ and $\widetilde{\Phi}$ on Ω and $\widetilde{\Omega}$ respectively. Define $\tau : \Omega \to \widetilde{\Omega}$ by

$$\tau(z) = \widetilde{\Phi}_j^{-1}(\widetilde{\sigma}(\sigma^{-1}(\Phi_j(z))))$$

where j is an integer large enough that $\Phi_j(z)$ is in $\sigma(W)$. The map τ is well defined since if $\Phi_j(z)$ is in $\sigma(W)$ and p is a positive integer then

$$\widetilde{\Phi}_{j+p}^{-1}(\widetilde{\sigma}(\sigma^{-1}(\Phi_{j+p}(z)))) = \widetilde{\Phi}_{j+p}^{-1}(\widetilde{\sigma}(\varphi_p(\sigma^{-1}(\Phi_j(z)))))$$

$$= \widetilde{\Phi}_j^{-1}(\widetilde{\sigma}(\sigma^{-1}(\Phi_j(z))))$$

One easily verifies that τ is a univalent map of Ω onto $\widetilde{\Omega}$, so Ω and $\widetilde{\Omega}$ are either both the plane or both a halfplane, τ is a linear fractional transformation and $\widetilde{\Phi} = \tau \circ \Phi \circ \tau^{-1}$. We note also that $\widetilde{\sigma} = \tau \circ \sigma$ on V. Taking $\Omega = \widetilde{\Omega}$, this establishes the uniqueness assertion of the theorem. ∎

COROLLARY 2.57
The map σ of the model is univalent on D if and only if φ is univalent on D.

PROOF   If φ is univalent, we may choose $V = D$, and σ was proved to be univalent on V. If σ is univalent, then φ is the composite of three univalent maps: $\varphi = \sigma^{-1} \circ \Phi \circ \sigma$. ∎

In order for Theorem 2.53 to be useful, we must be able to identify the Ω and Φ that correspond to a given φ. Clearly, when Ω is a halfplane, we can make a convenient choice for the halfplane and the uniqueness statement of the theorem allows us, using an automorphism of Ω, to make a convenient standard choice for Φ. It turns out that there are four distinct cases and we have picked a standard for each. Not surprisingly, the case depends on the behavior of φ near the Denjoy–Wolff point a. Since $\sigma(\varphi_n(z)) = \Phi_n(\sigma(z))$ for each z in the disk and each positive integer n, we will write $\sigma(a)$ for the Denjoy–Wolff point of Φ. If φ is a map satisfying the hypotheses, the model can be chosen from one of the following:

• **(plane/dilation)** $\Omega = C$, $\sigma(a) = 0$, and $\Phi(z) = sz$ where $0 < |s| < 1$.

- (**plane/translation**) $\Omega = C$, $\sigma(a) = \infty$, and $\Phi(z) = z + 1$.
- (**halfplane/dilation**) $\Omega = \{z : \text{Re } z > 0\}$, $\sigma(a) = 0$, and $\Phi(z) = sz$ where $0 < s < 1$.
- (**halfplane/translation**) $\Omega = \{z : \text{Im } z > 0\}$, $\sigma(a) = \infty$, and $\Phi(z) = z \pm 1$.

Examples will be given below (Example 2.58) to show that each of the four cases can occur, and the reader is invited to construct more examples in the exercises. The classification of maps φ is a non-trivial problem but is aided by the following facts

- $\varphi'(a) = \Phi'(\sigma(a))$
- φ is in the plane/dilation case if and only if $|a| < 1$.

In the plane/dilation case, the first of these assertions is easy because σ is analytic in a neighborhood of the Denjoy–Wolff point; in the other cases, it depends on quantifying the relationship of the pseudohyperbolic distances between the iterates $\varphi_n(0)$ to the growth of the map σ. The somewhat tedious proof of this fact will be omitted, but we will use it freely. (Proof may be found in [Co81], [Pom79], or [BaP79].) The second assertion follows from the fact that if φ has a fixed point in the fundamental set V, then Φ has a fixed point in the fundamental set $\sigma(V)$ and the plane/dilation case is the only case in which Φ has an interior fixed point. These facts do not enable one to distinguish the plane/translation case from the halfplane/translation case, but otherwise distinguish the cases. Distinguishing the plane/translation and halfplane/translation cases is quite difficult in practice. One theoretical tool uses *interpolating sequences* for $H^\infty(D)$ (see for example [Hof62, p. 194]). An interpolating sequence is a sequence $\{z_j\}$ in the disk such that for any bounded sequence of complex numbers $\{c_j\}$, there is a bounded analytic function f on D with $f(z_j) = c_j$. Clearly an interpolating sequence is a Blaschke sequence, but not all Blaschke sequences are interpolating (for example, a radial sequence is interpolating if and only if it approaches the boundary exponentially). We will see below that if Ω is a halfplane and z_0 is any point of D, the iterates $\varphi_j(z_0)$ are an interpolating sequence and it can be proved (see [Co81], [Pom79], and [BaP79]) that the converse is true as well. This clearly distinguishes the halfplane/translation and plane/translation cases, but it is ineffective in practice because it requires precise knowledge of the iterates of φ. A more practical, but not always effective, test is the second derivative test described in Exercise 2.4.10.

EXAMPLE 2.58

Examples of the four cases. The plane/dilation case arises from $\varphi(z) = sz$ where $|s| < 1$ (here $\varphi = \Phi$ and $\sigma(z) = z$). For the plane/translation case, we may take $\varphi(z) = (1 + z)(3 - z)^{-1}$ and $\sigma(z) = (1 + z)(1 - z)^{-1}$. We have seen (Example 2.55) that $\varphi(z) = .5z + .5$ leads to the halfplane/dilation case with $\sigma(z) = 1 - z$ and $\Phi(z) = .5z$. There are no linear fractional maps of the disk into, not onto, itself that lead to the halfplane/translation case (see Exercise 2.4.7). A

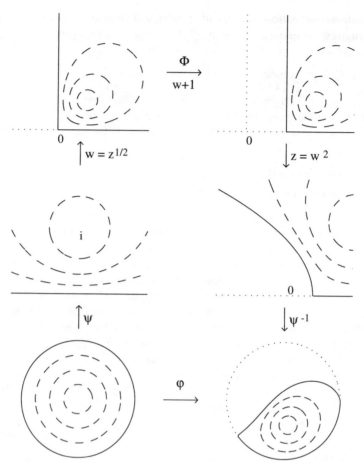

FIGURE 2.8

Constructing φ in the halfplane/translation case.

fairly simple example of a map in the halfplane/translation case can be constructed (see Figure 2.8) by letting $\psi(z) = i(1 + z)/(1 - z)$ be the map of the disk onto the upper halfplane and letting $\sigma(z) = \sqrt{\psi(z)}$ where the branch of the square root satisfies $\sqrt{2i} = 1 + i$ (and $\sqrt{1} = 1$). Then $\psi^{-1}(w) = (w - i)/(w + i)$ and $\sigma^{-1}(w) = (w^2 - i)/(w^2 + i)$. Defining $\varphi(z) = \sigma^{-1}(\sigma(z) + 1)$ gives (after some simplification)

$$\varphi(z) = \frac{(1 - i) + (1 + 3i)z + \sqrt{8(1 - z^2)}}{(3 + i) - (1 - i)z + \sqrt{8(1 - z^2)}}$$

This map is univalent on the disk and has Denjoy–Wolff point 1 with $\varphi'(1) = 1$.

In constructing a map φ using the model, it is important to notice that the intertwining $\Phi \circ \sigma = \sigma \circ \varphi$ implies $\Phi(\sigma(D)) \subset \sigma(D)$, that is, that $\sigma(D)$ is dilation or translation invariant. Indeed, if σ is univalent, the condition $\Phi(\sigma(D)) \subset \sigma(D)$ is enough to define $\varphi(z) = \sigma^{-1}(\Phi(\sigma(z)))$ to give a univalent map of the disk into itself and if $\sigma(D)$ is a fundamental set for Φ on Ω, then φ has σ, Φ, and Ω as its model.

The model also helps in understanding the fixed point structure of maps of the disk to itself. If b is a fixed point of φ, the intertwining of φ and Φ implies $\sigma(b)$ is a fixed point of Φ if $\sigma(b)$ exists in some sense. Since the fixed points of Φ are obvious, the fixed points of φ are the inverse images of these under σ. As in the construction of Example 2.58, the model makes it possible to construct and understand analytic selfmaps of the disk with specific fixed point properties.

EXAMPLE 2.59
In this example, we will construct a map in the plane/dilation case with one fixed point on the unit circle. Let $\psi(z) = (1 + z)/(1 - z)$ be the map of the unit disk onto the right halfplane that takes -1 to 0, 0 to 1, and 1 to infinity. As suggested by the proof of Corollary 2.57, a map of the disk into itself can be constructed by letting $\sigma(z) = \psi(z)^2 - 1$, multiplying by $1/2$, then going back by $\sigma^{-1}(w) = \psi^{-1}(\sqrt{w + 1})$ as is illustrated in Figure 2.9. The resulting φ is in the plane/dilation case because $\varphi(0) = 0$ and the corresponding automorphism of the plane to itself is $\Phi(w) = w/2$. Specifically,

$$\varphi(z) = \sigma^{-1}\left(\frac{1}{2}\sigma(z)\right) = \frac{\sqrt{1 + z^2} - 1 + z}{\sqrt{1 + z^2} + 1 - z}$$

where the branch of the square root satisfies $\sqrt{1} = 1$. It can be checked directly from the formula that $\varphi(0) = 0$ and $\varphi'(0) = 1/2$. Since 0 and ∞ are the only fixed points of $\Phi(w) = w/2$, the only fixed points of φ are $0 = \sigma^{-1}(0)$ and $1 = \sigma^{-1}(\infty)$. It is not difficult to check that $\varphi'(1) = \sqrt{2} > 1$. $\quad\square$

EXAMPLE 2.60
The model can be used to understand or construct non-univalent maps of the disk also! If the map φ is far from univalent, the intertwining map σ will of necessity be very complicated, for example, if $\varphi(z) = z(2z - 1)/(2 - z)$, which is a Blaschke product mapping the disk two-to-one onto itself, σ will be an infinite-to-one map of the disk into the plane. However, the model can be used to easily construct simple non-univalent maps of the disk with desired properties.

Let

$$S_1 = \{w = x + iy : x > 0, \ y > 0, \text{ and } w \neq 2^n + iy \text{ for } n = 0, 1, \ldots \text{ and } y \leq 2^n\}$$

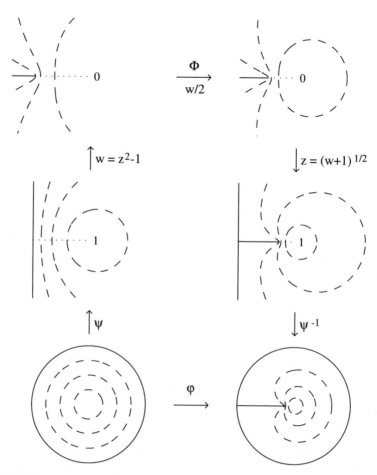

FIGURE 2.9

Constructing a map φ from the model.

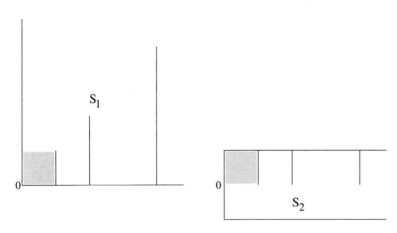

FIGURE 2.10

The sheets S_1 and S_2 of the Riemann surface.

let

$$S_2 = \{w = x+iy : x > 0, |y| < 1, \text{ and } w \neq 2^n+iy \text{ for } n = 0, 1, \ldots \text{ and } 0 \leq y\}$$

and let S be the Riemann surface obtained by identifying the points $\{x + iy : 0 < x < 1 \text{ and } 0 < y < 1\}$ of S_1 and S_2 (with the inherited analytic structure). The two sheets of the Riemann surface are shown in Figure 2.10 with the areas to be identified shaded. The surface S (see Figure 2.11) consists of two sheets over $\{x + iy : x > 1, 0 < y < 1, x \neq 2^n\}$ and one sheet elsewhere.

Speaking loosely, if q is a point of S, so is $q/2$. More precisely, if $q = x+iy$ is in S_1, then $(x+iy)/2$ is also in S_1 and we write $q/2 = (x+iy)/2$; similarly, if q is in S_2. If q is in the overlap, then so is $q/2$ and the equivalence class of $q/2$ is the same for both elements of the equivalence class of q, so the map $q \mapsto q/2$ is well defined on S. If ρ is the Riemann map of D onto S that takes 1 to "0" and 0 to "1/2", then we can define an analytic map φ of the disk into itself by $\varphi(z) = \rho^{-1}(\rho(z)/2)$. We see that φ is two-to-one on the subset $\rho^{-1}(\{x + iy : 1 < x < 2, 0 < y < 1\})$ because for such points q from S_1 and S_2, the points $q/2$ are identified.

Let π be the projection of S onto $S_1 \cup S_2$ in the right halfplane that acts by forgetting sheets and let σ be the analytic function mapping the unit disk into the halfplane by $\sigma = \pi \circ \rho$. Since $\pi(q/2) = \pi(q)/2$, the map σ intertwines φ and $\Phi(w) = w/2$ on the halfplane. Thus, φ is in the halfplane/dilation case and the attractive fixed point 0 of Φ corresponds to the Denjoy–Wolff point $a = 1$ of φ. Moreover, there are two other fixed points of φ: b_1 that corresponds to the fixed point ∞ of Φ by taking $w \to \infty$ in S_1 and b_2 that corresponds to the fixed point ∞ of Φ by taking $w \to \infty$ in S_2. Notice that φ maps the part of the unit circle between $a = 1$ and b_1 that corresponds to the positive imaginary axis in S onto itself because Φ maps that part of the boundary of $\sigma(D)$ onto itself. Moreover,

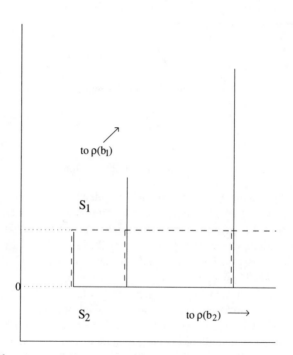

FIGURE 2.11

The Riemann surface $\mathcal{S}$.

there is a point c between 1 and b_2, namely $c = \sigma^{-1}(-i)$, so that φ maps the arc $[1, c]$ into but not onto itself and φ maps the arc (c, b_2) into the disk because Φ maps the segment $[0, -i]$ into but not onto itself and Φ maps the line $\{x - i : x > 0\}$ into $\sigma(D)$. Some points of the unit circle between b_1 and b_2 are mapped to the unit circle, but all such points are eventually mapped into the disk under the iteration because Φ maps that part of the boundary of $\sigma(D)$ partly into itself and partly into $\sigma(D)$. $\square$

Our principal interest in the model is that it enables us to solve Schroeder's functional equation.

LEMMA 2.61
Suppose φ, V, Φ, σ, and Ω are as in Theorem 2.53 and suppose $\lambda \neq 0$ is a complex number. If F is analytic on Ω and $F \circ \Phi = \lambda F$, then $f \circ \varphi = \lambda f$ where $f = F \circ \sigma$. Conversely, if f is analytic on D and $f \circ \varphi = \lambda f$, there is a function F analytic on Ω so that $F \circ \Phi = \lambda F$ and $f = F \circ \sigma$.

PROOF If $F \circ \Phi = \lambda F$, then

$$f \circ \varphi = F \circ \sigma \circ \varphi = F \circ \Phi \circ \sigma = \lambda F \circ \sigma = \lambda f$$

Conversely, if $f \circ \varphi = \lambda f$, since σ is univalent on V, we may define $\widetilde{F}$ on $\sigma(V)$ by $\widetilde{F} = f \circ \sigma^{-1}$ so that $\widetilde{F} \circ \Phi = \lambda \widetilde{F}$ on $\sigma(V)$. Since $\sigma(V)$ is a fundamental set for Φ on Ω, we may define F on Ω by $F(w) = \lambda^{-k}\widetilde{F}(\Phi_k(w))$ where k is an integer large enough that $\Phi_k(w)$ is in $\sigma(V)$. If k' is another such integer, say $k' = k + m$ with $m > 0$, then

$$\lambda^{-k'}\widetilde{F}(\Phi_{k'}(w)) = \lambda^{-k-m}\widetilde{F}(\Phi_{k+m}(w)) = \lambda^{-k-m}\widetilde{F}(\Phi_m(\Phi_k(w)))$$

$$= \lambda^{-k-m}\left(\lambda^m\widetilde{F}(\Phi_k(w))\right) = \lambda^{-k}\widetilde{F}(\Phi_k(w))$$

so that F is well defined. It clearly satisfies $F \circ \Phi = \lambda F$ and $f = F \circ \sigma$. $\blacksquare$

LEMMA 2.62
(1) *If $\Phi(z) = sz$ and $0 < |s| < 1$, then $F \circ \Phi = \lambda F$ has a non-zero solution F that is analytic in the plane if and only if $\lambda = s^j$ for $j = 0, 1, 2, \ldots$. Moreover, F is a solution of $F \circ \Phi = s^j F$ for some non-negative integer j if and only if $F(z) = cz^j$ for some constant c.*

(2) *If $\Phi(z) = z + 1$ and $\lambda \neq 0$, then F, analytic in the plane, the halfplane $\{z : \mathrm{Im}\, z > 0\}$, or the strip $\{z : |\mathrm{Im}\, z| < \kappa\}$, is a non-zero solution of $F \circ \Phi = \lambda F$ if and only if*

$$F(z) = e^{\gamma z}g(e^{2\pi i z}) \tag{2.4.1}$$

where $e^\gamma = \lambda$ and g is analytic in the punctured plane $C \setminus \{0\}$, the punctured disk $D \setminus \{0\}$, or the annulus $\{w : e^{-2\pi\kappa} < |w| < e^{2\pi\kappa}\}$ respectively.

PROOF   To prove (1), compare the Taylor series at $z = 0$ for $F(sz) = \sum(a_j s^j)z^j$ and $\lambda F(z) = \sum(\lambda a_j)z^j$.

Clearly functions of the form in Equation (2.4.1) satisfy the functional equation. Conversely, if F is analytic and satisfies the functional equation, define g by $g(w) = e^{-\gamma z}F(z)$ where $e^\gamma = \lambda$ and $w = e^{2\pi i z}$. Since F satisfies the functional equation, g is well defined and analytic.   ∎

THEOREM 2.63

Suppose φ is an analytic map of the unit disk into itself, not an automorphism, with Denjoy–Wolff point a.

(1)   If $\varphi'(a) = 0$, then the only non-zero solution of $f \circ \varphi = \lambda f$ is $\lambda = 1$ and f constant.

(2)   If $|a| < 1$ and $\varphi'(a) \neq 0$, then $f \circ \varphi = \lambda f$ has a non-zero solution if and only if $\lambda = \varphi'(a)^j$ for some non-negative integer j. Moreover, f is a non-zero solution of $f \circ \varphi = \varphi'(a)^j f$ for some non-negative integer j if and only if $f(z) = c\sigma(z)^j$ where σ is the map of the model (Theorem 2.53) and c is constant.

(3)   If $|a| = 1$, then the solution space of $f \circ \varphi = \lambda f$ is infinite dimensional for every $\lambda \neq 0$.

PROOF   (1) Clearly f constant and $\lambda = 1$ satisfies the functional equation. Since $f \equiv 0$ is the only solution for $\lambda = 0$, we assume $\lambda \neq 0$.

Suppose $f \circ \varphi = \lambda f$. Since $\varphi'(a) = 0$, the Julia–Carathéodory Theorem implies $|a| < 1$. Since a is a fixed point of φ, we have $f(a) = f(\varphi(a)) = \lambda f(a)$ and either $f(a) = 1$ or $f(a) = 0$. In any case, $f - f(a)$ satisfies the functional equation also, so we may assume $f(a) = 0$.

Let $f^{(j)}(a) = 0$ for $j = 0, 1, \ldots, k - 1$. Taking k^{th} derivatives on both sides of the equation $f \circ \varphi = \lambda f$ yields

$$\left(f^{(k)}(\varphi)\right)(\varphi')^k + \text{terms containing } f^{(j)} \text{ for } j < k = \lambda f^{(k)}$$

Now evaluating at a gives $\lambda f^{(k)}(a) = \left(f^{(k)}(a)\right)(0)^k + 0 = 0$ so $f^{(k)}(a) = 0$ also. Thus, if $\lambda \neq 1$, then $f \equiv 0$ is the only solution.

(2) Apply Lemma 2.61 and (1) of Lemma 2.62 recalling that $\varphi'(a) = \Phi'(\sigma(a))$.

(3) If $s = \varphi'(a) < 1$ so that φ is in the halfplane/dilation case with $\sigma_0 \circ \varphi = \Phi_0 \circ \sigma_0$ for σ_0 a map of the disk into the right halfplane and $\Phi_0(z) = sz$, then let $\sigma(z) = -(|\log s|)^{-1}\log(\sigma_0(z))$ which maps D into the strip $\{x + iy : |y| < \pi/(2|\log s|)\}$ and let $\Phi(z) = z + 1$. Then $\sigma \circ \varphi = \Phi \circ \sigma = \sigma + 1$. Thus, we have $\sigma \circ \varphi = \Phi \circ \sigma$ where $\Phi(z) = z + 1$ on the plane, the upper halfplane, or the strip

$\{x + iy : |y| < \pi/(2|\log s|)\}$ depending on whether φ is in the plane/translation case, the halfplane/translation case, or the halfplane/dilation case. Now, we apply Lemma 2.61 and (2) of Lemma 2.62 (or in the halfplane translation case $z - 1$, the obvious modification of this lemma). Since the sets of analytic functions g on the punctured plane, punctured disk, and the annulus are infinite dimensional, the solution space of $f \circ \varphi = \lambda f$ is infinite dimensional for every $\lambda \neq 0$. ∎

While this theorem gives the analytic solutions of the eigenvalue equation for the composition operator C_φ, it does not completely solve the problem because a function f satisfying $f \circ \varphi = \lambda f$ is an eigenvector for C_φ if and only if f is in the space of functions on which C_φ is defined. As we shall see in Chapter 7, answering this question can be delicate, but the following result gives growth estimates that are helpful in determining which eigenfunctions are actually in the space.

THEOREM 2.64

Suppose φ, an analytic map of the unit disk into itself, has Denjoy–Wolff point a with $|a| = 1$ and $\varphi'(a) < 1$. Suppose also that b is another fixed point of φ, that φ is analytic in a neighborhood of b, and that there is $\delta > 0$ for which $|\varphi(z)| < 1$ whenever $0 < |z - b| < \delta$ and $|z| \leq 1$. If σ is the map in the model that takes the disk into the right half plane and satisfies $\sigma \circ \varphi = \varphi'(a)\sigma$, then for every $p > |\log \varphi'(a)/\log \varphi'(b)|$, there is a constant M so that

$$|\sigma(z)| < M|z - b|^{-p}$$

in a neighborhood of b.

PROOF Let $r_0 = \varphi'(b)$, let $s = \varphi'(a)$, and let $r = \exp(|\log s|/p) < r_0$. Without loss of generality, we may assume δ is such that if $0 < |z - b| < \delta$ and $|z| \leq 1$ then $|\varphi'(b) - \varphi'(z)| < r_0 - r$ as well as $|\varphi(z)| < 1$. Let

$$K = \{w : |b - w| \geq \delta \text{ and } w = \varphi(z) \text{ for some } z \text{ with } |z| \leq 1 \text{ and } |b - z| \leq \delta\}$$

Now K is a compact subset of D with the property that if z is in the closed disk and $0 < |b - z| < \delta$ then either $|b - \varphi(z)| < \delta$ or $\varphi(z)$ is in K. In other words, if z is in the closed disk and $0 < |b - z| < \delta$ then $\varphi_k(z)$ is in K for some positive integer k. Let $\delta' = \sup\{|b - w| : w \in K\}$.

If n is a positive integer so that $\varphi_n(z)$ is in K, then $\sigma(\varphi_n(z)) = s^n \sigma(z)$, so $\sigma(z) = s^{-n}\sigma(\varphi_n(z))$ is in $s^{-n}\sigma(K)$. Since $\sigma(K)$ is a compact subset of the right half plane, this means $|\sigma(z)| \leq M_1 s^{-n}$, where $M_1 = \max\{|w| : w \in K\}$. We will estimate the integer n for which $\varphi_n(z)$ is in K.

If $|b - z| < \delta$, then

$$|b - \varphi(z)| = \left| \int_0^1 \varphi'(tb + (1 - t)z)(b - z) \, dt \right|$$

$$= |b - z| \left| \varphi'(b) - \int_0^1 \varphi'(b) - \varphi'(tb + (1 - t)z) \, dt \right|$$

$$\geq |b - z|(\varphi'(b) - (r_0 - r)) = r|b - z|$$

Similarly, if $|b - \varphi(z)| < \delta$ as well, then

$$|b - \varphi(\varphi(z))| \geq r|b - \varphi(z)| \geq r^2|b - z|$$

Thus, if n is the least integer so that $|b - \varphi_n(z)| \geq \delta$, which is the least integer with $\varphi_n(z)$ in K, we have $\delta' \geq |b - \varphi_n(z)| \geq r^n|b - z|$. Rewriting this, we find $n \leq \log(\delta'|b - z|^{-1})/\log r$. This means

$$|\sigma(z)| \leq M_1 s^{-n} \leq M_1 s^{-\log(\delta'|b-z|^{-1})/\log r} = M_2|b-z|^{\log s/\log r} = M_2|b-z|^{-p}$$

∎

We next show that if φ is in the halfplane/dilation case, then the iterates of a point form an interpolating sequence, a result that will be useful later. The proof of the corresponding result for the halfplane/translation case is left for the exercises.

THEOREM 2.65

Suppose φ, an analytic map of the unit disk into itself, has Denjoy–Wolff point a with $|a| = 1$ and $\varphi'(a) < 1$. If $\{z_j\}$ is a sequence in the disk with $\varphi(z_j) = z_{j+1}$, then $\{z_j\}$ is an interpolating sequence.

PROOF Let $\{z_j\}$ be a sequence in the disk with $\varphi(z_j) = z_{j+1}$. We may assume with no loss of generality that one of the points of the sequence is labeled z_0.

For $0 < s < 1$, the map $w \mapsto sw$ of the right halfplane onto itself is conjugate, using $w = (1 - z)/(1 + z)$, to

$$\Phi(z) = \frac{(1 + s)z + (1 - s)}{(1 - s)z + (1 + s)}$$

which is the automorphism of the unit disk that has fixed points ± 1 and $\varphi'(1) = s$. The map φ has a halfplane/dilation model, so using this conjugation and letting $s = \varphi'(a)$, we see there is σ, mapping the disk into the disk, so that $\Phi \circ \sigma = \sigma \circ \varphi$ and $\sigma(z_0) = ir$ is purely imaginary. (The normalization can be accomplished by replacing an arbitrary solution σ with $((1 + t)\sigma + (1 - t))/((1 - t)\sigma + (1 + t))$ for a suitable positive number t.)

Let $w_j = \sigma(z_j)$ so that

$$\Phi(w_j) = \Phi(\sigma(z_j)) = \sigma(\varphi(z_j)) = \sigma(z_{j+1}) = w_{j+1}$$

If $\{c_j\}$ is a bounded sequence of complex numbers and f is a bounded analytic function on the disk with $f(w_j) = c_j$, then $f \circ \sigma$ is a bounded analytic function on the disk with $f \circ \sigma(z_j) = f(w_j) = c_j$. Thus, it is enough to prove that the sequence w_j is an interpolating sequence.

Since Φ is an automorphism of the disk, using the relation $\Phi(w_j) = w_{j+1}$, we may extend the sequence w_j, if necessary, to include all integers j and an easy calculation shows

$$w_j = \frac{(1 + s^j)ir + (1 - s^j)}{(1 - s^j)ir + (1 + s^j)}$$

for all j. The condition of Carleson for a sequence to be an interpolating sequence is that there is δ such that for each $k = \ldots, -1, 0, 1, 2 \ldots$

$$\prod_{j \neq k} \left| \frac{w_j - w_k}{1 - \overline{w_k} w_j} \right| \geq \delta > 0$$

In our case, this product is

$$\prod_{j \neq k} \left| \frac{\left(\frac{(1+s^j)ir+(1-s^j)}{(1-s^j)ir+(1+s^j)}\right) - \left(\frac{(1+s^k)ir+(1-s^k)}{(1-s^k)ir+(1+s^k)}\right)}{1 - \left(\frac{(1+s^k)(-ir)+(1-s^k)}{(1-s^k)(-ir)+(1+s^k)}\right)\left(\frac{(1+s^j)ir+(1-s^j)}{(1-s^j)ir+(1+s^j)}\right)} \right|$$

$$= \prod_{j \neq k} \left| \frac{\begin{array}{c}\left((1+s^j)ir + (1-s^j)\right)\left((1-s^k)ir + (1+s^k)\right) \\ -\left((1-s^j)ir + (1+s^j)\right)\left((1+s^k)ir + (1-s^k)\right)\end{array}}{\begin{array}{c}\left((1-s^j)ir + (1+s^j)\right)\left((1-s^k)(-ir) + (1+s^k)\right) \\ -\left((1+s^j)ir + (1-s^j)\right)\left((1+s^k)(-ir) + (1-s^k)\right)\end{array}} \right|$$

$$= \prod_{j \neq k} \left| \frac{(s^k - s^j)(1 + r^2)}{(s^k + s^j)(1 - r^2) + 2ir(s^k - s^j)} \right|$$

Splitting the infinite product into the product for $j < k$ and $j > k$, dividing by s^j and s^k, respectively, and replacing $k - j$ and $j - k$, respectively, by m, this becomes

$$\prod_{m=1}^{\infty} \frac{\left|(1 - s^m)(1 + r^2)\right|^2}{\left|(1 + s^m)(1 - r^2) + 2ir(1 - s^m)\right|^2} = \prod_{m=1}^{\infty} \frac{1 - s^m(2 - s^m)}{1 + s^m\left(\frac{2(1-6r^2+r^4)}{(1+r^2)^2} + s^m\right)}$$

Since $2 - s^m$ and $(2(1 - 6r^2 + r^4)/(1 + r^2)^2) + s^m$ are bounded independent of m, the product converges to a positive number. Note that this product is independent of k, so the condition for being an interpolating sequence is satisfied. $\blacksquare$

Proofs of the Lemmas on iteration near the boundary

In the remainder of this section, we present the geometric results that lead up to Proposition 2.56 and give the proof of that proposition. These technical results from function theory are not critical for the remainder of the book and may omitted on a first reading.

LEMMA 2.66

If φ is an analytic mapping of the disk into itself with Denjoy–Wolff point a on the circle and $\varphi'(a) < 1$, then for any compact set K in D, there is a nontangential approach region containing all the iterates $\varphi_n(K)$.

PROOF   The proof is more convenient in the right halfplane, so we will change variables by letting $\psi(w) = (1+\overline{a}\varphi(z(w)))/(1-\overline{a}\varphi(z(w)))$ where $z(w) = a(w-1)/(w+1)$. Then ψ maps the right halfplane into itself with fixed point infinity and Julia's Lemma (2.41) guarantees $\mathrm{Re}\,\psi(w) > \kappa\,\mathrm{Re}\,(w)$ for $\kappa = \varphi'(a)^{-1} > 1$ for all w in the right halfplane (see Exercise 2.3.10). A nontangential approach region for the point infinity is a sector $\{w = x + iy : x > 0 \text{ and } |y| < \beta x\}$. Clearly, nontangential approach for iterates of ψ using the sectors is equivalent to nontangential approach for iterates of φ using the approach regions $\Gamma(1, \alpha)$ in the disk.

Given a compact set K in the halfplane, choose $r < 1$ so that for any point w in K, the pseudohyperbolic disk centered at w with radius r contains both K and $\psi(K)$. Let $x_0 = \min\{\mathrm{Re}\,w : w \in K\}$, let $w_0 = x_0 + iy_0$ be the point of K with real part x_0 and greatest imaginary part and let Δ_0 be the pseudohyperbolic disk with center w_0 and radius r. (Without loss of generality, we may assume $y_0 \geq 0$; if it is not, replace φ by $\widetilde{\varphi}(z) = \overline{\varphi(\overline{z})}$.) Since this is a pseudohyperbolic disk with respect to the halfplane (see Exercise 2.3.6), it is the set of w for which

$$\left| \frac{w - w_0}{w + \overline{w_0}} \right| \leq r$$

The Euclidean description of this disk is

$$\left(x - x_0 \frac{1 + r^2}{1 - r^2} \right)^2 + (y - y_0)^2 \leq \left(x_0 \frac{2r}{1 - r^2} \right)^2 \qquad (2.4.2)$$

Since $\kappa > 1$, the line $\mathrm{Re}\,w = \kappa x_0$ is to the right of x_0. Let $x_1 = \kappa x_0$ and let $w_1 = x_1 + iy_1$ be the upper point of intersection of the line $\mathrm{Re}\,(w) = x_1$ and the boundary circle of Δ_0 (see Figure 2.12). Now w_1 has pseudohyperbolic distance r from w_0, so

$$(y_1 - y_0)^2 = \left(x_0 \frac{2r}{1 - r^2} \right)^2 - \left(\kappa x_0 - x_0 \frac{1 + r^2}{1 - r^2} \right)^2$$

and

$$y_1 = y_0 + x_0 \sqrt{2\kappa \left(\frac{1 + r^2}{1 - r^2} \right) - (\kappa^2 + 1)}$$

Construct $w_2 = x_2 + iy_2$ by choosing $x_2 = \kappa x_1$ and letting w_2 be the upper intersection point of the line $\mathrm{Re}\,(w) = \kappa x_1$ and the boundary circle of the pseudo-

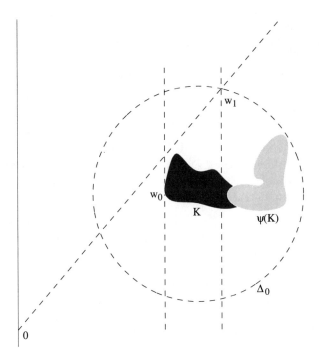

FIGURE 2.12
Nontangential iteration: $K, \psi(K), w_0, w_1$ and the disk Δ_0.

hyperbolic disk with center w_1 and radius r. It follows that

$$y_2 = y_1 + x_1 \sqrt{2\kappa \left(\frac{1 + r^2}{1 - r^2} \right) - \left(\kappa^2 + 1 \right)}$$

Writing d for the expression in the square root and continuing, we define $w_n = x_n + iy_n$ by $x_n = \kappa x_{n-1}$ and $y_n = y_{n-1} + dx_{n-1}$. Since $x_n = \kappa x_{n-1}$, we have

$$\frac{y_n}{x_n} = \frac{d}{\kappa} + \frac{1}{\kappa} \frac{y_{n-1}}{x_{n-1}}$$

By Julia's Lemma, $\psi(K)$ lies to the right of the line $\text{Re}\,(w) = x_1$. Since $\psi(K)$ is also in Δ_0, for every point $w = x + iy$ in $\psi(K)$, the ratio y/x is less than or equal to the ratio y_1/x_1. For w in $\psi(K)$, if $u + iv = \psi(w)$, then $u > \kappa x$ and the Schwarz–Pick Theorem (Exercise 2.3.6) implies that $\psi(K)$ and $\psi(\psi(K))$ both lie in a pseudohyperbolic disk of radius r with center at w. In particular, $v \leq y + xd$ and

$$\frac{v}{u} \leq \frac{d}{\kappa} + \frac{1}{\kappa} \frac{y}{x} \leq \frac{d}{\kappa} + \frac{1}{\kappa} \frac{y_1}{x_1} = \frac{y_2}{x_2}$$

In other words, for every point $u + iv$ of $\psi(\psi(K))$, the ratio v/u is no more than the ratio y_2/x_2. Continuing by induction, we see the ratio of the imaginary part of a point in $\psi_n(K)$ to its real part is no more than y_n/x_n.

Since

$$\frac{y_n}{x_n} = \frac{d}{\kappa} + \frac{1}{\kappa} \frac{y_{n-1}}{x_{n-1}} = d \sum_{j=1}^{n} \kappa^{-j} + \kappa^{-n} \frac{y_0}{x_0}$$

and κ is greater than 1 so that the infinite series converges, there is a ray in the right halfplane such that $\psi_n(K)$ is below the ray for all n.

Similarly, using the lower points of intersection, we see there is a ray in the right halfplane such that $\psi_n(K)$ is above the ray for all n. Thus, there is a nontangential approach region that contains all the iterates $\psi_n(K)$. ∎

In the case that $\varphi'(a) = 1$ at the Denjoy–Wolff point a on the circle, the situation is more difficult because the iterates do not have to converge nontangentially. We will state the results in terms of maps of the halfplane to itself as the geometry is clearer in this case. We begin with a lemma that gives the rate at which the real parts of the iterates of a point go to infinity.

LEMMA 2.67

Let ψ be an analytic mapping of the right halfplane into itself with

$$\kappa = \inf_{0 < x} \frac{\text{Re}\,(\psi(x + iy))}{x} = 1$$

If w_0 is a point of the halfplane and $w_n = x_n + iy_n = \psi_n(w_0)$ for $n = 1, 2, \ldots$, then

$$\lim_{n \to \infty} \frac{x_{n+1}}{x_n} = 1$$

PROOF The hypothesis on κ guarantees the ratios x_{n+1}/x_n are never less than 1. Let r_n be the pseudohyperbolic distance between the iterates,

$$r_n = \left| \frac{w_{n+1} - w_n}{w_{n+1} + \overline{w_n}} \right| \tag{2.4.3}$$

so the r_n form a decreasing sequence of positive numbers. The Euclidean description of this distance is

$$\left(x_{n+1} - x_n \frac{1 + r_n^2}{1 - r_n^2} \right)^2 + (y_{n+1} - y_n)^2 = \left(x_n \frac{2r_n}{1 - r_n^2} \right)^2$$

and we have

$$\left(\frac{x_{n+1}}{x_n} - \frac{1 + r_n^2}{1 - r_n^2} \right)^2 + \left(\frac{y_{n+1} - y_n}{x_n} \right)^2 = \left(\frac{2r_n}{1 - r_n^2} \right)^2 \tag{2.4.4}$$

In particular, the ratios x_{n+1}/x_n remain bounded since this inequality implies

$$\frac{x_{n+1}}{x_n} \leq \frac{1 + r_n^2}{1 - r_n^2} + \frac{2r_n}{1 - r_n^2} \leq \frac{(1 + r_0)^2}{1 - r_0^2}$$

Similarly the ratios $(y_{n+1} - y_n)/x_n$ remain bounded. If the sequence of distances r_n converges to zero, the right side of Inequality (2.4.4) is tending to zero, so the first term on the left tends to zero and we get

$$\lim_{n \to \infty} \frac{x_{n+1}}{x_n} = \lim_{n \to \infty} \frac{1 + r_n^2}{1 - r_n^2} = 1$$

in this case.

So we suppose the sequence of distances r_n is converging to a positive number and, thus, that $\lim_{n \to \infty} r_{n+1}/r_n = 1$. Define the maps f_n by

$$f_n(w) = \left(\frac{\psi_{n+1}(w) - w_{n+1}}{\psi_{n+1}(w) + \overline{w_{n+1}}} \right) \left(\frac{\psi_n(w) + \overline{w_n}}{\psi_n(w) - w_n} \right)$$

By the Schwarz–Pick Theorem for the halfplane (Exercise 2.3.6), $|f_n(w)| \leq 1$. Note that

$$f_n(w_0) = \frac{x_n}{x_{n+1}} \psi'(w_n) \quad \text{and} \quad |f_n(w_1)| = \frac{r_{n+1}}{r_n} \tag{2.4.5}$$

Since the sequence f_n is a normal family of maps into the closed disk and $|f_n(w_1)|$ converges to 1, Montel's theorem implies the sequence of functions $|f_n|$ converges to 1 uniformly on compact subsets of the halfplane. In particular,

$$\lim_{n \to \infty} \frac{x_n}{x_{n+1}} |\psi'(w_n)| = \lim_{n \to \infty} |f_n(w_0)| = 1$$

which implies, since the ratios x_{n+1}/x_n remain bounded, that

$$\lim_{n\to\infty} |\psi'(w_n)| - \frac{x_{n+1}}{x_n} = 0 \qquad (2.4.6)$$

Now let $\psi(w) = w + p(w)$ and let $u(w) = \operatorname{Re} p(w)$. From the Julia–Carathéodory Theorem (Exercise 2.3.10) and the hypothesis on κ, $u(w)$ is positive. For w and w' in the right halfplane, the Schwarz–Pick Theorem implies

$$\left| \frac{p(w') - p(w)}{w' - w} \right| \leq \left| \frac{p(w') + \overline{p(w)}}{w' + \overline{w}} \right|$$

so taking limits on both sides as w' approaches w gives

$$|p'(w)| \leq \frac{u(w)}{x} \qquad (2.4.7)$$

Let $x_{n+1}/x_n - 1 = \lambda_n$ and $p'(w_n) = a_n + ib_n$. Since $\psi'(w) = 1 + p'(w)$, Equation (2.4.6) gives

$$\lim_{n\to\infty} |1 + p'(w_n)|^2 - (1 + \lambda_n)^2 = 0$$

This means that there are numbers ϵ_n tending to zero so that

$$1 + 2\lambda_n + \lambda_n^2 - \epsilon_n = 1 + 2a_n + a_n^2 + b_n^2 \qquad (2.4.8)$$

But Inequality (2.4.7) implies

$$a_n^2 \leq a_n^2 + b_n^2 = |p'(w_n)|^2 \leq \left(\frac{\operatorname{Re}(p(w_n))}{x_n} \right)^2 = \left(\frac{x_{n+1}}{x_n} - 1 \right)^2 = \lambda_n^2 \quad (2.4.9)$$

Using Inequalities (2.4.8) and (2.4.9) we get

$$\lambda_n - \epsilon_n/2 \leq a_n \leq \lambda_n$$

Since the ϵ_n tend to zero, this means $\lim_{n\to\infty} \lambda_n - a_n = 0$, which in turn, by Inequality (2.4.9), implies $\lim_{n\to\infty} b_n = 0$. Thus,

$$\lim_{n\to\infty} \psi'(w_n) - \frac{x_{n+1}}{x_n} = \lim_{n\to\infty} (a_n - \lambda_n) + ib_n = 0$$

Combining this with Equation (2.4.5) we see that this means $f_n(w_0)$ converges to 1 and therefore that the sequence f_n converges uniformly to 1 on compact subsets of the halfplane.

Let g_n be the normalized functions

$$g_n(w) = \frac{\psi_n(w) - iy_n}{x_n}$$

which map the halfplane into itself and satisfy $g_n(w_0) = 1$. Now

$$g_n(\psi(w)) = \frac{\psi_{n+1}(w) - iy_n}{x_n} = \frac{x_{n+1}}{x_n} g_{n+1}(w) + i\frac{y_{n+1} - y_n}{x_n} \qquad (2.4.10)$$

Since the g_n's form a normal family, for any subsequence of the n, there is function g and a further subsequence n_j so that g_{n_j} converges uniformly on compact sets to g and there are c and d so that

$$\lim_{j\to\infty} \frac{x_{n_j+1}}{x_{n_j}} = c \quad \text{and} \quad \lim_{j\to\infty} \frac{y_{n_j+1} - y_{n_j}}{x_{n_j}} = d$$

Now

$$
\begin{aligned}
f_n(w) &= \left(\frac{\psi_{n+1}(w) - w_{n+1}}{\psi_{n+1}(w) + \overline{w_{n+1}}}\right)\left(\frac{\psi_n(w) + \overline{w_n}}{\psi_n(w) - w_n}\right) \\
&= \left(\frac{g_{n+1}(w) - 1}{g_{n+1}(w) + 1}\right)\left(\frac{g_n(w) + 1}{g_n(w) - 1}\right)
\end{aligned}
$$

so, since the f_n's converge to 1, the subsequence g_{n_j+1} also converges to g. Taking limits in Equation (2.4.10) on the subsequence n_j gives

$$g(\psi(w)) = cg(w) + id \tag{2.4.11}$$

We wish to show that $c = 1$.

To this end, we will show the ratio w_{n+1}/w_n tends to 1. If w_{n_k} is a subsequence so that $x_{n_k}/|w_{n_k}|$ is bounded away from zero, then the w_{n_k} converge to infinity in a sector and the Julia–Carathéodory Theorem would imply $\lim_{k\to\infty} w_{n_k+1}/w_{n_k} = 1$. On the other hand, if w_{n_k} is a subsequence so that $x_{n_k}/|w_{n_k}|$ converges to zero, then by Equation (2.4.3)

$$\frac{|w_{n_k+1} - w_{n_k}|}{|w_{n_k+1} - w_{n_k}| + 2x_{n_k}} \leq \left|\frac{w_{n_k+1} - w_{n_k}}{w_{n_k+1} + \overline{w_{n_k}}}\right| = r_{n_k} \leq r_0$$

and we have

$$\left|\frac{w_{n_k+1}}{w_{n_k}} - 1\right| \leq \frac{x_{n_k}}{|w_{n_k}|}\frac{2r_0}{1 - r_0}$$

which tends to zero. For every subsequence of the iterates w_n, either there is a sub-subsequence w_{n_k} on which $x_{n_k}/|w_{n_k}|$ is bounded away from zero or there is a sub-subsequence on which $x_{n_k}/|w_{n_k}|$ converges to zero and, in either case, $\lim_{k\to\infty} w_{n_k+1}/w_{n_k} = 1$. Thus, $\lim_{n\to\infty} w_{n+1}/w_n = 1$.

Putting $w = w_0$ in Equation (2.4.11) gives $\operatorname{Re} g(w_1) = \operatorname{Re} cg(w_0) = c$ and by induction, $\operatorname{Re} g(w_n) = c^n$. Since g is a map of the halfplane into itself, the Schwarz–Pick Theorem gives

$$\left|\frac{g(w_n) - g(w_0)}{g(w_n) + \overline{g(w_0)}}\right|^2 \leq \left|\frac{w_n - w_0}{w_n + \overline{w_0}}\right|^2$$

Clearing the denominators, we find

$$\left((\operatorname{Re}g(w_n) - \operatorname{Re}g(w_0))^2 + (\operatorname{Im}g(w_n) - \operatorname{Im}g(w_0))^2\right)\left((x_n + x_0)^2 + (y_n - y_0)^2\right)$$

is no more than

$$\left((Reg(w_n) + Reg(w_0))^2 + (Img(w_n) - Img(w_0))^2\right)\left((x_n - x_0)^2 + (y_n - y_0)^2\right)$$

Expanding each of these expressions, replacing the term

$$(Im\, g(w_n) - Im\, g(w_0))^2 (x_n - x_0)^2$$

by

$$(Im\, g(w_n) - Im\, g(w_0))^2 (x_n + x_0)^2$$

in the second expression and simplifying, we get

$$(Re\, g(w_n) - Re\, g(w_0))^2 (x_n + x_0)^2 + (Re\, g(w_n) - Re\, g(w_0))^2 (y_n - y_0)^2$$

$$\leq (Re\, g(w_n) + Re\, g(w_0))^2 (x_n - x_0)^2 + (Re\, g(w_n) + Re\, g(w_0))^2 (y_n - y_0)^2$$

This simplifies to

$$x_n x_0 \left((Re\, g(w_n))^2 + (Re\, g(w_0))^2\right)$$

$$\leq (Re\, g(w_n))(Re\, g(w_0)) \left(x_n^2 + x_0^2 + (y_n - y_0)^2\right)$$

which implies

$$x_n x_0 (Re\, g(w_n))^2 \leq (Re\, g(w_n))(Re\, g(w_0)) \left((x_n + x_0)^2 + (y_n - y_0)^2\right)$$

Thus, noting that $g(w_0) = 1$ and $x_n \geq x_0$, we have

$$c^n = Re\, g(w_n) \leq \frac{|w_n + \overline{w_0}|^2}{x_n x_0} Re\, g(w_0) \leq \frac{1}{x_0^2}|w_n + \overline{w_0}|^2$$

so

$$c \leq \limsup_{n\to\infty} |w_n + \overline{w_0}|^{2/n}$$

Now

$$\limsup_{n\to\infty} |w_n + \overline{w_0}|^{1/n} = \limsup_{n\to\infty} \left(\frac{|w_n + \overline{w_0}|}{|w_{n-1} + \overline{w_0}|} \cdots \frac{|w_1 + \overline{w_0}|}{|w_0 + \overline{w_0}|}\right)^{1/n}$$

$$\leq \limsup_{n\to\infty} \frac{|w_n + \overline{w_0}|}{|w_{n-1} + \overline{w_0}|}$$

$$= \limsup_{n\to\infty} \frac{|w_n|}{|w_{n-1}|} = 1$$

Since $c = \lim_{j\to\infty} x_{n_j+1}/x_{n_j} \geq 1$, we see $c = 1$. Thus, every subsequence of the x_n has a further subsequence such that $\lim_{j\to\infty} x_{n_j+1}/x_{n_j} = 1$ and we may conclude that $\lim_{n\to\infty} x_{n+1}/x_n = 1$. ∎

This having been accomplished, we can prove the main lemma about the iteration of maps with angular derivative 1 at the Denjoy–Wolff point on the unit circle.

LEMMA 2.68
Let ψ be an analytic mapping of the right halfplane into itself with

$$\kappa = \inf_{0 < x} \frac{\operatorname{Re}(\psi(x + iy))}{x} = 1$$

Given a compact subset K of the halfplane and w_0 in K, let $w_n = x_n + iy_n = \psi_n(w_0)$ for $n = 1, 2, \ldots$. There are numbers δ and λ, $0 < \delta < 1$ and $0 < \lambda < \infty$, so that if S is the sector $\{w = x + iy : x > 0 \text{ and } |y| < \beta x\}$ for some positive number β and

$$W = S \cup \bigcup_{n=1}^{\infty} \{x + iy : \delta x_n < x < \infty, \; |y - y_n| < \lambda x_n\}$$

we have

(1) $\psi_n(K) \subset W$ for all positive integers n.

(2)   For w approaching infinity in W,

$$\lim_{w \to \infty} \frac{\operatorname{Re}\psi(w)}{\operatorname{Re}w} = \lim_{w \to \infty} \frac{\psi(w)}{w} = \lim_{w \to \infty} \psi'(w) = 1$$

(3)   There is ρ so that ψ is univalent in $W \cap \{w : x > \rho\}$.

PROOF   (The region W together with the iterates w_n are illustrated in Figure 2.13.)
Given a compact set K in the halfplane and w_0 in K, choose r, δ, and λ so that $\{w : \delta x_0 < \operatorname{Re}(w) \text{ and } |\operatorname{Im}(w) - y_0| < \lambda x_0\}$ contains a pseudohyperbolic disk with radius r and center at w_0 that contains K. The Euclidean description of the pseudohyperbolic disks in the halfplane, Equation (2.4.2), shows that the Euclidean radius of the disk is proportional to the real part of the pseudohyperbolic center. Since the width of the strips used in the definition of W and the size of the extension of the strip to the left of w_n are also proportional to x_n, the domain W as defined in the statement of the lemma contains a pseudohyperbolic disk of radius r centered at each of the w_n. Since ψ is contractive in the pseudohyperbolic metric, W contains $\psi_n(K)$ for every integer n. This proves (1).
As in the proof of Lemma 2.67, let $\psi(w) = w + p(w)$ and let $u(w) = \operatorname{Re}p(w)$. From Inequality (2.4.7) we get $\operatorname{Re}(p'(w)) \leq u(w)/x$ which implies

$$\frac{\partial}{\partial x} \frac{u(w)}{x} = \frac{\operatorname{Re}(p'(w))}{x} - \frac{u(w)}{x^2} \leq 0$$

This means that $u(x + iy)/x$ is decreasing as a function of x for every y.

Again applying the Schwarz–Pick Theorem to p, we have, for $p(w) \neq p(w')$,

$$\left| \frac{p(w) + \overline{p(w')}}{p(w) - p(w')} \right|^2 \geq \left| \frac{w + \overline{w'}}{w - w'} \right|^2$$

that is,

$$\frac{|p(w) - p(w')|^2 + 4\operatorname{Re} p(w)\operatorname{Re} p(w')}{|p(w) - p(w')|^2} \geq \frac{|w - w'|^2 + 4\operatorname{Re} w\operatorname{Re} w'}{|w - w'|^2}$$

so

$$\frac{\operatorname{Re} p(w)\operatorname{Re} p(w')}{|p(w) - p(w')|^2} \geq \frac{\operatorname{Re} w\operatorname{Re} w'}{|w - w'|^2}$$

and

$$\frac{|p(w) - p(w')|^2}{\operatorname{Re} p(w)\operatorname{Re} p(w')} \leq \frac{|w - w'|^2}{\operatorname{Re} w\operatorname{Re} w'} \tag{2.4.12}$$

which holds for $p(w) = p(w')$ as well. It follows for any n that

$$u(w)^2 - 2u(w)u(w_n) \leq (u(w) - u(w_n))^2$$

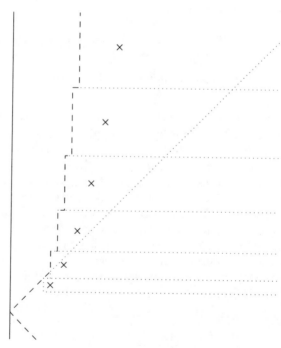

FIGURE 2.13

The region W with the points w_n.

$$\leq |p(w) - p(w_n)|^2 \leq \frac{|w - w_n|^2}{x x_n} u(w) u(w_n)$$

and

$$u(w) \leq \left(2 + \frac{|w - w_n|^2}{x x_n}\right) u(w_n)$$

Now if $|y - y_n| \leq \lambda x_n$ and $\delta x_n < x \leq x_n$ then

$$|w - w_n| \leq x_n - x + |y - y_n| \leq (\lambda + 1)x_n$$

and the inequality above implies

$$\frac{u(w)}{x} \leq \left(\frac{2}{\delta} + \frac{(\lambda + 1)^2}{\delta^2}\right) \frac{u(w_n)}{x_n} = \frac{2\delta + (\lambda + 1)^2}{\delta^2}\left(\frac{x_{n+1}}{x_n} - 1\right) \quad (2.4.13)$$

Moreover, this inequality holds for $x > x_n$ since u/x is decreasing as a function of x. Lemma 2.67 shows that $\lim_{n\to\infty} x_{n+1}/x_n = 1$ so Inequality (2.4.13) implies $\lim_{w\to\infty} u(w)/x = 0$ where the limit is taken in W *outside* the sector S. On the other hand, the Julia–Carathéodory Theorem implies $\lim_{w\to\infty} p(w)/x = 0$ in the sector S. Inequality (2.4.7) implies $\lim_{w\to\infty} p'(w) = 0$ for w in W. Inequality (2.4.12) implies

$$\frac{|p(w) - p(w')|^2}{|w - w'|^2} \leq \frac{\mathrm{Re}\,(p(w))\mathrm{Re}\,(p(w'))}{x x'} \quad (2.4.14)$$

Fixing w' and letting w go to infinity in W shows $\lim_{w\to\infty} p(w)/w = 0$ also. Recalling that $\psi(w) = w + p(w)$ gives the limits in (2).

Finally, for w and w' in W from Inequality (2.4.14)

$$\left|\frac{\psi(w) - \psi(w')}{w - w'}\right| = \left|1 + \frac{p(w) - p(w')}{w - w'}\right| \geq 1 - \left|\frac{p(w) - p(w')}{w - w'}\right|$$

$$\geq 1 - \left(\frac{\mathrm{Re}\,(p(w))\mathrm{Re}\,(p(w'))}{x x'}\right)^{1/2}$$

Since the final term goes to zero as w and w' go to infinity in W, there is ρ so that ψ is univalent in $W \cap \{w : x > \rho\}$ which is (3). ∎

We will put these lemmas to use in constructing the fundamental set V that used in the proof of Theorem 2.53. Recall that the goal of the proposition is to show that if φ is an analytic mapping of the disk into itself with Denjoy–Wolff point a and $\varphi'(a) \neq 0$, then there is a fundamental set for φ on D on which φ is univalent.

PROOF (of Proposition 2.56) If φ has a fixed point in the unit disk, using an automorphism of the disk, we can reduce to the case $a = 0$ which is covered by Exercise 2.4.3.

Suppose a is the Denjoy–Wolff point of φ and $|a| = 1$. From Corollary 2.47 and Lemmas 2.66 and 2.68, we see that for any compact set K in the disk, there is an integer M so that for $n \geq M$, the function φ is univalent on $\cup_{n \geq M} \varphi_n(K)$. For $k = 2, 3, \ldots$, let $K_k = \{z : |z| \leq 1 - k^{-1}\}$ with interior $K_k^\circ = \{z : |z| < 1 - k^{-1}\}$. Let m be the smallest integer such that $\varphi(0)$ is in K_m°.

We will construct the fundamental set by an inductive process, starting with the integer m. Let M_m be a positive integer large enough that φ restricted to $\cup_{n \geq M_m} \varphi_n(K_m)$ is one-to-one and let $U_m = \cup_{n \geq M_m} \varphi_n(K_m^\circ)$. By the choice of m, U_m is connected. Indeed, by continuity each $\varphi_n(K_m^\circ)$ is connected and $\varphi(0)$ is in K_m° and $\varphi(K_m^\circ)$ (since 0 is in K_m°). Similarly, $\varphi_2(0)$ is in $\varphi(K_m^\circ)$ and $\varphi_2(K_m^\circ)$, and so on. In this first step of the induction, we have found an integer M_m and a connected open set U_m in D so that $\varphi(U_m) \subset U_m$, φ is univalent on U_m, and $\cup_{n \geq M_m} \varphi_n(K_m^\circ) \subset U_m$.

Continuing with the induction, suppose integers $M_m \leq M_{m+1} \leq \cdots \leq M_{k-1}$ and open sets $U_m \subset U_{m+1} \subset \cdots \subset U_{k-1}$ have been chosen so that U_j is a connected subset of D, $\varphi(U_j) \subset U_j$, φ is univalent on U_j, and $\cup_{n \geq M_j} \varphi_n(K_j^\circ) \subset U_j$ for $j = m, m+1, \ldots, k-1$. Let M_k' be an integer so that $M_k' \geq M_{k-1}$ and φ restricted to $\cup_{n \geq M_k'} \varphi_n(K_k)$ is univalent. Now the set

$$L = \text{closure} \left(U_{k-1} \setminus \left(\cup_{n \geq M_k'} \varphi_n(K_k^\circ) \right) \right)$$

is a compact subset of D, so $\varphi(L)$ is also. Since φ_n converges uniformly to a on K_k and a is not in $\varphi(L)$, we can find $M_k \geq M_k'$ so that $\varphi(L) \cap \left(\cup_{n \geq M_k} \varphi_n(K_k^\circ) \right) = \emptyset$. Let $U_k = U_{k-1} \cup \left(\cup_{n \geq M_k} \varphi_n(K_k^\circ) \right)$. Clearly, U_k is a connected open subset of D, $\varphi(U_k) \subset U_k$ and $\cup_{n \geq M_k} \varphi_n(K_k^\circ) \subset U_k$. We claim φ is univalent on U_k. Suppose z and w are in U_k with $z \neq w$. If both z and w are in U_{k-1} or both in $\cup_{n \geq M_k'} \varphi_n(K_k^\circ)$, then clearly $\varphi(z) \neq \varphi(w)$, so we suppose z is in L and w is in $\cup_{n \geq M_k} \varphi_n(K_k^\circ)$. (Note the occurrences of M_k and M_k'!) But if z is in L, then $\varphi(L)$ is disjoint from $\cup_{n \geq M_k} \varphi_n(K_k^\circ)$, so $\varphi(z) \neq \varphi(w)$.

Thus, we construct the open sets U_k inductively for $k = m, m+1, \ldots$. Let $V' = \cup_{k \geq m} U_k$ and let $V = V' \cup (\text{holes of } V')$. By construction, V' is a connected open subset of D, $\varphi(V') \subset V'$, and φ is univalent on V'. By the maximum principle and the argument principle, $\varphi(V) \subset V$ and φ is univalent on V as well [Boa87, p. 115] (or see Exercise 2.4.4). It follows that V is a fundamental set for φ on D as we were required to show. ∎

It can be shown that in case the Denjoy–Wolff point a is on the unit circle, if the iterates of one point of the disk converge to a nontangentially, then the iterates of every point of the disk do. Moreover, in this case, V can be chosen to contain small sectors with vertex at a and opening θ for all $\theta < \pi$. We omit the proof since we will not need this result in our subsequent work.

Exercises

2.4.1 In the case of an interior fixed point, the function σ of Theorem 2.53 is sometimes called the **Koenigs' function** after G. Koenigs who proved [Koe84] that if φ is an analytic map of the disk into itself, not an automorphism, with $\varphi(0) = 0$, then

$$\sigma(z) = \lim_{j \to \infty} \frac{\varphi_j(z)}{s^j}$$

exists and satisfies $\varphi(\sigma(z)) = s\sigma(z)$ in the disk for $s = \varphi'(0)$. Prove Koenigs' theorem.

2.4.2 We say the functions φ and ψ **commute** with each other if $\varphi \circ \psi = \psi \circ \varphi$. Suppose φ is an analytic map of the unit disk into itself with $\varphi(0) = 0$ and $0 < |\varphi'(0)| < 1$.
   (a) Show that if ψ is an analytic map of the disk to itself and ψ commutes with φ, then $\psi(0) = 0$ also.
   (b) Show that if ψ is an analytic map of the disk to itself and ψ commutes with φ, then φ and ψ have the same Koenigs' function.
   (c) Show, conversely, that if ψ is an analytic map of the disk to itself and φ and ψ have the same Koenigs' function, then ψ commutes with φ.
   (d) Show that if σ denotes the Koenigs' function of φ, then

$$\{\psi : \psi \circ \varphi = \varphi \circ \psi\} = \{\psi : \sigma \circ \psi = \psi'(0)\sigma\}$$

   (e) Show that the operators C_φ and C_ψ commute if and only if the functions φ and ψ commute.

2.4.3 Suppose that φ is an analytic map of the disk into itself with $\varphi(0) = 0$ and $0 < |\varphi'(0)| < 1$. Show that there is $\delta > 0$ so that if $V = \{z : |z| < \delta\}$, then $\varphi(V) \subset V$, φ is univalent on V, and if K is a compact subset of the disk, there is a positive integer n so that if $j \geq n$ then $\varphi_j(K) \subset V$.

2.4.4 Suppose V' is a connected open subset of D and $V = V' \cup (\text{holes of } V')$.
   (a) Show that if φ is an analytic map of the disk into itself and $\varphi(V') \subset V'$, then $\varphi(V) \subset V$.
   (b) Show that if φ is an analytic map of the disk into itself and φ is univalent on V', then it is univalent on V also.

2.4.5 Show that if $\varphi(z) = \sum a_j z^j$ is an analytic map of the disk into itself, not an automorphism, with $\varphi(0) = 0$, then the Taylor coefficients of the Koenigs function σ can be determined recursively from those of φ.

2.4.6 Show that if φ is an analytic map of the disk into itself with $\varphi(0) = \varphi'(0) = 0$, then there is no non-constant analytic function σ and linear fractional transformation Φ so that $\Phi \circ \sigma = \sigma \circ \varphi$.

2.4.7 Prove the assertion of Example 2.58 that if φ is a linear fractional map of the disk into itself that has Denjoy–Wolff point 1 with $\varphi'(1) = 1$ then φ maps the disk onto itself.

2.4.8 Show that if φ has Denjoy–Wolff point 1 with $\varphi'(1) = 1$ and its model is $\Omega = \{z : \text{Im } z > 0\}$, and $\Phi(z) = z + 1$, then $\widetilde{\varphi}(z) = \overline{\varphi(\overline{z})}$ has model $\Omega = \{z : \text{Im } z > 0\}$, and $\Phi(z) = z - 1$.

2.4.9 (a) In the halfplane/translation case, show that the choices $\Phi(z) = z + 1$ and $\Phi(z) = z - 1$ are actually different.
   (b) Show that any automorphism Φ of a halfplane or the whole plane is conjugate to one of the four cases or else is an elliptic automorphism.

2.4.10 (a) Suppose φ has Denjoy–Wolff point 1 and is analytic in a neighborhood of 1, $\varphi'(1) = 1$ and $\mathrm{Re}\,\varphi''(1) \neq 0$. Show that the domain Ω of the model is the plane. (Caution: the second derivative at a fixed point is not invariant under conjugation by an automorphism, so $a = 1$ is important here (see Exercise 2.3.4).)

   (b) Give examples to show that both the halfplane/ and the plane/translation cases can occur when φ has Denjoy–Wolff point 1 and $\varphi'(1) = 1$ and $\mathrm{Re}\,\varphi''(1) = 0$.

2.4.11 Prove: If φ is an analytic map of the disk into itself with $\varphi(x)$ real for $-1 < x < 1$, then Ω, σ, and Φ can be chosen so that $\sigma(x)$ is real for $-1 < x < 1$.

2.4.12 Prove: If φ is an analytic map of the disk into itself with $\varphi(x)$ real for $-1 < x < 1$ and $\varphi(1) = \varphi'(1) = 1$, then φ is in the plane/translation case.

2.4.13 As in Example 2.55 let $\varphi(z) = sz + 1 - s$ for $0 < s < 1$ and $\Omega = \{z : \mathrm{Re}\,(z) > 0\}$, but instead of $\sigma(z) = 1 - z$ and $\Phi(z) = sz$, let $\sigma_1(z) = \sqrt{1 - z}$, and $\Phi_1(z) = \sqrt{s}z$. Check that Φ_1 is an automorphism of the halfplane Ω and $\Phi_1 \circ \sigma_1 = \sigma_1 \circ \varphi$. Clearly σ_1 and σ are not conjugate to each other. Why does this not contradict the uniqueness statement in Theorem 2.53?

2.4.14 Suppose φ is an analytic map of the unit disk into itself and σ is a map of the unit disk into the upper halfplane so that $\sigma(\varphi(z)) = \sigma(z) + 1$. (In other words, $\sigma \circ \varphi = \Phi \circ \sigma$ for $\Phi(w) = w + 1$.) Now suppose Ω_1 is the upper halfplane with some whiskers removed

$$\Omega_1 = \{w = x + iy : y > 0,\ w \neq n + iy \text{ for } 0 < y \leq 1, n \text{ an integer}\}$$

   (a) Show that there is a univalent map τ_1 of the upper halfplane onto Ω_1 so that $\tau_1(w + 1) = \tau_1(w) + 1$. (Hint: Let τ be the Riemann map of the upper halfplane onto Ω_1 with $\tau(2i) = 2i$ and $\tau(\infty) = \infty$. Consider the map $\tau_0(w) = \tau(x_0 + wy_0/2) - 1$ where $\tau(x_0 + iy_0) = 2i + 1$, then show $y_0 = 2$.)

   (b) Show that if $\sigma_1 = \tau_1 \circ \sigma$, then σ_1 is a map of the unit disk into the upper halfplane such that $\sigma_1(\varphi(z)) = \sigma_1(z) + 1$.

   (c) Show that σ_1 and σ are not conjugate to each other as maps of the disk into the upper halfplane. Why does this not contradict the uniqueness statement in Theorem 2.53?

2.4.15 (a) Construct a univalent map φ of the disk into itself with Denjoy–Wolff point 0 and $\varphi'(0) = .5$ and exactly three other fixed points such that $\varphi(\partial D) \supset \partial D$.

   (b) Construct a univalent map φ of the disk into itself with Denjoy–Wolff point 0 and $\varphi'(0) = .5$ such that $\varphi(\partial D) \cap \partial D$ consists of three boundary fixed points.

2.4.16 (a) Construct a univalent map φ of the disk into itself with Denjoy–Wolff point 1 and $\varphi'(1) = .5$ and exactly three other fixed points such that $\varphi(\partial D) \supset \partial D$.

   (b) Construct a univalent map φ of the disk into itself with Denjoy–Wolff point 1 and $\varphi'(1) = .5$ such that $\varphi(\partial D) \cap \partial D$ consists of four fixed points.

2.4.17 (a) Construct a univalent map φ of the disk into itself with Denjoy–Wolff point 1 and $\varphi'(1) = 1$ and exactly three other fixed points such that $\varphi(\partial D) \supset \partial D$.

   (b) Construct a univalent map φ of the disk into itself with Denjoy–Wolff point 1 and $\varphi'(1) = 1$ such that $\varphi(\partial D) \cap \partial D$ consists of four fixed points.

2.4.18 Suppose φ, an analytic map of the unit disk into itself, has Denjoy–Wolff point a with $|a| = 1$, $\varphi'(a) = 1$, and the model for φ is halfplane/translation. Show that if $\{z_j\}$ is a sequence in the disk with $\varphi(z_j) = z_{j+1}$, then $\{z_j\}$ is an interpolating sequence.

Notes

This circle of ideas has a very long history. In his paper of 1871 [Scr71], E. Schroeder considers the problem of fractional iteration and relates it to the solution of various functional equations including the functional equation $f \circ \varphi = sf$ that bears his name. In 1884, G. Koenigs [Koe84] constructed a solution of Schroeder's equation near a fixed point a when φ is analytic at a and $\varphi'(a) = s$, $|s| \neq 0, 1$. In the years since these classic works, much has been written; the interested reader might consult the bibliography of [Co81] or that of M. Kuczma [Kuc63]. For the most part, the literature has concentrated on cases when φ is analytic in a neighborhood of the fixed point or dropped analyticity completely and considered continuous functions in R^n.

Traditionally, the intertwining map σ is constructed by a limiting process from φ, as illustrated by Exercise 2.4.1. Ch. Pommerenke [Pom79] and I. N. Baker [BaP79] and Pommerenke used this technique for the case when the Denjoy–Wolff point is on the circle. The unified construction presented here is from C. Cowen's paper [Co81]. Although more abstract, it gives more insight into the problem of uniqueness than do the limiting arguments. Lemmas 2.67 and 2.68 that give information about the iteration and an approach region for the case $\varphi'(a) = 1$ are due to Pommerenke [Pom79] and the outline of his proofs have been followed.

Problems of identification of the case of the model given reasonable information about the map φ have been studied by several authors but there does not seem to be a completely satisfactory way to distinguish the two cases in which $\varphi'(a) = 1$. Some geometric information concerning the distinction is given in [BaP79] and [Co81]. The second derivative test and the symmetry condition of Exercises 2.4.10 and 2.4.12 are included in [Co81] and the second derivative test is refined in [BoS93, Theorems 4.4 and 6.2] of P. Bourdon and J. H. Shapiro. The connection between interpolating sequences and the case appeared in Cowen's paper [Co83].

The model for iteration has proved to be very useful in the study of composition operators because it relates information about the map φ to information about the simpler linear fractional map Φ. For example, we will use it to study spectra in Chapter 7, subnormality of adjoints of composition operators in Section 8.2 and unitary equivalence and similarity of composition operators in Section 9.2. P. Bourdon and J. H. Shapiro in [BoS90] and [BoS93] use the model for iteration to study cyclicity of composition operators. In [ShSS92], J. H. Shapiro, W. Smith, and D. Stegenga relate the geometric properties of the model to the compactness of the composition operator. Much of Shapiro's recent book [Sho93] is devoted to developing the model, especially in the plane/dilation case, and exploring its applications to cyclicity and compactness. Moreover, it is clear that much more can be done along these lines; the model does not depend on univalence of φ but, up to now, many applications of the model to composition operators have included this hypothesis.

The model for iteration is related to the theory of discrete branching processes (Galton–Watson processes) in probability theory through the probability generating function. The probability generating function for a Galton–Watson process is the function $\varphi(z) =$

$\sum_{k=0}^{\infty} p_k z^k$ where $p_k \geq 0$ and $\sum p_k = 1$. (T. E. Harris' book [Has63] begins with a very readable discussion of these ideas.) The coefficient p_k is to be interpreted as the probability that an individual will have k offspring and the kth coefficient of φ_n is interpreted as the probability that there will be k individuals in the nth generation. Questions concerning the eventual population size are related to the asymptotic behavior of the iterates of φ. We note that every probability generating function for a Galton–Watson process is an analytic function mapping D into itself, and if $p_0 + p_1 \neq 0$, φ satisfies the hypotheses of theorem 2.53. The bibliography of Harris [Has63] and K. Athreya and P. Ney [AtN72] are a guide to older work on branching processes.

In [Co81], Theorem 2.53 is used to study "fractional iterates" of φ, that is, a real semigroup of functions φ_t for $t \geq 0$, where $\varphi_1 = \varphi$ and $\varphi_{s+t} = \varphi_s \circ \varphi_t$. Since fractional iterates can be defined for linear fractional transformations, we try to define φ_t by $\Phi_t \circ \sigma = \sigma \circ \varphi_t$. But there are difficulties because, in general, σ is not univalent, and even if it were, σ^{-1} need not be defined on $\Phi_t(\sigma(D))$ unless $\sigma(D)$ meets some geometric condition. We do find, though, that for each z in D, $\varphi_t(z)$ can be defined for t sufficiently large. This is enough that we can find a meromorphic function on the disk that deserves to be called the infinitesimal generator of the semigroup. This point of view will become important in Section 7.7.

Exercise 2.4.2 was suggested by P. Bourdon. Much of this information is contained in Cowen's paper [Co84a]. The application of the model for iteration to the problem of commutation illustrates the use of the abstract formulation of the model and the resulting uniqueness. The problem of commutation of analytic functions mapping the disk to itself has also been studied by A. Shields [Shi64] and D. Behan [Beh73].

Iteration of functions for which $\varphi'(a) = 0$ is studied via Böttcher's functional equation, $f(\varphi(z)) = (f(z))^k$, in Cowen's paper [Co82].

2.5 The automorphisms of the ball

If $\psi = (\psi_1, \psi_2, \ldots, \psi_m)$ is an analytic map of an open set $\Omega \subset C^n$ into C^m, $\psi'(z) : C^n \rightarrow C^m$ is the linear operator which is represented (with respect to the standard basis on C^n) by the matrix (a_{jk}) where $a_{jk} = D_k \psi_j(z)$, for $1 \leq j \leq m$, $1 \leq k \leq n$ and $D_k = \frac{\partial}{\partial z_k}$. The derivative $\psi'(z)$ is the unique linear operator satisfying

$$\psi(z+h) - \psi(z) = \psi'(z)h + O(|h|^2)$$

for h near 0 in C^n. When $n = m$ the matrix (a_{jk}) is square and its determinant, the complex Jacobian, is denoted $J_\psi(z)$.

Before we can describe the automorphisms of B_N we need a several variable version of the Schwarz lemma, applicable to maps φ of B_N into B_N.

THEOREM 2.69 *(Schwarz Lemma in B_N)*
Suppose $\varphi : B_N \rightarrow B_N$ is analytic and $\varphi(0) = 0$. Then

(1) $|\varphi(z)| \leq |z|$ *for all z in B_N and*

(2) $\|\varphi'(0)\| \leq 1$.

PROOF For ζ, η any two points of ∂B_N define an analytic function of the disk D by

$$g(\lambda) = g_{\zeta,\eta}(\lambda) = \langle \varphi(\lambda\zeta), \eta \rangle$$

Note that g maps D into D and $g(0) = 0$. Thus the one variable Schwarz lemma implies that

$$|g(\lambda)| = |\langle \varphi(\lambda\zeta), \eta \rangle| \leq |\lambda|$$

for all ζ, η in ∂B_N. In particular, given z in B_N, write $z = r\zeta$ for some $r \geq 0$, ζ in ∂B_N and $\varphi(r\zeta) = s\eta$ for $s \geq 0$, η in ∂B_N. Thus $s \leq r$, that is $|\varphi(z)| \leq |z|$, giving (1).

Moreover, a computation shows that $g'(\lambda) = \langle \varphi'(\lambda\zeta)\zeta, \eta \rangle$ so again by the one variable Schwarz lemma we have

$$|g'(0)| = |\langle \varphi'(0)\zeta, \eta \rangle| \leq 1$$

for all ζ, η in ∂B_N. Thus $\|\varphi'(0)\| \leq 1$. ∎

Unlike the case $N = 1$ neither equality in (2) of Theorem 2.69, nor equality at some $z_0 \neq 0$ in (1) implies that φ is a linear map. To see this consider $\varphi(z_1, z_2) = (z_1 + \frac{1}{2}z_2^2, 0)$ mapping B_2 into B_2. Then $\varphi(z_1, 0) = (z_1, 0)$ and $\|\varphi'(0)\| = 1$, yet φ is not linear. Note that φ *is* a unitary map of the slice $[e_1]$ onto itself. Though we will not need it, we remark that this sort of partial linearity is a consequence of the equality cases of Theorem 2.69, see [Ab89b, p. 161] for the details. We will however have need of the next result, where we make the stronger assumption that $\varphi'(0)$ is an isometry of C^N into C^N.

PROPOSITION 2.70
Suppose $\varphi : B_N \rightarrow B_N$ is analytic with $\varphi(0) = 0$ and $\varphi'(0)$ unitary. Then $\varphi(z) = \varphi'(0)z$.

PROOF Fix ζ in ∂B_N and let $\eta = \varphi'(0)\zeta$. Define $g(\lambda) = \langle \varphi(\lambda\zeta), \eta \rangle$. Notice that $g'(\lambda) = \langle \varphi'(\lambda\zeta)\zeta, \eta \rangle$ and $g'(0) = 1$. By the Schwarz lemma in the disk $g(\lambda) = \lambda$, or $\langle \lambda^{-1}\varphi(\lambda\zeta), \eta \rangle = 1$. But

$$\frac{|\varphi(\lambda\zeta)|}{|\lambda|} \leq 1$$

so we must have $\varphi(\lambda\zeta) = \lambda\eta = \varphi'(0)(\lambda\zeta)$ for all $\lambda, |\lambda| < 1$. Since ζ is arbitrary this gives the conclusion. ∎

By an automorphism φ of B_N we mean a biholomorphic map of B_N onto B_N; that is, both φ and φ^{-1} are analytic maps of B_N onto B_N. It is true, but not obvious, that if φ maps B_N one-to-one and onto B_N then φ is an automorphism of B_N; see [Ru80, Theorem 15.1.8, p. 302].

We will first describe some automorphisms of B_N that are analogous to the disk automorphisms $(a - z)/(1 - \overline{a}z)$, for a in D. Let a be in B_N and set

$$P_a(z) = \frac{\langle z, a \rangle}{\langle a, a \rangle} a$$

so P_a is projection onto the subspace $[a]$ spanned by a, and $Q_a = I - P_a$, projection onto the orthogonal complement of $[a]$. To simplify notation write $s_a = \sqrt{1 - |a|^2}$. Define $\varphi_a(z)$ by

$$\varphi_a(z) = \frac{a - P_a(z) - s_a Q_a(z)}{1 - \langle z, a \rangle}$$

Clearly φ_a is analytic in $\overline{B_N}$, $\varphi_a(0) = a$ and $\varphi_a(a) = 0$. To obtain further properties of φ_a it is simplest to first consider the special case $a = (a_1, 0') \equiv (a_1, 0, \ldots, 0)$. Then one can easily compute the coordinate functions of φ_a to obtain

$$\varphi_a(z) = \varphi_a(z_1, z') = \left(\frac{a_1 - z_1}{1 - \overline{a_1} z_1}, \frac{-s_a z'}{1 - \overline{a_1} z_1} \right)$$

A direct computation will then verify that for $a = (a_1, 0')$

$$1 - \langle \varphi_a(z), \varphi_a(w) \rangle = \frac{(1 - |a|^2)(1 - \langle z, w \rangle)}{(1 - \langle z, a \rangle)(1 - \langle a, w \rangle)} \qquad (2.5.1)$$

for any z, w in the closed ball. We will refer to Equation (2.5.1) as the fundamental identity for automorphisms; our next goal is to extend it to any a in B_N. Note that consequences of Equation (2.5.1) include (setting $z = w$)

(i) $\varphi_a(B_N) \subset B_N$

(ii) $\varphi_a(\partial B_N) \subset \partial B_N$.

The verification of Equation (2.5.1) for arbitrary a in B_N will follow from the next result.

LEMMA 2.71
If U is a unitary map, then $\varphi_{Ua} = U \varphi_a U^{-1}$, for any a in B_N.

PROOF This is Exercise 2.5.1. ∎

Returning to Equation (2.5.1), let a be in B_N and find U unitary such that $Ua = (c, 0')$. Then

$$
\begin{aligned}
1 - \langle \varphi_{Ua}(z), \varphi_{Ua}(w) \rangle &= 1 - \langle U \varphi_a U^{-1}(z), U \varphi_a U^{-1}(w) \rangle \\
&= 1 - \langle \varphi_a U^{-1}(z), \varphi_a U^{-1}(w) \rangle
\end{aligned}
$$

Now replace z by Uz and w by Uw and use the fact that we have verified Equation (2.5.1) for $Ua = (c, 0')$. Thus Equation (2.5.1) holds for any a in B_N.

Next we compute, *for the special case* $a = (a_1, 0')$, the matrix of the derivative map $\varphi'_a(z)$, using the description of the coordinate functions of φ_a. Since the (j, k) entry of $\varphi'_a(z)$ is $D_k\varphi_j(z)$ we see that

$$\varphi'_a(0) = \operatorname{diag}\{-s_a^2, -s_a, \ldots, -s_a\}$$
$$\varphi'_a(a) = \operatorname{diag}\{-1/s_a^2, -1/s_a, \ldots, -1/s_a\}$$

and hence

$$\varphi'_a(a)\varphi'_a(0) = \varphi'_a(0)\varphi'_a(a) = I \tag{2.5.2}$$

Lemma 2.71 can be used again to extend Equation (2.5.2) from $a = (a_1, 0')$ to arbitrary a in B_N. The details are left to the reader.

For any a in B_N the analytic map $\varphi_a \circ \varphi_a$ of B_N into B_N satisfies

(i) $\varphi_a \circ \varphi_a(0) = 0$

(ii) $(\varphi_a \circ \varphi_a)'(0) = I$

Proposition 2.70 implies that $\varphi_a \circ \varphi_a(z) = z$ for all z in B_N. Hence φ_a is an automorphism of B_N, with $\varphi_a^{-1} = \varphi_a$. We can now give a complete description of the group of all automorphisms of B_N, $\operatorname{Aut}(B_N)$. This description should be compared with the description of the automorphisms of the unit disk given in Section 2.3.

THEOREM 2.72
Let ψ be an automorphism of B_N with $\psi^{-1}(0) = a$. There exists a unitary map U so that $\psi = U\varphi_a$.

PROOF   The map $\psi \circ \varphi_a$ is an automorphism of B_N fixing 0. We wish to show $\psi \circ \varphi_a$ is unitary. Write $\tau = \psi \circ \varphi_a$ and consider, for θ real,

$$\lambda(z) = \tau^{-1}(e^{-i\theta}\tau(e^{i\theta}z))$$

Check that $\lambda(0) = 0$ and $\lambda'(0) = I$. Thus, as before, an appeal to Proposition 2.70 implies $\lambda(z) = z$ and therefore $\tau(e^{i\theta}z) = e^{i\theta}\tau(z)$, for every real θ. If we write out the homogeneous expansion of τ, that is write $\tau(z) = Lz + \sum_{s=2}^{\infty} F_s(z)$ where L is linear and the components of F_s are homogeneous polynomials of degree s, we see that the homogeneity of τ, $\tau(e^{i\theta}z) = e^{i\theta}\tau(z)$, forces $F_s \equiv 0$ for $s \geq 2$, and we are done, since a linear map which is an automorphism of B_N is unitary.   ∎

With this classification theorem we can extend the fundamental identity (2.5.1) to all ψ in $\operatorname{Aut}(B_N)$. Set $\psi^{-1}(0) = a$. Then for z, w in $\overline{B_N}$

$$1 - \langle \psi(z), \psi(w) \rangle = \frac{(1 - |a|^2)(1 - \langle z, w \rangle)}{(1 - \langle z, a \rangle)(1 - \langle a, w \rangle)} \tag{2.5.3}$$

We will next study the fixed point sets of automorphisms of B_N. Our first result deals with the automorphisms φ_a.

THEOREM 2.73
For $a \neq 0$ the only point of $\overline{B_N}$ which is fixed by φ_a is $a/(1 + s_a)$ where $s_a = \sqrt{1 - |a|^2}$.

PROOF   If z is a fixed point of φ_a then

$$a - P_a z - s_a Q_a z = (1 - \langle z, a \rangle) z$$

Applying Q_a to both sides of this equation

$$-s_a Q_a(z) = (1 - \langle z, a \rangle) Q_a(z)$$

which can only hold if $Q_a(z) = 0$ since $\operatorname{Re}(1 - \langle z, a \rangle) \geq 0$. Thus any fixed point of $\varphi_a(z)$ is of the form λa for λ complex. Using the definition of φ_a we see that if $\varphi_a(\lambda a) = \lambda a$ then $\lambda^2 |a|^2 - 2\lambda + 1 = 0$ which has solutions

$$\lambda = \frac{1 \pm s_a}{|a|^2} = \frac{1}{1 \pm s_a}$$

Only $\lambda = 1/(1 + s_a)$ gives a point in $\overline{B_N}$; moreover $a/(1 + s_a)$ is in B_N.   ∎

Of course for a unitary map U, the fixed point set of U in $\overline{B_N}$ is the intersection of $\overline{B_N}$ with a (complex) subspace of C^N, either $\{0\}$ or the eigenspace of U corresponding to the eigenvalue 1. By an ***affine subset of*** B_N we mean the intersection of B_N with a translate of a subspace of C^N; its dimension is the dimension of this subspace. An affine subset of C^N is simply a translate of a subspace of C^N. It is easy to see that a set $E \subset C^N$ is affine if and only if whenever $z_1, \ldots, z_n$ in E and $\lambda_1, \ldots, \lambda_n$ in C satisfy $\sum \lambda_j = 1$, then $\sum \lambda_j z_j$ is in E. Write $A[z_1, \ldots, z_n]$ for the affine set generated by $z_1, z_2, \ldots, z_n$; that is, $A[z_1, \ldots, z_n] = \{\sum \lambda_j z_j : \sum \lambda_j = 1\}$. Suppose $z = \sum \lambda_j z_j$ where $\sum \lambda_j = 1$ and $\langle z, a \rangle \neq 1$. Then $\varphi_a(z) = \sum \mu_j \varphi_a(z_j)$ where $\mu_j = \lambda_j (1 - \langle z_j, a \rangle)/(1 - \langle z, a \rangle)$ so that $\sum \mu_j = 1$. This says

$$\varphi_a \{A[z_1, \ldots, z_n] \setminus V_a\} \subset A[\varphi_a(z_1), \ldots, \varphi_a(z_n)] \tag{2.5.4}$$

where V_a is the singularity set for φ_a; that is, $V_a = \{z : \langle z, a \rangle = 1\}$. This observation is the key step in proving the next result.

THEOREM 2.74
The image under ψ in $\operatorname{Aut}(B_N)$ of an affine subset E of $\overline{B_N}$ is an affine subset of $\overline{B_N}$.

PROOF   By Theorem 2.72 we may write $\psi = U\varphi_a$ for some unitary U. Thus it is enough to show $\varphi_a(E)$ is affine. Now $E = A[z_1, \ldots, z_m] \cap \overline{B_N}$ for some

$z_1, \ldots, z_m$ and $V_a \cap \overline{B_N} = \emptyset$, so Equation (2.5.4) shows

$$\varphi_a(E) \subset A[\varphi_a(z_1), \ldots, \varphi_a(z_m)] \cap \overline{B_N}$$

Since $\varphi_a(\varphi_a(z)) = z$ this also implies

$$\varphi_a\{A[\varphi_a(z_1), \ldots, \varphi_a(z_m)] \cap \overline{B_N}\} \subset A[z_1, \ldots, z_m] \cap \overline{B_N} = E$$

Thus $E = \varphi_a\{A[\varphi_a(z_1), \ldots, \varphi_a(z_m)] \cap \overline{B_N}\}$ and $\varphi_a(E)$ is the affine subset $A[\varphi_a(z_1), \ldots, \varphi_a(z_m)] \cap \overline{B_N}$. ∎

COROLLARY 2.75

If ψ in $\mathrm{Aut}(B_N)$ has a fixed point in B_N then the fixed point set of ψ in the open ball B_N is affine.

PROOF If ψ fixes a in B_N, consider $\varphi_a \circ \psi \circ \varphi_a$ which fixes 0. By Theorem 2.72, this automorphism is a unitary map whose fixed point set is a subspace V of C^N. It is easy to see that the fixed point set of ψ in B_N is $\varphi_a(V) \cap B_N$; by the last theorem this is an affine set. ∎

Perhaps surprisingly, this last result is actually true in much greater generality.

THEOREM 2.76

Let $\varphi : B_N \to B_N$ be analytic. Then the fixed point set of φ in B_N is affine or empty.

We will not give the proof of this result here, but instead refer the reader to [Ru80, p. 165] where it is shown that if $\varphi(0) = 0$ then φ and $\varphi'(0)$ have the same fixed points in B_N. The theorem then follows from the previous result for automorphisms, and the fact that automorphisms take affine sets to affine sets.

Exercises

2.5.1 Prove Lemma 2.71. (Hint: The goal is to show $\varphi_{Ua}(Uz) = U\varphi_a(z)$. Note that $s_a = s_{Ua}$.)

2.5.2 Use Theorem 2.69 and the fundamental identity, Equation (2.5.3), to prove the Schwarz–Pick Theorem for B_N: If $\varphi : B_N \to B_N$ is analytic and a in B_N then

$$\frac{|1 - \langle \varphi(z), \varphi(a) \rangle|^2}{(1 - |\varphi(z)|^2)(1 - |\varphi(a)|^2)} \leq \frac{|1 - \langle z, a \rangle|^2}{(1 - |z|^2)(1 - |a|^2)}$$

2.5.3 For a in B_N and $0 < r < 1$ set $\mathcal{E}(a, r) = \varphi_a(rB_N)$.
  (a) Show $U\mathcal{E}(a, r) = \mathcal{E}(Ua, r)$.

(b) Show that if $a = (a_1, 0')$ then $\mathcal{E}(a, r)$ is an ellipsoid and identify its center $c = (c_1, 0')$. Also show that $\mathcal{E}(a, r) \cap [e_1]$ is a disk of radius $\sim r(1 - |a|^2)$ while $\mathcal{E}(a, r) \cap \{z_1 = c_1\}$ is a ball of radius $\sim r\sqrt{1 - |a|^2}$ for r small. By virtue of (a) these observations give the shape of an arbitrary $\mathcal{E}(a, r)$.

(c) Interpret Exercise 2.5.2 in terms of these ellipsoids.

Notes

Early contributors to the study of the Schwarz Lemma in several variables include K. Reinhardt, C. Carathéodory and H. Cartan; see [Di89] for a discussion of some of the history of and references for generalizations of the Schwarz Lemma.

The automorphisms in B_2 were first described by H. Poincaré in [Poi07]. The discussion of Aut(B_N) given here was greatly influenced by the exposition in [Ru80].

Theorem 2.76 is due to M. Hervé [He63]; see also [Ru78].

2.6   Julia–Carathéodory theory in the ball

In this section we will discuss the analogues of the ideas of Section 2.3 for the ball. As much as possible our treatment here will parallel the treatment in the disk, although a few new technical details will be needed. For ζ in ∂B_N we will continue to use the notation $d(\zeta)$ for $\liminf_{z \to \zeta} (1 - |\varphi(z)|)/(1 - |z|)$ where φ is an analytic map of the ball B_N into itself. Julia's Lemma (Lemma 2.41) generalizes to the following:

LEMMA 2.77   (*Julia's Lemma in B_N*)
Suppose ζ is in ∂B_N with $d(\zeta) < \infty$. Suppose $a_n \to \zeta$ satisfies

$$\lim_{n \to \infty} \frac{1 - |\varphi(a_n)|}{1 - |a_n|} = d(\zeta)$$

and $\lim_{n \to \infty} \varphi(a_n) = \eta$ where η is in ∂B_N. Then for every z in B_N

$$\frac{|1 - \langle \varphi(z), \eta \rangle|^2}{1 - |\varphi(z)|^2} \leq d(\zeta) \frac{|1 - \langle z, \zeta \rangle|^2}{1 - |z|^2}$$

PROOF   By the Schwarz–Pick Theorem in B_N (Exercise 2.5.2) we have

$$\frac{|1 - \langle \varphi(z), \varphi(a_n) \rangle|^2}{(1 - |\varphi(z)|^2)(1 - |\varphi(a_n)|^2)} \leq \frac{|1 - \langle z, a_n \rangle|^2}{(1 - |z|^2)(1 - |a_n|^2)}$$

or

$$\frac{|1 - \langle \varphi(z), \varphi(a_n) \rangle|^2}{1 - |\varphi(z)|^2} \leq \frac{1 - |\varphi(a_n)|^2}{1 - |a_n|^2} \frac{|1 - \langle z, a_n \rangle|^2}{1 - |z|^2}$$

Letting $n \to \infty$ gives the conclusion. ∎

As in the one variable case, this lemma has an appealing geometric interpretation. Set $E(k, \zeta) = \{z \in B_N : |1 - \langle z, \zeta \rangle|^2 \leq k(1 - |z|^2)\}$. In the special case $\zeta = e_1$ a computation shows that this is equivalent to

$$\left| z_1 - \frac{1}{1+k} \right|^2 + \frac{k}{1+k} |z'|^2 \leq \left(\frac{k}{1+k} \right)^2$$

which is an ellipsoid tangent at e_1 with center $\frac{1}{1+k} e_1$. Its intersection with $[e_1]$ is a disk of radius $\frac{k}{1+k}$, while its intersection with $z_1 = \frac{1}{1+k}$ is a ball of radius $\sqrt{\frac{k}{1+k}}$. Because unitary maps preserve inner products, $U(E(k, \zeta)) = E(k, U\zeta)$ for any unitary U. Thus in general $E(k, \zeta)$ is an ellipsoid internally tangent to the unit sphere at ζ with center $\frac{1}{1+k}\zeta$. Julia's Lemma says φ maps each ellipsoid $E(k, \zeta)$ into the corresponding ellipsoid $E(d(\zeta)k, \eta)$.

One important technical complication for the Julia–Carathéodory theory in B_N arises from the appropriate analogue for nontangential approach regions when $N > 1$. For many of the basic function theory results in B_N, approach within a Koranyi approach region replaces nontangential approach. These regions are defined for $\alpha > 1$ by

$$\Gamma(\zeta, \alpha) = \{z \in B_N : |1 - \langle z, \zeta \rangle| < \frac{\alpha}{2}(1 - |z|^2)\}$$

Setting $N = 1$ we see that, since

$$\frac{\alpha}{2}(1 - |z|^2) \sim \alpha(1 - |z|)$$

for $|z|$ near 1, in essence, this gives our previous definition for a nontangential approach region in the disk, and we will use the same notation. However, when $N > 1$ these regions have a perhaps unexpected geometric property. Since $U\Gamma(\zeta, \alpha) = \Gamma(U\zeta, \alpha)$, we can concentrate on understanding $\Gamma(e_1, \alpha)$. An easy computation shows that while the intersection of $\Gamma(e_1, \alpha)$ with the complex line through e_1 is a standard nontangential approach region in the disk $B_N \cap [e_1]$, in other directions $\Gamma(e_1, \alpha)$ permits tangential approach to ∂B_N. In particular, the intersection of $\Gamma(e_1, \alpha)$ with $\{z : \operatorname{Im} z_1 = 0\}$ is the ball

$$(x_1 - 1/\alpha)^2 + |z'|^2 < (1 - 1/\alpha)^2$$

containing e_1 in its boundary. We say f has **admissible limit at** ζ if it has a limit $f^*(\zeta)$ along every curve lying in some Koranyi region $\Gamma(\zeta, \alpha)$.

In the definition of a Koranyi approach region one is comparing the distance (in the Euclidean metric) from z to $\zeta + T^C(\zeta)$, where $T^C(\zeta)$ is the maximal complex subspace of the tangent space to ∂B_N at ζ, with the distance from z to ∂B_N. In one variable this point of view degenerates, since the maximal complex subspace of the tangent space to ∂D at any point is just $\{0\}$, and the comparison then becomes simply that of the distances from z to ζ and from z to ∂D.

In the Julia–Carathéodory Theorem in the disk (Theorem 2.44), we made use of the fact that an analytic function in D that is bounded in every $\Gamma(\zeta, \alpha)$ and has radial limit at ζ in fact has nontangential limit at ζ. Since the analogue of this using Koranyi regions is false when $N > 1$, we need to use a smaller collection of approach sets.

DEFINITION 2.78   *We say f has **restricted limit at** ζ in ∂B_N if f has limit $f^*(\zeta)$ along every curve $\Lambda(t)$ approaching ζ that satisfies*

$$(1)\quad \lim_{t \to 1} \frac{|\Lambda(t) - \lambda(t)|^2}{1 - |\lambda(t)|^2} = 0$$

$$and\ (2)\quad \frac{|\lambda(t) - \zeta|}{1 - |\lambda(t)|} \le M < \infty\ for\ 0 \le t < 1$$

where $\lambda(t)$ is the orthogonal projection of $\Lambda(t)$ into the complex line $[\zeta]$ through 0 and ζ:

$$\lambda(t) = \langle \Lambda(t), \zeta \rangle \zeta$$

The meaning of (2) is that the projection $\lambda(t)$ lies in a nontangential approach region in the copy of the unit disk lying in $[\zeta]$. The notion of restricted limit arises because the full analogue of Lindelöf's Theorem, with admissible limits replacing nontangential limits, does not hold for $N > 1$. The following theorem, a version of Čirka's Theorem, provides an adequate substitute. A proof can be found in [Ru80, Theorem 8.4.8, p. 174].

THEOREM 2.79
Suppose f is analytic in B_N and bounded in every approach region $\Gamma(\zeta, \alpha)$. If $\lim_{r \to 1} f(r\zeta)$ exists, then f has restricted limit at ζ.

DEFINITION 2.80   *We say $\varphi : B_N \to B_N$ has **finite angular derivative** at ζ in ∂B_N if there exists η in ∂B_N so that*

$$\frac{\langle \eta - \varphi(z), \eta \rangle}{\langle \zeta - z, \zeta \rangle}$$

has finite restricted limit at ζ.

The following result is the Julia–Carathéodory Theorem in the ball. When η is in ∂B_N we use the notation φ_η for the coordinate of φ in the η–direction, i.e. $\varphi_\eta(z) = \langle \varphi(z), \eta \rangle$.

THEOREM 2.81 *(Julia–Carathéodory Theorem in B_N)*
For $\varphi : B_N \to B_N$ analytic and ζ in ∂B_N, the following are equivalent:

(1) $d(\zeta) = \liminf_{z \to \zeta}(1 - |\varphi(z)|)/(1 - |z|) < \infty$ where the limit is taken as z approaches ζ unrestrictedly in B_N.

(2) φ has finite angular derivative at ζ.

(3) φ has restricted limit η at ζ, where $|\eta| = 1$ and $D_\zeta \varphi_\eta(z) = \langle \varphi'(z)\zeta, \eta \rangle$ has finite restricted limit at ζ.

Moreover, when these conditions hold, $D_\zeta \varphi_\eta(z)$ has restricted limit $d(\zeta)$ at ζ.

The proof of this theorem parallels the proof of Theorem 2.44 in the disk and requires the following extension of Lemma 2.45 to Koranyi approach regions in B_N.

LEMMA 2.82
Let $1 < \alpha < \beta$. There exists $\delta = \delta(\alpha, \beta) > 0$ so that if (z_1, z') is in $\Gamma(e_1, \alpha)$ then $(z_1 + \lambda, z')$ is in $\Gamma(e_1, \beta)$ for all λ with $|\lambda| \leq \delta|1 - z_1|$.

PROOF If (z_1, z') is in $\Gamma(e_1, \alpha)$ then

$$1 - |z_1|^2 - |z'|^2 \geq \frac{2}{\alpha}|1 - z_1|$$

If $|\lambda| < 1$ we have

$$1 - |z_1 + \lambda|^2 \geq 1 - |z_1|^2 - 2|\lambda| - |\lambda|^2 \geq 1 - |z_1|^2 - 3|\lambda|$$

so that

$$1 - |z_1 + \lambda|^2 - |z'|^2 \geq \frac{2}{\alpha}|1 - z_1| - 3|\lambda|$$

and hence if $|\lambda| \leq \delta|1 - z_1|$ for some positive δ less than $1/2$

$$1 - |z_1 + \lambda|^2 - |z'|^2 \geq (\frac{2}{\alpha} - 3\delta)|1 - z_1|$$

But

$$\frac{2}{\beta}|1 - z_1 - \lambda| \leq \frac{2}{\beta}(|1 - z_1| + |\lambda|) \leq \frac{2}{\beta}(1 + \delta)|1 - z_1|$$

and $(z_1 + \lambda, z')$ will lie in $\Gamma(e_1, \beta)$ provided

$$\frac{2}{\alpha} - 3\delta \geq \frac{2}{\beta}(1 + \delta) \tag{2.6.1}$$

since $(z_1 + \lambda, z')$ is in $\Gamma(e_1, \beta)$ if

$$\frac{2}{\beta}|1 - z_1 - \lambda| < 1 - |z_1 + \lambda|^2 - |z'|^2$$

Inequality (2.6.1) holds provided δ is less than the positive number

$$\frac{2/\alpha - 2/\beta}{3 + 2/\beta}$$

the minimum of this value and $1/2$ is the δ we seek. ∎

PROOF (of Theorem 2.81)   We leave it as an exercise to check that it is sufficient to prove the theorem in the case $\zeta = \eta = e_1$, which will simplify notation.

For the implication (1) $\Rightarrow$ (2), assume $d \equiv d(e_1) < \infty$; our goal is to show $(1 - \varphi_1(z))/(1 - z_1)$ has restricted limit at e_1. By Julia's Lemma for the ball (Lemma 2.77)

$$\frac{|1 - \varphi_1(z)|^2}{1 - |\varphi(z)|^2} \leq d\frac{|1 - z_1|^2}{1 - |z|^2}$$

For $0 < r < 1$ we have

$$\frac{1 - |\varphi(re_1)|}{1 - r}\frac{1 + r}{1 + |\varphi(re_1)|} \leq \frac{1 - |\varphi_1(re_1)|}{1 - r}\frac{1 + r}{1 + |\varphi_1(re_1)|}$$

$$\leq \frac{|1 - \varphi_1(re_1)|^2}{1 - |\varphi_1(re_1)|^2}\frac{1 - r^2}{(1 - r)^2}$$

$$\leq \frac{|1 - \varphi_1(re_1)|^2}{1 - |\varphi(re_1)|^2}\frac{1 - r^2}{(1 - r)^2} \leq d$$

from which it follows (using (1)) that

$$\lim_{r \to 1} \frac{1 - |\varphi(re_1)|}{1 - r} = d$$

and

$$\lim_{r \to 1} \frac{1 - |\varphi_1(re_1)|}{1 - r} = d$$

and

$$\lim_{r \to 1} \frac{|1 - \varphi_1(re_1)|}{1 - r} = d$$

so that

$$\lim_{r \to 1} \frac{1 - \varphi_1(re_1)}{1 - r} = d$$

exactly as in the proof of Theorem 2.44. According to Theorem 2.79, restricted convergence will follow from this radial convergence if we can show that $(1 - \varphi_1(z))/(1 - z_1)$ is bounded in every $\Gamma(e_1, \alpha)$. To this end fix $\Gamma(e_1, \alpha)$ and z in $\Gamma(e_1, \alpha)$. Set $|1 - z_1| = \delta$ so that

$$\frac{|1 - z_1|^2}{1 - |z|^2} = \frac{|1 - z_1|}{1 - |z|^2}\delta \leq \alpha\delta/2$$

and by Julia's Lemma

$$\frac{|1 - \varphi_1(z)|^2}{1 - |\varphi(z)|^2} \leq \alpha\delta d/2$$

or

$$\frac{|1 - \varphi_1(z)|}{|1 - z_1|} \leq \frac{\alpha\delta d(1 - |\varphi_1(z)|^2)}{2\delta|1 - \varphi_1(z)|} \leq \alpha d$$

as desired.

Next we turn to (2) $\Rightarrow$ (3). The existence of a finite angular derivative at e_1 clearly implies that φ has restricted limit of norm 1 at e_1; our normalization is that the limit is e_1. We also need to show that $D_1\varphi_1$ has finite restricted limit at e_1. Fix $1 < \alpha < \beta$ and let $\delta = \delta(\alpha, \beta)$ be the δ of Lemma 2.82 so that (z_1, z') in $\Gamma(e_1, \alpha)$ implies $(z_1 + \lambda, z')$ is in $\Gamma(e_1, \beta)$ for all λ with $|\lambda| \leq \delta|1 - z_1|$. As in the proof of Theorem 2.44, the Cauchy integral formula applied to $D_1\varphi_1 - 1$ yields, for $r = r(z) = \delta(\alpha, \beta)|1 - z_1|$,

$$D_1\varphi_1(z) = D_1(\varphi_1 - 1)(z) = \int_0^{2\pi} \frac{(\varphi_1 - 1)(z_1 + re^{i\theta}, z')}{re^{i\theta}} \frac{d\theta}{2\pi}$$

$$= \int_0^{2\pi} \left(\frac{\varphi_1(z_1 + re^{i\theta}, z') - 1}{z_1 + re^{i\theta} - 1} \right) \left(\frac{z_1 + re^{i\theta} - 1}{re^{i\theta}} \right) \frac{d\theta}{2\pi} \quad (2.6.2)$$

If (2) holds, then $d(e_1) < \infty$ and the above argument shows that $(1 - \varphi_1(z))/(1 - z_1)$ is bounded in every Koranyi approach region. Since $(z_1 + re^{i\theta}, z')$ is in $\Gamma(e_1, \beta)$, the first factor in the integrand of Equation (2.6.2) is bounded as θ varies over $(0, 2\pi)$ and (z_1, z') varies in $\Gamma(e_1, \alpha)$. The second factor in the integrand is bounded by $1 + \delta^{-1}$. It is easy to see that $(t + re^{i\theta}, 0)$ approaches e_1 nontangentially as t approaches 1, where $r = r(te_1) = (\delta)(1 - t)$. Thus, if $z = te_1$, the first factor in the integrand of Equation (2.6.2) converges to the restricted limit of $(1 - \varphi_1)/(1 - z_1)$ as t approaches 1, while the second factor converges to $1 + \delta^{-1}e^{-i\theta}$. The dominated convergence theorem implies $D_1\varphi_1$ has radial limit equal to the restricted limit of $(1 - \varphi_1)/(1 - z_1)$ at e_1; since $D_1\varphi_1$ is also bounded in every Koranyi region, we may use Theorem 2.79 to conclude this is in fact the restricted limit of $D_1\varphi_1$ at e_1.

We finish by showing that (3) implies (1). Since $D_1\varphi_1(re_1)$ has a finite limit as $r \to 1$ there exists a finite number M so that

$$|1 - \varphi_1(re_1)| = \left| \int_r^1 D_1\varphi_1(te_1)dt \right| \leq M(1 - r)$$

Thus

$$\frac{1 - |\varphi(re_1)|}{1 - r} \leq \frac{|1 - \varphi_1(re_1)|}{1 - r} \leq M$$

and $d(\zeta) < \infty$. ∎

A version of the Denjoy–Wolff Theorem (Theorem 2.51) continues to hold in the ball B_N when $N > 1$. We state it here in the form that we will find most useful later on.

THEOREM 2.83   (Denjoy–Wolff Theorem in B_N)
If φ is an analytic map of the ball into itself with no fixed points in B_N, then there is a point ζ of norm 1 so that the iterates φ_n of φ converge to ζ uniformly on compact subsets of B_N.

The point ζ will be referred to as the Denjoy–Wolff point of φ. We will prove this theorem in a sequence of steps. Curiously, the most difficult step will be a verification of the theorem in the special case that φ is an automorphism of B_N.

Since the analytic maps of B_N into itself form a normal family, every sequence of iterates of some $\varphi : B_N \to B_N$ will have a convergent subsequence. A simple observation about the subsequential limits of $\{\varphi_n\}$ is contained in the next lemma.

LEMMA 2.84
If $\psi : B_N \to \overline{B_N}$ is analytic, then either $\psi(B_N) \subset B_N$ or $\psi(z) \equiv \zeta$, for some ζ in ∂B_N and all z in B_N.

PROOF   If ψ takes a point z_0 in B_N to a point ζ in ∂B_N, consider $G(z) \equiv (1 + \langle z, \zeta \rangle)/2$. Then G is in the ball algebra of functions analytic in B_N and continuous on $\overline{B_N}$, $G(\zeta) = 1$ and $|G(z)| < 1$ for $z \neq \zeta$ in $\overline{B_N}$. Apply the maximum modulus theorem (for analytic functions on B_N, see [Ru80, p. 5]) to $G \circ \psi$, noting that $G \circ \psi(z_0) = 1$ while $|G \circ \psi| \leq 1$ on B_N, to conclude that $G \circ \psi$ is identically 1 and hence $\psi(z) \equiv \zeta$.   ∎

We get our candidate for the limit of the iterates of an arbitrary fixed point free φ by a method reminiscent of the proof of Theorem 2.48: Choose positive numbers r_n increasing to 1 and find, using the Brouwer Fixed Point Theorem, fixed points a_n of the maps $r_n\varphi : r_n\overline{B_N} \to r_n\overline{B_N}$. Passing to a subsequence if necessary we may assume the a_n's have a limit ζ in $\overline{B_N}$. Just as in the proof of Theorem 2.48 in the disk, since φ has no fixed points in the open ball, ζ must be in the boundary of B_N. Without loss of generality we may suppose it to be e_1. Setting $f_n = r_n\varphi$ and considering f_n as a map of B_N into B_N, the Schwarz–Pick Theorem for the ball (Exercise 2.5.2) gives

$$\frac{|1 - \langle f_n(a_n), f_n(w) \rangle|^2}{(1 - |f_n(a_n)|^2)(1 - |f_n(w)|^2)} \leq \frac{|1 - \langle a_n, w \rangle|^2}{(1 - |a_n|^2)(1 - |w|^2)}$$

for all w in B_N. Since $f_n(a_n) = a_n$ this gives

$$\frac{|1 - \langle a_n, f_n(w) \rangle|^2}{1 - |f_n(w)|^2} \leq \frac{|1 - \langle a_n, w \rangle|^2}{1 - |w|^2}$$

Letting $n \to \infty$ and recalling that $a_n \to e_1$ and $f_n(w) \to \varphi(w)$ we obtain

$$\frac{|1 - \varphi_1(w)|^2}{1 - |\varphi(w)|^2} \leq \frac{|1 - w_1|^2}{1 - |w|^2}$$

We summarize this in geometric language:

PROPOSITION 2.85

If φ is a analytic map of the unit ball into itself with no fixed points in B_N, there is a unique ζ in ∂B_N so that each ellipsoid $E(k, \zeta)$ is mapped into itself by φ and every iterate of φ.

PROOF All that remains to be shown is the uniqueness of ζ and the argument follows the proof of the one variable case, Theorem 2.48. Suppose ζ_1 and ζ_2 both have the stated property. Choose k_1 and k_2 so that $\overline{E(k_1, \zeta_1)}$ and $\overline{E(k_2, \zeta_2)}$ are tangent to each other at w in B_N. Then $\varphi(w)$ is in $\overline{E(k_1, \zeta_1)} \cap \overline{E(k_2, \zeta_2)} = \{w\}$, contradicting the hypothesis that φ is fixed point free in B_N. ∎

We denote the mapping $\psi(z) \equiv \zeta$ of Proposition 2.85 by $\zeta(\varphi)$.

COROLLARY 2.86

If φ has no interior fixed points, then the only constant map which can appear as a subsequential limit of iterates of φ is $\zeta(\varphi)$.

PROOF The proof is identical to the proof of Lemma 2.49. ∎

For non-constant maps which are subsequential limits of $\{\varphi_n\}$ we have the following analogue of Lemma 2.50 in the disk. An idempotent is a map ψ with $\psi \circ \psi = \psi$. Idempotents are also called retractions.

PROPOSITION 2.87

If $\{\varphi_n\}$ has a non-constant limit, then there is a non-constant idempotent among the subsequential limits of $\{\varphi_n\}$.

PROOF Suppose $\varphi_{n_i} \to \tau$ where τ is not a constant map. Necessarily $\tau(B_N) \subset B_N$, by Lemma 2.84. Set $m_i = n_{i+1} - n_i$ and choose a convergent subsequence of $\{\varphi_{m_i}\}$, say $\varphi_{m_{i_k}} \to \psi$. On the one hand $\varphi_{m_{i_k}} \circ \varphi_{n_{i_k}} \to \psi \circ \tau$, while also $\varphi_{m_i} \circ \varphi_{n_i} = \varphi_{n_{i+1}} \to \tau$. So $\psi \circ \tau = \tau$, or ψ is the identity on the range of τ, which consists of more than one point. So, using Theorem 2.76, the fixed point set of ψ is an affine subset A of B_N with dimension at least 1. Clearly the range of ψ contains A; if this containment is proper the same argument, now applied to ψ instead of τ, produces another subsequential limit of $\{\varphi_n\}$ which is the identity on an affine subset A' of dimension strictly greater than the dimension of A. Since this process can only be repeated a finite number of times there must be a subsequential limit whose range in B_N is its fixed point set. This is an idempotent, and since the dimension of its fixed point set is at least one, it is non-constant. $\blacksquare$

Next we verify Theorem 2.83 when φ is an automorphism. Automorphisms of B_N with no interior fixed points will fix either exactly one point or exactly two points in ∂B_N (see Exercise 2.6.3 and Exercise 2.6.5). Those with two boundary fixed points are easy to handle.

PROPOSITION 2.88

Let φ be an automorphism of B_N which fixes precisely two points of ∂B_N. Then φ_n converges to one of these fixed points.

PROOF Suppose φ fixes ζ_1 and ζ_2 in the boundary. Consider the complex line L through ζ_1 and ζ_2. Since automorphisms take affine sets to affine sets (Theorem 2.74), φ maps $L \cap B_N$ onto $L \cap B_N$. The Denjoy–Wolff Theorem in one variable implies that the iterates of φ restricted to $L \cap B_N$ converge to one of the fixed points, say ζ_1. By Lemma 2.84 every convergent subsequence of $\{\varphi_n\}$ must converge to ζ_1. This implies $\varphi_n \to \zeta_1$ since the analytic self-maps of B_N form a normal family. $\blacksquare$

Clearly the limit identified in this last proposition must be the map $\zeta(\varphi)$ identified in Proposition 2.85. An alternate proof of Proposition 2.88 can be given using the result of Exercise 2.6.5.

The one boundary fixed point automorphisms require a more complicated analysis. It is conceptually simpler to discuss these automorphisms in an unbounded realization of B_N, the Siegel upper half space

$$\Omega = \{(w_1, w') \in C^N : \operatorname{Im} w_1 > |w'|^2 = |w_2|^2 + \cdots |w_N|^2\}$$

The Cayley transform $\Phi(z) = i(e_1 + z)/(1 - z_1)$, defined for $z_1 \neq 1$, is a biholomorphic map of B_N onto Ω that extends (with $\Phi(e_1) = \infty$) to a homeomorphism of $\overline{B_N}$ onto $\Omega \cup \partial\Omega \cup \{\infty\}$, the one point compactification of the closure of Ω. The automorphisms of B_N that fix only e_1 correspond, under conjugation by Φ, to automorphisms of Ω fixing only $\{\infty\}$. An example of a class of such automorphisms are the Heisenberg translations:

DEFINITION 2.89 *For each (b_1, b') in $\partial\Omega$ set*

$$h_b(w_1, w') = (w_1 + b_1 + 2i\langle w', b'\rangle, w' + b')$$

These "translations" are called the Heisenberg translations of Ω; they form a subgroup of Aut(Ω) and for each $b \neq 0$, h_b fixes ∞ only. The corresponding map in the disk, $\Phi^{-1} \circ h_b \circ \Phi$ will be called a Heisenberg translation of B_N.

Unlike the situation in one variable, these "translations" are *not* the complete set of all automorphisms fixing only e_1. For example, if λ_j is a complex number of modulus 1 for $j = 2, 3, \ldots, N$ and $b \neq 0$ is real, the map $(w_1, w') \rightarrow (w_1 + b, \lambda_2 w_2, \ldots, \lambda_N w_N)$ is an automorphism of Ω fixing ∞ only but is not a Heisenberg translation. Notice that this mapping fixes *as a set* the image under Φ of the complex line through 0 and e_1; that is the set $\{(w_1, w') \in \Omega : w' = 0\}$. That this is no accident is the content of the next theorem. We say φ *fixes A as a set* if $\varphi(A) \subset A$.

THEOREM 2.90
An automorphism of B_N which fixes e_1 only is either a Heisenberg translation of B_N or fixes as a set some non-empty proper affine subset of B_N.

Before we can prove this theorem we need to establish some other results about the automorphisms of B_N fixing e_1, or equivalently, the automorphisms of Ω fixing ∞. We will find it convenient to transfer back and forth between the ball and the upper half space by means of the Cayley transform Φ and temporarily adopt the convention of using lower case letters for automorphisms of B_N and the corresponding upper case letters for the associated automorphisms in Ω obtained by composition on the right and left by Φ and Φ^{-1} respectively.

LEMMA 2.91
Given any (a_1, a') in Ω with $\operatorname{Im} a_1 - |a'|^2 = 1$ there exists a Heisenberg translation h_b so that $h_b(a_1, a') = (i, 0')$.

PROOF Write $a_1 = c + i(1 + |a'|^2)$ with c real. Set $b = (-c + i|a'|^2, -a')$ so that b is in $\partial\Omega$. The Heisenberg translation h_b has the desired property. ∎

LEMMA 2.92
Suppose g in $\operatorname{Aut}(\Omega)$ fixes ∞ only. Then for every $w = (w_1, w')$ in Ω

$$\operatorname{Im} g_1(w) - |g'(w)|^2 = \operatorname{Im} w_1 - |w'|^2$$

where $g = (g_1, g_2, \ldots, g_N) = (g_1, g')$.

PROOF Set $g(i, 0') = (a_1, a')$ and let $t = \operatorname{Im} a_1 - |a'|^2$, a positive quantity since (a_1, a') is in Ω. For $s > 0$ define the *non-isotropic dilation* in $\operatorname{Aut}(\Omega)$ by

$$\delta_s(w_1, w') = (s^2 w_1, s w') \tag{2.6.3}$$

If $s \neq 1$ the fixed point set of δ_s is $\{0, \infty\}$. With $s = 1/\sqrt{t}$ we may use Lemma 2.91 to find a Heisenberg translation h_b so that $h_b \circ \delta_s \circ g$ fixes $(i, 0')$ and ∞. The corresponding automorphism of B_N fixes e_1 and 0, hence is unitary. Moreover, it must fix as a set the orthogonal complement of the complex line through 0 and e_1.

This implies

$$h_b \circ \delta_s \circ g(w_1, w') = (w_1, Uw')$$

for some unitary U on C^{N-1}. Equivalently

$$g(w_1, w') = \delta_{\sqrt{t}} \circ h_b^{-1}(w_1, Uw') = (t(w_1 - \overline{b_1} - 2i\langle Uw', b' \rangle), \sqrt{t}(Uw' - b'))$$

We claim that the hypothesis on g shows that t must be 1. If not, we may solve $\sqrt{t}(Uw' - b') = w'$ since $(U - t^{-1/2}I)$ is non-singular. Denote its solution by v'. Set $v_1 = \alpha + i|v'|^2$ for real α to be specified. We have (v_1, v') is in $\partial\Omega$ and therefore $g(v_1, v')$ is in $\partial\Omega$ since g is an automorphism. By our choice of v'

$$g(v_1, v') = (t(\alpha + i|v|^2 - \overline{b_1} - 2i\langle Uv', b' \rangle), v')$$

Since this is a point of the boundary of Ω we must have

$$\mathrm{Im}\,(t(\alpha + i|v'|^2 - \overline{b_1} - 2i\langle Uv', b' \rangle)) = |v'|^2 = \mathrm{Im}\,(\alpha + i|v'|^2)$$

Thus (v_1, v') will be a fixed point of g if α real is chosen to satisfy

$$\mathrm{Re}(t(\alpha + i|v'|^2 - \overline{b_1} - 2i\langle Uv', b' \rangle)) = \alpha$$

or

$$t\alpha + t\mathrm{Re}(-b_1 - 2i\langle Uv', b' \rangle) = \alpha$$

If $t \neq 1$ this may be solved for real α and then the fixed point set of g will contain more than one point, a contradiction. Thus t must be 1 and therefore

$$g(w_1, w') = (w_1 - \overline{b_1} - 2i\langle Uw', b' \rangle, Uw' - b')$$

from which it follows by direct calculation that

$$\mathrm{Im}\,g_1(w) - |g'(w)|^2 = \mathrm{Im}\,w_1 - |w'|^2$$

for all (w_1, w') in Ω. ∎

Note that the proof of Lemma 2.92 shows that any automorphism g of Ω fixing ∞ only can be written as $g(w_1, w') = h_c(w_1, Uw')$ for some unitary U on C^{N-1} and Heisenberg translation h_c. If $\Phi(z) = w$, where Φ is the Cayley transform, then $\mathrm{Im}\,w_1 - |w'|^2 = (1 - |z|^2)/(|1 - z_1|^2)$, so the geometric meaning of Lemma 2.92 is as follows: if an automorphism g of Ω fixes ∞ only, the corresponding automorphism $G = \Phi^{-1} \circ g \circ \Phi$ in $\mathrm{Aut}(B_N)$ maps the boundary of each ellipsoid $E(k, e_1)$ into (and therefore onto) itself. This should be compared with the result of Exercise 2.3.7 for parabolic automorphisms in the disk.

PROOF (of Theorem 2.90)   Let G be the given automorphism of B_N and let g be the corresponding automorphism of Ω. Set $g(i, 0') = (a_1, a')$. By Lemma 2.92 we know $\mathrm{Im}\,a_1 - |a'|^2 = 1$ so we may apply Lemma 2.91 to find a Heisenberg translation h_b so that $h_b \circ g$ fixes both ∞ and $(i, 0')$. As in the proof of Lemma 2.92 the corresponding ball automorphism $F = H_b \circ G$ is unitary and has the form $F(z_1, z') = (z_1, Uz')$ for some unitary U. Since

$$F \circ \Phi^{-1}(w_1, w') = \left(\frac{w_1 - i}{w_1 + i}, \frac{2}{w_1 + i} Uw' \right) = \Phi^{-1} \circ F(w_1, w')$$

on all points of C^N with first coordinate not equal to $-i$, we have $f(w_1, w') = (w_1, Uw')$ where $f = \Phi \circ F \circ \Phi^{-1}$ and (w_1, w') is in Ω.

At this point we distinguish two cases. If every eigenvalue of U is 1 then U, and hence F, is the identity and our original automorphism G is the Heisenberg translation H_b^{-1}.

On the other hand, if U has an eigenvalue $e^{i\theta} \neq 1$ find $0 \neq \Lambda = (\lambda_2, \lambda_3, \dots, \lambda_N)$ so that $\Lambda(U) = e^{i\theta} \Lambda$ where (U) is the matrix of the operator U with respect to the standard basis on C^{N-1}. Recall that $g = \Phi \circ G \circ \Phi^{-1} = \Phi \circ H_b^{-1} \circ F \circ \Phi^{-1} = h_b^{-1} \circ f$ where $h_b^{-1} = h_{\tilde{b}}, \tilde{b} = (-\overline{b_1}, -b')$. Let V be the column vector $(-b_2, \dots, -b_N)^t$ so that $\Lambda V = \sum_{j=2}^{N} -b_j \lambda_j$. Consider the set

$$\mathcal{A} = \left\{ (w_1, \dots, w_N) \in \Omega : \sum_{j=2}^{N} \lambda_j w_j = \frac{\Lambda V}{1 - e^{i\theta}} \right\}$$

which is a non-empty proper affine subset of Ω. We claim that g fixes $\mathcal{A}$ as a set. To see this, let $(w_1, w_2, \dots, w_N)$ be in $\mathcal{A}$. Now $g(w_1, w') = h_b^{-1} \circ f(w_1, w') = h_b^{-1}(w_1, Uw')$. Writing $W' = (w_2, \dots, w_N)^t$ we see that the last $N - 1$ coordinates of $h_b^{-1}(w_1, Uw')$ are $((U)W' + V)^t$. Since

$$\Lambda((U)W' + V) = e^{i\theta} \Lambda W' + \Lambda V = \frac{e^{i\theta} \Lambda V}{(1 - e^{i\theta})} + \Lambda V = \frac{\Lambda V}{1 - e^{i\theta}}$$

and $g(w_1, w')$ is in $\mathcal{A}$ as desired. Since Φ preserves affine sets (Exercise 2.6.4) the corresponding map G on B_N fixes as a set some non-empty, proper, affine subset of B_N. ∎

We can now prove Theorem 2.83 in the case that φ is an automorphism fixing one point of ∂B_N. We may normalize so that the fixed point is e_1.

THEOREM 2.93

Suppose φ is an automorphism of B_N fixing e_1 only. Then $\varphi_n \to e_1$, uniformly on compact subsets of B_N.

PROOF If φ is a Heisenberg translation $\Phi^{-1} \circ h_b \circ \Phi$ then the result is immediate from the easy observation that $(h_b)_n \to \infty$. The remainder of the proof is an inductive argument.

By Theorem 2.51 the result holds for $N = 1$. We assume it holds for $k < N$ and that φ is not a Heisenberg translation. Then by Theorem 2.90 φ fixes as a set some affine subset $\mathcal{A}$ of B_N with dimension k, $1 \leq k < N$. The restriction $\tilde{\varphi}$ of φ to $\mathcal{A} \cong B_k$ may be considered as an automorphism of B_k fixing e_1 only (it is easy to check from the description of $\mathcal{A}$ in the proof of Theorem 2.90 that e_1 is in $\partial\mathcal{A}$). By induction, $\tilde{\varphi}_n \to e_1$ and then by Lemma 2.84, $\varphi_n \to e_1$. $\blacksquare$

A final observation before we prove Theorem 2.83:

LEMMA 2.94
If some subsequence of $\{\varphi_n\}$ converges to the identity map, then φ must be an automorphism.

PROOF   Suppose $\varphi_{n_i} \to I$. By passing to a subsequence if necessary we may assume $\varphi_{n_i-1} \to \tau$. Then $\varphi_{n_i-1} \circ \varphi$ converges to $\tau \circ \varphi$ and also to I, so we must have $\tau \circ \varphi = I$. In particular, $\tau(B_N) \subset B_N$ and therefore $\varphi_{n_i} = \varphi \circ \varphi_{n_i-1} \to \varphi \circ \tau$ which implies that $\varphi \circ \tau = I$ as well. $\blacksquare$

PROOF (of Theorem 2.83.)   We have already established the result for an automorphism with no fixed point in B_N. Suppose now that φ is an arbitrary fixed point free self map of B_N. If every subsequential limit of $\{\varphi_n\}$ is constant, then $\varphi_n \to \zeta(\varphi)$ by Corollary 2.86, and we are done.

If there is a non-constant subsequential limit, then there is a non-constant idempotent ψ so that for some $\{n_i\}$, $\varphi_{n_i} \to \psi$. Let $\mathcal{A}$ be the fixed point set of ψ, an affine set (by Theorem 2.76) of dimension at least 1.

We claim that φ maps $\mathcal{A}$ into $\mathcal{A}$. To see this choose z_0 in $\mathcal{A}$. Since $\varphi_{n_i}(z_0) \to \psi(z_0) = z_0$ we have $\varphi(\varphi_{n_i}(z_0)) \to \varphi(z_0)$. But

$$\varphi(\varphi_{n_i}(z_0)) = \varphi_{n_i}(\varphi(z_0)) \to \psi(\varphi(z_0))$$

so $\psi(\varphi(z_0)) = \varphi(z_0)$ and $\varphi(z_0)$ is in $\mathcal{A}$. Moreover, since φ_{n_i} restricted to $\mathcal{A}$ converges to the identity on $\mathcal{A}$, Lemma 2.94 (with B replaces by $\mathcal{A}$) implies that φ restricted to $\mathcal{A}$ is an automorphism of $\mathcal{A}$, and is clearly a fixed point free automorphism. Thus by the result already established for automorphisms, the sequence of restrictions of φ_{n_i} to $\mathcal{A}$ converges to a constant, contradicting $\varphi_{n_i} \to \psi$. We conclude that all subsequential limits of $\{\varphi_n\}$ must be constant and we are done. $\blacksquare$

Exercises

2.6.1 Verify that the normalizations $\eta = \zeta = e_1$ cause no loss of generality in the proof of Theorem 2.81

2.6.2 Suppose that φ is an analytic map of B_N into B_N with finite angular derivative at e_1 and $\varphi(e_1) = e_1$. Show that for $j = 2, 3, \ldots, N$

(a)
$$\lim_{r \to 1} \frac{\varphi_j(re_1)}{(1 - r)^{1/2}} = 0$$

(b)
$$\frac{\varphi_j(z)}{(1 - z_1)^{1/2}}$$

is bounded in every region $\Gamma(e_1, \alpha)$.

Hence
$$\frac{\varphi_j(z)}{(1 - z_1)^{1/2}}$$

has restricted limit 0 at e_1.

2.6.3 If φ is an automorphism of B_N fixing at least three distinct points of ∂B_N then φ has a fixed point in B_N.

    Hints: Suppose $\zeta_1, \zeta_2, \zeta_3$ are fixed by φ. Use the fundamental identity for automorphisms, Equation (2.5.3), to show $\langle \zeta_1, a \rangle = \langle \zeta_2, a \rangle$ where $a = \varphi^{-1}(0)$. Then show that φ fixes $(\zeta_1 + \zeta_2)/2$.

2.6.4 The Cayley transform Φ sends affine sets in B_N to affine sets in Ω.

2.6.5 Recall that for $0 < s \neq 1$, the non-isotropic dilation $\delta_s(w_1, w') = (s^2 w_1, s w')$ is an automorphism of Ω fixing 0 and ∞ only.

  (a) Show that an arbitrary g in $\mathrm{Aut}(\Omega)$ fixing 0 and ∞ only has the form
$$g(w_1, w') = (s^2 w_1, s U w')$$
where s is positive and not equal to 1 and U is unitary on C^{N-1}.

  (b) Show that if φ is an automorphism of the ball fixing ζ_1, ζ_2 in the boundary of the ball and no other points then φ is conjugate to a map g as in (a); i.e.
$$\varphi = \psi^{-1} \circ g \circ \psi$$
for some biholomorphic map ψ of B_N onto Ω.

2.6.6 This exercise explores the nature of the subsequential limits of $\{\varphi_n\}$ when φ has fixed point(s) in B_N.

  (a) Show that if there is a constant function $g(z) \equiv z_0$, z_0 in B_N among the subsequential limits of $\{\varphi_n\}$, then in fact $\varphi_n \to g$.

  (b) Show that if case (a) does not occur, then φ must act as an automorphism on some affine subset of B_N of dimension at least 1.

  (c) If φ is not an automorphism, then every subsequential limit of $\{\varphi_n\}$ is degenerate in the sense that its range is contained in an affine subset of B_N of dimension less than N.

2.6.7 Give an example to show that the fixed point set of a subsequential limit of $\{\varphi_n\}$ may be strictly larger than the fixed point set of φ.

Notes

Parts of the Julia–Carathéodory Theorem in the ball are due to M. Hervé [He63]. An expanded version of the theorem was given by W. Rudin and the treatment given here follows [Ru80, §8.5] very closely. There are other extensions of Theorem 2.81 that give additional information about other directional derivatives of various coordinate functions of φ when φ has finite angular derivative at some point — see [Ru80, p. 178] where the result of Exercise 2.6.2 also appears. We have followed [Ru80] for the definition of restricted limit; see also [Cir73] for a generalization of this notion for more general domains.

The first proof of Theorem 2.83 goes back to M. Hervé [He63]; the result was rediscovered by B. D. MacCluer and the proof given here is that which appears in [Mc83]. There is related work by Y. Kubota on iteration of analytic self-maps of the ball in [Kub83], and by G. Chen in [Che84]. Lemma 2.92 appears in [Mc83] with thanks to D. Ullrich. Further information on the subsequential limits of $\{\varphi_n\}$ when φ has a fixed point in B_N can be found in [Ab89b, § 2.2.5].

The result of Exercise 2.6.3 is due to T. Hayden and T. Suffridge [HaS71]; the suggested proof outlined here appears in [Ru80, p. 33] where it is ascribed to D. Ullrich.

3

Norms

3.1 Boundedness in classical spaces on the disk

The question of the boundedness of composition operators is a basic but subtle one. Even among relatively benign spaces, it may happen that some natural composition operators are unbounded because the space is too small or because it is too big. We will see that in the most important classical spaces, all analytic maps of the disk into itself give bounded composition operators. In spaces of functions of several variables, the classical spaces have unbounded composition operators and even the general principles are unclear.

The Littlewood Subordination Theorem (Theorem 2.22), which applies to functions with $\varphi(0) = 0$, gives the boundedness of certain composition operators on a variety of spaces. If the space is invariant under automorphisms, that is, if all composition operators whose symbol is an automorphism are bounded, then the boundedness of many other composition operators follows immediately. Suppose φ_1 is an analytic map, ψ is an automorphism, and $\varphi_2 = \psi \circ \varphi_1$. This means $C_{\varphi_2} = C_{\varphi_1} C_\psi$ and $C_{\varphi_1} = C_{\varphi_2} C_\psi^{-1} = C_{\varphi_2} C_{\psi^{-1}}$, so if all automorphisms give continuous operators, then C_{φ_2} is bounded if and only if C_{φ_1} is. Choosing ψ with $\psi(\varphi_1(0)) = 0$ gives $\varphi_2(0) = 0$ so the Littlewood Subordination Theorem may give boundedness. Thus we will split the question of boundedness and prove boundedness theorems for functions that fix 0 and for automorphisms separately.

THEOREM 3.1

For $0 < p < \infty$ let μ be a finite positive measure on the closed unit interval and suppose $\mathcal{Y}$ is the space of all analytic functions on the disk for which

$$\|f\|^p \equiv \int_0^1 \int_0^{2\pi} |f(re^{i\theta})|^p \, \frac{d\theta}{2\pi} \, d\mu(r)$$

is finite. If φ is an analytic map of the unit disk into itself such that $\varphi(0) = 0$, then C_φ is a bounded operator on $\mathcal{Y}$ and $\|C_\varphi\| = 1$.

PROOF   Since $|f|^p$ is subharmonic, the Littlewood Subordination Theorem implies for each r,

$$\int_0^{2\pi} |f(\varphi(re^{i\theta}))|^p \frac{d\theta}{2\pi} \le \int_0^{2\pi} |f(re^{i\theta})|^p \frac{d\theta}{2\pi}$$

so

$$\|C_\varphi f\|^p = \int_0^1 \int_0^{2\pi} |f(\varphi(re^{i\theta}))|^p \frac{d\theta}{2\pi} d\mu(r)$$

$$\le \int_0^1 \int_0^{2\pi} |f(re^{i\theta})|^p \frac{d\theta}{2\pi} d\mu(r) = \|f\|^p$$

That is, $\|C_\varphi\| \le 1$. On the other hand, the constant function 1 is in $\mathcal{Y}$ and $C_\varphi 1 = 1$ so it follows that $\|C_\varphi\| = 1$.   ∎

This theorem applies to the Hardy spaces, the Bergman spaces, and the weighted Bergman spaces A_α^p, as well as other spaces. It does not apply to the Dirichlet space, and in fact there are maps φ with $\varphi(0) = 0$ that do not define bounded composition operators on the Dirichlet space. Indeed, since z is in the Dirichlet space and $C_\varphi(z) = \varphi$, φ must be in the Dirichlet space if C_φ is a bounded operator. Since the Dirichlet norm measures the area of the image counting multiplicity, it is easy to construct φ mapping the disk into the disk with $\varphi(0) = 0$ and φ not in the Dirichlet space.

A weighted Hardy space $H^2(\beta)$ is included in Theorem 3.1 if and only if the weight sequence $\beta(j)$ is a moment sequence. However, using the theory of Hadamard (Schur) products of matrices, the theorem can be extended to a wider class of weighted Hardy spaces. Since we do not wish to dwell on this technique from matrix analysis, we state the theorem without proof.

THEOREM 3.2
Suppose $H^2(\beta_1)$ and $H^2(\beta_2)$ are weighted Hardy spaces and

$$\frac{\beta_1(j+1)}{\beta_1(j)} \ge \frac{\beta_2(j+1)}{\beta_2(j)}$$

for $j = 0, 1, 2, \ldots$. If $\varphi(0) = 0$ and C_φ is bounded on $H^2(\beta_1)$, then C_φ is bounded on $H^2(\beta_2)$ and

$$\|C_\varphi\|_{\beta_1} \ge \|C_\varphi\|_{\beta_2}$$

By Theorem 3.1, every map φ that fixes 0 gives a bounded composition operator on $H^2(D)$. Applying Theorem 3.2 to the case $H^2(\beta_1) = H^2(D)$, that is, $\beta_1(j) \equiv 1$, we obtain the following interesting corollary.

COROLLARY 3.3

If $H^2(\beta)$ is a weighted Hardy space with $\beta(0) \geq \beta(1) \geq \beta(2) \geq \cdots$ and φ satisfies $\varphi(0) = 0$, then C_φ is bounded on $H^2(\beta)$ and $\|C_\varphi\| = 1$.

All the spaces in this corollary contain $H^2(D)$, but even among spaces that contain $H^2(D)$, some hypothesis like monotonicity is required for establishing boundedness.

EXAMPLE 3.4

We define a weighted Hardy space. Let $\beta(0) = \beta(1) = 1$. If $\beta(j - 1)$ has been defined and j satisfies $2^{2k} < j \leq 2^{2k+1}$ for some integer k, let $\beta(j) = \beta(j-1)/2$. If $\beta(j - 1)$ has been defined and j satisfies $2^{2k+1} < j \leq 2^{2k+2}$ for some integer k, let $\beta(j) = \sqrt{2}\beta(j - 1)$. It is not difficult to verify that $\beta(j) \leq 1$ for all positive integers j, that $\beta(2^{2k}) = 1$ for all k, and that $\beta(2^{2k+1}) = 2^{-4^k}$. To see that $H^2(\beta)$ is not entirely pathological, we note that multiplication by z on $H^2(\beta)$ is bounded ($\|zf\| \leq \sqrt{2}\|f\|$) and bounded below ($\|zf\| \geq \|f\|/2$). On the other hand, for $\varphi = z^2$ the resulting composition operator C_φ is not bounded. Since $C_\varphi(z^j) = z^{2j}$, we see that $\|C_\varphi(z^{2^{2k+1}})\| = \|z^{4^{k+1}}\| = 1$, but $\|z^{2^{2k+1}}\| = 2^{-4^k}$, so $\|C_\varphi\| \geq 2^{4^k}$ and C_φ is not bounded. $\quad\square$

Let us now consider invariance of spaces under automorphisms. One general argument, which depends on a change of variables, applies to weighted Dirichlet spaces.

THEOREM 3.5

Let ν be a positive function on the unit interval with $\int_D \nu(1 - |z|^2)\, dA(z) < \infty$ such that, for each $q > 1$, there is a constant $\kappa = \kappa(q)$ satisfying

$$\nu(s) \leq \kappa\nu(t) \quad \text{whenever } s \leq qt$$

For $1 \leq p < \infty$, suppose $\mathcal{Y}$ is the Banach space of all analytic functions on the disk for which the norm given by

$$\|f\|^p = |f(0)|^p + \int_D |f'(z)|^p \nu(1 - |z|^2)\, dA(z)/\pi$$

is finite. If ψ is an automorphism of the unit disk, then C_ψ is a bounded operator on $\mathcal{Y}$.

Taking $p = 2$ and $\kappa = q^\gamma$, we see this result applies to $\nu(t) = t^\gamma$ for $\gamma \geq 0$, which includes the classical Dirichlet space ($\gamma = 0$), the classical Hardy space $H^2(D)$ (by Theorem 2.30 and the observation that the weights $1 - |z|^2$ and $\log|z|^{-2}$ are interchangeable, $\gamma = 1$ gives an equivalent norm) and the weighted Bergman spaces $A_\alpha^2(D)$ (by Exercise 2.1.5, $\gamma = \alpha + 2$ gives an equivalent norm).

PROOF Let ψ be an automorphism of the disk, say $\psi(z) = \lambda(z + u)/(1 + \overline{u}z)$ where $|\lambda| = 1$ and $|u| < 1$. Since point evaluation at u is continuous on $\mathcal{Y}$, there is a constant κ_0 so that $|f(\psi(0))|^p \le \kappa_0^p \|f\|^p$.

Now

$$\int_D |(f \circ \psi)'(z)|^p \nu(1 - |z|^2) \, dA(z)/\pi$$

$$= \int_D |f'(\psi(z))|^p |\psi'(z)|^{p-2} \nu(1 - |z|^2) \, |\psi'(z)|^2 dA(z)/\pi$$

$$\le \left(\max |\psi'(z)|^{p-2}\right) \int_D |f'(\psi(z))|^p \nu(1 - |z|^2) \, |\psi'(z)|^2 \, dA(z)/\pi$$

An easy calculation gives

$$|\psi'(z)| = \frac{1 - |u|^2}{|1 + \overline{u}z|^2} \le \frac{1 + |u|}{1 - |u|}$$

and

$$|\psi'(z)| \ge \frac{1 - |u|}{1 + |u|}$$

for $|z| \le 1$.

We change variables in the integral by $w = \psi(z)$. Since

$$dA(w) = |\psi'(z)|^2 \, dA(z)$$

and $\psi(D) = D$, we get

$$\int_D |f'(\psi(z))|^p \nu(1 - |z|^2) \, |\psi'(z)|^2 \, dA(z)/\pi$$

$$= \int_D |f'(w)|^p \nu(1 - |\psi^{-1}(w)|^2) \, dA(w)/\pi$$

Now $\psi^{-1}(w) = \overline{\lambda}(w - \lambda u)/(1 - \overline{\lambda u}w)$ and Equation (2.3.1) gives

$$\frac{1 - |\psi^{-1}(w)|^2}{1 - |w|^2} = \frac{1 - |u|^2}{|1 - \overline{\lambda u}w|^2} \le \frac{1 + |u|}{1 - |u|}$$

that is, $(1 - |\psi^{-1}(w)|^2) \le q(1 - |w|^2)$ with $q = (1 + |u|)/(1 - |u|)$ and the hypothesis on ν guarantees $\nu(1 - |\psi^{-1}(w)|^2) \le \kappa\nu(1 - |w|^2)$. Therefore,

$$\int_D |f'(w)|^p \nu(1 - |\psi^{-1}(w)|^2) \, dA(w)/\pi \le \kappa \int_D |f'(w)|^p \nu(1 - |w|^2) \, dA(w)/\pi$$

Putting the inequalities together, we get the desired result

$$\|f \circ \psi\|^p = |f(\psi(0))|^p + \int_D |(f \circ \psi)'(z)|^p \nu(1 - |z|^2) \, dA(z)/\pi$$

$$\leq \kappa_0^p \|f\|^p + \kappa \left(\frac{1+|u|}{1-|u|}\right)^{|p-2|} \int_D |f'(z)|^p \nu(1-|z|^2)\, dA(z)/\pi$$

$$\leq \left(\kappa_0^p + \kappa \left(\frac{1+|u|}{1-|u|}\right)^{|p-2|}\right) \|f\|^p$$

∎

For the space $H^p(D)$ we can prove boundedness of these operators and calculate their norms by carrying out the change of variable explicitly. Note in the second part of the proof that follows, for an L^p space, the estimate would be carried out using characteristic functions on sets near where the derivative is biggest. Kernel functions replace the characteristic functions in the argument, both here and in other proofs.

THEOREM 3.6
If ψ is an automorphism of the disk, then on the Hardy space $H^p(D)$, $p \geq 1$,

$$\left(\frac{1-|\psi(0)|}{1+|\psi(0)|}\right)^{1/p} \|f\| \leq \|C_\psi(f)\| \leq \left(\frac{1+|\psi(0)|}{1-|\psi(0)|}\right)^{1/p} \|f\|$$

Moreover, these inequalities are best possible and

$$\|C_\psi\| = \left(\frac{1+|\psi(0)|}{1-|\psi(0)|}\right)^{1/p}$$

PROOF Let ψ be the automorphism

$$\psi(z) = \lambda \frac{z+u}{1+\overline{u}z}$$

where $|\lambda| = 1$ and $|u| < 1$. Let f be a polynomial. Then $f \circ \psi$ is a bounded analytic function on the disk and its norm in $H^p(D)$ satisfies

$$\|f \circ \psi\|^p = \int_0^{2\pi} |f(\psi(e^{i\theta}))|^p \frac{d\theta}{2\pi}$$

Since ψ is a smooth homeomorphism of the unit circle, we can change variables on the circle by

$$e^{it} = \psi(e^{i\theta}) = \lambda \frac{e^{i\theta} + u}{1 + \overline{u}e^{i\theta}}$$

or

$$e^{i\theta} = \psi^{-1}(e^{it}) = \overline{\lambda} \frac{e^{it} - \lambda u}{1 - \overline{\lambda u}e^{it}}$$

It follows that

$$ie^{i\theta} \frac{d\theta}{2\pi} = \overline{\lambda} \frac{1-|u|^2}{(1 - \overline{\lambda u}e^{it})^2} ie^{it} \frac{dt}{2\pi}$$

or

$$\frac{d\theta}{2\pi} = \frac{1 - |u|^2}{|1 - \overline{\lambda u} e^{it}|^2} \frac{dt}{2\pi}$$

and we have

$$\int_0^{2\pi} |f(\psi(e^{i\theta}))|^p \frac{d\theta}{2\pi} = \int_0^{2\pi} |f(e^{it})|^p \frac{1 - |u|^2}{|1 - \overline{\lambda u} e^{it}|^2} \frac{dt}{2\pi}$$

$$\leq \frac{1 + |u|}{1 - |u|} \int_0^{2\pi} |f(e^{it})|^p \frac{dt}{2\pi}$$

That is, since $|u| = |\psi(0)|$, we have

$$\|f \circ \psi\|^p \leq \frac{1 + |\psi(0)|}{1 - |\psi(0)|} \|f\|^p$$

which is the upper inequality. Since the polynomials are dense, we see that this inequality holds for all functions in $H^p(D)$ and, therefore, that C_ψ is bounded on $H^p(D)$.

To show that the upper inequality is sharp, consider the adjoint C_ψ^* acting on the evaluation kernels K_w in the dual of $H^p(D)$:

$$C_\psi^*(K_w) = K_{\psi(w)}$$

so by Corollary 2.14

$$\|C_\psi\| = \|C_\psi^*\| \geq \frac{\|K_{\psi(w)}\|}{\|K_w\|} = \left(\frac{1 - |w|^2}{1 - |\psi(w)|^2} \right)^{1/p}$$

Now if $u = se^{i\theta}$, we take $w = re^{i\theta}$ and

$$\|C_\psi\| \geq \lim_{r \to 1} \left(\frac{1 - |w|^2}{1 - |\psi(w)|^2} \right)^{1/p} = \left(\frac{1 + |u|}{1 - |u|} \right)^{1/p} = \left(\frac{1 + |\psi(0)|}{1 - |\psi(0)|} \right)^{1/p}$$

Note that

$$\|f\| = \|C_\psi^{-1} C_\psi f\| \leq \|C_\psi^{-1}\| \|C_\psi f\|$$

Since $|\psi(0)| = |\psi^{-1}(0)|$, this is

$$\|C_\psi f\| \geq \frac{1}{\|C_\psi^{-1}\|} \|f\| = \left(\frac{1 - |\psi(0)|}{1 + |\psi(0)|} \right)^{1/p} \|f\|$$

which is the lower inequality. Moreover, if g_n is a sequence that exhibits the norm of C_ψ^{-1}, then the substitution $f_n = C_\psi^{-1} g_n$ shows that the lower inequality is sharp also. ∎

This leads to a proof that all composition operators are bounded on $H^p(D)$ and gives an estimate for their norms. The resulting estimate is the best based only on $\varphi(0)$ (see Exercise 3.1.2 and Theorem 3.8) but apparently neither bound is achieved for many maps φ.

COROLLARY 3.7
If φ is an analytic map of the disk into itself, then on the Hardy space $H^p(D)$, $p \geq 1$,

$$\left(\frac{1}{1 - |\varphi(0)|^2}\right)^{1/p} \leq \|C_\varphi\| \leq \left(\frac{1 + |\varphi(0)|}{1 - |\varphi(0)|}\right)^{1/p}$$

PROOF For $\psi(z) = (\varphi(0) - z)/(1 - \overline{\varphi(0)}z)$ and $\varphi_0 = \psi \circ \varphi$, we have $\varphi_0(0) = 0$ and $C_\varphi = C_{\varphi_0} C_{\psi^{-1}}$. Since $\psi^{-1} = \psi$, by Corollary 2.24 and Theorem 3.6, we have

$$\|C_\varphi\| \leq \|C_{\varphi_0}\| \, \|C_{\psi^{-1}}\| = \left(\frac{1 + |\varphi(0)|}{1 - |\varphi(0)|}\right)^{1/p}$$

On the other hand, applying the adjoint to the linear functional for evaluation at 0, we get

$$\|C_\varphi\| \geq \frac{\|C_\varphi^*(K_0)\|}{\|K_0\|} = \frac{\|K_{\varphi(0)}\|}{1} = \left(\frac{1}{1 - |\varphi(0)|^2}\right)^{1/p}$$

∎

On $H^p(D)$, we get very specific information about the norms of composition with inner functions. The following proof is based on the observation that on $H^p(D)$, inner functions that vanish at 0 give isometries. In fact, $H^p = [1] \oplus (zH^p)$ is a decomposition of $H^p(D)$ and composition with φ is the identity on $[1]$ and a pure shift on zH^p. Other proofs can be given that emphasize the behavior of φ inside the disk, for example, using Theorem 2.29.

THEOREM 3.8
Let φ be an inner function on the disk. If f is in $H^p(D)$, $0 < p < \infty$, then

$$\left(\frac{1 - |\varphi(0)|}{1 + |\varphi(0)|}\right)^{1/p} \|f\| \leq \|C_\varphi f\| \leq \left(\frac{1 + |\varphi(0)|}{1 - |\varphi(0)|}\right)^{1/p} \|f\| \qquad (3.1.1)$$

Moreover, for each inner function, these inequalities are best possible.

PROOF Consider first the case $p = 2$ and $\varphi(0) = 0$. Since φ is inner, $\overline{\varphi(e^{i\theta})} = \varphi(e^{i\theta})^{-1}$ almost everywhere. Thus, $\|\varphi^j\| = 1$ for all j and if $j > k$,

$$\langle \varphi^j, \varphi^k \rangle = \int_0^{2\pi} \varphi^j(e^{i\theta})\overline{\varphi^k(e^{i\theta})} \frac{d\theta}{2\pi} = \int_0^{2\pi} \varphi^{j-k}(e^{i\theta}) \frac{d\theta}{2\pi} = \varphi^{j-k}(0) = 0$$

So for $f(z) = \sum c_j z^j$ in $H^2(D)$,

$$\|f \circ \varphi\|^2 = \langle \sum c_j \varphi^j, \sum c_k \varphi^k \rangle = \sum |c_j|^2 = \|f\|^2$$

For general p, as in the proof of Corollary 2.14, if F is in $H^p(D)$ with inner–outer factorization $F = uf$, then f is non-zero and $\|F\|_p = \|f\|_p$. Moreover, by Theorem 2.25, the inner–outer factorization of $F \circ \varphi$ is $(u \circ \varphi)(f \circ \varphi)$.

$$\|F \circ \varphi\|_p = \|f \circ \varphi\|_p = \|(f \circ \varphi)^{p/2}\|_2^{2/p} = \|f^{p/2}\|_2^{2/p} = \|f\|_p = \|F\|_p$$

Now for general inner functions φ, letting $\psi(z) = (\varphi(0) - z)/(1 - \overline{\varphi(0)}z)$ and $\varphi_0 = \psi \circ \varphi$, we have φ_0 is inner, $\varphi_0(0) = 0$, and $C_\varphi = C_{\varphi_0} C_{\psi^{-1}}$. Since $\psi^{-1} = \psi$, by Theorem 3.6 and Exercise 3.1.4

$$\|C_\varphi f\| = \|C_{\varphi_0} C_{\psi^{-1}} f\| = \|C_{\psi^{-1}} f\|$$

$$\leq \left(\frac{1 + |\psi^{-1}(0)|}{1 - |\psi^{-1}(0)|} \right)^{1/p} \|f\| = \left(\frac{1 + |\varphi(0)|}{1 - |\varphi(0)|} \right)^{1/p} \|f\| \quad (3.1.2)$$

and

$$\|C_\varphi f\| = \|C_{\varphi_0} C_{\psi^{-1}} f\| = \|C_{\psi^{-1}} f\|$$

$$\geq \left(\frac{1 - |\psi^{-1}(0)|}{1 + |\psi^{-1}(0)|} \right)^{1/p} \|f\| = \left(\frac{1 - |\varphi(0)|}{1 + |\varphi(0)|} \right)^{1/p} \|f\| \quad (3.1.3)$$

To see that both inequalities in Inequality (3.1.1) are best possible for each inner function, note that since (by Theorem 3.6 and Exercise 3.1.4)

$$\|C_{\psi^{-1}}\| = \left(\frac{1 + |\varphi(0)|}{1 - |\varphi(0)|} \right)^{1/p}$$

there is a sequence f_n with

$$\lim_{n \to \infty} \|C_{\psi^{-1}} f_n\| = \left(\frac{1 + |\varphi(0)|}{1 - |\varphi(0)|} \right)^{1/p} \|f_n\|$$

Inequality (3.1.2) shows that for this sequence

$$\lim_{n \to \infty} \|C_\varphi f_n\| = \left(\frac{1 + |\varphi(0)|}{1 - |\varphi(0)|} \right)^{1/p} \|f_n\|$$

Using Inequality (3.1.3) in the same way, since the lower inequality in Theorem 3.6 is sharp, there is a sequence g_n so that

$$\lim_{n \to \infty} \|C_\varphi g_n\| = \left(\frac{1 - |\varphi(0)|}{1 + |\varphi(0)|} \right)^{1/p} \|g_n\|$$

The previous few results depended on the observation that

$$\|C_\varphi\| \geq \|C_\varphi^*(K_w)\|/\|K_w\| = \|K_{\varphi(w)}\|/\|K_w\|$$

where K_w is the reproducing kernel for w in the disk. While this is a trivial inequality, it applies to all spaces and appears to be quite powerful. For example, this inequality is the principal ingredient in the computation below of the spectral radius of C_φ on $H^p(D)$. Furthermore, we believe the following interesting question is unsolved:

Is there an analytic map φ of the unit disk into itself such that, on $H^2(D)$,

$$\|C_\varphi\| > \sup_{w \in D} \|K_{\varphi(w)}\|/\|K_w\|$$

THEOREM 3.9
Suppose φ is an analytic map of the disk into itself with Denjoy–Wolff point a. Then the spectral radius of C_φ on $H^p(D)$, $p \geq 1$, is 1 when $|a| < 1$ and $\varphi'(a)^{-1/p}$ when $|a| = 1$.

PROOF The spectral radius of C_φ is

$$\lim_{k \to \infty} \|C_\varphi{}^k\|^{1/k} = \lim_{k \to \infty} \|C_{\varphi_k}\|^{1/k}$$

By Corollary 3.7

$$\left(\frac{1}{1 - |\varphi_k(0)|^2}\right)^{1/p} \leq \|C_{\varphi_k}\| \leq \left(\frac{1 + |\varphi_k(0)|}{1 - |\varphi_k(0)|}\right)^{1/p} \leq 2^{2/p}\left(\frac{1}{1 - |\varphi_k(0)|^2}\right)^{1/p}$$

so the spectral radius of C_φ is

$$\lim_{k \to \infty}\left(\left(\frac{1}{1 - |\varphi_k(0)|^2}\right)^{1/p}\right)^{1/k} = \left(\lim_{k \to \infty}\left(\prod_2^k \frac{1 - |\varphi_{j-1}(0)|^2}{1 - |\varphi_j(0)|^2}\right)^{1/k}\right)^{1/p}$$

$$= \left(\lim_{j \to \infty} \frac{1 - |\varphi_{j-1}(0)|^2}{1 - |\varphi_j(0)|^2}\right)^{1/p}$$

if the latter limit exists.

Since $a = \lim \varphi_j(0)$, we have $\lim(1 - |\varphi_{j-1}(0)|^2)/(1 - |\varphi_j(0)|^2) = 1$ when $|a| < 1$.

When $|a| = 1$ and $\varphi'(a) < 1$, the sequence $\{\varphi_j(0)\}$ converges to a nontangentially (Lemma 2.66) which means, by the Julia–Carathéodory Theorem (Theorem 2.44), that

$$\lim_{j \to \infty} \frac{1 - |\varphi_j(0)|^2}{1 - |\varphi_{j-1}(0)|^2} = \lim_{j \to \infty} \frac{|a - \varphi_j(0)|}{|a - \varphi_{j-1}(0)|} = \varphi'(a)$$

which gives the conclusion.

Suppose now that $|a| = 1$ and $\varphi'(a) = 1$. If z_j is a sequence in D converging to a such that $\varphi(z_j)$ tends to a and $s = \lim(1 - |\varphi(z_j)|)/(1 - |z_j|)$ exists then by the Julia–Carathéodory Theorem, $s \geq \varphi'(a) = 1$. It follows that the limit supremum of $(1 - |\varphi_{j-1}(0)|^2)/(1 - |\varphi_j(0)|^2)$ is no more than 1. On the other hand, since $1 - |\varphi_k(0)|^2 \leq 1$ for each k,

$$\lim_{k \to \infty} \left(\frac{1}{1 - |\varphi_k(0)|^2} \right)^{1/k} \geq 1$$

Therefore,

$$\lim_{k \to \infty} \left(\frac{1}{1 - |\varphi_k(0)|^2} \right)^{1/k} = 1 = \varphi'(a)$$

and the conclusion follows in this case also. ∎

COROLLARY 3.10
If φ satisfies the hypotheses of Theorem 3.9 and $|a| = 1$ then $\|C_\varphi\| \geq \varphi'(a)^{-1/p}$.

PROOF For any bounded operator the spectral radius is no more than the norm. ∎

The corollary is sometimes an improvement over the lower bound in Theorem 3.7. If $\varphi(z) = sz + (1 - s)$ for $0 < s < 1$, then the Denjoy–Wolff point is 1 and $\varphi'(1) = s$. We see that

$$\frac{1}{1 - |\varphi(0)|^2} = \frac{1}{2s - s^2} = \frac{1}{s}\frac{1}{2 - s} < \frac{1}{s} = \varphi'(a)^{-1}$$

(We will see later that, on $H^2(D)$, C_φ^* is subnormal, so the norm of C_φ is its spectral radius.)

Exercises

3.1.1 In Theorem 3.5 replace the hypothesis "for each $q > 1$ there exists $\kappa = \kappa(q)$ satisfying $\nu(s) \leq \kappa\nu(t)$ whenever $s \leq qt$" by the hypothesis "for each $0 < q < 1$ there exists $\kappa = \kappa(q)$ satisfying $\nu(s) \leq \kappa\nu(t)$ whenever $s \geq qt$" and show that the conclusion still holds. Note that this applies to $\nu(t) = t^\gamma$ for $-1 < \gamma < 0$.

3.1.2 Given b in D find a trivial analytic map φ of the unit disk into itself with $\varphi(0) = b$ and for which

$$\|C_\varphi\| = \left(\frac{1}{1 - |\varphi(0)|^2} \right)^{1/p}$$

as an operator on $H^p(D)$ to show that Corollary 3.7 is sharp.

3.1.3   (a)   Discover and prove a sharp norm inequality for the Bergman space $A^2(D)$, analogous to that of Corollary 3.7 for the Hardy space.

      (b)   Generalize Corollary 3.7 for the weighted Bergman spaces $A_\alpha^2(D)$, where $\alpha > -1$.

3.1.4   (a)   Extend Theorem 3.6 to the case $0 < p < 1$. Hint: To show that

$$\|C_\psi\| = \left(\frac{1 + |\psi(0)|}{1 - |\psi(0)|} \right)^{1/p}$$

      when $0 < p < 1$, use the corresponding result for $p = 2$ and a factorization argument.

      (b)   Extend Corollary 3.7 to the case $0 < p < 1$.

3.1.5   Consider the weighted Hardy space $H^2(\beta)$ where $\beta(j) = 2^j$. Show that this space is not invariant under automorphisms of the unit disk. Can you find a geometric reason why this should be so?

3.1.6   Let $\mathcal{H}$ be the space of Exercise 2.1.14. Show that $\varphi(z) = (3z + 3)/(z + 5)$ maps the unit disk into itself and that C_φ is an invertible composition operator but that C_φ^{-1} is not a composition operator; that is, show that there is no map ψ of the unit disk to itself so that $C_\psi C_\varphi = C_\varphi C_\psi = I$.

3.1.7   Use the description of $H^p(D)$ as those analytic functions whose p^{th} power has a harmonic majorant (Theorem 2.3) to show that for any analytic map φ of the disk to itself, C_φ is bounded $H^p(D)$ and

$$\|C_\varphi\| \leq \left(\frac{1 - |\varphi(0)|}{1 + |\varphi(0)|} \right)^{1/p}$$

3.1.8   Find the spectral radius of C_φ acting on the Bergman space $A^2(D)$.

3.1.9   For c in the unit disk, let

$$\varphi(z) = \lambda \frac{z - c}{1 - \bar{c}z} \quad \text{and} \quad \psi(z) = \left(\frac{1 - |c|^2}{(1 - \bar{c}z)^2} \right)^{1/p}$$

      Show that the weighted composition operator $(Wf)(z) = \psi(z)f(\varphi(z))$ is an invertible isometry on $H^p(D)$.

3.1.10   (a)   Show that the norm of a function $f = \sum_0^\infty a_n z^n$ in the Bergman space $A_\alpha^2(D)$ is *equal* (not just equivalent) to

$$\sum_0^\infty |a_n|^2 \beta(n)^2$$

      where

$$\beta(n)^2 = \frac{n!}{(\alpha + 1)(\alpha + 2) \cdots (\alpha + n + 1)}$$

      (b)   Use the hypergeometric function identity ([Er53, p. 56])

$$1 + \sum_{k=1}^\infty \binom{s}{k} \binom{t}{k} \frac{k!}{r(r+1) \cdots (r + k - 1)} = F(-s, -t; r; 1)$$

$$= \frac{\Gamma(r)\Gamma(r + s + t)}{\Gamma(r + s)\Gamma(r + t)}$$

      valid at least for $r > 0$ and $r + s + t > 0$ to show that there exist positive numbers a_u with

$$\langle (1 - z)^u, (1 - z)^v \rangle_\alpha = a_u a_v \langle (1 - z)^{u + (\alpha+1)/2}, (1 - z)^{v + (\alpha+1)/2} \rangle$$

for all $u, v > -(\alpha + 2)/2$, where the inner products are in $A_\alpha^2(D)$ and $H^2(D)$ respectively. Conclude there exists a unique unitary operator

$$U : A_\alpha^2(D) \to H^2(D)$$

with

$$U(1 - z)^u = a_u(1 - z)^{u+(\alpha+1)/2}$$

(c) Let $0 < s < 1$ and set $\varphi(z) = sz + 1 - s$. Show that U as above induces a unitary equivalence between C_φ acting on $A_\alpha^2(D)$ and $s^{-(\alpha+1)/2}C_\varphi$ acting on $H^2(D)$. Hint: Verify the equation $UC_\varphi = s^{-(\alpha+1)/2}C_\varphi U$ on the spanning set $\{(1 - z)^u\}$.

Notes

Theorem 3.2 and Corollary 3.3 are proved in [Co90b, pp. 30, 31]. Exercise 3.1.9 is a special case of the result of F. Forelli [Fo64] identifying the isometries of $H^p(D)$ for $p \neq 2$: they are weighted composition operators. Exercise 3.1.10 is taken from T. L. Kriete and H. C. Rhaly [KrR87].

3.2 Compactness and essential norms in classical spaces on the disk

A linear operator on a Banach space is compact if the image of the unit ball under the operator has compact closure. For composition operators on many of the spaces of interest to us this definition has a useful reformulation. We state this for the Hardy and standard weight Bergman spaces although it should be clear that the same characterization will hold for many other spaces, both in one and several variables (see for example Exercise 3.2.1). The statement of Proposition 3.11 includes the case $p < 1$; the definition of a compact operator on these spaces is the same.

PROPOSITION 3.11
Let $\mathcal{X}$ be $H^p(D)$ or $A_\alpha^p(D)$ for $0 < p \leq \infty$ and $\alpha > -1$. Then C_φ is compact on $\mathcal{X}$ if and only if whenever $\{f_n\}$ is bounded in $\mathcal{X}$ and $f_n \to 0$ uniformly on compact subsets of D then $f_n \circ \varphi \to 0$ in $\mathcal{X}$.

PROOF Assume C_φ is compact and suppose $\{f_n\}$ is bounded in $\mathcal{X}$ with $f_n \to 0$ uniformly on compact subsets of D. Then $\{f_n \circ \varphi\}$ has a subsequence which converges in $\mathcal{X}$; since $\{\varphi(z)\}$ is a compact set, $f_n(\varphi(z))$ converges to 0 for each z in the disk and the limit function is necessarily 0. Since this is true for any subsequence of the f_n's, we see that the limit of $f_n \circ \varphi$ in $\mathcal{X}$ is 0.

Conversely, let $\{g_n\}$ be any bounded sequence in $\mathcal{X}$. Since $\{g_n\}$ is a normal family we may extract a subsequence $\{g_{n_k}\}$ converging uniformly on compact sets to some function g. It is easy to see that g must be in $\mathcal{X}$. Then $\{g_{n_k} - g\}$ is a bounded sequence in $\mathcal{X}$ converging almost uniformly to 0 and the hypothesis guarantees that $g_{n_k} \circ \varphi \to g \circ \varphi$ in $\mathcal{X}$. Thus C_φ is compact. ∎

Much of the study of compact composition operators has been driven by a desire to relate compactness of C_φ to geometric properties of φ. The first and simplest result in this spirit is an observation which follows immediately from Proposition 3.11: On $H^p(D)$ or $A^p_\alpha(D)$ (as well as many other spaces) if $\overline{\varphi(D)} \subset D$ then C_φ is compact. For $p = \infty$ the converse is true as well: if C_φ is compact on $H^\infty(D)$ then $\overline{\varphi(D)} \subset D$ (Exercise 3.2.2). This provides an extreme example of the basic philosophy of compact composition operators: if $\varphi(D)$ is small in an appropriate sense (usually involving how the values of φ are allowed to approach ∂D) then C_φ will be compact.

When C_φ is compact on $H^p(D)$ the map φ cannot have radial limits of modulus one on a set of positive Lebesgue measure. To see this apply of characterization of Proposition 3.11 to the functions $f_n = z^n$, noting that $C_\varphi(z^n) = \varphi^n$.

Further information on compactness of composition operators will come from a Carleson measure characterization, which we turn to next. Given an analytic map φ of the disk into itself, define the pull-back measure μ on $\overline{D}$ by $\mu(E) = \sigma((\varphi^*)^{-1}(E))$, where σ is normalized Lebesgue measure on the circle and E is a subset of the closed disk.

THEOREM 3.12

Let $0 < p < \infty$.

(1) C_φ *is bounded on* $H^p(D)$ *if and only if* $\mu(\mathcal{S}(\zeta, h)) = O(h)$ *for all* ζ *in* ∂D *and* $0 < h < 1$.

(2) C_φ *is compact on* $H^p(D)$ *if and only if* $\mu(\mathcal{S}(\zeta, h)) = o(h)$ *as* $h \to 0$, *uniformly in* ζ *in* ∂D.

Since we already know that every C_φ is bounded on $H^p(D)$ (Corollary 3.7), the interest in (1) is that it tells us that $\mu(\mathcal{S}(\zeta, h)) = O(h)$ for every φ, a fact which will prove useful in the study of composition operators on $H^p(B_N)$ for $N > 1$. An important consequence of (2) is that compactness of C_φ on $H^p(D)$ is independent of p, $0 < p < \infty$.

PROOF   Let f be in $H^p(D)$. Using Proposition 2.25 and the change of variables formula from measure theory [Hal74, p. 163], we have

$$\int_{\partial D} |(f \circ \varphi)^*|^p \, d\sigma = \int_{\partial D} |f^* \circ \varphi^*|^p \, d\sigma = \int_{\overline{D}} |f^*|^p \, d\mu$$

Thus (1) follows immediately from Theorem 2.35.

For (2) first assume $\mu(\mathcal{S}(\zeta, h)) \neq o(h)$ so that we may find ζ_n in ∂D, positive numbers h_n decreasing to 0, and $\beta > 0$ with $\mu(\mathcal{S}(\zeta_n, h_n)) \geq \beta h_n$. Set $a_n = (1 - h_n)\zeta_n$ and $f_n = (1 - \overline{a_n}z)^{-4/p}$. Since $(1 - \overline{a_n}z)^{-2} = \sum k(\overline{a_n}z)^{k-1}$ we have

$$\|f_n\|_p^p \sim (1 - |a_n|^2)^{-3} \sim h_n^{-3}$$

as $n \to \infty$. (This estimate also follows from Exercise 2.1.4.) Thus if $g_n = f_n/\|f_n\|_p$ we have $g_n \to 0$ uniformly on compact subsets of D. But

$$\|g_n \circ \varphi\|_p^p = \int_{\partial D} |(g_n \circ \varphi)^*|^p \, d\sigma = \int_{\overline{D}} |g_n|^p \, d\mu \geq \|f_n\|_p^{-p} \int_{\mathcal{S}(\zeta_n, h_n)} |f_n|^p \, d\mu$$

If z is in $\mathcal{S}(\zeta_n, h_n)$

$$|1 - \overline{a_n}z| = |1 - (1 - h_n)\overline{\zeta_n}z| \leq |\overline{\zeta_n}(\zeta_n - z)| + |h_n\overline{\zeta_n}z| \leq 2h_n$$

so we have $|f_n|^p \geq (2h_n)^{-4}$ on $\mathcal{S}(\zeta_n, h_n)$. Thus $\|g_n \circ \varphi\|_p^p$ is bounded away from zero and C_φ cannot be compact by Proposition 3.11.

For the converse direction of (2) assume $\mu(\mathcal{S}(\zeta, h)) = o(h)$ uniformly in ζ. Then we also have $\mu(W(\zeta, h)) = o(h)$ where, here, $W(\zeta, h)$ are the Carleson windows in $\overline{D}$

$$W(\zeta, h) = \{re^{i\theta} \in \overline{D} : 1 - h < r \leq 1, |\theta - t| \leq h\}$$

where $\zeta = e^{it}$. Given any $\epsilon > 0$ we may find h_0 so that $\mu(W(\zeta, h)) \leq \epsilon h$ for all $h \leq h_0$. If μ' is the restriction of μ to the annulus $1 - h_0 < |z| \leq 1$, we claim that μ' is a Carleson measure with

$$\mu'(W(\zeta, h)) \leq 2\epsilon h \qquad (3.2.1)$$

Notice that Inequality (3.2.1) clearly holds for $h \leq h_0$; indeed $\mu'(W(\zeta, h)) \leq \epsilon h$ in this case. If $h > h_0$ we may cover $W(\zeta, h) \cap \{1 - h_0 < |z| \leq 1\}$ by at most $[h/h_0] + 1 < 2h/h_0$ windows $W(\eta_j, h_0)$ for appropriately chosen η_j in ∂D (see Figure 3.1). Then

$$\mu'(W(\zeta, h)) \leq \mu'(\cup_j W(\eta_j, h_0)) \leq \sum_j \mu'(W(\eta_j, h_0))$$

$$\leq \epsilon h_0 (2\frac{h}{h_0}) = 2\epsilon h$$

as desired.

We finish with an appeal to Proposition 3.11. Suppose $\{f_n\}$ is bounded in $H^p(D)$ and $f_n \to f$ uniformly on compact subsets of D. Then

$$\int_{\partial D} |(f_n - f)^* \circ \varphi^*|^p \, d\sigma = \int_{\overline{D}} |(f_n - f)^*|^p \, d\mu$$

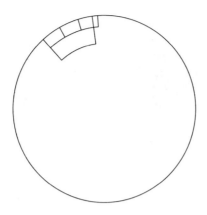

FIGURE 3.1

$$W(\zeta, h) \cap \{1 - h_0 < |z| \le 1\} \subset \cup_j W(\eta_j, h_0).$$

$$= \int_{\overline{D} \setminus (1-h_0)D} |(f_n - f)^*|^p \, d\mu + \int_{(1-h_0)\overline{D}} |(f_n - f)|^p \, d\mu \qquad (3.2.2)$$

The second integral on the right hand side of Equation (3.2.2) can be made arbitrarily small by choosing n sufficiently large. The first integral is

$$\int_{\overline{D}} |(f_n - f)^*|^p \, d\mu'$$

and by Inequality (3.2.1) and the remark following Theorem 2.37, it is bounded by $C\epsilon \|f_n - f\|_p^p$ for some absolute constant C. Since ϵ can be chosen arbitrarily small we are done. ∎

Analogous Carleson measure characterizations for the weighted Bergman spaces $A_\alpha^p(D)$ or weighted Dirichlet spaces $\mathcal{D}_\alpha$ are outlined in Exercises 3.2.6 and 3.2.9.

We can use this theorem to compare two examples. Fix $0 < s < 1$ and $0 < b < 1$ and let $\varphi(z) = sz + (1-s)$ and $\psi(z) = 1 - (1-z)^b$. Then φ and ψ are conformal maps of D onto regions whose closures intersect the unit circle only at 1. The image $\varphi(D)$ is a disk internally tangent to the circle at 1 while $\psi(D)$ is contained in a sector with vertex at 1 and angle $b\pi$ there (see Figure 3.2).

Theorem 3.12 provides one of several ways to see that on $H^p(D)$, C_ψ is compact while C_φ is not: since $\varphi^{-1}\mathcal{S}(1, h) = \mathcal{S}(1, h/s)$, condition (2) cannot hold. By comparison, $\psi^{-1}\mathcal{S}(1, h) = \mathcal{S}(1, h^{1/b})$ and thus $\mu\mathcal{S}(1, h) = o(h)$ since $b < 1$. Since $\psi(D)$ is contained in a nontangential approach region near 1 it is easy to extend this to see that $\mu(\mathcal{S}(\zeta, h)) = o(h)$ uniformly in ζ in ∂D. See Lemma 6.2 for the details to this argument.

As we will see, from the point of view of compactness the important feature distinguishing these two examples is the angular derivative. It is possible to show (see Exercise 3.2.10) that for a map with finite angular derivative at some point the

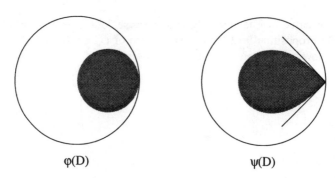

$$\varphi(D) \qquad\qquad\qquad \psi(D)$$

FIGURE 3.2

Maps inducing non-compact and compact operators.

pull-back measure can never satisfy the little-o Carleson condition of Theorem 3.12 (2). This is a special case of a more general principle which gives a lower estimate for the essential norm of a composition operator on the classical spaces. The *essential norm of an operator*, denoted $\| \cdot \|_e$, is its distance in the operator norm from the compact operators. The essential norm is 0 exactly when the operator is compact. The relation of the essential norm to the Calkin algebra is discussed some in Section 3.4.

PROPOSITION 3.13

Consider a weighted Hardy space $H^2(\beta)$ where $\sum \beta(n)^{-2} = +\infty$. Let K_w be the kernel function for evaluation at w. Then

$$\|C_\varphi\|_e \geq \limsup_{|w| \to 1} \frac{\|K_{\varphi(w)}\|}{\|K_w\|}$$

PROOF   Let w_j be in D tending to ∂D. The hypothesis on the weight sequence β ensures that $k_j = K_{w_j}/\|K_{w_j}\|$ tends to zero weakly as j approaches infinity (Theorem 2.17). If Q is an arbitrary compact operator on $H^2(\beta)$ we have $Q^*(k_j) \to 0$. Since $\|C_\varphi\|_e = \inf\{\|C_\varphi - Q\| : Q \text{ is compact}\}$ and for Q compact,

$$\|C_\varphi - Q\| \geq \limsup_{j \to \infty} \|(C_\varphi - Q)^* k_j\| = \limsup_{j \to \infty} \|C_\varphi{}^*(k_j)\| = \limsup_{j \to \infty} \frac{\|K_{\varphi(w_j)}\|}{\|K_{w_j}\|}$$

the result follows.   ∎

Of course $H^2(D)$ and $A_\alpha^2(D)$, as well as many other spaces of interest to us, satisfy the hypothesis of Proposition 3.13.

COROLLARY 3.14

If C_φ is compact on $H^p(D)$ or $A_\alpha^p(D)$ then φ has no finite angular derivative at any point of ∂D.

PROOF   By virtue of Theorem 3.12 and Exercise 3.2.6 it is enough to assume $p = 2$. By Proposition 3.13 the essential norm of C_φ on $H^2(D)$ or $A_\alpha^2(D)$ is bounded below by the limit supremum of a power of

$$\frac{1 - |w|^2}{1 - |\varphi(w)|^2}$$

as $|w| \to 1$, where the power depends on the particular space. This lower bound is zero only if φ has no finite angular derivative.   ∎

We next look at a method for obtaining *upper* estimates on the essential norm of a composition operator. The basic idea of this method is due to J. H. Shapiro, who developed it to study essential norms on $H^2(D)$ and $A_\alpha^2(D)$. We begin a bit more generally and consider any weighted Hardy space $\mathcal{H}$ of analytic functions on D that can be described as a weighted Dirichlet space. That is, we require that the norm of a function f in $\mathcal{H}$ be given by, or at least be equivalent to,

$$\left(|f(0)|^2 + \int_D |f'(z)|^2 H(|z|) \frac{dA(z)}{\pi} \right)^{1/2}$$

where $H(r)$ is non-negative, continuous on $(0, 1)$ and integrable on $[0, 1)$. *This will be our standing assumption on $\mathcal{H}$ for the remainder of this section.* Recall that, by Theorem 2.30 the choice $H(r) = |\log r^2|$ or $H(r) = 1 - r^2$ gives $H^2(D)$, up to an equivalent norm, while $H(r) = |\log r^2|^{\alpha+2}$ or $(1 - r^2)^{\alpha+2}$ gives the standard weighted Bergman space $A_\alpha^2(D)$ (Exercise 2.1.5). A general weighted Bergman space

$$A_G^2(D) \equiv \{ f \text{ analytic on } D : \int_D |f(z)|^2 G(|z|) \frac{dA}{\pi} < \infty \}$$

defined from an arbitrary positive, continuous function $G(r)$ with $\int_D G(|z|)\, dA$ finite can be described as a weighted Dirichlet space, with equivalent norm, by setting

$$H(r) = \int_r^1 \left(\int_x^1 G(s)\, ds \right) dx$$

for $0 \leq r < 1$. The details are left to the reader in Exercise 3.2.11.

Let R_n be the orthogonal projection of $\mathcal{H}$ onto $z^n \mathcal{H}$ and $Q_n = I - R_n$, that is, for $f = \sum_{k=0}^\infty a_k z^k$ in $\mathcal{H}$ let $(R_n f)(z) = \sum_{k=n}^\infty a_k z^k$ and set $(Q_n f)(z) = \sum_{k=0}^{n-1} a_k z^k$. We need two simple estimates.

PROPOSITION 3.15
For each r, $0 < r < 1$, and f in $\mathcal{H}$

(1) $|(R_n f)(w)| \leq \|f\| \left(\sum_{k=n}^\infty r^{2k} \beta(k)^{-2} \right)^{1/2}$ *for $|w| \leq r$*

(2) $|(R_n f)'(w)| \leq \|f\| \left(\sum_{k=n}^\infty k^2 r^{2k-2} \beta(k)^{-2} \right)^{1/2}$ *for $|w| \leq r$*

where $\beta(k) = \|z^k\|_{\mathcal{H}}$.

PROOF   If K_w is the kernel for evaluation at w in D,

$$|R_n f(w)| = |\langle R_n f, K_w \rangle| = |\langle f, R_n K_w \rangle|$$

$$\leq \|f\| \|R_n K_w\|$$

$$\leq \|f\| \left(\sum_{k=n}^{\infty} \frac{r^{2k}}{\beta(k)^2} \right)^{1/2}$$

which gives (1). The second inequality follows from a similar argument, using the kernel function for evaluation of the derivative.   ∎

Our interest in these projections comes from the next lemma.

LEMMA 3.16

If C_φ is bounded on $\mathcal{H}$, then $\|C_\varphi\|_e = \lim_{n \to \infty} \|C_\varphi R_n\|$.

PROOF   Note first that $\|C_\varphi R_n\|$ is a non-increasing sequence of non-negative numbers, so its limit exists. Since $(R_n + Q_n)f = f$ and Q_n is compact, we have for each n

$$\|C_\varphi\|_e \leq \|C_\varphi R_n + C_\varphi Q_n\|_e \leq \|C_\varphi R_n\|_e \leq \|C_\varphi R_n\|$$

so that

$$\|C_\varphi\|_e \leq \lim_{n \to \infty} \|C_\varphi R_n\|$$

For the other direction, let V be an arbitrary compact operator on $\mathcal{H}$. Then

$$\|C_\varphi - V\| \geq \|(C_\varphi - V)R_n\| = \|C_\varphi R_n - V R_n\| \geq \|C_\varphi R_n\| - \|V R_n\|$$

We claim that $\|V R_n\| \to 0$ as $n \to \infty$. Of course, $\|V R_n\| = \|(V R_n)^*\| = \|R_n V^*\|$. Since V^* is compact, the image of the unit ball of $\mathcal{H}$ under V^* is relatively compact. But $R_n V^*$ converges pointwise to 0 and $\|R_n\| = 1$, so $R_n V^*$ converges uniformly to zero on the unit ball of $\mathcal{H}$. This verifies the claim, and we have obtained

$$\|C_\varphi\|_e \geq \lim_{n \to \infty} \|C_\varphi R_n\|$$

which completes the proof.   ∎

Lemma 3.16 is our starting point for obtaining, on our general weighted Dirichlet space $\mathcal{H}$, an upper estimate on $\|C_\varphi\|_e$ when C_φ is bounded on $\mathcal{H}$. We have

$$\|C_\varphi\|_e = \lim_{n \to \infty} \left\{ \sup_{\|f\| \leq 1} \|C_\varphi R_n f\| \right\}$$

Taking f in the unit ball of $\mathcal{H}$, $\|C_\varphi R_n f\|^2$ is comparable to

$$|R_n f \circ \varphi(0)|^2 + \int_D |(R_n f \circ \varphi)'(z)|^2 H(|z|) \, \frac{dA(z)}{\pi}$$

The first term in this does not exceed

$$\sum_{k=n}^{\infty} \frac{1}{\beta^2(k)} |\varphi(0)|^{2k}$$

which tends to 0 as n approaches infinity. The Area Formula (Theorem 2.32) gives for the second term

$$\int_{\varphi(D)} |(R_n f)'(w)|^2 \left(\sum_{j \geq 1} H(|z_j(w)|) \right) \frac{dA(w)}{\pi}$$

where $\{z_j(w)\}$ are the zeros of $\varphi(z) - w$. If we fix an arbitrary r in $(0,1)$ this integral is dominated by

$$\sup_{|w| \leq r} |(R_n f)'(w)|^2 \int_{\varphi(D)} \sum_{j \geq 1} H(|z_j(w)|) \, \frac{dA(w)}{\pi}$$

$$+ \int_{\varphi(D) \setminus rD} |(R_n f)'(w)|^2 \sum_{j \geq 1} H(|z_j(w)|) \, \frac{dA(w)}{\pi}$$

Using the Area Formula again and Proposition 3.15, the first term of this sum is at most

$$\|f\|^2 \left(\sum_{k=n}^{\infty} \frac{k^2}{\beta^2(k)} r^{2k-2} \right) \int_D |\varphi'(z)|^2 H(|z|) \, \frac{dA(z)}{\pi}$$

which is dominated by a multiple of

$$\|f\|^2 \left(\sum_{k=n}^{\infty} \frac{k^2}{\beta^2(k)} r^{2k-2} \right) \|\varphi\|^2$$

As n approaches infinity, this tends to 0. We are left to consider the supremum of

$$\int_{\varphi(D) \setminus rD} |(R_n f)'(w)|^2 \left(\sum_{j \geq 1} H(|z_j(w)|) \right) \frac{dA(w)}{\pi}$$

as f ranges over the unit ball of $\mathcal{H}$. This will be dominated by the supremum of

$$\int_{\varphi(D) \setminus rD} |f'(w)|^2 \left(\sum_{j \geq 1} H(|z_j(w)|) \right) \frac{dA(w)}{\pi} \tag{3.2.3}$$

as f ranges over the same set, since in the second case the supremum is taken over a larger set. Set

$$\gamma_r = \sup_{r \le |w| < 1} \frac{\sum H(|z_j(w)|)}{H(|w|)}$$

Then the integral (3.2.3) is dominated by

$$\gamma_r \int_{\varphi(D) \setminus rD} |f'(w)|^2 H(|w|) \, dA(w)/\pi \le C\gamma_r \|f\|^2 \le C\gamma_r$$

where C is a constant, independent of f, which arises from the fact that

$$\left(|f(0)|^2 + \int |f'(z)|^2 H(|z|) \, dA(z)/\pi \right)^{1/2}$$

gives an *equivalent* norm for f in $\mathcal{H}$. Since this holds for arbitrary $0 < r < 1$ we have arrived at our desired estimate

$$\|C_\varphi\|_e^2 \le C \lim_{r \to 1} \gamma_r \tag{3.2.4}$$

where γ_r is as just defined.

Let us interpret this result in the case $\mathcal{H} = H^2(D)$. Here we may take $H(r) = |\log r^2|$ and $C = 1$, by the Littlewood–Paley identity (Theorem 2.30), and

$$\gamma_r = \sup_{r \le |w| < 1} \frac{\sum_{j \ge 1} -\log |z_j(w)|}{-\log |w|} = \sup_{r \le |w| < 1} \frac{N_\varphi(w)}{-\log |w|}$$

where $N_\varphi(w)$ is the Nevanlinna counting function. We state this result as a theorem.

THEOREM 3.17
If $\varphi : D \to D$ is analytic, then on $H^2(D)$

$$\|C_\varphi\|_e^2 \le \limsup_{|w| \to 1} \frac{N_\varphi(w)}{-\log |w|}$$

In the case of $H^2(D)$ we can improve the *lower* bound estimate of Proposition 3.13 to show that $\|C_\varphi\|_e \ge \limsup_{|w| \to 1} \frac{N_\varphi(w)}{-\log |w|}$. Once this is verified we can put it together with Theorem 3.17 to obtain a precise formula for essential norms on $H^2(D)$:

$$\|C_\varphi\|_e = \limsup_{|w| \to 1} \frac{N_\varphi(w)}{-\log |w|}$$

for $H^2(D)$. This will of course give an exact characterization of the compact composition operators on $H^2(D)$ as those with symbol φ satisfying

$$\lim_{|w| \to 1} \frac{N_\varphi(w)}{-\log |w|} = 0$$

To obtain our improved lower bound estimate on $\|C_\varphi\|_e$ we need a lemma that shows that the average value of the Nevanlinna counting function over disks dominates the value of the function at the center. The basic idea of this lemma is that the counting function $N_\varphi(w)$, while itself not necessarily subharmonic, is a pointwise limit of an increasing sequence of "partial" counting functions $N_\varphi(w, r)$ which are subharmonic in $C \setminus \{\varphi(0)\}$, and hence $N_\varphi(w)$ inherits from this sequence an area sub-mean-value property.

LEMMA 3.18
If Δ is a disk in C not containing $\varphi(0)$, with center a, then

$$N_\varphi(a) \leq \frac{1}{|\Delta|} \int_\Delta N_\varphi \, dA/\pi$$

where $|\Delta|$ is the normalized area measure of Δ, $|\Delta| = \int \chi_\Delta(z) \, dA(z)/\pi$.

PROOF Define, for $0 \leq r < 1$, $N_\varphi(w, r) = \sum \log(r/|z_j(w)|)$, where the sum is taken over all $z_j(w)$ in $\varphi^{-1}(w)$ with $|z_j(w)| < r$, including multiplicities. We consider $N_\varphi(w, r)$ to be defined on $C \setminus \{\varphi(0)\}$ by the understanding that $N_\varphi(w, r) = 0$ if w is not in $\varphi(rD)$. Of course $N_\varphi(w) = \lim_{r \to 1} N_\varphi(w, r)$. The crucial fact is that for each $0 \leq r < 1$, $N_\varphi(w, r)$ is subharmonic in $C \setminus \{\varphi(0)\}$. This follows from Jensen's formula: if $w \neq \varphi(0)$

$$N_\varphi(r, w) = \frac{1}{2\pi} \int_0^{2\pi} \log|\varphi(re^{i\theta}) - w| \, d\theta - \log|\varphi(0) - w|$$

$$= \int \log|\zeta - w| \, d\mu(\zeta) - \log|\varphi(0) - w|$$

where μ is the Borel probability measure supported on $\varphi_r(\partial D)$ defined by $\mu = \sigma \varphi_r^{-1}$ for $d\sigma = d\theta/(2\pi)$. From this last line we recognize that $N_\varphi(r, w)$ is subharmonic in $C \setminus \{\varphi(0)\}$. If Δ is any disk with center a not containing $\varphi(0)$ the Monotone Convergence Theorem and subharmonicity of $N_\varphi(r, w)$ give

$$N_\varphi(a) = \lim_{r \to 1} N_\varphi(r, a) \leq \lim_{r \to 1} \frac{1}{|\Delta|} \int_\Delta N_\varphi(r, w) \, dA(w)/\pi$$

$$= \frac{1}{|\Delta|} \int_\Delta N_\varphi(w) \, dA(w)/\pi$$

as desired. ∎

The same argument actually proves a more general result: if f is analytic in the disk and if Δ is any disk not containing $\{f^{-1}(\varphi(0))\}$ and centered at a, then

$$N_\varphi(f(a)) \leq \frac{1}{|\Delta|} \int_\Delta N_\varphi(f(w)) \, dA(w)/\pi \qquad (3.2.5)$$

We leave the details to the reader.

THEOREM 3.19
For C_φ acting on $H^2(D)$

$$\|C_\varphi\|_e^2 \geq \limsup_{|w|\to 1} \frac{N_\varphi(w)}{-\log|w|}$$

PROOF We use as test functions the family of normalized reproducing kernel functions $k_a(z) = (1-|a|^2)^{1/2}/(1-\bar{a}z)$. Since these unit vectors tend to 0 uniformly on compact subsets of D as $|a|\to 1$, if Q is an arbitrary compact operator on $H^2(D)$, $\|Qk_a\|\to 0$ as $|a|\to 1$. Thus

$$\|C_\varphi\|_e^2 \geq \limsup_{|a|\to 1}\|C_\varphi(k_a)\|^2 = \limsup_{|a|\to 1}\|k_a\circ\varphi\|^2$$

$$= \limsup_{|a|\to 1}\left(|k_a\circ\varphi(0)|^2 + \int_D |(k_a\circ\varphi)'(z)|^2\log\frac{1}{|z|^2}\,dA(z)/\pi\right)$$

$$= \limsup_{|a|\to 1}\int_{\varphi(D)}|k_a'(w)|^2(\sum\log\frac{1}{|z_j(w)|^2})\,dA(w)/\pi$$

$$= \limsup_{|a|\to 1}\left(2\int_D |k_a'(w)|^2 N_\varphi(w)\,dA(w)/\pi\right)$$

where we used the fact that $|k_a\circ\varphi(0)|\to 0$ as $|a|\to 1$ and the Area Formula.

Recall the automorphism $\psi_a(z) = (a-z)/(1-\bar{a}z)$ satisfies $\psi_a\circ\psi_a(z)=z$ and $|\psi_a'(z)|^2 = (1-|a|^2)^2/(|1-\bar{a}z|^4)$. Using this we may write

$$\int_D |k_a'(w)|^2 N_\varphi(w)\,dA(w)/\pi = \int_D |a|^2(1-|a|^2)\frac{1}{|1-\bar{a}w|^4}N_\varphi(w)\,dA(w)/\pi$$

$$= \frac{|a|^2}{(1-|a|^2)}\int_D |\psi_a'(w)|^2 N_\varphi(w)\,dA(w)/\pi$$

$$= \frac{|a|^2}{1-|a|^2}\int_D N_\varphi(\psi_a(w))\,dA(w)/\pi$$

Fix r with $0<r<1$. For a sufficiently close to ∂D (depending on r), $|\psi_a(\varphi(0))|>r$. For these values of a, we may use Inequality (3.2.5) to obtain

$$\frac{|a|^2}{(1-|a|^2)}\int_D N_\varphi(\psi_a(w))\,dA(w)/\pi \geq \frac{|a|^2}{(1-|a|^2)}\int_{rD} N_\varphi(\psi_a(w))\,dA(w)/\pi$$

$$\geq \frac{|a|^2}{(1-|a|^2)}r^2 N_\varphi(a)$$

So we have, for all a sufficiently close to ∂D

$$2\int_D |k_a'(w)|^2 N_\varphi(w)\,dA(w)/\pi \geq \frac{2|a|^2}{(1-|a|^2)}r^2 N_\varphi(a)$$

$$= |a|^2 r^2 \frac{(-\log|a|^2)}{(1-|a|^2)} \frac{N_\varphi(a)}{(-\log|a|)}$$

As $|a| \to 1$

$$|a|^2 \left(\frac{-\log|a|^2}{1-|a|^2} \right) \to 1$$

so

$$\limsup_{|a| \to 1} \left(2 \int_D |k_a'(w)|^2 N_\varphi(w)\, dA(w)/\pi \right) \geq r^2 \limsup_{|a| \to 1} \frac{N_\varphi(a)}{-\log|a|}$$

Since this is true for all $r < 1$, we have

$$\limsup_{|a| \to 1} \left(2 \int_D |k_a'(w)|^2 N_\varphi(w)\, dA(w)/\pi \right) \geq \limsup_{|a| \to 1} \frac{N_\varphi(w)}{-\log|a|}$$

from which the conclusion follows. ∎

We summarize the results of Theorem 3.17 and Theorem 3.19:

THEOREM 3.20
If φ is an analytic map of the unit disk into itself, then, for C_φ acting on $H^2(D)$, we have

$$\|C_\varphi\|_e^2 = \limsup_{|w| \to 1} \frac{N_\varphi(w)}{-\log|w|}$$

In particular C_φ is compact if and only if

$$\lim_{|w| \to 1} \frac{N_\varphi(w)}{-\log|w|} = 0$$

COROLLARY 3.21
If φ is univalent, or even just of bounded multiplicity, with no finite angular derivative at any point of ∂D, then C_φ is compact on $H^p(D)$.

PROOF It suffices to assume $p = 2$. Given w in $\varphi(D)$ let $z(w)$ denote a preimage of w of minimum modulus. As w approaches ∂D then so does $z(w)$. If M is an upper bound on the multiplicity of φ we have $N_\varphi(w) \leq M(-\log|z(w)|)$. Since $-\log|w|$ and $1-|w|$ are comparable for w near the unit circle there exists $M' < \infty$ so that

$$\frac{N_\varphi(w)}{-\log|w|} \leq M' \frac{1-|z(w)|}{1-|\varphi(z(w))|}$$

By the Julia–Carathéodory Theorem (Theorem 2.44), if φ has no finite angular derivative, the right hand side approaches zero as w tends to the circle. ∎

The conclusion of Corollary 3.21 does not hold if the hypothesis of bounded multiplicity is removed. Indeed, there exist inner functions with no finite angular derivative at any point of ∂D (see Exercise 3.2.7), but as we observed following Proposition 3.11, no map with radial limits of modulus one on a set of positive measure can induce a compact composition operator on $H^p(D)$. Exercise 3.2.8 elaborates on this example.

We finish this section by using our general upper estimate for $\|C_\varphi\|_e$ from Inequality (3.2.4) to connect the compactness question to angular derivatives for the standard weight Bergman spaces. We will use the notation $\|C_\varphi\|_{e,\alpha}$ for the essential norm on $A^2_\alpha(D)$. We have $\|C_\varphi\|^2_{e,\alpha} \leq C \lim_{r \to 1} \gamma_r$ where

$$\gamma_r = \sup_{r \leq |w| < 1} \frac{\sum_{j \geq 1} H(|z_j(w)|)}{H(|w|)}$$

and we may take $H(r) = |\log r^2|^{\alpha+2}$. For w in $\varphi(D)$, let $z(w)$ be a point of $\{z_j(w)\}$ with minimum modulus. Choose $\{w_n\}$ so that

$$\lim_{r \to 1} \gamma_r = \lim_{n \to \infty} \frac{\sum_{j \geq 1} H(|z_j(w_n)|)}{H(|w_n|)}$$

For each j,

$$H(|z_j(w_n)|) = 2^{\alpha+2} \left(\log \frac{1}{|z_j(w_n)|} \right)^{\alpha+1} \log \frac{1}{|z_j(w_n)|}$$

$$\leq 2^{\alpha+2} \left(\log \frac{1}{|z(w_n)|} \right)^{\alpha+1} \log \frac{1}{|z_j(w_n)|}$$

and hence

$$\sum_{j \geq 1} H(|z_j(w_n)|) \leq 2^{\alpha+2} \left(\log \frac{1}{|z(w_n)|} \right)^{\alpha+1} \sum_{j \geq 1} \log \frac{1}{|z_j(w_n)|}$$

$$= 2^{\alpha+2} \left(\log \frac{1}{|z(w_n)|} \right)^{\alpha+1} N_\varphi(w_n)$$

So

$$\frac{\sum_{j \geq 1} H(|z_j(w_n)|)}{H(|w_n|)} \leq \left(\frac{-\log(|z(w_n)|)}{-\log(|w_n|)} \right)^{\alpha+1} \frac{N_\varphi(w_n)}{-\log(|w_n|)}$$

By Littlewood's Inequality (Theorem 2.29)

$$N_\varphi(w_n) \leq \log \left| \frac{1 - \overline{\varphi(0)}w_n}{\varphi(0) - w_n} \right|$$

and an application of L'Hôpital's Rule shows

$$\limsup_{n \to \infty} \frac{N_\varphi(w_n)}{-\log|w_n|} \leq \frac{1 + |\varphi(0)|}{1 - |\varphi(0)|}$$

The factor

$$\left(\frac{-\log(|z(w_n)|)}{-\log(|w_n|)}\right)^{\alpha+1}$$

is comparable, as $n \to \infty$, to

$$\left(\frac{1-|z(w_n)|}{1-|w_n|}\right)^{\alpha+1}$$

since $|z(w_n)| \to 1$ as $|w_n| \to 1$. Letting $\beta(\varphi) = \liminf_{|z|\to 1}(1-|\varphi(z)|)/(1-|z|)$ we see that

$$\limsup_{n\to\infty}\left(\frac{1-|z(w_n)|}{1-|w_n|}\right)^{\alpha+1} \leq \left(\frac{1}{\beta(\varphi)}\right)^{\alpha+1}$$

or

$$\|C_\varphi\|_e^2 \leq c(\alpha)\frac{1+|\varphi(0)|}{1-|\varphi(0)|}\left(\frac{1}{\beta(\varphi)}\right)^{\alpha+1}$$

In particular we get a simple characterization of the compact composition operators on $A_\alpha^p(D)$:

THEOREM 3.22
On $A_\alpha^p(D)$ C_φ is compact if and only if φ has no finite angular derivative.

PROOF The only if direction is Corollary 3.14. Conversely if φ has no finite angular derivative, then $\beta(\varphi) = +\infty$ and the above argument shows that on $A_\alpha^2(D)$, $\|C_\varphi\|_{e,\alpha} = 0$ and C_φ is compact. By Exercise 3.2.6 the conclusion holds for all $0 < p < \infty$. ∎

A different approach to this result is outlined in Exercise 3.2.9.

Exercises

3.2.1 Let $H^2(\beta)$ be any weighted Hardy space in D. Show that C_φ is compact on $H^2(\beta)$ if and only if whenever $\{f_n\}$ is a bounded sequence in $H^2(\beta)$ and $f_n \to 0$ uniformly on compact subsets of D then $f_n \circ \varphi \to 0$ in the norm of $H^2(\beta)$.

3.2.2 Show that C_φ is compact on $H^\infty(D)$ if and only if $\overline{\varphi(D)} \subset D$.

3.2.3 Construct a map φ with $\varphi(D) = D$ yet C_φ is compact on $H^2(D)$.

3.2.4 Show that if φ is an inner function, the norm and the essential norm of C_φ on $H^2(D)$ are the same.

3.2.5 Suppose φ is analytic in a neighborhood of the closed unit disk. Show that on $H^2(D)$, the essential norm of C_φ is given by

$$\|C_\varphi\|_e^2 = \max\{\sum_{\varphi(e^{i\theta})=w}|\varphi'(e^{i\theta})|^{-1} : |w| = 1\}$$

3.2.6 Formulate and prove an analogue of Theorem 3.12 for the weighted Bergman spaces $A_\alpha^p(D)$. Conclude that boundedness and compactness of C_φ on $A_\alpha^p(D)$ are independent of p, $0 < p < \infty$. Hint:
$$\int_{S(\zeta,h)} (1 - |z|^2)^\alpha \, dA(z) \sim h^{\alpha+2}$$

3.2.7 P. Ahern and D. Clark [AhC74, Theorem 2] show that the singular inner function $\varphi(z) = \exp \int_{\partial D} (z + \zeta)/(z - \zeta) \, d\mu(\zeta)$ for μ singular to Lebesgue measure has finite angular derivative at η in ∂D if and only if
$$\int_{\partial D} \frac{d\mu(\zeta)}{|\zeta - \eta|^2} < \infty$$
Use this to show that there exist inner functions that induce compact composition operators on $A_\alpha^p(D)$ for $\alpha > -1$ and $0 < p < \infty$.

3.2.8 Give an example of a mapping $\varphi : D \to D$ satisfying
- φ has no finite angular derivative
- $\lim_{r \to 1} |\varphi(r\zeta)| < 1$ for almost every ζ in ∂D
- C_φ is not compact on $H^p(D)$

Hint: Consider $\varphi = \tau \circ \varphi_a \circ \psi$ where ψ is an inner function as in the previous exercise, φ_a is an automorphism of D taking $a = \psi(0)$ to 0, and $\tau(z) = (1+z)/2$.

3.2.9 Recall from Exercise 2.1.5 the one parameter family of weighted Dirichlet spaces $\mathcal{D}_\alpha$, $\alpha > -1$, defined by
$$\mathcal{D}_\alpha = \{f \text{ analytic} : \|f\|^2 = |f(0)|^2 + \int_D |f'(z)|^2 (1 - |z|^2)^\alpha \frac{dA(z)}{\pi} < \infty\}$$
Define the measure μ_α on D by
$$d\mu_\alpha = |\varphi'(z)|^2 (1 - |z|^2)^\alpha \frac{dA(z)}{\pi}$$

(a) Show C_φ is bounded on $\mathcal{D}_\alpha$ if and only if
$$\mu_\alpha \varphi^{-1} S(\zeta, h) = O(h^{\alpha+2})$$
while C_φ is compact on $\mathcal{D}_\alpha$ if and only if
$$\mu_\alpha \varphi^{-1} S(\zeta, h) = o(h^{\alpha+2})$$
as $h \to 0$, uniformly in ζ.

(b) Suppose $-1 < \alpha < \beta$. Show if C_φ is bounded (respectively, compact) on $\mathcal{D}_\alpha$ then C_φ is bounded (compact) on $\mathcal{D}_\beta$.

(c) Again let $-1 < \alpha < \beta$. Suppose φ has no finite angular derivative at any point of ∂D and C_φ is bounded on $\mathcal{D}_\alpha$. Show that C_φ is compact on $\mathcal{D}_\beta$.

(d) Show that (c) gives an alternate proof of Theorem 3.22 and Corollary 3.21.

3.2.10 Suppose φ has finite angular derivative at ζ in ∂D. Show directly that there exists a constant c, independent of h so that
$$\mu_\alpha \varphi^{-1}(S(\zeta, h)) \geq ch^{\alpha+2}$$
where μ_α is the measure defined in the previous problem. This gives an alternate proof that if C_φ is compact on $\mathcal{D}_\alpha$ then φ cannot have finite angular derivative at any point.

3.2.11 Verify that a general weighted Bergman space $A_G^2(D)$ can be described as a weighted Dirichlet space with weight function $H(r) = \int_r^1 (\int_x^1 G(s) \, ds) \, dx$.

3.2.12 Suppose C_φ is compact on $H^p(D)$, or $A_\alpha^p(D)$, or $H^2(\beta)$ where $\sum \beta(n)^{-2} = \infty$. Show that φ has a unique fixed point in D.

Notes

The study of compact composition operators on $H^p(D)$ was begun by H. J. Schwartz in his thesis [Scz69]. He proved Proposition 3.11 and noted as a consequence that if C_φ is compact on $H^p(D)$ then $|\varphi^*(e^{i\theta})| < 1$ almost everywhere. The example $\varphi(z) = (1+z)/2$, giving a non-compact C_φ with $|\varphi^*(e^{i\theta})| < 1$ except at the single point 1 also appears in [Scz69], with a proof based on Proposition 3.11 directly. J. H. Shapiro and P. D. Taylor [ShT73] were the first to note explicitly the role of the angular derivative in studying compactness of C_φ in $H^p(D)$. In particular, the $H^p(D)$ case of Corollary 3.14 appears in [ShT73] with a proof that applies Proposition 3.11 directly. Many of the results in [ShT73] were later extended to the Bergman space $A^2(D)$ by D. M. Boyd [Boy74].

The Carleson criteria for boundedness and compactness on Hardy spaces were first explicitly written down in [Mc85] for the case $N > 1$. The Carleson point of view in general is a cornerstone for the paper [McS86] of B. D. MacCluer and J. H. Shapiro in which Theorem 3.22 and Corollary 3.21, with proofs as outlined in Exercise 3.2.9 appear. The results of Exercises 3.2.3, 3.2.6, 3.2.7, 3.2.8, 3.2.9 and 3.2.10 are all taken from [McS86].

Theorem 3.20, giving the essential norm of C_φ on $H^p(D)$ in terms of the Nevanlinna counting function, is due to J. H. Shapiro [Sho87a]. Theorem 2.4 of [Co83] is a precursor of this result and is closely related to Exercise 3.2.5. A similar, though somewhat less precise formula for $\|C_\varphi\|_e$ on $A_\alpha^2(D)$, using generalized Nevanlinna counting functions, can also be found in [Sho87a]. The generalization of Shapiro's method for obtaining upper estimates on the essential norm on general weighted Bergman or Dirichlet spaces appears in the work of T. Kriete and B. MacCluer in [KrM92].

Exercise 3.2.12 is proved later as Theorem 7.14 but is presented here as an exercise because it is a good application of the ideas of this section.

3.3   Hilbert–Schmidt operators

The spectral theorem implies that every positive compact operator P on a Hilbert space can be written as $P = \sum \lambda_n \langle \cdot, v_n \rangle v_n$ where $\{\lambda_n\}$ is the sequence of eigenvalues of P, with multiplicities, and $\{v_n\}$ is an orthonormal set so that $P v_n = \lambda_n v_n$. (For proofs of this and other facts noted below see [Con91, Chap. I] or [Zh90b, Chap. 1].) A bounded operator T on $\mathcal{H}$ has polar decomposition $T = UP$ where P is the positive operator $(T^*T)^{1/2}$ and U is a partial isometry with $\|Ux\| = \|x\|$ on $(\ker T)^\perp$. Since P is compact if and only if T is, when T is compact, we may write $P = \sum \lambda_n \langle \cdot, v_n \rangle v_n$ as above, where the eigenvalues $\{\lambda_n\}$ of P are arranged in non-increasing order. In this case, the vectors $u_n = U v_n$ also form an orthonormal set and we have

$$T = \sum \lambda_n \langle \cdot, v_n \rangle u_n \qquad (3.3.1)$$

The numbers $\{\lambda_n\}$ are called the singular values of T and we refer to Equation (3.3.1) as a singular value decomposition of the compact operator T. A compact operator T is said to be in the Schatten p–class $\mathcal{S}_p$ for $0 < p < \infty$ if

$\sum_{n=1}^{\infty} \lambda_n^p < \infty$. The class $\mathcal{S}_2$, called the Hilbert–Schmidt class, is of particular importance. It is clear from the definition that if T has polar decomposition $T = UP$, then T is a Hilbert–Schmidt operator if and only if P is, but this class can also be usefully characterized in ways that do not explicitly involve the polar decomposition.

THEOREM 3.23

Let T be an operator on a separable Hilbert space $\mathcal{H}$ and let $T = UP$ be its polar decomposition. The following are equivalent:

(1) T is Hilbert–Schmidt.

(2) P is Hilbert–Schmidt.

(3) $\sum_{m=1}^{\infty} \|Te_m\|^2 < \infty$ for all orthonormal bases $\{e_m\}$ of $\mathcal{H}$.

(4) $\sum_{m=1}^{\infty} \|Te_m\|^2 < \infty$ for some orthonormal basis $\{e_m\}$ of $\mathcal{H}$.

PROOF   As noted above, the equivalence of (1) and (2) is a direct consequence of the definition. Suppose T is a compact operator with singular value decomposition $T = \sum \lambda_n \langle \cdot, v_n \rangle u_n$. If $\{e_m\}$ is any orthonormal basis for $\mathcal{H}$ we have

$$T(e_m) = \sum_{n=1}^{\infty} \lambda_n \langle e_m, v_n \rangle u_n$$

so that

$$\sum_{m=1}^{\infty} \|Te_m\|^2 = \sum_{m=1}^{\infty} \sum_{n=1}^{\infty} |\lambda_n|^2 |\langle e_m, v_n \rangle|^2$$

$$= \sum_{n=1}^{\infty} |\lambda_n|^2 \sum_{m=1}^{\infty} |\langle v_n, e_m \rangle|^2 = \sum_{n=1}^{\infty} |\lambda_n|^2 \|v_n\|^2 = \sum_{n=1}^{\infty} |\lambda_n|^2$$

This calculation shows that (1) implies (3). Trivially (3) implies (4). The same calculation will show that (4) implies (1) provided we know that if (4) holds then T must be compact. If $\sum \|Te_m\|^2 < \infty$ for some orthonormal basis $\{e_m\}$ of $\mathcal{H}$, define finite rank operators T_k by

$$T_k \left(\sum_{m=1}^{\infty} a_m e_m \right) = \sum_{m=1}^{k} a_m Te_m$$

For x an arbitrary vector in $\mathcal{H}$ with $x = \sum_{m=1}^{\infty} a_m e_m$, we have

$$(T - T_k)x = \sum_{m=k+1}^{\infty} a_m Te_m$$

from which it follows that

$$\|T - T_k\|^2 \leq \sum_{m=k+1}^{\infty} \|Te_m\|^2$$

Thus if (4) holds, T is the norm limit of finite rank operators and is therefore compact. ∎

Set $\|T\|_{HS}^2 = \sum \|Te_m\|^2$ where $\{e_m\}$ is any orthonormal basis. The proof of Theorem 3.23 shows $\sum \|Te_m\|^2 = \sum |\lambda_n|^2$ so $\|T\|_{HS}$ does not depend on the choice of orthonormal basis. Although we do not need it here, we note that $\| \cdot \|_{HS}$ gives a norm on $S_2(\mathcal{H})$ and under this norm $S_2(\mathcal{H})$ is a Hilbert space; more generally, $S_p(\mathcal{H})$ is a Banach space for $p \geq 1$ with $\|T\|_p = (\sum \lambda_n^p)^{1/p}$. It is easy to see from (4) of Theorem 3.23 that the Hilbert–Schmidt operators form a left ideal in $\mathcal{B}(\mathcal{H})$. In addition, condition (4) shows that an operator unitarily equivalent to a Hilbert–Schmidt operator is also in the class and since T with polar decomposition $T = UP$ is Hilbert–Schmidt if and only if P is, the equality

$$U^*T^*U = U^*(UP)^*U = U^*P$$

shows that T is Hilbert–Schmidt if and only if T^* is. Since a left ideal that is closed under adjoints is also a right ideal, the Hilbert–Schmidt class is a two-sided ideal in $\mathcal{B}(\mathcal{H})$.

The Schatten class $S_1(\mathcal{H})$ is also called the **trace class** of $\mathcal{H}$. It is not difficult to use the definition of the Schatten classes to show that an operator is in the trace class if and only if it is the product of two operators in the Hilbert–Schmidt class. In Chapter 4, we will have occasion to use the following partial analogue of Theorem 3.23 for trace class operators.

THEOREM 3.24
If $\sum \|Te_n\| < \infty$ for some orthonormal basis $\{e_n\}$ of a Hilbert space $\mathcal{H}$ then T is in $S_1(\mathcal{H})$.

PROOF By a modification of the proof of Theorem 3.23 (or see [Con91, p. 1]), if $T = UP$ is the polar decomposition of T, the finite or infinite trace of P, the sum $\sum \langle Pw_n, w_n \rangle = \sum \lambda_n$, is independent of the orthonormal basis $\{w_n\}$. Thus, the operator T is trace class if this sum is finite for any basis and our hypothesis implies that it is for $\{e_n\}$

$$\sum \langle Pe_n, e_n \rangle \leq \sum \|Pe_n\|\|e_n\| = \sum \|Te_n\|$$

∎

Using the orthonormal basis $\{z^n\}$ for $H^2(D)$, Theorem 3.23 shows that C_φ is

Hilbert–Schmidt on $H^2(D)$ if and only if

$$\sum_{n=0}^{\infty} \|C_\varphi(z^n)\|^2 = \sum_{n=0}^{\infty} \|\varphi^n\|_2^2 = \sum_{n=0}^{\infty} \int_{\partial D} |\varphi^n|^2 \, d\sigma$$

$$= \int_{\partial D} \frac{1}{1 - |\varphi|^2} \, d\sigma(\zeta) < \infty \qquad (3.3.2)$$

Alternately, using Theorem 2.31 to compute $\|\varphi^n\|_2^2$ we see that (Exercise 3.3.2)

$$C_\varphi \text{ is in } \mathcal{S}_2(H^2(D)) \text{ if and only if } \frac{N_\varphi(z)}{-\log|z|} \text{ is in } L^1\left(\frac{dA}{(1-|z|^2)^2}\right)$$

By using the orthonormal basis $\{z^n/\|z^n\|_\alpha\}$ on the weighted Bergman space $A_\alpha^2(D)$ we see that C_φ is in $\mathcal{S}_2(A_\alpha^2(D))$ if and only if

$$\int_D \frac{1}{(1 - |\varphi(z)|^2)^{\alpha+2}} (1 - |z|^2)^\alpha \, dA(z) < \infty$$

Equivalently,

$$\|C_\varphi\|_{HS}^2 = \int_D \|K_{\varphi(z)}\|^2 (1 - |z|^2)^\alpha \, dA(z)/\pi$$

where K_w denotes the kernel function for evaluation at w in $A_\alpha^2(D)$. Exercise 3.3.1 generalizes this to Bergman spaces defined from arbitrary positive, continuous, circularly symmetric weight functions.

We can use the characterization of Hilbert–Schmidt composition operators on $H^2(D)$ by means of the integral condition of Equation (3.3.2) to give some interesting examples. We consider maps φ for which $\varphi(D)$ is contained in a nontangential approach region.

PROPOSITION 3.25
Suppose φ is an analytic map of the unit disk into itself. If there is a point ζ on the circle and an $\alpha > 1$ such that $\varphi(D)$ is contained in the nontangential approach region $\Gamma(\zeta, \alpha)$, then C_φ is Hilbert–Schmidt on $H^2(D)$.

PROOF Without loss of generality, we may assume $\zeta = 1$. Recalling that $\Gamma(1, \alpha) = \{z : |1 - z| < \alpha(1 - |z|)\}$, find $\beta > 1$ so that

$$\Gamma(1, \alpha) \subset \Gamma \equiv \{z : |1 - z| < \frac{\beta}{2}(1 - |z|^2)\}$$

Thus, if $\varphi(D)$ is contained in $\Gamma(1, \alpha)$ we have

$$\frac{1}{1 - |\varphi(z)|^2} < \frac{\beta}{2} \frac{1}{|1 - \varphi(z)|}$$

and by Equation (3.3.2), it suffices to show that $g(z) \equiv (1 - \varphi(z))^{-1}$ is in $H^1(D)$. By Exercise 3.3.3 the mapping $\lambda(z) = \frac{1}{1-z}$ takes $\Gamma(1, \alpha)$ into the sector S contained in the right halfplane and defined by

$$S = \{z : \text{Re } z > 0 \text{ and } |\text{Arg} z| < \cos^{-1}(1/\beta)\}$$

Set $b = \cos^{-1}(1/\beta)$. The function

$$F(z) = \left(\frac{1+z}{1-z}\right)^{2b/\pi}$$

is a univalent map of D onto S and is in $H^1(D)$ since $2b/\pi < 1$. Moreover, $F^{-1} \circ g$ is well-defined on D, since $\varphi(D) \subset \Gamma(1,\alpha)$, and maps D into D. Thus $g = F \circ (F^{-1} \circ g)$ is in $H^1(D)$ since F is. ∎

The point of Proposition 3.25 is to show that one of the simple ways of constructing non-trivial examples of compact composition operators actually gives Hilbert–Schmidt operators. To give an example of a composition operator which is compact but not Hilbert–Schmidt on $H^2(D)$ requires some work. The rest of this section is devoted to the construction of such an example.

We define a map φ by $\varphi(z) = 1 - f(g(z))$ where $f(z) = -z \log z$ (principal branch) and $g(z)$ is a univalent map of D onto $H_r = \{z : \text{Re } z > 0 \text{ and } |z| < r\}$ for small r to be specified (see Figure 3.3). Note that such a g can be obtained as

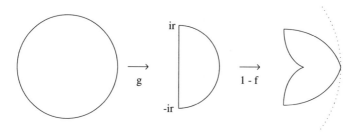

FIGURE 3.3

$\varphi(D)$ **giving** C_φ **compact but not Hilbert–Schmidt.**

the composition of the Cayley transform of D onto the upper halfplane, followed by a suitable branch of $\sqrt{z}$, the inverse Cayley transform, and a suitable rotation and dilation. By preceding this by a rotation of D we can assume $g(1) = 0$. Also notice that g is conformal in a neighborhood of 1, so that $g'(1) \neq 0$.

We show next that if $r < 1/e$ then φ as defined above is univalent in D. Since g is univalent we need only check that f is univalent in H_r. Since $f'(z) = -1 - \log z$ has real part $-1 - \log|z|$ which is positive on the convex set H_r if $r < 1/e$, Exercise 2.3.14 shows that f is one-to-one on $\overline{H_r} \setminus \{0\}$. Indeed, since $f(z) = 0$ only if $z = 0$, f is one-to-one on $\overline{H_r}$.

Next we claim that $\varphi(D)$ is a subset of D if r is chosen sufficiently small. By the definition of φ it is enough to check that f maps H_r into $\{z : |1 - z| < 1\}$.

Consider $f(se^{i\theta})$ for $0 < s \leq r$ and $\theta \in [-\pi/2, \pi/2]$, where $r = e^{-2}$. A computation shows that

$$|f(se^{i\theta}) - 1|^2 = 1 + s^2 \log^2 s + s^2\theta^2 - 2s\theta \sin\theta + 2s\cos\theta \log s$$

Consider s fixed and maximize this over $\theta \in [-\pi/2, \pi/2]$. A calculus argument shows that, provided $s \leq e^{-2}$, the maximum occurs at $\theta = \pi/2$ and is

$$1 + s^2 \log^2 s + s^2\pi^2/4 - \pi s < 1$$

since $s \log^2 s < \pi(1 - \pi s/4)$ on $(0, 1)$.

We have shown that $\varphi(D) \subset D$ and φ has radial limit of modulus 1 only at 1. Since φ is univalent, C_φ will be compact on $H^2(D)$ if φ does not have finite angular derivative at $z = 1$, by Theorem 3.21. Since $\varphi'(r) = [1 - \log(g(r))]g'(r)$ where $g'(1)$ is bounded and non-zero and $g(r) \to 0$ as $r \to 1$, $|\varphi'(1)| = \infty$ as desired.

Lastly we claim that C_φ is not Hilbert–Schmidt on $H^2(D)$. This will be the case if $\int_{\partial D}(1 - |\varphi|^2)^{-1} d\sigma = \infty$. Since g extends conformally to a neighborhood of 1 and $g(1) = 0$, g^{-1} extends conformally to a neighborhood of 0. In particular, $(g^{-1})'$ and its reciprocal are bounded on a small interval $[-i\delta, i\delta]$ on the imaginary axis contained in ∂H_r. By a change of variables $g(\zeta) = x + iy$ we see that C_φ fails to be Hilbert–Schmidt if

$$\int_0^\delta \frac{1}{1 - |1 - f(iy)|^2}\, dy = \infty \tag{3.3.3}$$

But

$$|1 - f(iy)|^2 = |1 + iy\log(iy)|^2 = (1 - y\pi/2)^2 + y^2 \log^2 y$$

so

$$1 - |1 - f(iy)|^2 = y\pi - y^2\pi^2/4 - y^2 \log^2 y < y$$

for y sufficiently small and the integral in (3.3.3) must diverge.

Exercises

3.3.1 Let $G(r)$ be a positive, continuous function on $(0, 1)$ with $\int_0^1 G(r)r\, dr < \infty$. Define the Bergman space $A_G^2(D)$ by

$$A_G^2(D) \equiv \{f \text{ analytic in } D : \|f\|_G^2 \equiv \int_D |f|^2 G(|z|)\, dA(z)/\pi < \infty\}$$

with kernel functions

$$K_w(z) = \sum_{n=0}^\infty \frac{\overline{w}^n z^n}{\beta(n)^2}$$

where $\beta(n)^2 = \|z^n\|_G^2$. Show that C_φ on $A_G^2(D)$ is Hilbert–Schmidt if and only if

$$\int_D \|K_{\varphi(z)}\|^2 G(|z|)\, dA < \infty$$

3.3.2   (a)   Show that $C_\varphi \in \mathcal{S}_2(H^2(D))$ if and only if

$$\frac{N_\varphi(z)}{-\log(|z|)} \in L^1((1-|z|^2)^{-2}\, dA)$$

     (b)   Show that $C_\varphi \in \mathcal{S}_2(A_\alpha^2(D))$ if and only if

$$\left(\frac{1-|z|^2}{1-|\varphi(z)|^2}\right)^{\alpha+2} \in L^1((1-|z|^2)^{-2}\, dA)$$

     (c)   Show that if

$$\int_{\varphi(D)} (1-|z|^2)^{-2}\, dA(z) < \infty$$

     then C_φ is Hilbert–Schmidt on $H^2(D)$.

3.3.3   Given $\beta > 1$, show that

$$\Gamma = \{z : |1-z| < \frac{\beta}{2}(1-|z|^2)\}$$

is contained in the cone with vertex 1 and angular opening $2\cos^{-1}(1/\beta)$ defined by

$$|\mathrm{Arg}(1-z)| < \cos^{-1}(1/\beta)$$

Show that the map $\lambda(z) = \frac{1}{1-z}$ takes Γ into the sector S in the right halfplane defined by

$$S = \{z : \mathrm{Re}\, z > 0 \text{ and } |\mathrm{Arg} z| < \cos^{-1}(1/\beta)\}$$

3.3.4   (a)   Extend the ideas of Proposition 3.25 to show that if $\varphi(D)$ is contained in a polygon inscribed in the unit circle, then C_φ is Hilbert–Schmidt on $H^2(D)$.

     (b)   Show that if $\varphi(D)$ is contained in a polygon inscribed in the unit circle, then C_φ is trace class on $H^2(D)$. Hint: If P is a polygon inscribed in the unit circle and Q is a polygon inscribed in the unit circle whose vertices include the vertices of P and a vertex between each of the vertices of P, show that the Riemann map φ of the disk onto P can be expressed as a composition $\varphi = \varphi_1 \circ \varphi_2$ where φ_1 maps the disk onto Q and $\varphi_1^{-1}(P)$ lies in a polygon inscribed in the disk. It follows that $C_\varphi = C_{\varphi_2} C_{\varphi_1}$ is the product of two Hilbert–Schmidt operators.

Notes

The characterization of Hilbert–Schmidt composition operators on $H^2(D)$ given by Equation (3.3.2) appears in J. H. Shapiro and P. D. Taylor's paper [ShT73]. The consequences of these ideas developed in Exercise 3.3.4 are Corollary 3.2 and Theorem 5.1 of [ShT73]. The analogous condition on $A^2(D)$ can be found in [Boy75].

Exercise 3.3.2 is an improvement of a result due to J. H. Shapiro and P. D. Taylor. The results in Exercise 3.3.2 are special cases of general characterization of the composition operators belonging to $\mathcal{S}_p(H^2)$ for $0 < p < \infty$ due to D. H. Luecking and K. Zhu ([LuZ92]):

$$C_\varphi \in \mathcal{S}_p(H^2) \Leftrightarrow \frac{N_\varphi(z)}{-\log|z|} \in L^{p/2}(d\lambda)$$

where $d\lambda(z) = (1 - |z|^2)^{-2}\,dA(z)$. To prove this Luecking and Zhu make use of an identification of $C_\varphi^* C_\varphi$ with a certain Toeplitz-type operator; previous work ([Lue87]) on the membership of these Toeplitz-type operators in the Schatten p–classes can be invoked. An analogous result for membership in $\mathcal{S}_p(A_\alpha^2)$ involving generalized Nevanlinna counting functions is also given. Since C_φ is compact on $H^2(D)$ if and only if

$$\limsup_{|w|\to 1} \frac{N_\varphi(w)}{-\log|w|} = 0$$

these characterizations suggest that there should exist compact C_φ on $H^2(D)$ not lying in any Schatten p–class. The problem of constructing such an example is considered by T. Carroll and C. Cowen in [CaC91]. The construction given here of a compact C_φ that is not Hilbert–Schmidt on $H^2(D)$ is adapted from [ShT73, Theorem 4.2].

Conditions for C_φ to be in a Schatten p–class for the Bergman space of a smoothly bounded strictly convex domain in C^N (analogous to the Luecking-Zhu criterion) were worked out by S. Y. Li in [Lis94].

3.4   Composition operators with closed range

We saw in Chapter 1 that on an algebraically consistent functional Banach space of analytic functions on a set X, invertible composition operators arise only when the symbol map is an automorphism of the underlying set X. For spaces of functions analytic on the disk, at least, the composition operator C_φ is one-to-one provided that φ is non-constant, so C_φ will have a continuous left inverse when it has closed range. In this section, we consider the question "When does C_φ have closed range?" and we also look at closely related questions concerning the essential spectrum. *Throughout this section we assume that φ is not a constant map.*

If H is an (infinite dimensional) Hilbert space, the collection of compact operators on H, denoted $\mathcal{K}(H)$, is a closed ideal in the collection of bounded operators $\mathcal{B}(H)$. The C*–algebra $\mathcal{B}(H)/\mathcal{K}(H)$, called the **Calkin algebra**, is important in the study of those properties of operators that are preserved under compact perturbations, that is, that are *essential* properties of the operator. A good reference for the information we will need about the Calkin algebra and related ideas is Chapter XI of Conway's book *A Course in Functional Analysis* [Con90]. Because the Calkin algebra is a Banach algebra, norm and spectrum are defined; moreover, the spectral mapping theorem and the formula relating the spectral radius to the norm are true when applied to this algebra. If A is a bounded operator on H, the norm, spectrum, etc., of the equivalence class in the Calkin algebra that contains A are called the **essential norm** of A, the **essential spectrum** of A, and so on. (Since the compact operators form the equivalence class 0 in the Calkin algebra, the essential norm of A is the distance of A from the compact operators as we defined it in Section 3.2.) In order for the equivalence class of A to be invertible in the Calkin algebra, there must be a bounded operator B for which $I - BA$ and $I - AB$ are both compact. If the equivalence class of A in the Calkin algebra is invertible, A is called a ***Fredholm operator***; such operators are characterized as the operators with closed range for which the operator and its adjoint have finite dimensional kernel [Con90, pp. 350–352]. The spectrum of A contains its essential spectrum and points on the boundary of the spectrum of A are either in its essential spectrum or they are eigenvalues of A and A^* of finite, equal multiplicity [Con90, p. 362]. We look first at conditions under which composition operators acting on various Hilbert spaces are Fredholm.

LEMMA 3.26
Let $\mathcal{H}$ be a Hilbert space of analytic functions in D that contains the polynomials. If C_φ is a bounded composition operator on $\mathcal{H}$ and the kernel of C_φ^* is finite dimensional, then φ is univalent.

PROOF Suppose φ is non-constant and not univalent. Then we may find distinct points a and b in the disk with $\varphi(a) = \varphi(b)$. Let $\mathcal{O}_a$ and $\mathcal{O}_b$ be disjoint open sets containing a and b respectively. Since φ is an open map, $\varphi(\mathcal{O}_a) \cap \varphi(\mathcal{O}_b)$ is open and non-empty. Choose a sequence $\{c_n\}$ of distinct points in $\varphi(\mathcal{O}_a) \cap \varphi(\mathcal{O}_b)$. For each n let a_n be in $\mathcal{O}_a$ with $\varphi(a_n) = c_n$ and b_n be in $\mathcal{O}_b$ with $\varphi(b_n) = c_n$. The sets $\{a_n\}$ and $\{b_n\}$ are of course infinite and disjoint, with $\varphi(a_n) = \varphi(b_n)$. Consider $K_{a_n} - K_{b_n}$ where K_w denotes the kernel function for evaluation at w. We have $C_\varphi^*(K_{a_n} - K_{b_n}) = 0$. Since $\mathcal{H}$ contains the polynomials, no kernel function can be a linear combination of other kernel functions (Exercise 1.1.10). Thus $\{K_{a_n} - K_{b_n}\}$ is a linearly independent set in the kernel of C_φ^* and we have our desired contradiction. $\blacksquare$

While our standing hypothesis in this section is that φ is not a constant map, we note that if φ is constant, then a similar argument shows that the kernel of C_φ^* is infinite dimensional. We note that essentially the same argument can be made to work for reproducing kernel Hilbert spaces of analytic functions in B_N, for $N > 1$. If φ is an analytic map of the ball into itself and $\varphi^{-1}(w)$ is infinite for some w in B_N, then by considering $C_\varphi^*(K_{z_j} - K_{z_k})$ for $z_j \neq z_k$ in $\varphi^{-1}(w)$ we see that the dimension of the kernel of C_φ^* is infinite. If $\varphi^{-1}(w)$ is finite for every w in B_N it follows that φ is an open map [Ru80, Theorem 15.1.6] and the argument proceeds as in Lemma 3.26.

Recall that $N_\varphi(w)$ is the Nevanlinna counting function

$$N_\varphi(w) = \sum \log \frac{1}{|z_j(w)|}$$

where $\{z_j(w)\}$ are the pre-images of w under φ, with multiplicities; $N_\varphi(w) = 0$ if w is not in the range of φ. If φ is any analytic map of the disk into itself we always have Littlewood's Inequality (Theorem 2.29):

$$N_\varphi(w) \leq \log \left| \frac{1 - \overline{\varphi(0)}w}{\varphi(0) - w} \right|$$

for w in $D \setminus \{\varphi(0)\}$. When φ is inner the inequality in Littlewood's Inequality becomes equality, almost everywhere:

LEMMA 3.27
If φ is an inner function then

$$N_\varphi(w) = \log \left| \frac{1 - \overline{\varphi(0)}w}{\varphi(0) - w} \right|$$

for all w in D outside a set of area measure 0.

PROOF   If φ is inner and $\varphi(0) = 0$ then Theorem 3.8 and Theorem 2.31 give for any f in $H^2(D)$

$$\|f\|_2^2 = \|f \circ \varphi\|_2^2 = |f(0)|^2 + 2\int_D |f'(w)|^2 N_\varphi(w) \frac{dA(w)}{\pi}$$

By the Littlewood–Paley identity (Theorem 2.30)

$$\|f\|_2^2 = |f(0)|^2 + 2\int_D |f'(w)|^2 \log \frac{1}{|w|} \frac{dA(w)}{\pi}$$

Thus $N_\varphi(w) = -\log |w|$ for almost all $[dA]$ w in D.

If φ is inner and $\varphi(0) = u \neq 0$ then for $\varphi_u(z) = (u - z)/(1 - \overline{u}z)$ the map $\psi = \varphi_u \circ \varphi$ is inner with value 0 at 0. Thus the above argument gives

$$N_\psi(w) = \log \frac{1}{|w|}$$

off a set of area 0. Since $N_\psi(w) = N_\varphi(\varphi_u(w))$ we have

$$N_\varphi(w) = \log \frac{1}{|\varphi_u(w)|} = \log \left| \frac{1 - \overline{\varphi(0)}w}{\varphi(0) - w} \right|$$

for almost all $[dA]$ w in D.   ∎

COROLLARY 3.28
If φ is a univalent inner function then φ is an automorphism of D.

PROOF   First suppose φ is a univalent inner function with $\varphi(0) = 0$. By Lemma 3.27 we may find $w_0 \neq 0 = \varphi(0)$ in $\varphi(D)$ so that

$$N_\varphi(w_0) = -\log |w_0|$$

By the Schwarz Lemma we have

$$\log |w_0| = \log |\varphi(\varphi^{-1}(w_0))| \le \log |\varphi^{-1}(w_0)| = \log |w_0|$$

Thus we must have $|\varphi(\varphi^{-1}(w_0))| = |\varphi^{-1}(w_0)|$ and the Schwarz Lemma implies φ is a rotation.

If $\varphi(0) = u \ne 0$ apply the above argument to the univalent inner function $\psi = \varphi_u \circ \varphi$ where $\varphi_u(z) = (u - z)/(1 - \overline{u}z)$ to get the desired conclusion. ∎

It is also the case that if $N_\varphi(w) = \log |1 - \overline{\varphi(0)}w|/|\varphi(0) - w|$ holds for some $w = w_0 \ne \varphi(0)$ in D, then φ must be inner. Since we will not need this result here, we will not give the proof but instead refer the interested reader to [Sho87a, p. 388].

We can now characterize the Fredholm composition operators on a wide variety of spaces.

THEOREM 3.29
Let $\mathcal{H}$ be any Hilbert space of analytic functions on D which contains the Dirichlet space $\mathcal{D}$. If C_φ is Fredholm then φ is an automorphism of D.

PROOF Suppose we can find an infinite set $\{\zeta_j\}$ of distinct points in the unit circle so that $\lim_{r \to 1} \varphi(r\zeta_j)$ exists and has modulus less than 1. Fix an unbounded simply connected region Ω of finite area as shown in Figure 3.4. For each j let ψ_j be the biholomorphic map of D onto Ω with $\psi_j(0) = 0$ and $\psi_j(\zeta_j) = \infty$. Notice that each ψ_j is in the Dirichlet space $\mathcal{D}$ and hence in $\mathcal{H}$, and for $e^{i\theta} \ne \zeta_j$, $\lim_{r \to 1} \psi_j(re^{i\theta})$ exists and is finite.

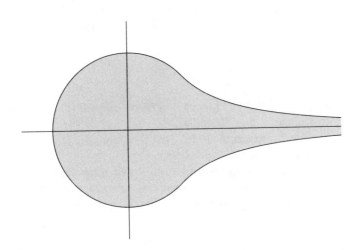

FIGURE 3.4

The region Ω.

We claim that no non-zero linear combination of the ψ_j's is in the range of C_φ. To see this consider any sum $\sum_{j=1}^N c_j \psi_j$ with $c_N \neq 0$ and suppose that $\sum_{j=1}^N c_j \psi_j = f \circ \varphi$ for some f in $\mathcal{H}$. Since $\lim_{r \to 1} f \circ \varphi(r\zeta_N)$ exists and is finite by assumption, while $\lim_{r \to 1} \sum_{j=1}^N c_j \psi_j(r\zeta_N) = \infty$ by the construction of the ψ_j's, we have a contradiction and the claim is verified.

Since C_φ is Fredholm, the range of C_φ is closed and we may write $\mathcal{H} = \operatorname{ran} C_\varphi \oplus M$ where $M = (\operatorname{ran} C_\varphi)^\perp$ is finite dimensional. If $P : \mathcal{H} \to M$ is the orthogonal projection onto M, then M contains $\{P\psi_j\}$. But $\{P\psi_j\}$ is a linearly independent set since

$$0 = \sum c_j P(\psi_j) = P(\sum c_j \psi_j)$$

would mean $\sum c_j \psi_j$ is in the range of C_φ which, by the argument above, implies $c_j = 0$ for all j. Thus we must conclude that φ has radial limits of modulus 1 almost everywhere so that φ is an inner function. By Lemma 3.26, φ must also be univalent and by Corollary 3.28 φ is therefore an automorphism of D. ∎

If the automorphisms of C_φ give bounded composition operators on $\mathcal{H}$ then the converse to this theorem holds trivially; if φ is an automorphism, then C_φ is in fact invertible, with inverse $C_{\varphi^{-1}}$.

We turn next to the problem of identifying the composition operators on $H^2(D)$ with closed range. The closed range composition operators are precisely those operators that are bounded below.

PROPOSITION 3.30
A bounded composition operator C_φ acting on a Hilbert space $\mathcal{H}$ of analytic functions in the disk has closed range if and only if there exists $\epsilon > 0$ so that

$$\|C_\varphi f\| \geq \epsilon \|f\|$$

for all f in $\mathcal{H}$.

PROOF   Suppose $\|C_\varphi f\| \geq \epsilon \|f\|$ for all f in $\mathcal{H}$ and suppose $\{C_\varphi f_n\}$ is a Cauchy sequence in the range of C_φ. Then $\|C_\varphi f_n - C_\varphi f_m\| \geq \epsilon \|f_n - f_m\|$ implies that $\{f_n\}$ is a Cauchy sequence in $\mathcal{H}$. Set $f = \lim_{n \to \infty} f_n$. Then $C_\varphi f = \lim_{n \to \infty} C_\varphi f_n$ and the range of C_φ is closed.

The converse direction follows readily from closed graph theorem and the observation that C_φ is one-to-one for any composition operator (with non-constant symbol) on a space of analytic functions in D, so C_φ is a bijective map of $\mathcal{H}$ onto the range of C_φ. ∎

Recall that Theorem 3.8 showed that if φ is inner, then C_φ is bounded below on $H^2(D)$, and therefore C_φ has closed range. This result, which will also be obtained as a special case Theorem 3.32 below, is instructive for our point of view here. We begin by noting another consequence of Lemma 3.27.

COROLLARY 3.31
If φ is inner, then there exist constants $c > 0$ and $\rho < 1$ so that for w in $\rho < |w| < 1$ outside a set of area 0

$$\frac{N_\varphi(w)}{-\log|w|} > c$$

PROOF If $\varphi(0) = 0$ the result is immediate from Lemma 3.27, with $\rho = 0$ and $c = 1/2$, say.

If $\varphi(0) = u \neq 0$ and $\psi = \varphi_u \circ \varphi$ as in the proof of Lemma 3.27 we have

$$N_\varphi(w) = \log\frac{1}{|\varphi_u(w)|}$$

for almost every $[dA]$ w in D. But

$$\log\frac{1}{|\varphi_u(w)|} \sim 1 - |\varphi_u(w)|^2 = \frac{1 - |u|^2}{|1 - \bar{u}w|^2}(1 - |w|^2)$$

$$\geq \frac{1 - |u|^2}{4}(1 - |w|^2) \sim \frac{1 - |u|^2}{4}\log\frac{1}{|w|}$$

where $\sim$ indicates that the ratio of the two quantities is bounded above and below by positive constants for $|w|$ near 1. This gives the desired result. ∎

As we will see, for general φ the operator C_φ will have closed range on $H^2(D)$ provided there is a positive constant c so that the set

$$\left\{w \in D : \frac{N_\varphi(w)}{-\log|w|} > c\right\}$$

is sufficiently spread out near ∂D in a sense that will be made precise presently. Define

$$\tau_\varphi(w) \equiv \frac{N_\varphi(w)}{-\log|w|}$$

and set $G_c = \{w \in D : \tau_\varphi(w) > c\}$. Of course, τ_φ is a non-negative measurable function. We can now state a theorem which characterizes the closed range composition operators on $H^2(D)$. The area of a set A in D will be denoted $|A|$.

THEOREM 3.32
The operator C_φ acting on $H^2(D)$ has closed range if and only if there are positive constants c and δ so that if $G_c = \{w \in D : \tau_\varphi(w) > c\}$ then

$$|G_c \cap S(\zeta, h)| \geq \delta h^2 \tag{3.4.1}$$

for all Carleson sets $S(\zeta, h)$ with $|\zeta| = 1$ and $0 < h < 1$.

The right hand side of Inequality (3.4.1) can, of course, equivalently be written as $\delta|S(\zeta, h)|$. Thus Inequality (3.4.1) can be viewed as a reverse Carleson measure inequality. Moreover, Theorem 3.32 has several equivalent formulations where other sets play the role of the Carleson sets $S(\zeta, h)$. Particularly useful is a version using pseudohyperbolic disks, defined for b in D and $0 < \epsilon < 1$ by

$$D(b, \epsilon) = \{z : |\varphi_b(z)| < \epsilon\} = \{\varphi_b(z) : |z| < \epsilon\}$$

where $\varphi_b(z) = (b - z)/(1 - \bar{b}z)$ (see Section 2.2). Careful attention to the geometry of these disks shows that the following are equivalent (Exercise 3.4.3) for positive constant c:

(1) There exists a positive constant δ so that

$$|G_c \cap S(\zeta, h)| \geq \delta h^2$$

for all $|\zeta| = 1$ and $0 < h < 1$.

(2) There exists $0 < \epsilon < 1$ and $\delta' > 0$ so that

$$|G_c \cap D(b, \epsilon)| \geq \delta'(1 - |b|)^2$$

for all b in D.

A key ingredient in the proof of Theorem 3.32 is the following result due to D. H. Luecking:

THEOREM 3.33
Suppose G is a measurable subset of D. For each $\alpha > -1$ the following are equivalent.

(1)   There exists $C > 0$ such that for every g in $A_\alpha^2(D)$

$$\int_D |g(z)|^2 (1 - |z|^2)^\alpha dA(z) \leq C \int_G |g(z)|^2 (1 - |z|^2)^\alpha dA(z)$$

(2)   There exists $\delta > 0$ so that

$$|G \cap S(\zeta, h)| \geq \delta h^2$$

for every $|\zeta| = 1$ and $0 < h < 1$.

A proof of the easier direction ((1)$\Rightarrow$(2)) of this theorem is outlined in Exercise 3.4.4. To characterize the closed range composition operators on $H^2(D)$ we use the $\alpha = 1$ version of this theorem in a modified form.

COROLLARY 3.34
If $\tau(z)$ is a bounded, measurable, non-negative function on D the following are equivalent.

1. There exists $K > 0$ such that for every f in $H^2(D)$

$$\int_D |f'(z)|^2 \tau(z) \log \frac{1}{|z|^2} dA(z) \geq K \int_D |f'(z)|^2 \log \frac{1}{|z|^2} dA(z)$$

2. There exist $c > 0$ and $\delta > 0$ so that

$$|G_c \cap S(\zeta, h)| \geq \delta h^2$$

for every $|\zeta| = 1$ and $0 < h < 1$ where $G_c = \{z \in D : \tau(z) > c\}$.

To see that the corollary follows from Theorem 3.33 first suppose that

$$\int_D |f'(z)|^2 \tau(z) \log \frac{1}{|z|^2} dA(z) \geq K \int_D |f'(z)|^2 \log \frac{1}{|z|^2} dA(z) \qquad (3.4.2)$$

for all f in $H^2(D)$. Since f in $H^2(D)$ if and only if f' is in $A_1^2(D)$ and $\log \frac{1}{|z|^2} \sim 1 - |z|^2$ near ∂D, Inequality (3.4.2) is equivalent to

$$\int_D |g(z)|^2 \tau(z)(1 - |z|^2) dA(z) \geq K' \int_D |g(z)|^2 (1 - |z|^2) dA(z)$$

for some $K' > 0$ and all g in $A_1^2(D)$. Choose $c < K'/2$ and set $G = \{\tau(z) > c\}$. We have

$$K' \int_D |g(z)|^2 (1 - |z|^2) dA(z) \leq \int_D |g(z)|^2 \tau(z)(1 - |z|^2) dA(z)$$

$$\leq \int_G |g(z)|^2 \tau(z)(1 - |z|^2) dA(z) + \int_{D\backslash G} |g(z)|^2 \tau(z)(1 - |z|^2) dA(z)$$

$$\leq \|\tau\|_\infty \int_G |g(z)|^2 (1 - |z|^2) dA(z) + c \int_D |g(z)|^2 (1 - |z|^2) dA(z)$$

from which we have

$$\frac{K' - c}{\|\tau\|_\infty} \int_D |g(z)|^2 (1 - |z|^2) dA(z) \leq \int_G |g(z)|^2 (1 - |z|^2) dA(z)$$

By Theorem 3.33 there exists $\delta > 0$ so that $|G \cap S(\zeta, h)| \geq \delta h^2$ for all $|\zeta| = 1$ and $0 < h < 1$. The derivation of $(2) \Rightarrow (1)$ of the corollary is essentially immediate from $(2) \Rightarrow (1)$ of Theorem 3.33.

PROOF (of Theorem 3.32) We first prove the theorem under the additional hypothesis that $\varphi(0) = 0$ and then show how the result follows in general. Let T denote the restriction of C_φ to the subspace $H_0^2(D) = \{f \in H^2(D) : f(0) = 0\}$. Note that $H_0^2(D)$ is invariant under C_φ. First suppose that there exists positive constants δ and c so that

$$|G_c^\varphi \cap S(\zeta, h)| \geq \delta h^2$$

for all $|\zeta| = 1$ and $0 < h < 1$, where $G_c^\varphi = \{w \in D : \tau_\varphi(w) > c\}$. Then, using Theorem 2.31 we have

$$\|Tf\|^2 = \|f \circ \varphi\|^2 = \int_D |f'(\varphi(z))|^2 |\varphi'(z)|^2 \log \frac{1}{|z|^2} \, dA(z)/\pi$$

$$= 2 \int_D |f'(w)|^2 N_\varphi(w) \, dA(w)/\pi$$

$$= \int_D |f'(w)|^2 \tau_\varphi(w) \log \frac{1}{|w|^2} \, dA(w)/\pi$$

Since $\varphi(0) = 0$, Littlewood's Inequality (Theorem 2.29) implies $\tau_\varphi(w) \le 1$ on $D \setminus \{0\}$ so τ_φ is a bounded, non-negative, measurable function and by Corollary 3.34 there exists $k > 0$ so that for all f in $H^2(D)$

$$\int_D |f'(z)|^2 \tau_\varphi(z) \log \frac{1}{|z|^2} \, dA(z)/\pi \ge k \int_D |f'(z)|^2 \log \frac{1}{|z|^2} \, dA(z)/\pi$$

In particular, for f in $H_0^2(D)$, we have

$$\|Tf\|^2 \ge k\|f\|^2$$

and T is bounded below on $H_0^2(D)$. From this it follows easily that C_φ is bounded below on $H^2(D)$ since an arbitrary f in $H^2(D)$ can be written as $f = g + f(0)$ where g is in $H_0^2(D)$ and $f \circ \varphi = g \circ \varphi + f(0)$.

Conversely, if $\|f \circ \varphi\|^2 \ge k\|f\|^2$ for all f in $H^2(D)$, we have

$$\int_D |f'(w)|^2 \tau_\varphi(w) \log \frac{1}{|w|^2} dA(w)/\pi \ge k \int_D |f'(w)|^2 \log \frac{1}{|w|^2} dA(w)/\pi$$
$$(3.4.3)$$

Since $\tau_\varphi(w) \le 1$ for $w \ne 0$, Corollary 3.34 guarantees that there exist positive numbers c and δ so that if $G_c^\varphi = \{\tau_\varphi(z) > c\}$,

$$|G_c^\varphi \cap S(\zeta, h)| \ge \delta h^2$$

for all $|\zeta| = 1$ and $0 < h < 1$. This completes the proof in the case $\varphi(0) = 0$.

If $\varphi(0) = u \ne 0$ let $\varphi_u(z) = (u - z)/(1 - \bar{u}z)$. Set $\psi = \varphi_u \circ \varphi$ so that $\psi(0) = 0$ and $C_\psi = C_\varphi C_{\varphi_u}$. We claim

$$C_\varphi \text{ has closed range } \Leftrightarrow C_\psi \text{ has closed range}$$

$$\Leftrightarrow \exists c_1, \delta_1 > 0 \text{ such that } |G_{c_1}^\psi \cap S(\zeta, h)| \ge \delta_1 h^2$$

$$\Leftrightarrow \exists c_2, \delta_2 > 0 \text{ such that } |G_{c_2}^\varphi \cap S(\zeta, h)| \ge \delta_2 h^2$$

for all $|\zeta| = 1$ and $0 < h < 1$. The first equivalence follows from the fact that $\{C_\varphi(f) : f \in H^2(D)\} = \{C_\psi(f) : f \in H^2(D)\}$ since C_{φ_u} is invertible on $H^2(D)$, with inverse C_{φ_u}. The second equivalence is what was just proven, since $\psi(0) = 0$. The final equivalence, the details of which are left to the reader in Exercise 3.4.5, exploits properties of the involution automorphism φ_u. ∎

As an easy application of this theorem, suppose that $\varphi(D) \cap S(\zeta, h) = \emptyset$ for some ζ in the circle and $h > 0$. Then C_φ cannot have closed range on $H^2(D)$, since $\tau_\varphi(z) = 0$ on $S(\zeta, h)$ and hence for every $c > 0$, $|G_c^\varphi \cap S(\zeta, h)| = 0$.

Exercises

3.4.1 Suppose φ is univalent and C_φ does not have closed range on $H^2(D)$ or $A_\alpha^2(D)$ for some $\alpha > -1$. Show that if $\psi(D) \subset \varphi(D)$ then C_ψ does not have closed range either.

3.4.2 Suppose that C_φ has closed range on $H^2(D)$ or $A_\alpha^2(D)$ for some $\alpha > -1$. If $\psi(D) \supset \varphi(D)$ where ψ is univalent, then C_ψ has closed range.

3.4.3 Let G be any measurable subset of the disk. Show that the following are equivalent:
  (1) There exists $\delta > 0$ so that
$$|G \cap S(\zeta, h)| \geq \delta h^2$$
  for all $|\zeta| = 1$ and $0 < h < 1$.
  (2) There exists $\epsilon < 1$ and $\delta' > 0$ so that
$$|G \cap D(b, \epsilon)| \geq \delta'(1 - |b|)^2$$
  for all b in the disk.
  Hints: The easy direction is (2) $\Rightarrow$ (1). Exercise 2.2.6 is relevant here. For (1) $\Rightarrow$ (2) the geometry is a little more complicated. As an intermediate step show that if (1) holds then there exists $\delta'' > 0$ and $0 < s < 1$ so that
$$|G \cap \Delta(b, s)| \geq \delta''(1 - |b|)^2$$
  for all b in D, where $\Delta(b, s)$ is the disk $\{z : |z - b| < s(1 - |b|)\}$.

3.4.4 Show that (1) $\Rightarrow$ (2) in Theorem 3.33.
  Hints: By the previous exercise it is enough to show that
$$|G \cap D(b, \epsilon)| \geq \delta(1 - |b|)^2$$
  for some $\epsilon, \delta > 0$ and all b in D. Use the test functions
$$\frac{(1 - |b|^2)^{(\alpha+2)/2}}{(1 - \bar{b}z)^{\alpha+2}}$$

3.4.5 Suppose $\psi = \varphi_u \circ \varphi$ where $\varphi_u = (u - z)/(1 - \bar{u}z)$ and $\varphi(0) = u$. Complete the proof of Theorem 3.32 by showing that there exist $c_1, \delta_1 > 0$ such that $|G_{c_1}^\psi \cap S(\zeta, h)| \geq \delta_1 h^2$ for all $|\zeta| = 1$ and $0 < h < 1$ if and only if there exists $c_2, \delta_2 > 0$ such that $|G_{c_2}^\varphi \cap S(\zeta, h)| \geq \delta_2 h^2$ for all $|\zeta| = 1$ and $0 < h < 1$.
  Hint: By the Exercise 3.4.3 it is enough to verify this equivalence with the Carleson sets $S(\zeta, h)$, for $|\zeta| = 1$ and $0 < h < 1$ replaced by pseudohyperbolic disks $D(b, \epsilon)$ for b in D and some fixed $0 < \epsilon < 1$. Automorphisms of the disk permute the pseudohyperbolic disks, and for any measurable subset E of D we have $|\varphi_u(E)| \sim |E|$ where $\sim$ indicates that the ratio of the two quantities is bounded above and below by constants depending only on u.

3.4.6 If φ is inner, must C_φ be bounded below on the Bergman spaces $A_\alpha^2(D)$ for $\alpha > -1$?

Notes

The Fredholm composition operators on $H^2(D)$ were first identified by J. Cima, J. Thomson and W. Wogen ([CiTW74]) and later by a different and more general method by P. Bourdon in [Bou90] where Lemma 3.26 appears. Fredholm composition operators on the Dirichlet space $\mathcal{D}$ in the disk were characterized by J. Cima ([Cim77]) and the proof given here of Theorem 3.29 uses some ideas from that paper.

In [CiTW74] the closed range composition operators on $H^2(D)$ are characterized in terms of boundary behavior of φ: C_φ has closed range if and only if there exists $M < \infty$ such that $\sigma(E) \leq M\sigma(\varphi^{-1}(E) \cap \partial D)$ for every Borel measurable set E in ∂D; i.e. $\frac{d(\sigma\varphi^{-1})}{d\sigma}$ is essentially bounded away from 0.

Theorem 3.32 can be extended to the Bergman spaces $A_\alpha^2(D)$, for $\alpha > -1$. The result in this context is: C_φ has closed range on $A_\alpha^2(D)$ if and only if there exist positive numbers c and δ such that

$$|G_c^{\varphi,\alpha} \cap S(\zeta,h)| \geq \delta h^2$$

for $|\zeta| = 1$ and $0 < h < 1$ where

$$G_c^{\varphi,\alpha} \equiv \{z : \tau_{\varphi,\alpha+2}(z) > c\}$$

and $\tau_{\varphi,\alpha+2}$ is defined in terms of the generalized Nevanlinna counting function:

$$\tau_{\varphi,\beta}(w) \equiv \frac{N_{\varphi,\beta}(w)}{(-\log|w|)^\beta}$$

and

$$N_{\varphi,\beta}(w) = \sum_{\varphi(z_j(w))=w} (-\log|z_j(w)|)^\beta$$

Both Theorem 3.32 and this generalization are due to N. Zorboska [Zo94b], as is the result of Exercise 3.4.1. Theorem 3.33 is due to D. Luecking([Lue81]).

3.5   Boundedness on $H^p(B_N)$

In this section we will take our first look at the question of boundedness of C_φ on $H^p(B_N)$ when $N > 1$ and $\varphi : B_N \to B_N$ is analytic. An indication that this may present more difficulty than the $N = 1$ case comes from considering the $N = 1$ boundedness proof outlined in Exercise 3.1.7: $H^p(D)$ can be described as the set of all analytic f in D for which there exists a harmonic u satisfying $|f|^p \leq u$. Then if φ is an analytic map of the disk into itself, the composition $u \circ \varphi$ is harmonic and $|f \circ \varphi|^p \leq u \circ \varphi$ so $f \circ \varphi$ is also in $H^p(D)$. This works because in one variable the harmonic functions on D are exactly the real parts of analytic functions. Moving to $N > 1$, the spaces $H^p(B_N)$ can still be described [Ru80] in terms of harmonic or $\mathcal{M}$–harmonic majorants where a function is called **harmonic** it is annihilated by the ordinary Laplacian Δ where $\Delta g = 4\sum_{i=1}^N D_i \overline{D}_i g$ and $\mathcal{M}$–harmonic if it is annihilated by the invariant Laplacian $\widetilde{\Delta}$ where $\widetilde{\Delta}g(a) = \Delta(g \circ \varphi_a)(0)$ and φ_a is the automorphism defined in Section 2.5. However, when $N > 1$ a harmonic or $\mathcal{M}$–harmonic function need not be the real part of an analytic function. The reader is invited to also look again at the proof of Theorem 3.1 with an eye to finding a similar problem in attempting to generalize it to the case $N > 1$. In spite of these difficulties in extending the proof of boundedness, it is also not obvious that there exist *unbounded* composition operators when $N > 1$. For $N = 2$ the first examples to be discovered included the following maps:

(i) $\varphi(z_1, z_2) = (2z_1 z_2, 0)$

(ii) $\varphi(z_1, z_2) = (z_1^2 + z_2^2, 0)$

(iii) $\varphi(z_1, z_2) = (A\varphi_1, B\varphi_2)$ where $A, B \geq 0$, $A^2 + B^2 = 1$ and φ_i are *inner* functions on B_2.

There are analogous examples for $N > 2$. The class of examples in (iii) requires some comment. An inner function on B_N is a function f in $H^\infty(B_N)$ whose radial limits have modulus 1 almost everywhere with respect to normalized surface area measure σ_N. There are no higher dimensional analogues of the finite Blaschke products and the existence of non-constant inner functions when $N > 1$ was a long standing open problem; the first proofs of their existence were given by A. Aleksandrov [Alv82] and E. Løw [Løw82]. While the original arguments to show that the maps in examples (i)-(iii) gave unbounded operators used ad hoc methods, we will treat these examples in a unified way shortly. Notice however that all of these maps exhibit a certain degeneracy. In (i) and (ii) the image of B_2 under φ is contained in the complex line $[e_1] = \{\lambda e_1 : \lambda \in C\}$, and certain arcs in ∂B_2 are collapsed to single points in ∂B_2. The maps in (iii) collapse a set of full measure in ∂B_2 into a set of measure 0 (either a torus if $A, B > 0$ or a circle if either A or $B = 0$).

The first attempt at a generalized understanding of boundedness of composition operators when $N > 1$ involved a Carleson measure criterion. Recall from Section 2.2 our notation for Carleson sets in $\overline{B_N}$. For ζ in ∂B_N and $h > 0$ we have

$$\mathcal{S}(\zeta, h) = \{z \in \overline{B_N} : |1 - \langle z, \zeta \rangle| < h\}$$
$$Q(\zeta, h) = \mathcal{S}(\zeta, h) \cap \partial B_N$$

In the next theorem φ^* denotes the radial limit of the mapping φ considered as a map of $\partial B_N \to \overline{B_N}$.

THEOREM 3.35

Let $p < \infty$ and suppose $\varphi : B_N \to B_N$ is analytic. Define a Borel measure μ on $\overline{B_N}$ by $\mu(A) = \sigma_N(\varphi^{*-1}(A))$. Then

(1) C_φ *is bounded on* $H^p(B_N)$ *if and only if there exists $C < \infty$ so that*

$$\mu(\mathcal{S}(\zeta, h)) \leq Ch^N \text{ for } \zeta \in \partial B_N, \ h > 0.$$

(2) C_φ *is compact on* $H^p(B_N)$ *if and only if*

$$\mu(\mathcal{S}(\zeta, h)) = o(h^N) \text{ as } h \to 0, \text{ uniformly in } \zeta.$$

As we observed in Section 2.2, the quantity h^N appearing on the right hand side is best interpreted as comparable to $\sigma_N(Q(\zeta, h))$. Part (1) of this theorem is a direct consequence of Theorem 2.37. In the first part of the proof of Theorem 3.35 below, we will, as promised in Section 2.2, give the argument for the easier direction of Theorem 2.37.

Note that an immediate and useful consequence of Theorem 3.35 is the following corollary.

COROLLARY 3.36

If C_φ is bounded (respectively compact) on $H^p(B_N)$ for some finite value of p, then C_φ is bounded (compact) on $H^p(B_N)$ for all p, $0 < p < \infty$.

PROOF (of Theorem 3.35) First suppose that C_φ is bounded on $H^p(B_N)$ and fix any η in ∂B_N and $0 < h < 1$. Consider the test functions

$$f_w(z) = (1 - \langle z, w \rangle)^{-4N/p}$$

where $w = (1 - h)\eta$. For these functions we have

$$\|f_w \circ \varphi\|_p^p = \int_{\partial B_N} |f_w \circ \varphi^*|^p \, d\sigma_N = \int_{\overline{B_N}} |f_w|^p \, d\mu$$

where $\mu = \sigma_N \varphi^{*-1}$ and where we have used the relationship $(f_w \circ \varphi)^*(\zeta) = f_w \circ \varphi^*(\zeta)$ a.e.$[\sigma_N]$, which is obvious since f_w extends continuously to $\overline{B_N}$. Our assumption that C_φ is bounded guarantees that there exists C depending only on φ such that

$$\int_{\partial B_N} |f_w \circ \varphi^*|^p \, d\sigma_N \leq C \int_{\partial B_N} |f_w|^p \, d\sigma_N$$

Thus

$$\int_{\mathcal{S}(\eta, h)} |f_w|^p \, d\mu \leq \int_{\overline{B_N}} |f_w|^p \, d\mu \leq C \int_{\partial B_N} |f_w|^p \, d\sigma_N$$

A computation (Exercise 3.5.2) shows that $|f_w|^p \geq (2h)^{-4N}$ on $\mathcal{S}(\eta, h)$. As outlined in Exercise 2.1.4, we may use the binomial series expansion, the orthogonality of $\langle w, \zeta \rangle^k$ and $\langle w, \zeta \rangle^m$ for $k \neq m$, and Equation (2.1.1) to see that $\|f_w\|_p^p \sim h^{-3N}$; see [Ru80, p. 18] for the details of this calculation. Thus

$$\mu(\mathcal{S}(\eta, h))(2h)^{-4N} \leq C_1 h^{-3N}$$

and therefore $\mu(\mathcal{S}(\eta, h)) \leq C_2 h^N$. This gives the "only if" direction of (1).

Conversely, if $\mu(\mathcal{S}(\eta, h)) \leq K h^N$ for all η in ∂B_N and $h > 0$ we apply the harder direction ((1) implies (2)) of Theorem 2.37 to see that

$$\int_{\overline{B_N}} |f|^p \, d\mu \leq C \int_{\partial B_N} |f^*|^p \, d\sigma_N$$

for some $C < \infty$ and every f in $H^p(B_N)$. In particular if f is in the ball algebra $A(B_N)$ of functions analytic in B_N and continuous in $\overline{B_N}$

$$\|f \circ \varphi\|_p^p = \int_{\partial B_N} |(f \circ \varphi)^*|^p \, d\sigma_N = \int_{\partial B_N} |f \circ \varphi^*|^p \, d\sigma_N =$$

$$= \int_{\overline{B_N}} |f|^p \, d\mu \le C \int_{\partial B_N} |f|^p \, d\sigma_N$$

Since $A(B_N)$ is dense in $H^p(B_N)$ this completes the proof of (1).

For (2) first assume $\mu(\mathcal{S}(\zeta, h)) \ne o(h^N)$, and choose sequences ζ_n in ∂B_N and h_n decreasing to zero and $\beta > 0$ so that $\mu(\mathcal{S}(\zeta_n, h_n)) \ge \beta h_n^N$. For $w_n = (1 - h_n)\zeta_n$ we use again as test functions $f_n(z) = (1 - \langle z, w_n \rangle)^{-4N/p}$, for which $\|f_n\|_p^p \sim h_n^{-3N}$. Let $g_n = f_n / \|f_n\|_p$. Note that g_n tends to 0 uniformly on compact subsets of B_N and

$$\|g_n \circ \varphi\|_p^p = \int_{\partial B_N} |(g_n \circ \varphi)^*|^p \, d\sigma_N = \int_{\partial B_N} |g_n \circ \varphi^*|^p \, d\sigma_N$$

$$= \int_{\overline{B_N}} |g_n|^p \, d\mu \ge \frac{1}{\|f_n\|_p^p} \int_{\mathcal{S}(\zeta_n, h_n)} |f_n^*|^p \, d\mu.$$

Again we use the estimates $|f_n|^p \ge (2h_n)^{-4N}$ and $\|f_n\|_p^p \sim h_n^{-3N}$ to see that this last integral is bounded away from zero as n tends to ∞.

Conversely, assume $\mu(\mathcal{S}(\zeta, h)) = o(h^N)$, uniformly in ζ in ∂B_N. It is easier, and equivalent, to replace the sets $S(\zeta, h)$ by the windows

$$W(\zeta, h) \equiv \{ z \in \overline{B_N} : 1 - |z| < h, \frac{z}{|z|} \in Q(\zeta, h) \}.$$

Given ϵ choose h_0 small enough so that $\mu(W(\zeta, h)) \le \epsilon h^N$ for $h \le h_0$ and all ζ in ∂B_N. Let μ' be the restriction of μ to $\overline{B_N} \setminus (1 - h_0)\overline{B_N}$. We claim μ' is a Carleson measure, with $\mu'(W(\zeta, h)) \le C\epsilon h^N$ for some constant C depending only on N. This is clearly true, with $C = 1$, for $h \le h_0$. Suppose $h > h_0$. Now $Q(\zeta, h)$ can be covered by a finite collection $\{Q(w_j, h_0/3)\}$ where the w_j are in $Q(\zeta, h)$. Since $Q(w_j, h_0/3)$ and $Q(\zeta, h)$ intersect, and $h > h_0/3$, the absorption property described in Exercise 3.5.7 guarantees that there is a constant $s > 0$ independent of h so that

$$Q(\zeta, sh) \supset Q(w_j, h_0/3)$$

for each j. Furthermore, there is (see [Ru80, p. 68]) a disjoint subcollection Γ of $\{Q(w_j, h_0/3)\}$ so that

$$Q(\zeta, h) \subset \bigcup_\Gamma Q(w_j, h_0)$$

We can obtain an upper estimate on card Γ:

$$\sigma_N(\bigcup_\Gamma Q(w_j, h_0/3)) \ge (\text{card } \Gamma) c_1 (h_0/3)^N$$

since $\sigma_N(Q(w_j, h_0/3)) \geq c_1(h_0/3)^N$ for some constant c_1 depending only on N. Also

$$\sigma_N(\bigcup_\Gamma Q(w_j, h_0/3)) \leq \sigma_N(Q(\zeta, sh)) \leq c_2 s^N h^N$$

for some c_2 depending only on N, so

$$\text{card } \Gamma \leq \frac{C_2 s^N h^N}{C_1 (h_0/3)^N} = C\left(\frac{h}{h_0}\right)^N$$

From this it follows that

$$\mu'(W(\zeta, h)) \leq \sum_\Gamma \mu'(W(w_j, h_0)) \leq C\left(\frac{h}{h_0}\right)^N \epsilon h_0^N = C \epsilon h^N$$

as desired.

Finally, suppose $\{f_n\}$ is a bounded sequence in $H^p(B_N)$ with f_n converging to 0 uniformly on compact subsets of B_N. By the first part of the theorem, C_φ is bounded on $H^p(B_N)$ and by Exercise 3.5.1, this implies $(f_n \circ \varphi)^* = f_n^* \circ \varphi^*$ almost everywhere on ∂B_N. Thus

$$\int_{\partial B_N} |(f_n \circ \varphi)^*|^p \, d\sigma_N = \int_{\partial B_N} |f_n^* \circ \varphi^*|^p \, d\sigma_N = \int_{B_N} |f_n^*|^p \, d\mu$$

$$= \int_{\overline{B_N} \setminus (1-h_0)\overline{B_N}} |f_n^*|^p \, d\mu + \int_{(1-h_0)\overline{B_N}} |f_n^*|^p \, d\mu = I + II.$$

Integral (II) can be made as small as desired by choosing n large. By Theorem 2.37 and the remark following it, integral (I) is less than $K\epsilon(\sup \|f_n\|_p^p)$ for some finite constant K. Since ϵ is arbitrary we are done. ∎

As expected, there is a completely analogous version of Theorem 3.35 for the weighted Bergman spaces $A_\alpha^p(B_N)$:

THEOREM 3.37
Let $p < \infty$ and suppose $\varphi : B_N \to B_N$ is analytic. Define a Borel measure μ on B_N by $\mu(A) = \nu_\alpha \varphi^{-1}(A)$ where $d\nu_\alpha = (1 - |z|^2)^\alpha \, d\nu_N$. Then

(1) C_φ is bounded on $A_\alpha^p(B_N)$ if and only if there exists $C < \infty$ so that

$$\mu(S(\zeta, h)) \leq C h^{N+\alpha+1} \text{ for } \zeta \in \partial B_N, \ h > 0.$$

(2) C_φ is compact on $A_\alpha^p(B_N)$ if and only if

$$\mu(S(\zeta, h)) = o(h^{N+\alpha+1}) \text{ as } h \to 0, \text{ uniformly in } \zeta.$$

The proof is left to the reader in Exercise 3.5.8. As a corollary, we see that boundedness and compactness of C_φ on $A_\alpha^p(B_N)$ is independent of $p, p < \infty$.

We will finish this section by examining some other easy consequences of Theorem 3.35, while Chapter 6 will explore some of its deeper consequences. We begin with a result that prevents the symbol map of a bounded operator from having a particularly strong kind of degeneracy.

COROLLARY 3.38
If C_φ is bounded on $H^p(B_N)$ then φ^* cannot carry a set of positive Lebesgue measure in ∂B_N into a set of Lebesgue measure 0.

PROOF Suppose $\varphi^*(A) \subset E \subset \partial B_N$ where $\sigma_N(E) = 0$. For any $\epsilon > 0$ there exists Carleson sets $Q(\zeta_k, h_k)$ with

$$E \subset \bigcup_1^\infty Q(\zeta_k, h_k)$$

and

$$\sum_1^\infty \sigma_N(Q(\zeta_k, h_k)) < \epsilon$$

In particular, $\sum_1^\infty h_k^N < c\epsilon$ for some positive constant c depending only on N. If C_φ is bounded then there exists $C < \infty$ so that

$$\sigma_N \varphi^{*-1} Q(\zeta_k, h_k) \leq C h_k^N$$

for each $k = 1, 2, \ldots$. Thus

$$A \subset \varphi^{*-1}(E) \subset \varphi^{*-1}(\bigcup_1^\infty Q(\zeta_k, h_k))$$

and

$$\sigma_N \varphi^{*-1}(E) \leq \sum_1^\infty \sigma_N \varphi^{*-1} Q(\zeta_k, h_k) \leq C \sum_1^\infty h_k^N \leq Cc\epsilon$$

Since ϵ is arbitrary, we are done. ∎

This corollary explains the previously mentioned examples of unbounded operators induced by maps of the form $\varphi(z_1, z_2) = (A\varphi_1, B\varphi_2)$ (for $A, B \geq 0$, $A^2 + B^2 = 1$ and φ_i inner), since φ^* carries ∂B_N into either the torus $\{|z_1| = A, |z_2| = B\}$ or a circle (if A or B is 0), which are sets of Lebesgue measure 0.

On a positive note, Theorem 3.35 can be used to show that the automorphisms give bounded composition operators on $H^p(B_N)$ (see Exercise 3.5.4). Alternately, this result can be obtained from the $\mathcal{M}$–harmonic majorant description of $H^p(B_N)$; see [Ru80, Theorem 5.6.3].

The examples $\varphi(z) = (2z_1 z_2, 0)$ and $\varphi(z) = (z_1^2 + z_2^2, 0)$ can also be understood in the context of Theorem 3.35. One way to do this is to explicitly compute

$\sigma_N \varphi^{*-1}(S(e_1, h))$. We will instead give a theorem which applies to these maps as special cases. Note that the map $(2z_1 z_2, 0)$ collapses the arc $\{(e^{i\theta}/\sqrt{2}, e^{-i\theta}/\sqrt{2}) : \theta \text{ real}\}$ to the point e_1, and the map $(z_1^2 + z_2^2, 0)$ collapses the circle $\{(z_1, z_2) \in \partial B_2 : \text{Im } z_1 = \text{Im } z_2 = 0\}$ to the point e_1. Our next result gives a condition for C_φ to be unbounded involving the cardinality of inverse images of points in ∂B_N.

THEOREM 3.39

Suppose $\varphi : B_N \to B_N$ is analytic such that the derivative map φ' is uniformly bounded on B_N. If

$$\sup(card\{\varphi^{*-1}(\zeta) : \zeta \in \partial B_N\}) = \infty$$

then C_φ is unbounded on $H^p(B_N)$.

The proof of this theorem uses the following lemma.

LEMMA 3.40

If $\varphi : B_N \to B_N$ is analytic and φ' is uniformly bounded, then there exist $A, \delta > 0$ so that if $\varphi(\zeta) = \eta$ (where $|\zeta| = |\eta| = 1$) then

$$\varphi(Q(\zeta, h)) \subset S(\eta, Ah)$$

for all $h < \delta$.

PROOF Since φ' is uniformly bounded, φ extends to be Lipschitz on $\overline{B_N}$; we will call its extension φ also. First suppose $\zeta = n = e_1$. Set

$$d = \liminf_{z \to e_1} \frac{1 - |\varphi(z)|}{1 - |z|}$$

By Theorem 2.81, $\lim_{r \to 1} D_1 \varphi_1(re_1) = d < \infty$. By Julia's Lemma, Lemma 2.77, φ maps each ellipsoid

$$E(k, e_1) \equiv \left\{ z : \frac{|1 - z_1|^2}{1 - |z|^2} \leq k \right\}$$

into the corresponding ellipsoid $E(dk, e_1)$. Straightforward calculations will verify that for $h < 1/2$

(i) If $z = (z_1, z')$ is in $Q(e_1, h)$ then $(1 - 2h, z')$ is in $E(\frac{2h}{1-2h}, e_1)$.

(ii) $E(\frac{h}{1-h}, e_1) \subset S(e_1, \frac{2h}{1-h})$.

Thus if $z = (z_1, z') \in Q(e_1, h)$ we have

$$\varphi(1 - 2h, z') \subset E(\frac{2dh}{1 - 2h}, e_1) \subset S(e_1, \frac{4dh}{1 - 2h})$$

If M is the Lipschitz constant for φ then if $z = (z_1, z') \in Q(e_1, h)$ we have

$$|\varphi(z) - \varphi(1 - 2h, z')| \leq M|z_1 - 1 + 2h| \leq M(3h)$$

Putting it all together for $z = (z_1, z')$ in $Q(e_1, h)$ and $h < 1/4$ we obtain

$$|1 - \varphi_1(z)| \leq |1 - \varphi_1(1 - 2h, z')| + |\varphi_1(1 - 2h, z') - \varphi_1(z)|$$
$$\leq \frac{4dh}{1 - 2h} + |\varphi(1 - 2h, z') - \varphi(z)|$$
$$\leq \frac{4dh}{1 - 2h} + 3Mh \leq h(8d + 3M)$$

which gives the conclusion that $\varphi(z)$ is in $\mathcal{S}(e_1, Ah)$ for $A = 8d + 3M$, provided $h \leq \delta \equiv 1/4$. While superficially A appears to depend on d, and hence on the normalization $n = \zeta = e_1$ the hypothesis that φ' is uniformly bounded implies that $d \leq \sup \|\varphi'\|$ and can be taken to be a uniform constant. The removal of the normalization then proceeds in the usual way: find unitary maps U, V so that $Ue_1 = \zeta$, $V\eta = e_1$ and apply the above argument to $V \circ \varphi \circ U$. Recall that $\mathcal{S}(e_1, Ah) \supset V \circ \varphi \circ U(Q(e_1, h)) = V \circ \varphi(Q(\zeta, h))$ so $V^{-1}\mathcal{S}(e_1, Ah) = \mathcal{S}(\eta, Ah) \supset \varphi(Q(\zeta, h))$ provided $h \leq 1/4$. ∎

PROOF (of Theorem 3.39)   For an arbitrary given positive integer m we can find η in ∂B_N and distinct $\zeta_1, \zeta_2, \ldots, \zeta_m$ in ∂B_N with $\varphi(\zeta_i) = \eta$. By the previous lemma, there exist finite positive constants A and δ_0 so that for all $h < \delta_0$, $\varphi(Q(\zeta_i, h)) \subset \mathcal{S}(\eta, Ah)$. By choosing h sufficiently small the sets $Q(\zeta_i, h)$ can be made disjoint. Then

$$\sigma_N \varphi^{*-1} \mathcal{S}(\eta, Ah) \geq \sum_1^m \sigma_N Q(\zeta_i, h) \geq mch^N$$

for some constant c depending only on the dimension N and all h sufficiently small. Since m is arbitrary, $\sigma_N \varphi^{*-1}$ is not a Carleson measure and C_φ is unbounded by Theorem 3.35. ∎

In Chapter 6, we will see that in contrast to this sufficient condition for unboundedness, even the strong non-degeneracy requirement that φ be univalent together with the smoothness requirement that φ be analytic in a neighborhood of the closed ball is not sufficient to guarantee that C_φ be bounded.

It is possible to give a sufficient condition for C_φ to be bounded if a little more than univalence is required. Define

$$\Omega(z) = \frac{\|\varphi'(z)\|^2}{|J_\varphi(z)|^2}$$

where $\varphi'(z)$ is the derivative of φ at z, $\| \cdot \|$ denotes its norm as a linear transformation of C^N, and $J_\varphi = \det(\varphi')$ is its Jacobian. Since φ is univalent, $J_\varphi(z)$ is non-zero in B_N ([Ru80, 15.1.8, p. 302]). Of course in one variable $\Omega(z) \equiv 1$.

THEOREM 3.41
Let $\varphi : B_N \to B_N$ *be univalent with* $\Omega(z)$ *bounded in* B_N. *Then* C_φ *is bounded on* $H^p(B_N)$ *for all* $p < \infty$.

Of course it is enough to prove this for the case $p = 2$. It will be convenient to work with a different description of $H^2(B_N)$, analogous to the weighted Dirichlet description of $H^2(D)$ which first appeared in Section 2.2. This alternate description motivates the definition of a one parameter family of weighted Dirichlet spaces in B_N which will be given in Exercise 3.5.9. The complex gradient of an analytic function f on B_N is defined by

$$\nabla f(z) = (D_1 f(z), D_2 f(z), \dots, D_N f(z))$$

LEMMA 3.42
For f analytic in the ball, f is in $H^2(B_N)$ if and only if

$$\int_{B_N} |\nabla f(z)|^2 (1 - |z|^2)\, d\nu_N(z) < \infty$$

where ν_N is normalized volume measure in the ball. Furthermore, $\|f\|_{H^2}^2$ and $|f(0)|^2 + \int_{B_N} |\nabla f|^2 (1 - |z|^2)\, d\nu_N$ are comparable in the sense that there are finite positive constants k_1 and k_2 with

$$\|f\|_{H^2}^2 \le k_1 \left(|f(0)|^2 + \int_{B_N} |\nabla f|^2 (1 - |z|^2)\, d\nu_N \right)$$

and

$$k_2 \left(|f(0)|^2 + \int_{B_N} |\nabla f|^2 (1 - |z|^2)\, d\nu_N \right) \le \|f\|_{H^2}^2$$

for all f analytic on B_N.

PROOF Let $\sum_{s=0}^{\infty} f_s$ be the homogeneous expansion of f in B_N. Since the polynomials f_s are pairwise orthogonal in $L^2(\sigma_N)$ we have $\|f\|_2^2 = \sum_{s=0}^{\infty} \|f_s\|_2^2$. We begin by showing that $\sum_{s=0}^{\infty} \|f_s\|_2^2$ and $|f(0)|^2 + \int_B |\mathcal{R}f(z)|^2 (1 - |z|^2)\, d\nu_N$ are comparable, where $\mathcal{R}f(z)$ is the radial derivative defined by $\mathcal{R}f = \sum_{s=1}^{\infty} s f_s$. Changing to polar coordinates yields

$$\int_{B_N} |\mathcal{R}f(z)|^2 (1 - |z|^2)\, d\nu_N = 2N \int_0^1 r^{2N-1}(1 - r^2)\, dr \int_{\partial B_N} |\mathcal{R}f(r\zeta)|^2\, d\sigma_N(\zeta)$$

Using the orthogonality of the monomials z^α and the definition of $\mathcal{R}f$ we see that the inner integral is

$$\sum_{s=1}^{\infty} r^{2s} s^2 \|f_s\|_2^2$$

from which it follows that

$$\int_{B_N} |\mathcal{R}f(z)|^2 (1 - |z|^2)\, d\nu_N = 2N \sum_{s=1}^{\infty} \frac{2}{(2N + 2s)(2N + 2s + 2)} s^2 \|f_s\|_2^2$$

The latter sum is comparable to $\sum_{s=1}^{\infty} \|f_s\|_2^2$ since

$$\lim_{s \to \infty} \frac{2s^2}{(2N + 2s)(2N + 2s + 2)} = \frac{1}{2}$$

This gives the desired equivalence when ∇f is replaced by $\mathcal{R}f$.

To finish we need to show that

$$\int_{B_N} |\mathcal{R}f(z)|^2 (1 - |z|^2) \, d\nu_N \quad \text{and} \quad \int_{B_N} |\nabla f(z)|^2 (1 - |z|^2) \, d\nu_N$$

are comparable. Since $\mathcal{R}f(z) = \langle \nabla f(z), \bar{z} \rangle$ we have $|\mathcal{R}f(z)|^2 \le |\nabla f(z)|^2$, so one of the two needed inequalities is trivial. For the other inequality we begin by comparing $\int_{\partial B_N} |\nabla f(r\zeta)|^2 \, d\sigma_N(\zeta)$ and $\int_{\partial B_N} |\mathcal{R}f(r\zeta)|^2 \, d\sigma_N(\zeta)$ when f is any monomial z^α with $|\alpha| \ge 1$ and $r \ge 1/2$. For $f(z) = z^\alpha$ we compute using Equation (2.1.1) for the $L^2(\sigma_N)$ norm of a monomial from Section 2.1:

$$\int_{\partial B_N} |\nabla f(r\zeta)|^2 \, d\sigma_N(\zeta) = \frac{(N-1)! r^{2(|\alpha|-1)}}{(N-2+|\alpha|)!} \alpha! |\alpha|$$

and

$$\int_{\partial B_N} |\mathcal{R}f(r\zeta)|^2 \, d\sigma_N(\zeta) = \frac{(N-1)! r^{2|\alpha|}}{(N-1+|\alpha|)!} \alpha! |\alpha|^2$$

Thus for $r \ge 1/2$ and $f(z) = z^\alpha$ we have

$$\int_{\partial B_N} |\nabla f(r\zeta)|^2 \, d\sigma_N(\zeta) \le 4 \frac{(N-1+|\alpha|)}{|\alpha|} \int_{\partial B_N} |\mathcal{R}f(r\zeta)|^2 \, d\sigma_N(\zeta)$$

$$\le 4N \int_{\partial B_N} |\mathcal{R}f(r\zeta)|^2 \, d\sigma_N(\zeta) \tag{3.5.1}$$

To pass from monomials to arbitrary analytic f in B_N note that if $f = \sum_\alpha c(\alpha) z^\alpha$ then $\mathcal{R}f(r\zeta) = \sum_\alpha c(\alpha) |\alpha| r^{|\alpha|} \zeta^\alpha$ and the orthogonality of the monomials gives

$$\|\mathcal{R}f(r\zeta)\|_2^2 = \sum_\alpha |c(\alpha)|^2 |\alpha|^2 r^{2|\alpha|} \|\zeta^\alpha\|_2^2$$

while

$$\nabla f(r\zeta) = \sum_\alpha c(\alpha)(\nabla z^\alpha)(r\zeta)$$

so

$$\int_{\partial B_N} |\nabla f(r\zeta)|^2 \, d\sigma_N(\zeta) = \int_{\partial B_N} \sum_\alpha |c(\alpha)|^2 |\nabla z^\alpha(r\zeta)|^2 \, d\sigma_N(\zeta)$$

Using Inequality (3.5.1), it follows that for all f analytic in the ball and $r \ge 1/2$

$$\int_{\partial B_N} |\nabla f(r\zeta)|^2 \, d\sigma_N(\zeta) \le 4N \int_{\partial B_N} |\mathcal{R}f(r\zeta)|^2 \, d\sigma_N(\zeta)$$

and therefore if R denotes the subset of B_N consisting of those points z with $|z| \geq 1/2$ we have

$$\int_R |\nabla f(z)|^2(1 - |z|^2)\, d\nu_N = 2N \int_{1/2}^1 r^{2N-1}(1 - r^2) \int_{\partial B_N} |\nabla f(r\zeta)|^2\, d\sigma_N(\zeta)$$

$$\leq 8N^2 \int_{1/2}^1 r^{2N-1}(1 - r^2) \int_{\partial B_N} |\mathcal{R}f(r\zeta)|^2\, d\sigma_N$$

$$= 4N \int_R |\mathcal{R}f(z)|^2(1 - |z|^2)\, d\nu_N$$

Since it is easily checked that there exists a constant c independent of g so that

$$\int_{B_N} |g(z)|^2(1 - |z|^2)\, d\nu_N \leq c \int_R |g(z)|^2(1 - |z|^2)\, d\nu_N$$

for all g analytic in B_N we are done.   ∎

PROOF (of Theorem 3.41)   It is enough to assume $p = 2$ and $\varphi(0) = 0$. By the previous lemma and the closed graph theorem, it is enough to show that

$$\int_{B_N} |\nabla(f \circ \varphi)|^2(1 - |z|^2)\, d\nu_N < \infty$$

whenever

$$\int_{B_N} |\nabla f(z)|^2(1 - |z|^2)\, d\nu_N < \infty$$

By the Chain Rule

$$\int_{B_N} |\nabla(f \circ \varphi)|^2(1 - |z|^2)\, d\nu_N$$

$$\leq \int_{B_N} |\nabla f(\varphi(z))|^2 \frac{\|\varphi'(z)\|^2}{|J_\varphi(z)|^2}|J_\varphi(z)|^2(1 - |z|^2)\, d\nu_N$$

$$\leq C \int_{B_N} |\nabla f(\varphi(z))|^2|J_\varphi(z)|^2(1 - |z|^2)\, d\nu_N$$

where $C = \sup \|\varphi'(z)\|^2/|J_\varphi(z)|^2$ which is finite by hypothesis. Now a change of variables in the last integral gives

$$\int_{B_N} |\nabla(f \circ \varphi)|^2(1 - |z|^2)\, d\nu_N \leq C \int_{B_N} |\nabla f(\varphi(z))|^2|J_\varphi(z)|^2(1 - |z|^2)\, d\nu_N$$

$$= C \int_{\varphi(B_N)} |\nabla f(w)|^2(1 - |\varphi^{-1}(w)|^2)\, d\nu_N(w)$$

$$\leq C \int_{\varphi(B_N)} |\nabla f(w)|^2 (1 - |w|^2) \, d\nu_N(w)$$

where the last inequality follows from the Schwarz Lemma $|\varphi(z)| \leq |z|$, since we are assuming $\varphi(0) = 0$. $\quad \blacksquare$

The requirement that $\Omega(z) = \|\varphi'(z)\|^2 / |J_\varphi(z)|^2$ be bounded in B_N holds whenever φ is univalent and the derivative of φ^{-1} is bounded on $\varphi(B_N)$. To see this let $w = \varphi(z)$ and write $\psi = \varphi^{-1}$ so $z = \psi(w)$. Since $\psi \circ \varphi = I$ we have $\psi'(w)\varphi'(z) = I$ so $(\psi'(w))^{-1} = \varphi'(z)$ and $J_\psi(w) = 1/(J_\varphi(z))$. Thus $\Omega(z) = |J_\psi(w)|^2 \|\psi'(w)^{-1}\|^2$. Write $A = (a_{ij})$ for the matrix of $\psi'(w)$ with respect to the standard basis on C^N. The $(i,j)^{th}$ entry of A^{-1} is

$$\frac{1}{\det A}(-1)^{i+j} \det M_{ji}$$

where M_{ji} is the $(j,i)^{th}$ minor of A. To show that $\Omega(z)$ is bounded it is enough to show that $\sum_{i,j=1}^{N} |\det M_{ji}|^2$ is bounded, independent of w. But $|\det M_{ji}| \leq (N-1)! \max |a_{ij}|^{N-1}$ so

$$\sum_{i,j=1}^{N} |\det M_{ji}|^2 \leq (N!)^2 \max\{|a_{ij}|^{2(N-1)} : i, j = 1, \dots, N\}$$

$$\leq (N!)^2 \|A\|^{2(N-1)}$$

Thus if ψ', the derivative of φ^{-1}, is bounded on $\varphi(B_N)$, $\Omega(z)$ is bounded in B_N.

Using ideas of the proof of Theorem 3.41 (see Exercise 3.5.10) it can be shown that if φ is univalent and $\Omega(z)$ is bounded in B_N, then C_φ will be compact on $H^p(B_N)$ if φ has no finite angular derivative at any point of ∂B_N. The same result holds in the weighted Bergman spaces $A^p_\alpha(B_N)$ (see Exercises 3.5.9 and 3.5.10). The converse, that C_φ compact implies φ has no finite angular derivative at any point of ∂B_N, holds for any φ (not necessarily univalent) on a wide variety of spaces. We prove this for the Hardy and weighted Bergman spaces, where the idea of the proof is the same as in the disk; other generalizations will be considered later.

THEOREM 3.43
If C_φ is compact on $H^p(B_N)$ or $A^p_\alpha(B_N)$ for some $p < \infty$ and $\alpha > -1$ then φ cannot have finite angular derivative at any point of ∂B_N.

PROOF It is enough to consider the case $p = 2$. We give the argument first for the Hardy space. Let K_w be the kernel function for evaluation at w in $H^2(B_N)$. Since $K_w = (1 - \langle z, w \rangle)^{-N}$ we have $\|K_w\|^2 = (1 - |w|^2)^{-N}$. The same argument as in Proposition 3.13 shows that

$$\|C_\varphi\|_e \geq \limsup_{|w| \to 1} \frac{\|K_{\varphi(w)}\|}{\|K_w\|}$$

which implies that for C_φ compact

$$\left(\frac{1 - |w|^2}{1 - |\varphi(w)|^2} \right)^N \to 0$$

for any sequence $w \to \zeta$ in ∂B_N, and hence φ cannot have finite angular derivative at ζ.

The argument is entirely analogous in the weighted Bergman space $A_\alpha^2(B_N)$ except now the kernel functions have norms whose squares are comparable to $(1 - |w|^2)^{-(N+\alpha+1)}$ (see Exercise 3.5.9). ∎

Exercises

3.5.1 As a corollary of Theorem 3.35 show that if C_φ is bounded on $H^p(B_N)$, then $(f \circ \varphi)^*(\zeta) = f^*(\varphi^*(\zeta))$ for f in $H^p(B_N)$ and $[\sigma_N]$–almost every ζ in ∂B_N. Here * denotes the radial limit function.

3.5.2 Let $f_w(z) = (1 - \langle z, w \rangle)^{-4N/p}$. Show that $|f_w|^p \geq (2h)^{-4N}$ on $S(\eta, h)$ where $w = (1 - h)\eta$.

3.5.3 Let $\varphi(z) = (c(\alpha) z^\alpha, 0')$ be a map on B_N where $\alpha = (\alpha_1, \ldots, \alpha_N)$ is a multi-index with at least two non-zero entries and

$$c(\alpha) = \frac{|\alpha|^{|\alpha|/2}}{\prod_{\alpha_i \neq 0} \alpha_i^{\alpha_i/2}}$$

   (a) Show $\varphi(B_N) \subset B_N$.
   (b) Show C_φ is unbounded on $H^p(B_N)$.

3.5.4 Use Theorem 3.35 to show that any automorphism of B_N gives a bounded composition operator on $H^p(B_N)$. Hint: Its enough to consider the special automorphisms φ_a defined in Section 2.5. Show

$$\varphi_a(S(\zeta, \delta)) \subset S\left(\varphi_a(\zeta), \frac{1 + |a|}{1 - |a|} \delta\right)$$

3.5.5 Consider any $\varphi : B_2 \to B_2$ with φ' uniformly bounded on B_2 and suppose $\varphi^{-1}(e_1)$ is the one-dimensional torus $\{(\frac{1}{\sqrt{2}} e^{i\theta}, \frac{1}{\sqrt{2}} e^{-i\theta}) : 0 \leq \theta \leq 2\pi\}$. Examples include $\varphi(z_1, z_2) = (2z_1 z_2, 0)$ and $\varphi(z_1, z_2) = (1/2 + z_1 z_2, z_2(1/2 - z_1 z_2))$.
   Consider $Q(\zeta_k, \delta) = \{\eta \in \partial B : |1 - \langle \zeta_k, \eta \rangle| < \delta\}$ where

$$\zeta_k = (\frac{1}{\sqrt{2}} e^{ik/\sqrt{n}}, \frac{1}{\sqrt{2}} e^{-ik/\sqrt{n}}) \text{ for } k = 1, 2, \ldots, [\sqrt{n}]$$

and $\delta = \frac{1}{2n}$.
   (a) Show that the $Q(\zeta_k, \delta)$ are pairwise disjoint.
   (b) Show that $\sigma_N \varphi^{-1} S(e_1, \delta) \geq C \delta^{3/2}$.

3.5.6 Show C_φ is Hilbert–Schmidt on $H^2(B_N)$ if and only if

$$\int_{\partial B_N} (1 - |\varphi(\zeta)|)^{-N} d\sigma_N(\zeta) < \infty$$

3.5.7 For z, w in $\overline{B_N}$ set $d(z, w) \equiv |1 - \langle z, w \rangle|$. Show that $d(\zeta, \eta)^{1/2}$ is a metric on ∂B_N and satisfies the triangle inequality in $\overline{B_N}$. From this it follows that $d(\zeta, \eta)$ is a quasi-metric on $\overline{B_N}$: Given z, w, and v in $\overline{B_N}$ we have

$$d(z, v) \leq 3\{d(z, w) + d(w, v)\}$$

Also show that the "balls" $Q(\zeta, \delta)$ have the following absorption property: There exists a constant $s > 0$ so that if $Q(\zeta, \delta_1)$ and $Q(\eta, \delta_2)$ intersect, where $\delta_1 > \delta_2$ then $Q(\zeta, s\delta_1) \supset Q(\eta, \delta_2)$.

3.5.8 Prove Theorem 3.37. Hints: Use the test functions

$$f(z) = (1 - \langle z, w \rangle)^{-4(N+\alpha+1)/p}$$

where $w = (1 - h)\zeta$. As in the proof of Theorem 3.35 we have $|1 - \langle z, w \rangle| \leq 2h$ on $\mathcal{S}(\zeta, h)$. You may use the estimate

$$\int_{B_N} \frac{(1 - |z|^2)^\alpha}{|1 - \langle z, w \rangle|^{N+1+\alpha+c}} d\nu_N(z) \sim (1 - |w|^2)^{-c}$$

for $c > 0$ from Exercise 2.1.4(c).

3.5.9 For $\alpha > -1$ set

$$\mathcal{D}_\alpha(B_N) = \{ \text{analytic } f = \sum_{s=0}^{\infty} f_s : \sum_{s=0}^{\infty} \|f_s\|_2^2 (s+1)^{1-\alpha} < \infty \}$$

normed by $\|f\|_{\mathcal{D}_\alpha}^2 = \sum \|f_s\|_2^2 (s+1)^{1-\alpha}$. Recall that $D_1(B_N)$ is $H^2(B_N)$.

(a) Show

$$f \in \mathcal{D}_\alpha(B_N) \Leftrightarrow \int_{B_N} |\nabla f(z)|^2 (1 - |z|^2)^\alpha \, d\nu_N(z) < \infty$$

with the series norm and $|f(0)|^2 + \int_{B_N} |\nabla f(z)|^2 (1 - |z|^2)^\alpha \, d\nu_N(z)$ being comparable.

(b) Show $\mathcal{D}_{\alpha+2}(B_N) = A_\alpha^2(B_N)$, again with an equivalence of norms.

(c) Extend Theorem 3.41 to $\mathcal{D}_\alpha(B_N)$, $\alpha \geq 0$.

(d) Show that the reproducing kernel functions K_w in $A_\alpha^2(B_N)$ have norms whose squares are comparable to $(1 - |w|^2)^{-(N+\alpha+1)}$.

3.5.10 Suppose φ is univalent in B_N and suppose

$$\Omega(z) \equiv \frac{\|\varphi'(z)\|^2}{|J_\varphi(z)|^2}$$

is bounded in B_N. Show that if φ has no finite angular derivative at any point of ∂B_N, then C_φ is compact on $H^2(B_N)$ and also on $\mathcal{D}_\alpha(B_N)$, for $\alpha \geq 0$.

3.5.11 Use slice integration (Equation (2.1.6)) to show that if $\varphi(z_1, z_2) = g(z_1, z_2)(z_1, z_2)$ where $g : B_2 \to D$ is analytic, then C_φ is bounded on $H^p(B_N)$ and $\|f \circ \varphi\| \leq \|f\|$. If g is chosen to be an inner function of B_2 this gives an example of an inner mapping on B_2 that is not an automorphism but for which C_φ is bounded.

3.5.12 Show that the mapping $\varphi(z_1, z_2) = (z_1, z_1)$ from the polydisk D^2 into itself does not induce a bounded composition operator on $H^2(D^2)$.

3.5.13 Let M_n be the subspace in $H^2(B_2)$ spanned by $\{z_1^k z_2^n\}_{k=0}^{\infty}$ for each $n = 0, 1, \ldots$, and note $H^2(B_2) = \sum_{n=0}^{\infty} \oplus M_n$.

(a) Show that $W_n(f(z)) = f(z_1)z_2^n$ defines a unitary operator from the Bergman space $A_\alpha^2(D)$ with $\alpha = n$ onto M_n. Hint: W_n carries the orthogonal basis $\{z^k\}$ in $A_n^2(D)$ onto the orthogonal basis $\{z_1^k z_2^n\}_{k=0}^{\infty}$ for M_n. Show W_n preserves norms on this basis and extend by linearity and continuity.

(b) Let $\varphi(z) = sz + (1 - s)$ for $0 < s < 1$ and write T_n for C_φ acting on $A_n^2(D)$. Let ψ map B_2 into B_2 by $\psi(z) = sz + (1 - s)e_1$ where $e_1 = (1, 0)$ and consider C_ψ acting on $H^2(B_2)$. Show that $C_\psi W_n = s^n W_n T_n$ for each n and thus that M_n reduces C_ψ.

(c) Conclude that C_ψ acting on $H^2(B_2)$ is unitarily equivalent to

$$\sum_{n=0}^{\infty} \oplus s^{(n-1)/2} C_\varphi$$

acting on

$$H^2(D) \oplus H^2(D) \oplus H^2(D) \oplus \cdots$$

Hint: See part (c) of Exercise 3.1.10.

Notes

Among the early examples of unbounded composition operators on $H^p(B_2)$, the example $\varphi(z_1, z_2) = (2z_1 z_2, 0)$ is due to J. H. Shapiro and the examples $\varphi(z_1, z_2) = (\psi(z_1 z_2), 0)$ for ψ inner were given by B. D. MacCluer [Mc84a] and J. A. Cima, C. S. Stanton, and W. R. Wogen [CiSW84].

Theorem 3.35 and Corollary 3.36 were noticed by B. D. MacCluer [Mc85]. Corollary 3.38, which also appears in [Mc85] with another proof, was suggested by D. H. Lueck-ing. Theorem 3.39 is due to J. A. Cima and W. R. Wogen [CiW87]. Theorem 3.41 and the related results of Exercises 3.5.10 and 3.5.9 were given by B. D. MacCluer and J. H. Shapiro in [McS86], where Lemma 3.42 was used implicitly, though without the details of the proof given. An earlier theorem of J. A. Cima, C. S. Stanton and W. R. Wogen showed by a quite different argument that if φ is univalent with $|J_\varphi(z)| \geq c > 0$ on B_2 and $|D_i \varphi_j(z)| \leq M < \infty$ for $i, j = 1, 2$, then C_φ is bounded on $H^2(B_2)$. Of course, under such hypotheses $\Omega(z)$ is bounded in B_2.

Theorem 3.43 can be proved in a variety of ways. In addition to the Hilbert space proof given here, it can be obtained from direct calculation with Carleson measures ([McS86]) or by an argument which parallels J. H. Shapiro and P. D. Taylor's original proof in the disk (see [Mc84a, Lemma 1.6 and Theorem 1.7]).

The maps of Exercise 3.5.3, which, of course, generalize the map $(2z_1 z_2, 0)$, are ubiquitous in the literature as a source of examples and counterexamples — see [Mc85], [McS86], [McM93], and [Lis94]. The example of Exercise 3.5.11 is due to J. A. Cima, C. S. Stanton, and W. R. Wogen [CiSW84]. The example of Exercise 3.5.12 is due to S. Sharma and R. Singh [SiS81].

The proof that automorphisms give bounded composition operators on $H^p(B_N)$ outlined in Exercise 3.5.4 uses the Carleson criterion of Theorem 3.35. Other arguments are possible; see for example [Ru80, p. 85].

The result of Exercise 3.5.13 is due to T. L. Kriete and B. D. MacCluer (unpublished).

Appropriately defined analogues $B(P, h)$ of the non-isotropic balls $Q(\zeta, h)$ (and associated admissible convergence regions) can be defined for arbitrary C^2 bounded domains in C^N, having $(N - 1)$-dimensional area measure $\sim h^N$ and satisfying the crucial absorption property of Exercise 3.5.7; see [Kra92, §8.6] by S. Krantz for details.

4

Small Spaces

4.1 Compactness on small spaces

We next consider composition operators acting on fairly general spaces of *bounded* analytic functions on B_N for $N \geq 1$ (of course, $B_1 = D$). To begin with we will simply suppose that $(\mathcal{Y}, \| \cdot \|_{\mathcal{Y}})$ is a Banach space of analytic functions on D or B_N satisfying

(1) Every function in $\mathcal{Y}$ is bounded on B_N.

(2) $\mathcal{Y}$ contains the polynomials.

(3) Evaluation at each point of B_N is a bounded linear functional.

(4) $\mathcal{Y}$ is $\mathcal{U}$–invariant; that is, if f is in $\mathcal{Y}$ and U is a unitary transformation of B_N then $f \circ U$ is also in $\mathcal{Y}$.

We refer to such spaces as "small" spaces. The first and third assumptions guarantee that convergence in the norm of $\mathcal{Y}$ implies convergence in the sup norm: the identity map from $(\mathcal{Y}, \| \cdot \|_{\mathcal{Y}})$ to $(\mathcal{Y}, \| \cdot \|_{\infty})$ is continuous by the closed graph theorem. Moreover, another closed graph theorem argument using (3) shows that (4) implies that C_U is bounded on $\mathcal{Y}$ whenever U is unitary.

Some examples of small spaces in one and several variables are:

- $H^{\infty}(B_N)$ and the ball algebra $A(B_N)$ for $N \geq 1$, both with the supremum norm.

- The analytic Lipschitz spaces $\text{Lip}_{\alpha}(B_N)$ for $0 < \alpha \leq 1$ and $N \geq 1$. The space $\text{Lip}_{\alpha}(B_N)$ is the set

$$\{f \text{ analytic in } B_N : |f(z) - f(w)| = O(|z - w|^{\alpha}) \text{ for all } z, w \in \overline{B_N}\}$$

Norms are computed by

$$\|f\|_{\alpha} = |f(0)| + \sup \left\{ \frac{|f(z) - f(w)|}{|z - w|^{\alpha}} : z \neq w \in \partial B_N \right\}$$

- $S^p(D) = \{f \text{ analytic in } D : f' \in H^p(D)\}$ for $p \geq 1$, with

$$\|f\|_{S^p} = |f(0)| + \|f'\|_p$$

As with the Lipschitz spaces, functions in these spaces extend to be continuous on the closed disk (see [Dur70, p. 42]). Analogous spaces exist for $N > 1$ using, for example, the radial derivative $\mathcal{R}f$ of f; however the hypothesis "$\mathcal{R}f$ belongs to $H^p(B_N)$" will only guarantee that f is in $H^\infty(B_N)$ when $p > N$ ([Kra92, p. 368].

- Weighted Hardy spaces $H^2(\beta, D)$, where the weight sequence satisfies

$$\sum_{n=1}^{\infty} \frac{1}{\beta(n)^2} < \infty$$

The analytic Lipschitz spaces also have a characterization in terms of pointwise growth of derivatives. In one variable this takes the following form:

THEOREM 4.1
If f is analytic in D, then f is in $Lip_\alpha(D)$ for some $0 < \alpha < 1$ if and only if

$$|f'(z)| = O\left(\frac{1}{1-|z|^2}\right)^{1-\alpha}$$

Furthermore,

$$\sup\{(1-|z|^2)^{1-\alpha}|f'(z)| : z \in D\}$$

and the Lipschitz constant

$$\sup\left\{\frac{|f(z)-f(w)|}{|z-w|^\alpha} : z \neq w \in \partial B_N\right\}$$

are comparable.

PROOF   We refer the reader to [Dur70, p. 74] for the details of the proof. If f is in $Lip_\alpha(D)$ the Lipschitz condition and the Cauchy integral formula for f' show that

$$|f'(z)| \leq C\left(\frac{1}{1-|z|^2}\right)^{1-\alpha}$$

where C can be taken to be a constant multiple of the Lipschitz constant. For the converse direction assume

$$|f'(z)| \leq C\left(\frac{1}{1-|z|^2}\right)^{1-\alpha}$$

and note that

$$f(e^{i\theta}) - f(e^{i\tau}) = \int_\Gamma f'(\zeta)d\zeta$$

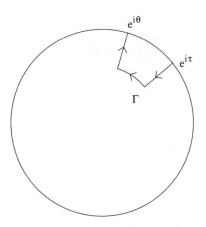

FIGURE 4.1

Contour for the proof of Theorem 4.1.

where Γ is the contour consisting of 2 radial segments and a part of the circle of radius $1 - (\theta - \tau)$ as shown in Figure 4.1. Estimating the integral over each of the three pieces of this contour gives $|f(e^{i\theta}) - f(e^{i\tau})| \leq C'|\theta - \tau|^\alpha$ where C' is a constant multiple of C. ∎

A similar result holds in several variables: For $0 < \alpha < 1$

$$f \in \mathrm{Lip}_\alpha(B_N) \Leftrightarrow \sup\{(1 - |z|^2)^{1-\alpha}|\mathcal{R}f(z)| : z \in B_N\} < \infty$$

where $\mathcal{R}f$ is the radial derivative of f

$$(\mathcal{R}f)(\lambda\zeta) = \lambda f'_\zeta(\lambda)$$

for ζ in ∂B_N, λ in D, and $f_\zeta(\lambda) = f(\lambda\zeta)$. For a proof, the reader is referred to [Ru80, §6.4].

The functions in many of the small spaces of interest to us, such as the Lipschitz spaces, $S^p(D)$ (for $p \geq 1$), $S^p(B_N)$ (for $p > N > 1$), and $H^2(\beta, D)$ when $\sum \beta(n)^{-2} < \infty$, all extend continuously to the closure of the disk or the ball. Such spaces will be referred to as **boundary-regular** small spaces.

We begin with a very general theorem that limits the possible eigenvalues of a composition operator induced by a map that fixes 0, acting on any space of analytic functions in B_N.

THEOREM 4.2
Suppose $\varphi : B_N \to B_N$ is analytic with $\varphi(0) = 0$ and suppose f, not the zero function, is analytic on B_N with $f \circ \varphi = \lambda f$. Then λ is either 0, 1, or a product of the eigenvalues of $\varphi'(0)$.

When $N = 1$ this is essentially part of Theorem 2.63 and Lemma 2.62. Alternately, a proof can be obtained by taking k^{th} derivatives on both sides of the equation $f \circ \varphi = \lambda f$ and evaluating at 0. A proof for the case $N > 1$ is outlined in Exercise 4.1.1. This is the easiest case of the very general argument that is used in the proof of Theorem 7.20 to find the spectrum of compact composition operators.

An analytic map $\varphi : B_N \to B_N$ is said to be ***unitary on a slice*** if there exist ζ and η in ∂B_N so that $\varphi(\lambda \zeta) = \lambda \eta$ for all λ in D. Of course, when $N = 1$ this simply means φ is a rotation. Furthermore, if $\varphi(0) = 0$ and φ is not unitary on any slice, then the eigenvalues of $\varphi'(0)$ all have modulus strictly less than 1, and for any z_0 in B_N, the iterates $\varphi_n(z_0)$ converge to 0. Note this last observation implies that if $f \circ \varphi = f$, then f must be constant.

THEOREM 4.3

For $N \geq 1$ let φ be an analytic map of B_N into B_N with $\varphi(0) = 0$. Suppose C_φ is compact on the small Banach space $\mathcal{Y}$. Then $\langle \varphi_n, \zeta \rangle$, the coordinate of the nth iterate of φ in the $\zeta-$ direction, satisfies

$$\| \langle \varphi_n, \zeta \rangle \|_\infty \to 0$$

as $n \to \infty$ for every ζ in ∂B_N.

PROOF   First we show that φ cannot be unitary on any slice. When $N = 1$ this is obvious: if φ were a rotation then C_φ would be invertible and hence not compact. If $N > 1$ we may use the unitary invariance of $\mathcal{Y}$ to assume that $\varphi(\lambda e_1) = \lambda e_1$ for all λ in D. The adjoint of C_φ satisfies

$$C_\varphi^*(K_{\lambda e_1}) = K_{\varphi(\lambda e_1)} = K_{\lambda e_1}$$

where $K_{\lambda e_1}$ is the linear functional of evaluation at λe_1. Thus C_φ^* has eigenvalue 1 with eigenvectors $\{K_{\lambda e_1} : \lambda \in D\}$. Since $\mathcal{Y}$ contains the polynomials these evaluation functionals are all linearly independent (as λ ranges over D), so C_φ^* and hence also C_φ cannot be compact.

Next we consider the subspace

$$\mathcal{Y}_0 = \{f \in \mathcal{Y} : f(0) = 0\}$$

which is a closed subspace of $\mathcal{Y}$. Since $\varphi(0) = 0$, $\mathcal{Y}_0$ is invariant under C_φ. Thus the restriction T of C_φ to $\mathcal{Y}_0$ is also compact. If $\lambda \neq 0$ is in the spectrum of T it must be an eigenvalue and there is f in $\mathcal{Y}_0$ so that $Tf = \lambda f$. Note $\lambda \neq 1$ since in this case f would be constant and hence the zero function (consider $f \circ \varphi_n = f$ and let n go to infinity). By Theorem 4.2 and the assumption that φ is not unitary on any slice we have $|\lambda| < 1$. In particular the spectral radius of T, $\rho(T)$, must be strictly less than 1. But $\rho(T) = \lim_{n \to \infty} \|T^n\|^{1/n}$ so that $\|T^n\| \to 0$ as n tends to infinity. Pick any ζ in ∂B_N and consider the polynomial p_ζ in $\mathcal{Y}_0$ defined by $p_\zeta(z) = \langle z, \zeta \rangle$. We have

$$\| \langle \varphi_n, \zeta \rangle \|_{\mathcal{Y}} = \| T^n p_\zeta \|_{\mathcal{Y}} \leq \| T^n \| \| p_\zeta \|_{\mathcal{Y}} \to 0$$

as n tends to infinity. But the topology in $\mathcal{Y}$ is stronger than the sup-norm topology so also

$$\|\langle \varphi_n, \varsigma \rangle\|_\infty \to 0$$

as n tends to infinity, and we are done. $\blacksquare$

For boundary-regular small spaces Theorem 4.3 immediately yields the following striking compactness theorem.

THEOREM 4.4
Let $\mathcal{Y}$ be a boundary-regular small space. Suppose $\varphi(0) = 0$ and C_φ is compact on $\mathcal{Y}$. Then $\|\varphi\|_\infty < 1$.

PROOF Since $\mathcal{Y}$ is boundary-regular, φ must extend continuously to $\overline{B_N}$. If $\|\varphi\|_\infty = 1$, we may use unitary invariance to suppose $\varphi(e_1) = e_1$ hence also $\varphi_n(e_1) = e_1$. Thus $\|\langle \varphi_n(z), e_1 \rangle\|_\infty = 1$ for all n, in contradiction to Theorem 4.3.
$\blacksquare$

Exercises 4.1.2 and 4.1.3 explore further consequences of Theorem 4.3 for small spaces which are not necessarily boundary-regular.

Perhaps surprisingly, the hypothesis $\varphi(0) = 0$ in Theorem 4.4 cannot be removed unless an additional condition is placed on the space $\mathcal{Y}$. To see this, take $\mathcal{Y} = H^2(\beta, D)$ where $\beta(n) = 2^n$, a boundary-regular small space with orthonormal basis $\{z^n/\beta(n)\}$ (see Exercise 3.1.5). Let $\varphi(z) = (1+z)/2$ so that $\varphi(1) = 1$ and $\varphi'(1) = 1/2$. Since

$$\sum_{n=0}^\infty \left\| C_\varphi \left(\frac{z^n}{\beta(n)} \right) \right\| = \sum_{n=0}^\infty \frac{\|\varphi^n\|}{\beta(n)} = \sum_{n=0}^\infty \frac{\|(1+z)^n\|}{4^n}$$

$$\leq \sum_{n=0}^\infty \frac{1}{4^n} \left\| \sum_{k=0}^n C(n,k)z^k \right\| \leq \sum_{n=0}^\infty \frac{1}{4^n} \sum_{k=0}^n C(n,k)2^k$$

$$= \sum_{n=0}^\infty (3/4)^n < \infty$$

where $C(n,k)$ is the binomial coefficient $n!/(k!(n-k)!)$, C_φ is trace class on $H^2(\beta)$ by Theorem 3.24 (and thus of course compact). This example is somewhat unconvincing because the space is not algebraically consistent: the functions in this space all extend to be analytic on the disk of radius 4 and the map φ takes that disk properly into itself, but a more convincing example of this type is given below (Example 4.6).

Examples of this sort can be avoided by replacing the condition that $\mathcal{Y}$ be unitarily invariant by the stronger requirement that $\mathcal{Y}$ be automorphism invariant, that is, whenever f is in $\mathcal{Y}$ and φ is in $\mathrm{Aut}(B_N)$, then $f \circ \varphi$ is in $\mathcal{Y}$. Note that C_φ

will then be bounded on $\mathcal{Y}$ for all φ in $\text{Aut}(B_N)$. The ball algebra, the analytic Lipschitz spaces, S^p, $H^2(\beta, D)$ with $\beta(n) = (n+1)^a$ for $a > 1/2$ are all examples of boundary-regular automorphism invariant spaces.

THEOREM 4.5
Suppose $\mathcal{Y}$ is an automorphism invariant boundary-regular small space. If C_φ is compact on $\mathcal{Y}$, then $\|\varphi\|_\infty < 1$.

PROOF   If $\varphi(0) = u$, let φ_u in $\text{Aut}(B_N)$ satisfy $\varphi_u^{-1} = \varphi_u$ and $\varphi_u(u) = 0$. If C_φ is compact, so is C_ψ where $\psi = \varphi_u \circ \varphi$. Since $\psi(0) = 0$ we have $\|\psi\|_\infty < 1$. But $\varphi = \varphi_u \circ \psi$ so $\|\varphi\|_\infty < 1$.   ∎

Another interesting example of a boundary-regular small space is $\mathcal{AC}$, the space of all analytic functions in D with Taylor series coefficients that are summable, normed by $\|f\|_{\mathcal{AC}} = \sum |a_n|$ where $f = \sum a_n z^n$. When considered as a space of functions on $\overline{D}$, this space is algebraically consistent. Though not obvious, $\mathcal{AC}$ is not automorphism invariant [Alp60]. But by part (a) of Exercise 4.1.8, if C_φ is compact on $\mathcal{AC}$, then $\|\varphi\|_\infty < 1$. Thus automorphism invariance is a sufficient, but not a necessary condition for a boundary-regular small space to fail to support a compact C_φ with $\|\varphi\|_\infty = 1$.

While the role that automorphism invariance plays in the proof is transparent, its role in the statement of the theorem is more mysterious. To explore this a little further we next consider any weighted Hardy space $H^2(\beta)$ where $\{\beta(n)\}$ is an increasing sequence satisfying

$$\sum_{n=0}^{\infty} \frac{\beta([\alpha n])}{\beta(n)} < \infty \qquad (4.1.1)$$

for some α, $1 > \alpha > 1/2$ where $[x]$ denotes the greatest integer less than or equal to x. Some remarks may help put Condition (4.1.1) in perspective. If $\beta(n) = R^n$ where $R > 1$ then Condition (4.1.1) is satisfied and all functions in the corresponding space $H^2(\beta)$ are actually analytic in the disk of radius R. However if $\beta(n) = \exp(n^a)$ for any $0 < a < 1$ then Condition (4.1.1) is still satisfied (see Exercise 4.1.4) but it follows from a basic result about random series [Kah85, p. 40] that the space $H^2(\beta)$ contains functions which are not continuable across any point of ∂D. We will see later (Corollary 7.15), however, that for any sequence $\beta(n)$ satisfying Condition (4.1.1), all functions in $H^2(\beta)$ are in $C^\infty(\overline{D})$. Finally, the weight sequences $\beta(n) = (n + 1)^a$ for $a > 1/2$ give rise to boundary-regular small spaces which fail to satisfy Condition (4.1.1).

Our interest in Condition (4.1.1) is in the following example.

EXAMPLE 4.6
Whenever an increasing sequence $\beta(n)$ satisfies Condition (4.1.1) then the function $\varphi(z) = (z + 1)/2$ gives rise to a compact operator on $H^2(\beta)$, and consequently

$H^2(\beta)$ can not be automorphism invariant. We will verify that C_φ is compact by doing a calculation which actually shows that it is trace class. To this end, begin with the orthonormal basis $\{e_n\}$ where $e_n(z) = z^n/\beta(n)$. We have

$$\|C_\varphi e_n\| = \frac{\|\varphi^n\|}{\beta(n)} = \frac{1}{\beta(n)} \frac{1}{2^n} \|(1+z)^n\|$$

$$= \frac{1}{\beta(n)} \frac{1}{2^n} \left\| \sum_{k=0}^{n} C(n,k) z^k \right\|$$

$$\leq \frac{1}{\beta(n)} \frac{1}{2^n} \sum_{k=0}^{n} C(n,k) \beta(k)$$

where $C(n,k)$ are the appropriate binomial coefficients. Our goal is to show that

$$\sum_{n=0}^{\infty} \frac{1}{\beta(n)} \frac{1}{2^n} \sum_{k=0}^{n} C(n,k) \beta(k) \qquad (4.1.2)$$

converges so C_φ satisfies the hypotheses of Theorem 3.24. We concentrate first on the inner sum, remembering that $\beta(n)$ is increasing so we may write, for our fixed $\alpha > 1/2$,

$$\sum_{k=0}^{n} C(n,k) \beta(k) = \sum_{k=0}^{[\alpha n]} C(n,k) \beta(k) + \sum_{k=[\alpha n]+1}^{n} C(n,k) \beta(k) \qquad (4.1.3)$$

$$\leq \beta([\alpha n]) 2^n + \sum_{k=[\alpha n]+1}^{n} C(n,k) \beta(k)$$

Since

$$\sum_{n=0}^{\infty} \frac{1}{\beta(n)} \frac{1}{2^n} \sum_{k=0}^{[\alpha n]} C(n,k) \beta(k) \leq \sum_{n=0}^{\infty} \frac{\beta([\alpha n])}{\beta(n)} < \infty$$

by hypothesis, it is only necessary to consider further that part of Expression (4.1.2) corresponding to the second sum in Equation (4.1.3),

$$\sum_{n=0}^{\infty} \frac{1}{\beta(n)} \frac{1}{2^n} \sum_{k=[\alpha n]+1}^{n} C(n,k) \beta(k)$$

The binomial coefficients $C(n,k)$ are non-increasing for k in the range of summation, since $\alpha > 1/2$. Thus we can make the simple estimate

$$\sum_{k=[\alpha n]+1}^{n} C(n,k) \beta(k) \leq n C(n, [\alpha n]+1) \beta(n) \leq n C(n, [\alpha n]) \beta(n)$$

For n sufficiently large, say $n \geq N_0$, use Stirling's formula

$$\lim_{n\to\infty} \frac{\sqrt{2n\pi}(n/e)^n}{n!} = 1$$

and the bounds $[\alpha n] > \alpha n - 1$ and $n - [\alpha n] \geq n - \alpha n$ to obtain the estimate

$$C(n, [\alpha n]) \leq \frac{c(\alpha)\sqrt{n}}{(\alpha^\alpha (1-\alpha)^{1-\alpha})^n}$$

where $c(\alpha)$ is a constant depending only on α. Since the function $x^x(1-x)^{1-x}$ is strictly increasing on $[1/2, 1)$ with value $1/2$ when $x = 1/2$, the assumption $\alpha > 1/2$ yields $\alpha^\alpha(1-\alpha)^{1-\alpha} > 1/2$. Thus

$$\sum_{n=N_0}^\infty \frac{1}{\beta(n)} \frac{1}{2^n} \sum_{k=[\alpha n]+1}^n C(n,k)\beta(k) \leq \sum_{n=N_0}^\infty \frac{1}{\beta(n)} \frac{1}{2^n} nC(n, [\alpha n])\beta(n)$$

$$\leq \sum_{n=N_0}^\infty \frac{nc(\alpha)\sqrt{n}}{2^n(\alpha^\alpha(1-\alpha)^{1-\alpha})^n} < \infty$$

since $(\alpha^\alpha(1-\alpha)^{1-\alpha}) > \frac{1}{2}$. ⬚

So for boundary-regular spaces that are so small as to fail to be automorphism invariant there can exist compact C_φ with $\|\varphi\|_\infty = 1$ and finite angular derivative at some point of ∂D.

We conclude this section by looking at the question of whether φ in the small space $\mathcal{Y}$ and $\|\varphi\|_\infty < 1$ implies C_φ compact on $\mathcal{Y}$. We will confine our attention to the case of weighted Hardy spaces $H^2(\beta)$. For the choice $\beta(n) = R^n$ where $R > 1$ it is an easy computation to show that $\varphi(z) = z^k/R^{k-1}$, k an integer, $k \geq 2$, gives rise to an isometry on $H^2(\beta)$ even though $\|\varphi\|_\infty = 1/(R^{k-1}) < 1$. This example, however, seems to depend on a pathology of the space $H^2(\beta)$, namely that all functions in the space are in fact analytic in the disk RD and so the natural domain of $H^2(\beta)$ is RD not D, that is, the space is not algebraically consistent. We pursue this observation a little further.

It is natural at this point to distinguish two different cases: $H^2(\beta)$ contains all functions analytic in a neighborhood of the closed unit disk, and there are functions analytic in a neighborhood of the closed disk not contained in $H^2(\beta)$. In the first case we have the following satisfying result.

THEOREM 4.7
Suppose $H^2(\beta)$ contains all functions analytic in a neighborhood of the closed unit disk and let φ be analytic in a neighborhood of the closed disk with $\|\varphi\|_\infty < 1$. Then C_φ acting on $H^2(\beta)$ is trace class. In particular then, C_φ is compact.

PROOF Set $\|\varphi\|_\infty = r$ and pick $\tilde{r}$ with $r < \tilde{r} < 1$. It is easy to check that for $\psi(z) = \tilde{r}z$, C_ψ is trace class — simply compute $\sum \|C_\psi(e_n)\|$ for the orthonormal

basis $e_n = z^n/\beta(n)$. Now $\varphi = \psi \circ \widetilde{\varphi}$ where $\widetilde{\varphi}(z) = \varphi(z)/\widetilde{r}$, so we will be done if we can show $C_{\widetilde{\varphi}}$ is bounded on $H^2(\beta)$. Using $\|\widetilde{\varphi}\|_\infty < 1$ find a disk $\widetilde{D} \supset \overline{D}$ with $\widetilde{\varphi}$ analytic on $\widetilde{D}$ and $\widetilde{\varphi}(\widetilde{D}) \subset D$. Then for any f in $H^2(\beta)$, $f \circ \widetilde{\varphi}$ is analytic in the neighborhood $\widetilde{D}$ of the closed disk and so is contained in $H^2(\beta)$ as desired.
∎

An informal statement of Theorem 4.7 is "for good spaces and good symbols, $\|\varphi\|_\infty < 1$ implies C_φ is compact." An informal contrapositive of this is "if φ is a good symbol with $\|\varphi\|_\infty < 1$ but C_φ is not compact, then $H^2(\beta)$ is a bad space." This contrapositive can be made precise for a large class of weighted spaces.

Consider an arbitrary space $H^2(\beta)$ where the weight sequence is regular in the sense that $\lim_{n\to\infty} \sqrt[n]{\beta(n)}$ exists, either finite or infinite. Examples of sequences with such regularity include $\beta(n) = n^a$, $\beta(n) = \exp(n^a)$, $\beta(n) = \exp((\log n)^q)$, as well as products and quotients (defined in the obvious way) of regular sequences. By comparison, if $\beta(n) = 2^{2^k}$ for $2^k \le n < 2^{k+1}$ then $\beta(n)$ is a monotone sequence that is not regular in this sense.

If the weight sequence β is regular and $H^2(\beta)$ supports a non-compact composition operator C_φ with φ analytic in a neighborhood of the closed unit disk and $\|\varphi\|_\infty < 1$ then all functions in $H^2(\beta)$ are, in fact, analytic in RD for some fixed R with $R > 1$.

Indeed, if $H^2(\beta)$ does not include every function analytic in a neighborhood of the disk, find $f = \sum a_n z^n$ that is analytic in a neighborhood of $R\overline{D}$ for some $R > 1$ but which is not in $H^2(\beta)$, so that $\sum |a_n|^2 \beta(n)^2 = \infty$. Now if $|a_n|\beta(n) \ge 1$ for infinitely many n, then $\sum |a_n|\beta(n) = \infty$. On the other hand, if $|a_n|\beta(n) < 1$ for all but finitely many n, then since $|a_n|^2\beta(n)^2 < |a_n|\beta(n)$ for all but finitely many n, the divergence of $\sum |a_n|^2\beta(n)^2$ implies the divergence of $\sum |a_n|\beta(n)$ and the Cauchy–Schwarz inequality implies

$$\infty = \sum_{n=0}^{\infty} |a_n|\beta(n) \le \left(\sum_{n=0}^{\infty} |a_n|^2 R^{2n} \right)^{1/2} \left(\sum_{n=0}^{\infty} \frac{\beta(n)^2}{R^{2n}} \right)^{1/2}$$

Since f is analytic in a disk larger than $R\overline{D}$, the series $\sum |a_n| R^n$ converges. In particular, for all but finitely many terms, $|a_n| R^n < 1$ which implies the series $\sum |a_n|^2 R^{2n}$ converges and we must have

$$\sum_{n=0}^{\infty} \frac{\beta(n)^2}{R^{2n}} = \infty$$

The divergence of this sum implies

$$\limsup_{n\to\infty} \frac{\sqrt[n]{\beta(n)^2}}{R^2} \ge 1 \tag{4.1.4}$$

Our regularity hypothesis lets us replace the limit supremum by the limit in Inequality (4.1.4). Fix any s with $1 < s < R$ and consider the kernel functions

$$K_w(z) = \sum_{n=0}^{\infty} \frac{\overline{w}^n z^n}{\beta(n)^2}$$

For z and w in sD

$$\sum_{n=0}^{\infty} \left| \frac{\overline{w}^n z^n}{\beta(n)^2} \right| \leq \sum_{n=0}^{\infty} \frac{s^{2n}}{\beta(n)^2}$$

Because

$$\lim_{n \to \infty} \frac{s^2}{\sqrt[n]{\beta(n)^2}} \leq \frac{s^2}{R^2} < 1$$

by Inequality (4.1.4), we know the kernel functions K_w and hence all functions in
$H^2(\beta)$ are analytic in sD for any $s < R$, and thus also in RD.

Exercises

4.1.1 Prove Theorem 4.2 for the case $N > 1$.

   Hints: Since $\varphi'(0)$ is unitarily equivalent to an upper triangular matrix there is
no loss of generality in assuming $\varphi'(0)$ is upper triangular at the outset; i.e. assume
$D_k \varphi_j(0) = 0$ for $j > k$. Suppose $f \circ \varphi = \lambda f$ where $\lambda \neq 0, 1$ or a product of the
eigenvalues $\{D_k \varphi_k(0)\}$ and where f has homogeneous expansion $\sum_{s=0}^{\infty} F_s(z)$.
Show inductively that $F_s \equiv 0$ for $s = 0, 1, 2, \ldots$.

4.1.2 Suppose $\mathcal{Y}$ is small but not necessarily boundary-regular. Assume φ maps D into
D with $\varphi(0) = 0$. Show that if φ has finite angular derivative at any point of ∂D
then C_φ is not compact on $\mathcal{Y}$.

   Hint: If φ has finite angular derivative at 1 with nontangential limit 1 at 1, then
$\{\varphi(r) : 0 < r < 1\}$ is a nontangential curve at 1. Apply Lindelöf's Theorem.

4.1.3 Re-do the previous exercise for $\mathcal{Y}$ a small space on the ball B_N, for $N > 1$. Hints:
The following consequence of Theorem 2.79 is relevant.

   Theorem ([Ru80, p. 171])
   *Suppose f is in $H^\infty(B_N)$, $|\zeta| = 1$ and $\lim_{r \to 1} f(r\zeta) = L$. Then f has restricted
limit L at ζ.*
   (The definition of restricted limit is Definition 2.78.) The exercise can then be
done for $N > 1$ by showing

   - If φ has finite angular derivative at ζ in ∂B_N, then $\Lambda(r) = \varphi(r\zeta)$ satisfies
     properties (1) and (2) in Definition 2.78.
   - If φ has finite angular derivative at e_1 with $\varphi^*(e_1) = e_1$, then $\varphi \circ \varphi$ also has
     finite angular derivative at e_1, with radial limit e_1 there.

   Exercise 2.6.2 will be helpful here.

4.1.4 Let $\beta(n) = \exp(n^a)$ for $0 < a < 1$. Show that

$$\sum_{n=0}^{\infty} \frac{\beta([\alpha n])}{\beta(n)} < \infty$$

for all α, $0 < \alpha < 1$.

4.1.5 Repeat the previous exercise for $\beta(n) = \exp((\log n)^q), q > 1$ except now the range of allowable α will depend on q.

4.1.6 Let $\beta(n) = (n+1)^a$. Show that $H^2(\beta)$ is boundary-regular if $a > 1/2$ and automorphism invariant for all a. Conclude that for all $a > 1/2$, if C_φ is compact then $\|\varphi\|_\infty < 1$.

4.1.7 If $H^2(\beta)$ supports a non-compact C_φ with $\|\varphi\|_\infty < 1$, then $H^2(\beta)$ supports an unbounded C_φ with $\varphi \in H^2(\beta)$ and $\|\varphi\|_\infty < 1$.

4.1.8 (a) Note that Theorem 4.4 does not apply to the small space $H^\infty(D)$, but recall from Exercise 3.2.2 that if C_φ is compact on $H^\infty(D)$ then $\|\varphi\|_\infty < 1$. Generalize this to any small space in which the functions z^n are a bounded set.

(b) Show that if C_φ is compact on $H^\infty(B_N), N > 1$ then $\overline{\varphi(B_N)} \subset B_N$.

4.1.9 Show that if there are functions analytic in a neighborhood of the closed unit disk not in $H^2(\beta)$, then $H^2(\beta)$ is not automorphism invariant.

4.1.10 Let $\varphi(z) = z^2$. Show that C_φ is unbounded on $H^2(\beta)$ for
(a) $\beta(n) = \exp(n^a), a > 0$
(b) $\beta(n) = \exp((\log n)^q), q > 1$

4.1.11 Give an example of a weighted Hardy space $H^2(\beta)$ and an analytic map φ of the disk into itself such that

$$\sup_{w \in D} \frac{\|C_\varphi^*(K_w)\|}{\|K_w\|} < \infty$$

yet C_φ is not bounded on $H^2(\beta)$.

Notes

R. Roan made the first study of composition operators on analytic Lipschitz spaces in the disk in [Roa80a] and on $S^p(D)$ in [Roa78b]. An error in Proposition 7 of [Roa78b] is corrected by B. D. MacCluer in [Mc87], where a Carleson criterion (Theorems 4.11 and 4.12 of the next section) is given for boundedness and compactness of C_φ on $S^p(D)$. This Carleson criterion is used in [Mc87] to prove Theorem 4.5 for the particular choices $\mathcal{Y} = S^p(D)$, $p \geq 1$. The general version of Theorem 4.5 presented here is due to J. H. Shapiro [Sho87b] who also first noted that some boundary-regular non-automorphism invariant small spaces could support compact C_φ with $\|\varphi\|_\infty = 1$. Our examples are a simplification of Theorem 3.4 in that paper.

P. Turan [Tu58] investigated the question of automorphism invariance of convergence of power series on the unit circle and found that, surprisingly, if the series for f converges at ζ_0 and φ is an automorphism of the disk, the series for $f \circ \varphi$ need not converge at $\varphi(\zeta_0)$! In that paper, he raised the question of automorphism invariance of the class $\mathcal{AC}$ and L. Alpar [Alp60] showed that it is not. In the possibly more accessible reference, G. Halasz [Haz67] strengthened Alpar's result.

Theorem 4.7 is essentially due to N. Zorboska [Zo89a]; where the hypothesis that $H^2(\beta)$ contains all functions analytic in a neighborhood of the unit disk is not explicitly stated, although it is assumed in the proof. Zorboska has also made a study of the spaces $H^2(\beta)$ with $\beta(n) = (n + 1)^a$ in [Zo90a], where various Carleson measure results are given for boundedness on C_φ on these spaces for a in the range $1 < a \le 3/2$ and $a > 3/2$.

The proof outlined for Exercise 4.1.1 can be found in [Mc84a]. Exercise 4.1.8 is due to H. J. Schwartz [Scz69] in the case of $H^\infty(D)$.

4.2 Boundedness on small spaces

In this section we look at boundedness of C_φ on certain boundary-regular small spaces. Theorem 4.4 in the last section has an interesting, and perhaps counterintuitive, consequence for the study of boundedness on certain small spaces. We will illustrate it first on the boundary-regular spaces $H^2(\beta)$ where $\beta(n) = (n + 1)^a$, for $1/2 < a < 1$, spaces which will be temporarily denoted by $H_a^2(\beta)$. Notice that $H_a^2(\beta)$ is also the weighted Dirichlet space $\mathcal{D}_\alpha$ for $\alpha = 1 - 2a$; see Exercise 2.1.5.

THEOREM 4.8

If C_φ is bounded on $H_a^2(\beta)$ for some $1/2 < a < 1$ then φ must have finite angular derivative at every point ζ in ∂D where $|\varphi(\zeta)| = 1$.

PROOF We first argue that φ must have finite angular derivative at *some* point of ∂D if $\|\varphi\|_\infty = 1$. Pick b with $1/2 < b < a$. If φ has no finite angular derivative at any point of ∂D then C_φ is compact on $H_b^2(\beta) = \mathcal{D}_{1-2b}$ by Exercise 3.2.9(c). Now use Theorem 4.4 to conclude that $\|\varphi\|_\infty < 1$.

To finish the proof suppose $|\varphi(\zeta)| = 1$. Let $\psi(z) = (\zeta + z)/2$. Since $H_a^2(\beta) = \mathcal{D}_{1-2a}$ it is easy to see that C_ψ is bounded on $H_a^2(\beta)$; so also is $C_{\varphi \circ \psi} = C_\psi C_\varphi$. The first part of the proof, applied now to $\varphi \circ \psi$ shows that $\varphi \circ \psi$ must have finite angular derivative at some point, since $\|\varphi \circ \psi\|_\infty = 1$. The only candidate for this point is the point ζ and hence $(\varphi \circ \psi)'$ must have finite radial limit at ζ. But $(\varphi \circ \psi)'(r\zeta) = \frac{1}{2}\varphi'(\psi(r\zeta))$ so φ' has finite radial limit at ζ and we are done. ∎

We will see that the same phenomenon persists in many other, although not all, boundary-regular spaces. The rest of this section will look in some detail at boundedness on the analytic Lipschitz spaces and the spaces S^p of functions with H^p derivative. For simplicity we work with the one variable spaces.

The spaces $\mathrm{Lip}_\alpha(D)$, $0 < \alpha \le 1$, were defined in the previous section. When $\alpha = 1$ the question of when C_φ is bounded is easy to dispense with: φ in $\mathrm{Lip}_1(D)$ implies φ' is in H^∞ which implies C_φ is bounded on $\mathrm{Lip}_1(D)$, where the first implication follows from [Dur70, p. 74]. Since the converse is automatically true, this gives a complete characterization.

The more interesting case is $0 < \alpha < 1$. In the following result, which makes use of certain test functions in $\text{Lip}_\alpha(D)$, we use Theorem 4.1 to assert that the Lipschitz constant of a function in $\text{Lip}_\alpha(D)$ is equivalent to

$$\sup\{(1 - |z|^2)^{1-\alpha}|f'(z)| : z \in D\}$$

THEOREM 4.9
For $0 < \alpha < 1$, *C_φ is bounded on $\text{Lip}_\alpha(D)$ if and only if*

$$\sup_{z \in D}\left\{\left(\frac{1 - |z|^2}{1 - |\varphi(z)|^2}\right)^{1-\alpha}|\varphi'(z)|\right\} < \infty \qquad (4.2.1)$$

PROOF First assume Condition (4.2.1) holds. Let f be in $\text{Lip}_\alpha(D)$, we wish to show $f \circ \varphi$ is in $\text{Lip}_\alpha(D)$. To this end, it suffices to argue that

$$\sup_{z \in D}\{|(f \circ \varphi)'(z)|(1 - |z|^2)^{1-\alpha}\} < \infty$$

But this supremum is bounded above by

$$\sup_{z \in D}\{|f'(\varphi(z))|(1 - |\varphi(z)|^2)^{1-\alpha}\}\sup_{z \in D}\left\{|\varphi'(z)|\frac{(1 - |z|^2)^{1-\alpha}}{(1 - |\varphi(z)|^2)^{1-\alpha}}\right\}$$

which gives the desired result, since $\sup\{|f'(w)|(1 - |w|^2)^{1-\alpha} : w \in D\}$ is bounded above by a constant multiple of $\|f\|_\alpha$ by Theorem 4.1.

For the converse, assume that C_φ is bounded on $\text{Lip}_\alpha(D)$, with $\|f \circ \varphi\|_\alpha \le C\|f\|_\alpha$ for all f in $\text{Lip}_\alpha(D)$. We will make use of a family of test functions $\{f_w : w \in D\}$ in $\text{Lip}_\alpha(D)$ defined as follows: For $w \ne 0$, let f_w be the antiderivative of

$$\left(1 - \frac{\overline{w}^2}{|w|^2}z^2\right)^{\alpha-1}$$

with $f_w(0) = 0$. Since for $w \ne 0$

$$\sup\left\{(1 - |z|^2)^{1-\alpha}\left|1 - \frac{\overline{w}^2}{|w|^2}z^2\right|^{\alpha-1} : z \in D\right\}$$

is bounded above by

$$\sup\{(1 - |z|^2)^{1-\alpha}(1 - |z|^2)^{\alpha-1} : z \in D\} = 1$$

the functions $\{f_w\}$ are a bounded family in $\text{Lip}_\alpha(D)$, again by Theorem 4.1. Thus there exists $C' < \infty$ so that $\|f_w \circ \varphi\|_\alpha \le C'$ for all w in D, $w \ne 0$. In particular, there exists $M < \infty$ such that

$$\sup\{|(f_w \circ \varphi)'(z)|(1 - |z|^2)^{1-\alpha} : z \in D\} \le M$$

Equivalently, for all z in D and $w \neq 0$ in D we have

$$(1 - |\varphi(z)|^2)^{1-\alpha} |f'_w(\varphi(z))| |\varphi'(z)| \left(\frac{1 - |z|^2}{1 - |\varphi(z)|^2} \right)^{1-\alpha} \leq M \qquad (4.2.2)$$

Fix an arbitrary z in D with $\varphi(z) \neq 0$ and set $w = \varphi(z)$ in Inequality (4.2.2) to obtain

$$(1 - |\varphi(z)|^2)^{1-\alpha} (1 - |\varphi(z)|^2)^{\alpha-1} |\varphi'(z)| \left(\frac{1 - |z|^2}{1 - |\varphi(z)|^2} \right)^{1-\alpha} \leq M$$

which shows that for all z in D with $\varphi(z) \neq 0$ we have

$$|\varphi'(z)| \left(\frac{1 - |z|^2}{1 - |\varphi(z)|^2} \right)^{1-\alpha} \leq M$$

But if $\varphi(z) = 0$ then clearly

$$\left(\frac{1 - |z|^2}{1 - |\varphi(z)|^2} \right)^{1-\alpha} |\varphi'(z)| = (1 - |z|^2)^{1-\alpha} |\varphi'(z)| \leq K \|\varphi\|_\alpha$$

for some constant $K < \infty$, which completes the proof, since if C_φ is bounded on $\mathrm{Lip}_\alpha(D)$ we must have φ in $\mathrm{Lip}_\alpha(D)$. ∎

It is easy to obtain from this theorem a comparison result: if C_φ is bounded on $\mathrm{Lip}_\alpha(D)$ and if $\beta < \alpha$ then C_φ must be bounded on $\mathrm{Lip}_\beta(D)$. The details are left to the reader in Exercise 4.2.1. We can also use Theorem 4.9 to give an analogue of Theorem 4.8 for the Lipschitz spaces.

COROLLARY 4.10
If C_φ is bounded on $\mathrm{Lip}_\alpha(D)$, then φ has finite angular derivative at every ζ in ∂D with $|\varphi(\zeta)| = 1$.

PROOF As in the proof of Theorem 4.8, we first show that φ must have finite angular derivative at some point of the unit circle if $\|\varphi\|_\infty = 1$. Fix $0 < \beta < \alpha$. We have

$$\left(\frac{1 - |z|^2}{1 - |\varphi(z)|^2} \right)^{1-\beta} |\varphi'(z)| = \left(\frac{1 - |z|^2}{1 - |\varphi(z)|^2} \right)^{1-\alpha} |\varphi'(z)| \left(\frac{1 - |z|^2}{1 - |\varphi(z)|^2} \right)^{\alpha-\beta}$$

If φ has no finite angular derivative, then

$$\liminf_{|z| \to 1} \frac{1 - |\varphi(z)|}{1 - |z|} = \infty$$

Since

$$\left(\frac{1 - |z|^2}{1 - |\varphi(z)|^2} \right)^{1-\alpha} |\varphi'(z)|$$

is bounded by Theorem 4.9, we must have

$$\lim_{|z|\to 1} \left(\frac{1 - |z|^2}{1 - |\varphi(z)|^2} \right)^{1-\beta} |\varphi'(z)| = 0$$

We will show next that this implies C_φ is compact on $\text{Lip}_\beta(D)$.

By Exercise 4.2.3, C_φ will be compact on $\text{Lip}_\beta(D)$ provided whenever $\{f_n\}$ is a bounded sequence in $\text{Lip}_\beta(D)$ with f_n converging to 0 uniformly on compact subsets of D then $\|f_n \circ \varphi\|_\beta$ converges to 0. Suppose $\{f_n\}$ is such a sequence. Since $|f_n(\varphi(0))|$ approaches 0 as n tends to infinity, we need only consider further $|(f_n \circ \varphi)'(z)|(1 - |z|^2)^{1-\beta}$. Given $\epsilon > 0$ choose $r < 1$ such that

$$|z| > r \text{ implies } \left(\frac{1 - |z|^2}{1 - |\varphi(z)|^2} \right)^{1-\beta} |\varphi'(z)| < \epsilon$$

Then

$$\sup_{z \in D} |(f_n \circ \varphi)'(z)|(1 - |z|^2)^{1-\beta}$$

$$\leq \sup_{|z| \leq r} |f_n'(\varphi(z))||\varphi'(z)|(1 - |z|^2)^{1-\beta}$$

$$+ \sup_{|z| > r} |f_n'(\varphi(z))|(1 - |\varphi(z)|^2)^{1-\beta}|\varphi'(z)| \left(\frac{1 - |z|^2}{1 - |\varphi(z)|^2} \right)^{1-\beta}$$

$$\leq K\|\varphi\|_\beta \sup_{|z| \leq r} |f_n'(\varphi(z))| + K\epsilon\|f_n\|_\beta$$

where the constant K comes from the equivalence of Theorem 4.1. Since f_n, and hence also f_n', tend to 0 uniformly on compact subsets of D as n tends to infinity,

$$\sup_{|z| \leq r} |f_n'(\varphi(z))| \to 0$$

as n tends to infinity. Thus $\|f_n \circ \varphi\|_\beta$ can be made arbitrarily small by choosing n large, and C_φ is compact on $\text{Lip}_\beta(D)$. But by Theorem 4.5 if C_φ is compact on $\text{Lip}_\beta(D)$ then $\|\varphi\|_\infty < 1$, and we conclude that φ must have finite angular derivative at some point of ∂D if $\|\varphi\|_\infty = 1$.

Since for each ζ in ∂D the map $\psi(z) = (\zeta + z)/2$ gives a bounded composition operator on $\text{Lip}_\alpha(D)$, we may argue exactly as in the proof of Theorem 4.8 to see that φ must have finite angular derivative at *every* ζ with $|\varphi(\zeta)| = 1$. ∎

Next we turn to a study of bounded composition operators on the spaces $S^p(D)$ for $p \geq 1$. Our approach will be to develop appropriate Carleson measure conditions for boundedness, using the familiar Carleson sets

$$S(\zeta, h) = \{z \in \overline{D} : |z - \zeta| < h\}$$

Since if φ is in $S^p(D)$ for some $p \geq 1$, φ must be continuous on the closed disk, we will regard φ as a map of the closed disk into itself.

THEOREM 4.11

Let φ mapping D into itself be in $S^p(D)$. Then C_φ is bounded on $S^p(D)$ if and only if there is a constant C so that

$$\mu(\varphi^{-1}\mathcal{S}(\zeta, h) \cap \partial D) \leq Ch$$

for all ζ in ∂D and $h > 0$ where μ is the measure defined on ∂D by

$$\mu(E) = \int_E |\varphi'|^p \, d\sigma$$

PROOF   First assume C_φ is bounded on $S^p(D)$ so

$$\|f \circ \varphi\|_{S^p} \leq C'\|f\|_{S^p}$$

for all f in $S^p(D)$. If $f(0) = 0$ this implies

$$\|(f \circ \varphi)'\|_{H^p} \leq C'\|f'\|_{H^p}$$

or

$$\int_{\partial D} |f'(\varphi(\zeta))|^p |\varphi'(\zeta)|^p \, d\sigma(\zeta) \leq C' \int_{\partial D} |f'(\zeta)|^p \, d\sigma(\zeta) \qquad (4.2.3)$$

where Proposition 2.25 is applicable since f' is in $H^p(D)$. By the definition of μ we have

$$\int_{\partial D} |f' \circ \varphi|^p \, d\mu \leq C' \int_{\partial D} |f'|^p \, d\sigma$$

Now considering φ as a map from ∂D to $\overline{D}$, a change of variables gives

$$\int_{\overline{D}} |f'|^p \, d(\mu\varphi^{-1}) \leq C' \int_{\partial D} |f'|^p \, d\sigma$$

for all f in $S^p(D)$ with $f(0) = 0$. Since $\{f' : f$ is in $S^p(D)$ and $f(0) = 0\}$ is $H^p(D)$, Theorem 2.35 gives the desired conclusion: There exists $C < \infty$ so that

$$\mu(\varphi^{-1}\mathcal{S}(\zeta, h) \cap \partial D) \leq Ch$$

for all ζ in the unit circle and $h > 0$.

For the converse we use Theorem 2.35 again to conclude that if $\mu(\varphi^{-1}\mathcal{S}(\zeta, h) \cap \partial D) \leq C''h$ for all ζ and h then

$$\int_{\overline{D}} |g|^p \, d(\mu\varphi^{-1}) \leq C \int_{\partial D} |g|^p \, d\sigma \qquad (4.2.4)$$

for all g in $H^p(D)$ and therefore for $g = f'$ where f is in $S^p(D)$. If f is a polynomial, the left side of Inequality (4.2.4) is equal to $\int_{\partial D} |(f \circ \varphi)'|^p \, d\sigma$ and thus for any polynomial we have

$$\|(f \circ \varphi)'\|_{H^p} \leq C^{1/p}\|f'\|_{H^p}$$

Therefore,

$$\|f \circ \varphi\|_{S^p} = \|(f \circ \varphi)'\|_{H^p} + |f(\varphi(0))|$$
$$\leq C^{1/p}\|f'\|_{H^p} + K\|f\|_{S^p} \leq (C^{1/p} + K)\|f\|_{S^p}$$

which, since the polynomials are dense in $S^p(D)$, is the desired result. ∎

By Theorem 4.5 and Exercise 4.2.8 the compact composition operators on $S^p(D)$ for $p \geq 1$ are characterized as those C_φ for which the symbol φ is in $S^p(D)$ and $\|\varphi\|_\infty < 1$. We will also have a use for the expected "little o" Carleson condition for compactness on $S^p(D)$.

THEOREM 4.12
Let φ mapping D into D be in $S^p(D)$. Then C_φ is compact on $S^p(D)$ if and only if

$$\mu(\varphi^{-1}\mathcal{S}(\zeta, h) \cap \partial D) = o(h) \qquad (4.2.5)$$

as h tends to 0, uniformly for ζ in the unit circle.

PROOF We give the proof for the direction which will be used in Theorem 4.13 below and leave the proof of the other direction for the reader as Exercise 4.2.10. Suppose φ is in $S^p(D)$ and Equation (4.2.5) holds. We wish to show C_φ is compact on $S^p(D)$. Since a bounded sequence in $S^p(D)$ is a normal family, it suffices, by an argument identical to that of Proposition 3.11, to show that if $\{f_n\}$ is bounded in $S^p(D)$ with $f_n \to 0$ uniformly on compact subsets of D, then $\|f_n \circ \varphi\|_{S^p} \to 0$. Since $\|f_n \circ \varphi\|_{S^p} = |f_n(\varphi(0))| + \|(f_n \circ \varphi)'\|_{H^p}$ we need only show $\|(f_n \circ \varphi)'\|_{H^p} \to 0$. By Theorem 4.11 we know $f_n \circ \varphi$ is in $S^p(D)$ for every n so

$$\|(f_n \circ \varphi)'\|_{H^p}^p = \int_{\partial D} |f_n' \circ \varphi|^p |\varphi'|^p \, d\sigma = \int_{\partial D} |f_n' \circ \varphi|^p \, d\mu = \int_{\overline{D}} |f_n'|^p \, d(\mu\varphi^{-1})$$

where in the first equality we have tacitly used Proposition 2.25. Using the hypothesis Equation (4.2.5) and replacing the sets $\mathcal{S}(\zeta, h)$ by the equivalent windows

$$W(\zeta, h) = \{re^{i\theta} \in \overline{D} : 1 - h < r \leq 1, |\theta - t| < h\}$$

where $\zeta = e^{it}$, we know that given any $\epsilon > 0$ there exists $h_0 > 0$ so that

$$\mu(\varphi^{-1}W(\zeta, h) \cap \partial D) \leq \epsilon h$$

for all ζ in the circle and $h \leq h_0$. Just as in the proof of Theorem 3.12 we consider the measure $\mu\varphi^{-1}$ restricted to $\{1 - h_0 \leq |z| \leq 1\}$ to see that

$$\int_{\overline{D}\backslash(1-h_0)\overline{D}} |f_n'|^p \, d(\mu\varphi^{-1}) \leq C\epsilon \int_{\partial D} |f_n'|^p \, d\sigma$$

$$\leq C\epsilon \|f_n\|_{S^p}^p$$

$$\leq CM\epsilon$$

where $\|f_n\|_{S^p}^p \leq M < \infty$ for all n. Since $f_n' \to 0$ uniformly on $(1 - h_0)\overline{D}$ we have for sufficiently large n

$$\int_{\overline{D}} |f_n'|^p \, d(\mu\varphi^{-1}) \leq CM\epsilon + \epsilon\mu(\varphi^{-1}(\overline{D})) = CM\epsilon + \epsilon\|\varphi'\|_{H^p}^p$$

Since ϵ is arbitrary, we are done. ∎

We use this result in the following.

THEOREM 4.13
Suppose C_φ is bounded on $S^p(D)$ for some $p \geq 2$. Then φ must have finite angular derivative at every point ζ in the unit circle with $|\varphi(\zeta)| = 1$.

PROOF We assume $\|\varphi\|_\infty = 1$, else there is nothing to prove. Just as in the proof of Theorem 4.8 it suffices to show that φ has finite angular derivative at some point of ∂D, since $\psi(z) = (\zeta + z)/2$ gives a bounded operator on $S^p(D)$.

Suppose φ has infinite angular derivative at every point of ∂D. If we knew C_φ was compact on $H^2(D)$ we could then write, for arbitrary Carleson set $\mathcal{S}(\zeta, h)$ and conjugate exponents p and q

$$\int_{\varphi^{-1}\mathcal{S}(\zeta,h)\cap\partial D} |\varphi'| \, d\sigma \leq \left(\int_{\varphi^{-1}\mathcal{S}(\zeta,h)\cap\partial D} |\varphi'|^p \, d\sigma\right)^{1/p} \left(\int_{\varphi^{-1}\mathcal{S}(\zeta,h)\cap\partial D} d\sigma\right)^{1/q}$$

where compactness of C_φ on $H^p(D)$ would give $\int_{\varphi^{-1}\mathcal{S}(\zeta,h)\cap\partial D} d\sigma \leq \epsilon(h)h$ where $\epsilon(h)$ approaches 0 as h tends to 0. Then by Theorem 4.11

$$\int_{\varphi^{-1}\mathcal{S}(\zeta,h)\cap\partial D} |\varphi'| \, d\sigma \leq (Ch)^{1/p}(\epsilon(h)h)^{1/q} = o(h)$$

as h goes to 0. This is sufficient to show that C_φ is compact on $S^1(D)$, by Theorem 4.12. But Theorem 4.5 says that all compact C_φ on $S^p(D)$, $p \geq 1$, have $\|\varphi\|_\infty < 1$, contradicting our initial assumption.

It remains to show that if C_φ is bounded on $S^p(D)$ for some $p \geq 2$ where φ has no finite angular derivative, then C_φ is compact on $H^2(D)$. By Exercise 3.2.9 we can accomplish this by showing that C_φ is bounded on $\mathcal{D}$ the (unweighted) Dirichlet space consisting of those analytic functions f on D for which $\int_D |f'|^2 \, dA < \infty$.

Since both $\mathcal{D}$ and $S^p(D)$ are automorphism invariant it is enough to consider the case $\varphi(0) = 0$. If C_φ is bounded on $S^p(D)$, $p \geq 2$ then it follows from Theorem 4.11 that C_φ is bounded on $S^2(D)$. (See Exercise 4.2.4). Since $\mathcal{D}$ is equivalent to the weighted Hardy space $H^2(\beta_2)$ with $\beta_2(j) = \sqrt{j+1}$ and $S^2(D)$ is equivalent to $H^2(\beta_1)$ with $\beta_1(j) = (j+1)$ the desired conclusion follows from Theorem 3.2. Alternately (see Exercise 4.2.17), the result can be obtained by making direct estimates to show

$$\int_{\varphi^{-1}S(\zeta,h)} |\varphi'(z)|^2 \, dA(z) = O(h^2)$$

and appealing to Exercise 3.2.9(a). ∎

This result is false in $S^1(D)$. Consider the mapping $\varphi(z) = 1 - (1-z)^b$ where $0 < b < 1$. Note that $\varphi(1) = 1$ and φ is in $S^q(D)$ for all $q < (1-b)^{-1}$. Since $\varphi'(r) = b(1-r)^{b-1}$ which tends to infinity as r goes to 1, φ does not have finite angular derivative at 1. But C_φ is bounded on $S^1(D)$. The verification of this fact is outlined in Exercise 4.2.5.

Exercises

4.2.1 Show that if C_φ is bounded on $\text{Lip}_\alpha(D)$ and $\beta < \alpha$ then C_φ is bounded on $\text{Lip}_\beta(D)$.

4.2.2 Suppose φ is in $\text{Lip}_\alpha(D)$. Show that if there exists $r < 1$ and $M < \infty$ so that $|\varphi(z)| \geq r$ implies $|\varphi'(z)| \leq M$, then C_φ is bounded on $\text{Lip}_\alpha(D)$.

4.2.3 Suppose that whenever $\{f_n\}$ is a bounded sequence in $\text{Lip}_\alpha(D)$ converging to 0 uniformly on compact subsets of D, then $\lim \|f_n \circ \varphi\|_\alpha = 0$. Show that C_φ is compact on $\text{Lip}_\alpha(D)$.

4.2.4 Show that for $1 \leq p \leq q$, C_φ bounded on $S^q(D)$ implies C_φ is bounded on $S^p(D)$.

4.2.5 Show that $\varphi(z) = 1 - (1-z)^b$ with $0 < b < 1$ gives a bounded operator on $S^1(D)$. Hint: Argue first that since $\varphi(D)$ is contained in a nontangential approach region at 1, it suffices to check the requisite Carleson condition only for Carleson sets based at 1. Then compute $\varphi^{-1}S(1,h) \cap \partial D$ explicitly.

4.2.6 In this problem $p > 1$ and $\alpha = 1 - 1/p$. For $0 < \alpha < 1$, define $\text{lip}_\alpha(D)$ to be those functions in $\text{Lip}_\alpha(D)$ for which

$$\frac{|f(z) - f(w)|}{|z - w|^\alpha} = o(1)$$

as w converges to z with $|w| = |z| = 1$.

(a) Show that $S^p(D)$ is continuously contained in $\text{lip}_\alpha(D)$.

(b) Show that $S^p(D)$ is dense in $\text{lip}_\alpha(D)$.

(c) Show that $\text{lip}_\alpha(D)$ can be alternately described as the closure of the polynomials in $\text{Lip}_\alpha(D)$.

(d) Suppose φ is in $\mathrm{lip}_\alpha(D)$. Show that C_φ is bounded on $\mathrm{lip}_\alpha(D)$ if and only if

$$\sup_{z \in D} \left\{ |\varphi'(z)| \left(\frac{1 - |z|^2}{1 - |\varphi(z)|^2} \right)^{1-\alpha} \right\} < \infty$$

4.2.7 Suppose C_φ is bounded on $\mathrm{Lip}_\alpha(D)$, $0 < \alpha < 1$ and $\{z_n\}$ is any sequence in D such that $\lim_{n \to \infty} |\varphi(z_n)| = 1$. Show that

$$\lim_{n \to \infty} |\varphi'(z_n)|(1 - |z_n|^2)^{1-\alpha} = 0$$

4.2.8 If X is $\mathrm{Lip}_\alpha(D)$ or $S^p(D)$ show that φ in X and $\|\varphi\|_\infty < 1$ implies C_φ is compact on X.

4.2.9 Show that Theorem 4.13 holds for $1 < p < 2$ in the case that φ is univalent, or even just has finite multiplicity.

4.2.10 Complete the proof of Theorem 4.12 by showing that Equation (4.2.5) holds if C_φ is compact on $S^p(D)$.

4.2.11 (a) Show directly from the Carleson criterion for compactness on $S^p(D)$ that if C_φ is compact on $S^p(D)$ for some $p > 1$ then C_φ is compact on $S^1(D)$.

(b) Now use Theorem 4.12 to give an alternate proof that if C_φ is compact on $S^p(D)$, $p \geq 1$ then $\|\varphi\|_\infty < 1$. Hint: If φ is in $S^1(D)$ then φ is absolutely continuous on the circle and for $E \subset \partial D$, $\int_E |\varphi'| \, d\sigma$ gives the arc length of the image of E under φ.

4.2.12 Suppose $\varphi : D \to D$ with $\varphi(1) = 1$ and infinite angular derivative at 1. For $-1 < \alpha < 0$ let $d\mu_\alpha = |\varphi'(z)|^2(1 - |z|^2)^\alpha dA$. Show directly that $\mu_\alpha \varphi^{-1} S(1, \delta) \neq O(\delta^{\alpha+2})$. This gives an alternate proof of Theorem 4.8.

4.2.13 On the unit disk the Bloch space $\mathcal{B}$ is defined to be those analytic f in D with

$$\sup_{z \in D} \left\{ (1 - |z|^2)|f'(z)| \right\} < \infty$$

With the norm

$$\|f\|_{\mathcal{B}} = |f(0)| + \sup_{z \in D} \left\{ (1 - |z|^2)|f'(z)| \right\}$$

$\mathcal{B}$ is a Banach space. The little Bloch space $\mathcal{B}_0$ is the closed subspace of $\mathcal{B}$ consisting of those f for which

$$\lim_{|z| \to 1} (1 - |z|^2)|f'(z)| = 0$$

(a) Show $H^\infty(D)$ is properly contained in $\mathcal{B}$. (Hint: Consider $f(z) = \log(1 - z)$.) Less immediate is that $\mathcal{B}_0$ neither contains, nor is contained in, $H^\infty(D)$. Show this by considering the singular inner function $f(z) = \exp[(z+1)/(z-1)]$ on the one hand, and any unbounded Dirichlet space function on the other. ([Zh90b])

(b) Show that every analytic $\varphi : D \to D$ induces a bounded composition operator on $\mathcal{B}$ and that C_φ is bounded on $\mathcal{B}_0$ if and only if φ is in $\mathcal{B}_0$.

4.2.14 Show that $\varphi : D \to D$ gives a compact composition operator on $\mathcal{B}_0$ if and only if

$$\lim_{|z| \to 1} \frac{1 - |z|^2}{1 - |\varphi(z)|^2} |\varphi'(z)| = 0$$

(Hint: For the only if direction, use the test functions $\log(1 - \overline{w_n}z)$.)

4.2.15 Let $\varphi(z) = 1 - (1 - z)^b$ where $0 < b < 1$. Recall that φ is a conformal map of D onto a nontangential approach region at 1. Notice that φ is in $\mathcal{B}_0$ so that C_φ is bounded on $\mathcal{B}_0$ by Exercise 4.2.13. Use Exercise 4.2.14 to show that C_φ is not compact on $\mathcal{B}_0$.

4.2.16 Let φ be a univalent map of D onto a region G in D satisfying
  (i) $\overline{G} \cap \partial D = 1$
  (ii) $\text{dist}(w, \partial G) = o(|1 - w|)$ as $w \to 1$ in G
  (iii) $G \cap \{|1 - z| < \delta\} \subset \Gamma(1)$ for some $\delta > 0$ and some nontangential approach region $\Gamma(1) = \{z : (|1 - z|)/(1 - |z|^2) \leq M < \infty$
  (iv) φ is C^1 on $\overline{D} \setminus \{1\}$
  By the Koebe distortion theorem we have
$$\frac{1}{4}(1 - |z|^2)|\varphi'(z)| \leq \text{dist}(\varphi(z), \partial G) \leq (1 - |z|^2)|\varphi'(z)|$$
  for all z in D. Show that C_φ is compact on $\mathcal{B}_0$. Thus there do exist compact C_φ on $\mathcal{B}_0$ where $\|\varphi\|_\infty = 1$.

4.2.17 Suppose $\varphi(0) = 0$ and φ is bounded on $S^2(D)$. Show by direct estimates that
$$\int_{\varphi^{-1}S(\zeta,h)} |\varphi'(z)|^2 \, dA(z) = O(h^2)$$
  for all ζ in the unit circle and $h > 0$. Hint: Note that $\|C_{\varphi_r}\|_{S^p} \leq \|C_\varphi\|_{S^p}$ where $\varphi_r(z) = \varphi(rz)$, $r < 1$.

4.2.18 The goal of this problem is to extend Theorem 3.29 to the small spaces $\mathcal{D}_\alpha(D)$, where $-1 < \alpha < 0$. Recall that the functions in these spaces all extend continuously to ∂D so that if C_φ is bounded on $\mathcal{D}_\alpha(D)$ we may regard φ as defined on the closed unit disk. Assume that C_φ is Fredholm on $\mathcal{D}_\alpha(D)$ for some $-1 < \alpha < 0$.
  (a) Show that if C_φ has closed range then $\varphi(\partial D) \supset \partial D$. Hint: Suppose not, then without loss of generality $\partial D \setminus \varphi(\partial D)$ contains an open arc centered at η in ∂D. Consider f_n in $\mathcal{D}_\alpha$ with
$$f_n'(z) = \frac{(1 - r_n^2)^{(\alpha+2)/2}}{(1 - r_n\overline{\eta}z)^{(\alpha+2)}}$$
  where $r_n \uparrow 1$ and $f_n(0) = 0$. Show that C_φ is not bounded below on $\{f_n\}$.
  (b) Using Theorem 4.8 show that φ is a one-to-one map of ∂D onto ∂D. Hint: By [CoP82, Lemma 8.2] if $\varphi : D \to D$ is univalent and $\varphi(\zeta_1) = \varphi(\zeta_2) = \lambda$ where $|\zeta_1| = |\zeta_2| = |\lambda| = 1$ then φ cannot have finite angular derivative at both ζ_1 and ζ_2.
  (c) Show that C_φ is Fredholm on $\mathcal{D}_\alpha(D)$ for some $-1 < \alpha < 0$ if and only if φ is in $\text{Aut}(D)$.

4.2.19 Modify the argument in the previous problem to identify the Fredholm composition operators on $S^2(D)$.

Notes

Theorem 4.9 is due to K. Madigan [Mad93b]. The proof given here, using explicit test functions, is a modification of Madigan's original argument. Theorem 4.8 was essentially noted by J. H. Shapiro in [Sho87b]; the analogous results in Theorem 4.10 and Theorem 4.13 appear to be new. Theorems 4.11 and 4.12 are due to B. D. MacClucr [Mc87], as is Exercise 4.2.11. The result of Exercise 4.2.12 was shown to us by T. Carroll. Exercises 4.2.13, 4.2.14, 4.2.15, and 4.2.16 are all taken from K. M. Madigan and A. Matheson's study of compact composition operator on the Bloch and little Bloch spaces in the disk [MaM93].

The result of Exercise 4.2.2 appears in the work of R. Roan [Roa80a], where the main result is the converse statement to this exercise. However, the proof given there appears to be in error (the estimates in (8) on p. 375 are not correct). We believe it is an open question as to whether or not the result still holds.

The results of Exercise 4.2.18 are in the paper of M. Jovovic and B. MacCluer [JoM95]. The conclusion of part (a) of Exercise 4.2.18 on closed range operators on the small spaces $\mathcal{D}_\alpha(D)$ as well as the analogous result on $S^2(D)$ (Exercise 4.2.19) can be obtained by a quite different argument that makes use of the fact that these spaces are, under equivalent norms, Banach algebras; see R. Roan [Roa78b] for this point of view.

5

Large Spaces

5.1 Boundedness on large spaces

In contrast to the function spaces considered in the last chapter, here we consider weighted Bergman or Hardy spaces that include functions which grow much more rapidly than functions in $H^2(D)$ or the standard weighted Bergman spaces $A_\alpha^2(D)$. We will confine our attention to function spaces on the disk, and concentrate on spaces $A_G^2(D)$ defined for weights G positive, continuous, and non-increasing on $(0, 1)$ with $\int_0^1 G(r) r \, dr < \infty$ so that $G(|z|) \, dA(z)$ is a positive, circularly symmetric finite measure on D. Set

$$A_G^2(D) \equiv \{f \text{ analytic} : \int_D |f(z)|^2 G(|z|) \frac{dA(z)}{\pi} < \infty\}$$

with

$$\|f\|_G^2 = \int_D |f(z)|^2 G(|z|) \frac{dA(z)}{\pi}$$

The space $A_G^2(D)$ is equivalent to the weighted Hardy space $H^2(\beta)$ where $\beta(n)^2 = 2p_n/c$ where $\{p_n\}$ are the moments

$$p_n = \int_0^1 r^{2n+1} G(r) \, dr$$

and $c = 2 \int_0^1 G(r) r \, dr$, chosen so that $\beta(0) = 1$. If $G(r) = (1 - r)^\alpha$ (for $\alpha > -1$) or $(-\log r)^\alpha$, we have the standard weighted Bergman spaces, with equivalent norms. The large spaces of principal interest here will arise from "fast weights"; functions $G(r)$ that decrease more rapidly than any power of $(1 - r)$.

DEFINITION 5.1 We say $G(r)$ is a **fast weight** if

$$\lim_{r \to 1} \frac{G(r)}{(1 - r)^\alpha} = 0 \qquad (5.1.1)$$

*for all $\alpha > 0$. If, in addition, the quantity in Equation (5.1.1) is decreasing for r near 1, for all $\alpha > 0$, we say that G is **fast** and **regular**.*

Throughout this first section our standing hypothesis on G is that it is positive, continuous, and non-increasing on $(0, 1)$ with $\int_0^1 G(r)r \, dr < \infty$. Specific examples of such G that are, in addition, both fast and regular include

$$G(r) = \exp\left\{ -c \left(\log \frac{1}{1-r} \right)^\beta \right\}$$

for $c > 0$ and $\beta > 1$ and

$$G(r) = \exp\left\{ -c \left(\frac{1}{1-r} \right)^\beta \right\}$$

for $c > 0$ and $\beta > 0$. The spaces $A_G^2(D)$ all have reproducing kernel functions that are obtained from the orthonormal basis $\{z^n/\sqrt{2p_n}\}_{n=0}^\infty$ in the usual way:

$$K_w(z) = \frac{1}{2} \sum_{n=0}^\infty \frac{1}{p_n} \overline{w}^n z^n$$

Since the moments p_n are bounded, we see that $\|K_w\|$ tends to infinity as w tends to 1.

By multiplying the integral inequality of Littlewood's Subordination Principle (Theorem 2.22) by $rG(r)$ and integrating with respect to r from 0 to 1 we see that whenever $\varphi(0) = 0$, C_φ will be bounded on $A_G^2(D)$ with $\|C_\varphi\| = 1$. When we remove the restriction $\varphi(0) = 0$ we will see that C_φ may fail to be bounded. In particular, our first result will have the important consequence that no non-rotation automorphisms give bounded composition operators on $A_G^2(D)$ if G is fast. While this fact can be verified directly (see Exercise 5.1.5) we prefer to obtain it as a corollary of a more general result. Note that any automorphism $\psi(z) = e^{i\theta}\frac{z-a}{1-\overline{a}z}$ has derivative $\psi'(z) = e^{i\theta}\frac{1-|a|^2}{(1-\overline{a}z)^2}$ and if $a \neq 0$, $|\psi'(z)| < 1$ on the arc of ∂D satisfying $\mathrm{Re}\,\overline{a}z < |a|^2$. As usual we use $|\varphi'(\zeta)|$ for the (modulus of the) angular derivative at ζ on the unit circle with the understanding that this is $+\infty$ in the case that the angular derivative does not exist at ζ. Our first theorem is in a setting somewhat more general than that of fast weight Bergman spaces.

THEOREM 5.2
Suppose $H^2(\beta)$ is a weighted Hardy space such that $\lim_{n\to\infty} n^\alpha \beta(n) = 0$ for every $\alpha > 0$. If $\varphi : D \to D$ is analytic and satisfies $|\varphi'(\zeta)| < 1$ at some ζ in the circle then C_φ does not map $H^2(\beta)$ into itself.

Every $A_G^2(D)$ space for which G is a fast weight is equivalent to an $H^2(\beta)$ space satisfying the hypotheses of Theorem 5.2; see Exercise 5.1.1.

PROOF (of Theorem 5.2)   We first use the hypothesis on $\beta(n)$ to get a lower estimate on the norms of the kernel functions K_w for $H^2(\beta)$. The kernel function for evaluation at w in D is

$$K_w(z) = \sum_{n=0}^{\infty} \frac{1}{\beta(n)^2} \overline{w}^n z^n$$

for which

$$\|K_w\|^2 = \sum_{n=0}^{\infty} \frac{1}{\beta(n)^2} |w|^{2n}$$

Now consider the Bergman space $A_W^2(D)$ where

$$W(r) = \frac{2^\alpha}{\Gamma(\alpha+1)} (-\log r)^\alpha$$

for $\alpha > 0$. Denoting the kernel function for this space K_w^α we have

$$\|K_w^\alpha\|^2 = \sum_{n=0}^{\infty} (n+1)^{\alpha+1} |w|^{2n}$$

which is comparable to $(1-|w|)^{-(\alpha+2)}$ on D. For any $\alpha > 0$ the hypothesis guarantees that $(n+1)^{\alpha+1}\beta(n)^2$ is bounded and hence that there is a finite constant C so that for w in the disk

$$\frac{1}{(1-|w|)^{\alpha+2}} \sim \|K_w^\alpha\|^2 \leq C\|K_w\|^2 \tag{5.1.2}$$

Suppose ζ is a point with $|\varphi'(\zeta)| < 1$ and assume C_φ maps $H^2(\beta)$ into itself. By the closed graph theorem C_φ acts boundedly on $H^2(\beta)$; set $\|C_\varphi\| = M$. We have

$$M \geq \frac{\|C_\varphi^* K_{r\zeta}\|}{\|K_{r\zeta}\|} = \frac{\|K_{\varphi(r\zeta)}\|}{\|K_{r\zeta}\|} = \frac{\|K_{|\varphi(r\zeta)|}\|}{\|K_r\|} \tag{5.1.3}$$

As a consequence of the Julia–Carathéodory theory there exist $\delta < 1$ and $r_0 < 1$ so that $r \geq r_0$ implies $1 - |\varphi(r\zeta)| < \delta(1-r)$. Our goal is to use this and Inequality (5.1.3) to obtain an upper estimate on $\|K_w\|$ which is incompatible with the estimates obtained in Inequality (5.1.2).

We will simplify notation by writing $g(t) = \|K_{1-t}\|$; since $\|K_z\|$ increases with $|z|$, $g(t)$ is decreasing and Inequality (5.1.3) shows $g(\delta(1-r)) \leq Mg(1-r)$ for all $r \geq r_0$, or $g(\delta t) \leq Mg(t)$ for all $t \leq t_0 = 1 - r_0$. By induction we have $g(\delta^n t_0) \leq M^n g(t_0)$. For $0 < s \leq t_0$, find n so that $\delta^{n+1}t_0 < s \leq \delta^n t_0$ and hence

$$g(s) \leq g(\delta^{n+1}t_0) \leq M^{n+1}g(t_0)$$

Since M is a fixed finite value and $\delta < 1$ we may write $M = \delta^{-A}$ for some $A \geq 0$. This means that for $0 < s \leq t_0$

$$g(s) \leq \delta^{-An} M g(t_0) \leq M g(t_0) \left(\frac{t_0}{s} \right)^A$$

or

$$\| K_{1-s} \| \leq C' \left(\frac{1}{s} \right)^A$$

for $s \leq t_0$, where $C' = M g(t_0) t_0^A$. Thus we have proved

$$\| K_w \| \leq C' \left(\frac{1}{1 - |w|} \right)^A$$

for all w with $|w| \geq 1 - t_0$. By adjusting C' this estimate holds for all z in D, which contradicts the estimates obtained in Inequality (5.1.2) when α is chosen so $\alpha + 2 > 2A$. ∎

Just as we have done with the standard weighted Bergman spaces, we will sometimes find it convenient to think of $A_G^2(D)$ as a weighted Dirichlet space, in the following way. Set

$$H(r) = \int_r^1 \int_x^1 G(s) \, ds \, dx \qquad (5.1.4)$$

for $0 \leq r < 1$. The space $A_G^2(D)$ consists of those f analytic in D for which

$$\left(|f(0)|^2 + \int_D |f'(z)|^2 H(|z|) \frac{dA(z)}{\pi} \right)^{1/2}$$

is finite; moreover this expression gives a norm equivalent to the original norm.

LEMMA 5.3
Suppose f is analytic in D with $f(0) = 0$. Then

$$\int_D |f'|^2 H \, dA \leq \int_D |f|^2 G \, dA \leq 6 \int_D |f'|^2 H \, dA$$

where we allow the possibility that the integrals are infinite.

PROOF Set $p_n = \int_0^1 r^{2n+1} G(r) \, dr$ and $q_n = \int_0^1 r^{2n+1} H(r) \, dr$. Integrating by parts twice shows that $p_n = 2n(2n+1) q_{n-1} \leq 6n^2 q_{n-1}$. If $f = \sum_{n=1}^{\infty} a_n z^n$ we have

$$\int_D |f|^2 G \, dA = 2\pi \sum_1^{\infty} |a_n|^2 p_n \leq 2\pi \sum_1^{\infty} 6n^2 |a_n|^2 q_{n-1} = 6 \int_D |f'|^2 H \, dA$$

while

$$\int_D |f'|^2 H \, dA = 2\pi \sum_1^\infty n^2 |a_n|^2 q_{n-1} \le 2\pi \sum_1^\infty |a_n|^2 p_n = \int_D |f|^2 G \, dA$$

∎

We will write $\|f\|_0^2$ for $\int_D |f'(z)|^2 H(|z|) \, dA(z)/\pi$; note that $\|f\|_0$ is a semi-norm on $A_G^2(D)$.

To investigate further the boundedness of C_φ on $A_G^2(D)$ define F on the positive real axis by

$$F(-\log r) = H(r) \tag{5.1.5}$$

and set

$$E(t) = F(t)/t \tag{5.1.6}$$

LEMMA 5.4

For $0 < t < 1/2$, $E(t)$ is increasing, and $\lim_{t \to 0} E(t) = 0$.

PROOF Since G is non-increasing, we have

$$H(r) = \int_r^1 \int_x^1 G(s) \, ds \, dx$$

$$\le \int_r^1 (1 - x) G(x) \, dx$$

$$= (1 - r) \int_r^1 G(x) \, dx - H(r)$$

where the last equality follows from an integration by parts. Thus

$$F(-\log r) = H(r) \le \frac{1 - r}{2} \int_r^1 G(x) \, dx$$

and so $E(t) = F(t)/t \to 0$ as $t \to 0$.

To see that $E(t)$ is increasing on $(0, \frac{1}{2})$, note that $E'(t) = [tF'(t) - F(t)]/t^2$. It suffices to show that $F(t) < tF'(t)$ for $t < \frac{1}{2}$. We have $F(t) = H(e^{-t})$,

$$F'(t) = e^{-t} \int_{e^{-t}}^1 G(x) \, dx$$

and by the above estimates

$$F(t) \le \frac{1 - e^{-t}}{2} \int_{e^{-t}}^1 G(x) \, dx$$

The result then follows since $e^t - 1 < 2t$ when $0 < t < \frac{1}{2}$. ∎

We emphasize that Lemma 5.4 holds for arbitrary non-increasing weight functions G, not just for fast weights. A stronger version of this lemma is given in Exercise 5.1.7.

The next result may seem technical, as the hypothesis involves a condition on the function E rather than the more natural G, but we will find it useful nevertheless. For $\varphi : D \to D$ analytic, recall the definition

$$M(r) = \max_{\theta} |\varphi(re^{i\theta})|$$

for $0 < r < 1$.

THEOREM 5.5

If G is a positive, continuous, non-increasing weight function such that

$$\limsup_{r \to 1} \frac{E(-\log r)}{E(-\log M(r))} < \infty$$

then C_φ is bounded on $A_G^2(D)$.

In preparation for the proof of this theorem, the next result gives a useful method for estimating $\|f \circ \varphi\|_0^2$.

LEMMA 5.6

Let $\varphi : D \to D$ be analytic with $|\varphi(0)| = s$ and suppose $N(r)$ is a continuous, strictly increasing function mapping $[0, 1]$ onto $[0, 1]$. If

(1) there is r_0 in $(0, 1)$ such that $r_0 \le |z| < 1$ implies $|\varphi(z)| \le N(|z|)$

(2) and there is ρ in $(0, 1)$ such that

$$\frac{E(-\log r)}{E(-\log N(r))} \le R < \infty$$

for $\rho \le N(r) < 1$,

then there exists c depending on r_0, ρ, s, and G but not on R or f, so that

$$\|f \circ \varphi\|_0^2 \le cR\|f\|_0^2$$

for all f analytic in D.

PROOF We may assume $\rho > s$. By Lemma 5.4, $E(t)$ is increasing for small $t > 0$. Altering $E(t)$ away from 0 amounts to altering H away from 1; this produces a semi-norm equivalent to $\| \cdot \|_0$. Hence we may assume that $E(t)$ is increasing for *all* $t > 0$. By Exercise 5.1.2, there is a constant c_1 depending only on r_0 and H (or equivalently, on r_0 and G) so that

$$\|f \circ \varphi\|_0^2 = \int_D |f'(\varphi(z))|^2 |\varphi'(z)|^2 H(|z|) \, dA(z)/\pi$$

$$\leq c_1 \int_S |f'(\varphi(z))|^2 |\varphi'(z)|^2 H(|z|)\, dA(z)/\pi \qquad (5.1.7)$$

where $S = \{z : r_0 < |z| < 1\}$. Using the Area Formula (Theorem 2.32) with

$$W(z) = \chi_S(z) H(|z|) = \chi_S(z) E(-\log|z|)(-\log|z|)$$

the right hand side of Inequality (5.1.7) is seen to be equal to

$$c_1 \int_{\varphi(D)} |f'(w)|^2 \left(\sum_{j\geq 1} \chi_S(z_j(w)) E(-\log|z_j(w)|)(-\log|z_j(w)|) \right) dA(w)/\pi$$

$$(5.1.8)$$

where the sum is over all preimages $z_j(w)$ of w under φ, listed with multiplicities. Now if $z_j(w)$ is in S, then $|w| = |\varphi(z_j(w))| \leq N(|z_j(w)|)$ so $N^{-1}(|w|) \leq |z_j(w)|$ which gives

$$E(-\log|z_j(w)|) \leq E(-\log N^{-1}(|w|)) \qquad (5.1.9)$$

Also recall Littlewood's Inequality (Theorem 2.29):

$$\sum_{j\geq 1}(-\log|z_j(w)|) \leq \log \left| \frac{1 - \overline{\varphi(0)}w}{w - \varphi(0)} \right|$$

Set

$$Y(r) = \max \left\{ \log \left| \frac{1 - \overline{\varphi(0)}w}{w - \varphi(0)} \right| : |w| = r \right\}$$

Since $Y(r) \leq \log|2/(r - |\varphi(0)|)|$, $Y(r)$ is integrable on $(0,1)$. It is straightforward to verify that for $r > |\varphi(0)|$, the maximum in the definition of $Y(r)$ is attained at $w = r\varphi(0)/|\varphi(0)|$ (see Exercise 5.1.6).

Set $y(s,r) = \log\left(\frac{1-sr}{r-s}\right)$ for $s < r < 1$. For fixed r, $y(s,r)$ increases with s, as is easily seen from

$$\frac{\partial}{\partial s}\left(\frac{1-sr}{r-s}\right) = \frac{1-r^2}{(r-s)^2} > 0$$

For fixed s, L'Hôpital's rule shows $y(s,r)/(-\log r)$ converges to $(1+s)/(1-s)$ as r tends to 1. If $|\varphi(0)| = s < \rho < r$ we have $Y(r) = y(s,r)$. Thus

$$Y(r) \leq c_2 \frac{1+s}{1-s}(-\log r) \quad \text{for } \rho < r < 1$$

for some constant c_2 depending on ρ and s.

It follows from Inequality (5.1.9) and Littlewood's Inequality that Expression (5.1.8) is bounded above by

$$c_1 \int_D |f'(w)|^2 E\left(-\log N^{-1}(|w|)\right) Y(|w|)\, dA(w)/\pi$$

Again we may replace the region of integration D by the annulus $\{\rho < |w| < 1\}$ at the expense of making c_1 larger, since the integrand, except for $|f'(w)|^2$, is rotation invariant. On $\{\rho < |w| < 1\}$ the estimates

$$E(-\log N^{-1}(|w|)) \leq RE(-\log|w|)$$

and

$$Y(|w|) \leq c_2 \frac{1+s}{1-s}(-\log|w|)$$

hold. Thus we have

$$\|f \circ \varphi\|_0^2 \leq c_3 \frac{1+s}{1-s} R \int_D |f'(w)|^2 E(-\log|w|)(-\log|w|)\, dA(w)/\pi$$

$$= cR \int_D |f'(w)|^2 F(-\log|w|)\, dA(w)/\pi$$

$$= cR\|f\|_0^2$$

where c is independent of f and R. ∎

PROOF (of Theorem 5.5)   The weighted Dirichlet description of $A_G^2(D)$ tells us that for f in $A_G^2(D)$, $\|f \circ \varphi\|_G^2$ is dominated by a constant times

$$|f(\varphi(0))|^2 + \|f \circ \varphi\|_0^2$$

Since $|f(\varphi(0))| \leq \|K_{\varphi(0)}\|\|f\|$ we need only show that $\|f \circ \varphi\|_0^2$ is bounded by a constant multiple of $\|f\|^2$. To do this we apply the previous lemma with $N(r)$ defined as follows

(i)   If $\|\varphi\|_\infty < 1$ then $N(r) = r$.

(ii) If $\varphi(0) = 0$ and $\|\varphi\|_\infty = 1$ then $N(r) = M(r)$.

(iii) If $\|\varphi\|_\infty = 1$ and $|\varphi(0)| = s > 0$ let $N(r) = M(r)$ on $[1/2, 1]$, $N(0) = 0$, and define $N(r)$ to be linear on $[0, 1/2]$.

Notice that in each case $N(r)$ is a continuous, strictly increasing map of $[0, 1]$ onto $[0, 1]$. Furthermore, there exists $r_0 < 1$ so that $|\varphi(z)| \leq N(|z|)$ for $r_0 \leq |z| < 1$: we choose $r_0 = \|\varphi\|_\infty$ in (i), $r_0 = 0$ in (ii), and $r_0 = 1/2$ in (iii).

The hypothesis on $E(-\log r)/E(-\log M(r))$ and the definition of $N(r)$ guarantee that there exists $R < \infty$ and $\rho < 1$ so that

$$\frac{E(-\log r)}{E(-\log N(r))} \leq R \text{ whenever } \rho < N(r) < 1$$

By Lemma 5.6 there exists c independent of f with

$$\|f \circ \varphi\|_0^2 \leq cR\|f\|_0^2 \leq cR\|f\|_G^2$$

for all f analytic in D. ∎

As an application of Theorem 5.5, recall that if $|\varphi'(\zeta)| > 1$ for all ζ in ∂D then $\inf\{|\varphi'(\zeta)| : \zeta \in \partial D\} > 1$ since, by Proposition 2.46, this infimum is actually attained. By Exercise 2.3.17

$$\lim_{r \to 1} \frac{\log M(r)}{\log r} > 1$$

Thus, ignoring the trivial case $\lim_{r \to 1} M(r) < 1$ from which it easily follows that C_φ is compact on $A_G^2(D)$, for r sufficiently close to 1, $E(-\log M(r)) > E(-\log r)$ and Theorem 5.5 guarantees that C_φ is bounded on $A_G^2(D)$:

COROLLARY 5.7
If $|\varphi'(\zeta)| > 1$ for every ζ in ∂D, then C_φ is bounded on $A_G^2(D)$, where G is any positive, continuous, non-increasing weight function.

Much more is actually true, as we shall see in the next section.

A theorem analogous to Theorem 5.5, but with a hypothesis involving the function $G(r)$ directly, gives a sufficient condition for C_φ to be bounded.

THEOREM 5.8
If G is a non-increasing weight function and

$$\limsup_{r \to 1} \frac{G(r)}{G(M(r))} < \infty$$

then C_φ is bounded on $A_G^2(D)$.

The proof of this theorem, which we will not give here, requires only a technical lemma that shows that this assumption on G implies that either the hypothesis of Theorem 5.5 holds or that $\lim_{r \to 1} M(r) < 1$ (so that C_φ is in fact compact on $A_G^2(D)$).

The next theorem provides almost the converse to Theorem 5.8.

THEOREM 5.9
Suppose G is a non-increasing weight function and

$$\liminf_{r \to 1} \frac{G(r)}{G(M(r))} = \infty$$

then

$$\limsup_{r \to 1} \frac{\|K_{M(r)}\|}{\|K_r\|} = \infty$$

Our desired corollary is immediate:

COROLLARY 5.10

If G is a non-increasing weight function and

$$\liminf_{r \to 1} \frac{G(r)}{G(M(r))} = \infty$$

then C_φ is unbounded on $A_G^2(D)$.

PROOF (of Theorem 5.9) If $|\varphi'(\zeta)| > 1$ for all ζ on the unit circle, then for r sufficiently close to 1, $M(r) \leq r$. Since G is non-increasing this says

$$\limsup_{r \to 1} \frac{G(r)}{G(M(r))} \leq 1$$

Thus for the hypothesis to hold we must have a point ζ_0 in ∂D with $|\varphi'(\zeta_0)| \leq 1$. By Julia's Lemma

$$M(r) \geq |\varphi(r\zeta_0)| \geq r \quad \text{for } 0 < r < 1 \tag{5.1.10}$$

Let

$$B = \limsup_{r \to 1} \frac{\|K_{M(r)}\|^2}{\|K_r\|^2}$$

Assume, for a contradiction, that B is finite. There exists $r_1 < 1$ so that

$$\frac{\|K_{M(r)}\|^2}{\|K_r\|^2} \leq 2B$$

if $r_1 \leq r < 1$. Choose $A > 2B$ and use the hypothesis on G to find $r_2 < 1$ with

$$\frac{G(r)}{G(M(r))} \geq A$$

if $r_2 \leq r < 1$. Set $r_0 = \max\{r_1, r_2\}$.

Let $h(r) = \|K_r\|^2 G(r)$. Since the identity map does not induce a Hilbert–Schmidt composition operator we have, by Exercise 3.3.1 of Section 3.3,

$$\int_D \|K_z\|^2 G(|z|) \, dA(z) = \infty$$

so that $h(r)$ must be unbounded as r approaches 1. Now for $r_0 \leq r < 1$

$$A > 2B \geq \frac{\|K_{M(r)}\|^2}{\|K_r\|^2} = \frac{h(M(r))}{h(r)} \frac{G(r)}{G(M(r))} \geq \frac{h(M(r))}{h(r)} A$$

so that

$$\frac{h(M(r))}{h(r)} < 1 \tag{5.1.11}$$

for $r_0 \leq r < 1$. We claim that Inequality (5.1.11) cannot hold. To see this, fix $N > \sup\{h(r) : 0 \leq r \leq M(r_0)\}$ and let $t_1 = \inf\{t : h(t) \geq N\}$. Then

$h(t_1) = N$ and $h(s) < N$ for all $s < t_1$. Let r^* be such that $M(r^*) = t_1$; by the definition of t_1 it follows that $r^* > r_0$. Also, by Inequality (5.1.10) we have $r^* \le M(r^*)$. If $r^* = M(r^*)$ then

$$\frac{h(M(r^*))}{h(r^*)} = 1$$

contradicting Inequality (5.1.11). If $r^* < M(r^*)$ then $h(r^*) < N$ by the definition of t_1 while $h(M(r^*)) = h(t_1) = N$, so again Inequality (5.1.11) is contradicted. Thus B is infinite and we are done. ∎

Although the main interest in Theorems 5.5, 5.8, and 5.9 and Lemma 5.6 is for fast weights G, the only requirement on G in these results is that it is a positive, continuous, non-increasing weight function. The next result, another application of Lemma 5.6, also holds whenever G is positive, continuous, and non-increasing, and is of interest even in the case of the standard weights. The functions H, F, and E continue to be defined from G via Equations (5.1.4), (5.1.5), and (5.1.6).

THEOREM 5.11

Let $\varphi : D \to D$ be analytic and suppose there exist $r_0 < 1$ and $A > 0$ with $|\varphi(z)| \le |z|^A$ for $r_0 \le |z| < 1$. Then for any $t_0 > 0$, there exists c independent of f so that

$$\|f \circ \varphi\|_0^2 \le c \left(\sup_{0 < t \le t_0} \frac{A F(t/A)}{F(t)} \right) \|f\|_0^2$$

for all f analytic in D.

PROOF We will apply Lemma 5.6 with $N(r) = r^A$. Set $\rho = \exp(-t_0)$ and $t = -\log r^A$. Using $E(t) = F(t)/t$ and the definition of $N(r)$ we have

$$\sup_{\rho \le N(r) < 1} \frac{E(-\log r)}{E(-\log N(r))} = \sup_{0 < t \le t_0} \frac{A F(t/A)}{F(t)} \equiv R$$

Thus by Lemma 5.6 $\|f \circ \varphi\|_0^2 \le cR\|f\|_0^2$ as desired. ∎

Notice that if $A \ge 1$ this guarantees that C_φ will be bounded on $A_G^2(D)$.

Whenever $\varphi(0) = 0$ the subspace $\mathcal{M} \equiv \{f \in A_G^2(D) : f(0) = 0\}$ is a reducing subspace for C_φ. Writing $A_G^2(D) = C \oplus \mathcal{M}$ the matrix of C_φ with respect to this decomposition has the form

$$\begin{pmatrix} 1 & 0 \\ 0 & C_\varphi|_{\mathcal{M}} \end{pmatrix}$$

If $Pf = f(0)$ so that P is orthogonal projection onto the constants then $C_\varphi - P$ has matrix

$$\begin{pmatrix} 0 & 0 \\ 0 & C_\varphi|_{\mathcal{M}} \end{pmatrix}$$

On $\mathcal{M}$ the $A_G^2(D)$ norm is equivalent to $\|\cdot\|_0$ so Theorem 5.11 says

$$\|C_\varphi|_{\mathcal{M}}\|^2 = \|C_\varphi - P\|^2 \leq c \left(\sup_{0 < t \leq t_0} \frac{AF(t/A)}{F(t)} \right)$$

which can be quite small, in contrast to $\|C_\varphi\| = 1$. Even when $\varphi(0) \neq 0$ it still makes sense to look at $\|C_\varphi - P\|$. The next theorem will be used in Section 9.4.

THEOREM 5.12
Suppose $G(r)$ is non-increasing on $[0, 1)$ and $\{\varphi_n\}$ is any sequence of maps of D into D. If $\lim_{n \to \infty} \varphi_n(w) = 0$ for every w in D, then C_{φ_n} is bounded on $A_G^2(D)$ for large n and $\|C_{\varphi_n} - P\|$, the norm of $C_{\varphi_n} - P$ as an operator on $A_G^2(D)$, tends to 0 as n approaches infinity.

PROOF Since $\|\varphi_n\|_\infty \leq 1$, the convergence of φ_n to 0 is actually uniform on compact subsets of D. In particular,

$$\delta_n \equiv M_n(.5) = \max\{|\varphi_n(z)| : |z| = .5\}$$

approaches 0 as n tends to infinity. Set $A_n = (\log \delta_n)/(-\log 2)$ so that A_n tends to infinity. By Exercise 2.3.18

$$|\varphi_n(z)| \leq |z|^{A_n}$$

for $\frac{1}{2} \leq |z| < 1$.
Pick any $0 < \delta < 1$. By Exercise 5.1.7

$$\frac{F(t)}{t^{1+\delta}}$$

is non-decreasing for t sufficiently small, say for $0 < t \leq t_0$. By the previous theorem there exists a constant c so that

$$\|C_{\varphi_n} f\|_0^2 \leq c \left(\sup_{0 < t \leq t_0} \frac{A_n F(t/A_n)}{F(t)} \right) \|f\|_0^2$$

Note that $A_n \geq 1$ for large n, so C_{φ_n} is bounded on $A_G^2(D)$ for such n. Write

$$\frac{A_n F(t/A_n)}{F(t)} = \frac{F(t/A_n)}{(t/A_n)^{1+\delta}} \frac{t^{1+\delta}}{F(t)} \frac{1}{A_n^\delta}$$

Since A_n approaches infinity as n tends to infinity, for n sufficiently large and all $t < t_0$

$$\frac{F(t/A_n)}{(t/A_n)^{1+\delta}} \leq \frac{F(t)}{t^{1+\delta}}$$

so that for such n

$$\frac{A_n F(t/A_n)}{F(t)} \leq \frac{1}{A_n^\delta}$$

and

$$\|C_{\varphi_n} f\|_0^2 \leq c \frac{1}{A_n^\delta} \|f\|_0^2$$

Since $(C_{\varphi_n} - P)f = f \circ \varphi_n - f(0)$ the description of $A_G^2(D)$ as a weighted Dirichlet space gives

$$
\begin{aligned}
\|(C_{\varphi_n} - P)f\|^2 &\leq C \left(|f \circ \varphi_n(0) - f(0)|^2 + \|f \circ \varphi_n\|_0^2 \right) \\
&= C \left(|\langle f, K_{\varphi_n(0)} - K_0 \rangle|^2 + \|f \circ \varphi_n\|_0^2 \right) \\
&\leq C \left(\|f\|^2 \|K_{\varphi_n(0)} - K_0\|^2 + c \frac{1}{A_n^\delta} \|f\|_0^2 \right)
\end{aligned}
$$

for n sufficiently large. Now let n go to infinity and use the fact that $\varphi_n(0)$ converges to 0 to get the desired conclusion. ∎

Exercises

5.1.1 Show that if G is fast and regular, then $A_G^2(D)$ is equivalent to $H^2(\beta)$ for some sequence $\{\beta(n)\}$ satisfying $\lim_{n \to \infty} n^\alpha \beta(n) = 0$ for every $\alpha > 0$.

5.1.2 Assume $G(r)$ is a positive, continuous function on $(0, 1)$ with $\int_0^1 G(r) r\, dr$ finite. Let $R = \{z \in D : r_0 \leq |z| < 1\}$ where $0 < r_0 < 1$. Show there exists $C < \infty$ depending only on G and r_0 so that

$$\int_D |f|^2 G\, dA \leq C \int_R |f|^2 G\, dA$$

for all f analytic in D.

5.1.3 Suppose that G_1 and G_2 are as in the previous problem. Show that if $G_1(r)/G_2(r)$ is bounded above and away from 0 for $r_0 \leq r < 1$ then $A_{G_1}^2(D) = A_{G_2}^2(D)$ with equivalent norms.

5.1.4 Suppose $\{K_w^1\}$ and $\{K_w^2\}$ are the reproducing kernel functions for the weighted Bergman spaces $A_{G_1}^2(D)$ and $A_{G_2}^2(D)$. Show that if $G_1(r)/G_2(r)$ is bounded above for r near 1, then

$$\frac{\|K_w^2\|_{G_2}}{\|K_w^1\|_{G_1}}$$

is bounded above in D.

5.1.5 Suppose that $\lim_{n \to \infty} n^\alpha \beta(n) = 0$ for all $\alpha > 0$.
 (i) Show that $f_k = [(1+z)/(1-z)]^k$ is in $H^2(\beta)$ for $k = 1, 2, 3, \dots$.
 (ii) Check that for r in $(0, 1)$ and $\varphi(z) = (z + r)/(1 + rz)$ then

$$f_k \circ \varphi = \left(\frac{1+r}{1-r} \right)^k f_k$$

and conclude that C_φ is not bounded on $H^2(\beta)$.

5.1.6 Recall the definition

$$Y(r) = \max\left\{\log\left|\frac{1 - \overline{\varphi(0)}w}{w - \varphi(0)}\right| : |w| = r\right\}$$

Show that if $r > |\varphi(0)|$, $Y(r)$ is attained at $w = r\varphi(0)/|\varphi(0)|$

5.1.7 If $G(r)$ is positive, continuous, and non-increasing on $(0, 1)$ then for $0 < \delta < 1$

$$\frac{F(t)}{t^{1+\delta}}$$

is non-decreasing for t small and tends to 0 as $t \to 0$. (F is defined in Equation (5.1.5)).

5.1.8 Let $\varphi(z) = z + t(1 - z)^\beta$ and $G(r) = \exp[-B\frac{1}{(1-r)^\alpha}]$, for $\alpha > 0$ and $B > 0$ where $1 < \beta < 3$ and $0 < t < 2^{1-\beta}$.

(a) Show that φ maps D into D. Note that $\varphi(1) = 1$ and φ has angular derivative 1 at 1. Show that $|\varphi(e^{i\theta})| < 1$ if $e^{i\theta} \neq 1$ so that φ has finite angular derivative only at 1.

(b) If $\{z_n\}$ is a sequence in D with $z_n \to 1$ tangentially then $|\varphi(z_n)| < |z_n|$ for n large.

(c) Use Theorem 5.8 and Corollary 5.10 to show that C_φ is bounded on $A_G^2(D)$ if and only if $\beta \geq \alpha + 1$.

5.1.9 Set $G(r) = (1 - r)^\alpha$ where $\alpha \geq 0$ and let φ be an analytic map of D into D with $\varphi(0) = 0$. Let $\mathcal{M} = \{f \in A_G^2(D) : f(0) = 0\}$.

(a) Given r_0 with $0 < r_0 < 1$, show that there exists $c < \infty$ so that

$$\|C_\varphi|_\mathcal{M}\|^2 \leq cA^{-(\alpha+1)}$$

provided $|\varphi(z)| \leq |z|^A$ on $r_0 < |z| < 1$.

(b) Let $\varphi(z) = z^n$. Part (a) shows that $n^{\alpha+1}\|C_{z^n}|_\mathcal{M}\|^2$ is bounded as n goes to infinity. Is it also bounded away from zero?

(c) Suppose $\{w_j\}$ are the zeros of φ repeated with multiplicities. Show that there exists $b < \infty$ so that

$$\|C_\varphi|_\mathcal{M}\|^2 \leq b\left(\frac{1}{\sum(1 - |w_j|)}\right)^{\alpha+1}$$

where b depends only on α.

5.1.10 Suppose $\{\varphi_n\}$ is a sequence of analytic self-maps of D such that C_{φ_n} is bounded on $A_G^2(D)$ for some positive, continuous decreasing weight function G.

(a) Let P be the orthogonal projection of $A_G^2(D)$ onto the constants defined by $P(f) = f(0)$. Prove the converse of Theorem 5.12; that is, show that if $\|C_{\varphi_n} - P\| \to 0$ then $\varphi_n(z) \to 0$ for each z in D.

(b) Define a not necessarily orthogonal projection of $A_G^2(D)$ onto the constants by $P_a(f) = f(a)$ for a in D. Show that if φ_n is a sequence of maps of the disk into itself such that $\|C_{\varphi_n} - P_a\|$ converges to 0, then $\varphi_n(z) \to a$ for every z in D.

(c) Suppose further that $\psi_a(z) = (z - a)/(1 - \overline{a}z)$ induces a bounded composition operator on $A_G^2(D)$. Show that if $\varphi_n(z) \to a$ pointwise in D then $\|C_{\varphi_n} - P_a\|$ converges to 0.

Notice that the conclusion of (c) holds whenever $A_G^2(D)$ is automorphism invariant and φ_n is the nth iterate of φ provided φ, not an elliptic automorphism of D, has fixed point a in D.

Notes

The study of composition operators on large spaces was begun by T. L. Kriete and B. D. Mac-Cluer in [KrM92] and most of the results of this section, as well as most of the exercises, are contained in, or essentially contained in, that paper. The proof of Theorem 5.8 can be found there as well. Exceptions are Theorem 5.9 which was recently obtained by T. L. Kriete ([Kri94]), and Theorem 5.12 and the related Exercise 5.1.10, also due to T. L. Kriete (private communication).

Exercise 5.1.8 is due to T. L. Kriete and B. D. MacCluer [KrM94] and [KrM92]. Note that a consequence of this exercise is that $|\varphi'(\zeta)| \geq 1$ does not characterize boundedness on large spaces. The results of this exercise remain true for $\beta = 3$ as well. In particular $\varphi(z) = z + t(1-z)^3$, with $0 < t < 1/4$ give bounded operators on $A_G^2(D)$ for $G(r) = \exp[-B(\frac{1}{1-r})^2]$. As we will see in the next section, C_φ is not compact since $\varphi'(1) = 1$. The main theorem of [KrM94] shows that for weights that decay more rapidly than $G(r) = \exp[-B(\frac{1}{1-r})^2]$ there are no bounded, non-compact operators, other than those induced by rotations. The $\beta = 3$ example shows that this result is sharp.

5.2 Compactness on large spaces

Recall that by Proposition 3.13, in any weighted Hardy space $H^2(\beta)$ where $\sum \beta(n)^{-2} = \infty$ the essential norm satisfies

$$\|C_\varphi\|_e \geq \limsup_{|w| \to 1} \frac{\|K_{\varphi(w)}\|}{\|K_w\|}$$

This easily yields a necessary condition for C_φ to be compact.

PROPOSITION 5.13
Suppose C_φ is bounded on $H^2(\beta)$ where $\sum \beta(n)^{-2} = \infty$. If there exists ζ in ∂D with $|\varphi'(\zeta)| \leq 1$ then $\|C_\varphi\|_e \geq 1$; in particular C_φ is not compact.

PROOF By Julia's Lemma if $|\varphi'(\zeta)| \leq 1$ then $1 - |\varphi(r\zeta)| \leq 1 - r$ for all $r > 0$ and hence $|\varphi(r\zeta)| \geq r$. Since $\|K_z\|$ increases with $|z|$, we have

$$\|C_\varphi\|_e \geq \limsup_{r \to 1} \frac{\|K_{\varphi(r\zeta)}\|}{\|K_{r\zeta}\|} \geq 1$$

A similar result, under the weaker hypothesis that

$$\sum_{n=0}^{\infty} \frac{n^{2k}}{\beta(n)^2} = \infty$$

for some non-negative integer k is discussed later, in Chapter 7.

Of course, Proposition 5.13 applies to the spaces $A_G^2(D)$ where G is a fast weight and shows that a necessary condition for C_φ to be compact on $A_G^2(D)$ is that $|\varphi'(\zeta)| > 1$ at all ζ in the unit circle. A result which lies deeper than this is that, for fast weights, this condition is *sufficient* for compactness. Before proceeding to a proof of this, we need to assemble several relevant ideas.

In Section 3.2, we described a procedure for estimating essential norms from above which worked in a fairly general setting including that of weighted Bergman spaces $A_G^2(D)$ with G fast and regular. If C_φ is bounded on $A_G^2(D)$ and $H(r) = \int_r^1 \int_x^1 G(s)\, ds\, dx$ we have by Inequality (3.2.4)

$$\|C_\varphi\|_e^2 \leq C \lim_{r \to 1} \gamma_r$$

where

$$\gamma_r = \sup_{r \leq |w| < 1} \frac{\sum H(|z_j(w)|)}{H(|w|)}$$

and $\{z_j(w)\} = \varphi^{-1}(w)$. We will interpret this, for arbitrary fast regular weights G, in a way that parallels the treatment for the standard weight spaces $A_\alpha^2(D)$ at the end of Section 3.2. Assume that $\|\varphi\|_\infty = 1$ for otherwise $\|C_\varphi\|_e$ is trivially seen to be 0. As in Section 3.2, let $z(w)$ be a point of $\{z_j(w)\}$ with minimum modulus, and choose w_n tending to ∂D so that

$$\lim_{r \to 1} \gamma_r = \lim_{n \to \infty} \frac{\sum H(|z_j(w_n)|)}{H(|w_n|)}$$

It is easy to see that $|z(w_n)| \to 1$. Recall from the last section the function $E(t)$ defined on $(0, \infty)$ by $E(-\log r) = -H(r)/\log r$.

Now

$$H(|z_j(w_n)|) = E(-\log|z_j(w_n)|)(-\log|z_j(w_n)|)$$

$$\leq E(-\log|z(w_n)|)(-\log|z_j(w_n)|)$$

for n large enough, since E is increasing on $(0, 1/2)$. So

$$\sum_{j \geq 1} H(|z_j(w_n)|) \leq E(-\log|z(w_n)|) \sum_{j \geq 1} (-\log|z_j(w_n)|)$$

$$= E(-\log|z(w_n)|)N_\varphi(w_n)$$

Using Littlewood's Inequality

$$N_\varphi(w_n) \leq \log\left|\frac{1 - \overline{\varphi(0)}w_n}{\varphi(0) - w_n}\right|$$

exactly as we did in Section 3.2 gives

$$\|C_\varphi\|_e^2 \leq C\frac{1 + |\varphi(0)|}{1 - |\varphi(0)|} \limsup_{n \to \infty} \frac{E(-\log|z(w_n)|)}{E(-\log|w_n|)}$$

$$\leq C \frac{1 + |\varphi(0)|}{1 - |\varphi(0)|} \limsup_{r \to 1} \frac{E(-\log r)}{E(-\log M(r))}$$

since $\|\varphi\|_\infty = 1$ implies $M(r)$ tends to 1 as r approaches 1.

Our task is now to use this general, but somewhat unintuitive formula, to derive some strikingly simple characterizations of the compact C_φ on $A_G^2(D)$ when G is fast and regular.

THEOREM 5.14

Let G be a fast regular weight and let $\varphi : D \to D$ be analytic. The following are equivalent:

(1) C_φ *is compact on* $A_G^2(D)$

(2) $|\varphi'(\zeta)| > 1$ *for all ζ in* ∂D

(3) *There exist r_0, $0 < r_0 < 1$, and $A > 1$ so that $|\varphi(z)| \leq |z|^A$ for $r_0 \leq |z| < 1$.*

PROOF We have just observed in Proposition 5.13 that (1) implies (2). For (2) implies (3) let $A_0 = \inf\{|\varphi'(\zeta)| : \zeta \in \partial D\}$; since this infimum is attained by Proposition 2.46, $A_0 > 1$. Choose A with $1 < A < A_0$. By Exercise 2.3.17 $\lim_{r \to 1} \log M(r)/\log r = A_0$ and hence there is $r_0 < 1$ so that $r \geq r_0$ implies

$$\frac{\log M(r)}{\log r} \geq A$$

which is equivalent to (3).

Finally assume that (3) holds. The definition of the angular derivative $|\varphi'(\zeta)|$ shows that (3) $\Rightarrow$ (2) and thus by Corollary 5.7 we know that C_φ is bounded on $A_G^2(D)$. This puts us in a position to apply the essential norm estimate

$$\|C_\varphi\|_e^2 \leq C \frac{1 + |\varphi(0)|}{1 - |\varphi(0)|} \limsup_{r \to 1} \frac{E(-\log r)}{E(-\log M(r))}$$

For $r \geq r_0$, $M(r) \leq r^A$ and, letting $\delta = 1/A < 1$, we have $-\log r \leq -\delta \log M(r)$. We may assume $M(r)$ approaches 1 as r tends to 1, otherwise C_φ is trivially compact. Thus, since $E(t)$ is increasing on $(0, 1/2)$, for r sufficiently close to 1 we have

$$E(-\log r) \leq E(-\delta \log M(r))$$

and

$$\frac{E(-\log r)}{E(-\log M(r))} \leq \frac{E(-\delta \log M(r))}{E(-\log M(r))}$$

We will have our desired result if we can argue that

$$\lim_{t \to 0} \frac{E(\delta t)}{E(t)} = 0$$

This requires tracing through the relationship between E, H, and G. In particular

$$\frac{E(\delta t)}{E(t)} = \frac{H(e^{-\delta t})}{\delta t} \frac{t}{H(e^{-t})} = \frac{1}{\delta} \frac{H(e^{-\delta t})}{H(e^{-t})}$$

Use L'Hôpital's Rule twice, and the relationship $H'' = G$ to obtain

$$\lim_{t \to 0} \frac{E(\delta t)}{E(t)} = \lim_{t \to 0} \frac{G(1 - \delta t)}{G(1 - t)}$$

Regularity of G guarantees that this latter limit is zero; see Exercise 5.2.2. Thus, $\|C_\varphi\|_e = 0$ and C_φ is compact, establishing (1). ∎

It is possible to obtain a characterization of compactness of C_φ on $A_G^2(D)$ in terms of the fast, regular weight function G, specifically

$$C_\varphi \text{ is compact on } A_G^2(D) \text{ if and only if } \lim_{|z| \to 1} \frac{G(r)}{G(M(r))} = 0$$

and similarly to obtain an upper estimate on $\|C_\varphi\|_e$ expressed in terms of the function G instead of E:

$$\|C_\varphi\|_e^2 \le C \frac{1 + |\varphi(0)|}{1 - |\varphi(0)|} \limsup_{|z| \to 1} \frac{G(r)}{G(M(r))}$$

Since the arguments needed for a complete proof of these results are quite technical, we will not give them here. They can be found in [KrM92]. Exercise 5.2.3 contains a partial result along these lines.

Exercises

5.2.1 Show that if $\varphi(0) = 0$ and φ is not a rotation of D, then C_φ is compact on $A_G^2(D)$, where G is fast and regular.

5.2.2 Show that if G is fast and regular, then

$$\lim_{t \to 0} \frac{G(1 - \delta t)}{G(1 - t)} = 0$$

for every $0 < \delta < 1$.

5.2.3 Show that if G is fast and regular and $|\varphi'(\zeta)| > 1$ for all ζ in the unit circle then

$$\lim_{r \to 1} \frac{G(r)}{G(M(r))} = 0$$

5.2.4 Suppose that $G(r)$ is any non-increasing weight function. Show that if C_φ is bounded on $A_G^2(D)$ then

$$\liminf_{r \to 1} \frac{G(r)}{G(M(r))} \le \|C_\varphi\|_e^2$$

Hint: Modify the proof of Theorem 5.9.

5.2.5 Let $G(r) = \exp[-B/(1-r)]$ where $B > 0$ and set $\varphi(z) = z^n/n + (n-1)/n$. Show that C_φ is bounded on $A_G^2(D)$ and find a lower bound for $\|C_\varphi\|_e$.

Notes

Theorem 5.14 is due to T. L. Kriete and B. D. MacCluer[KrM92] and the proof given here is taken from that paper.

5.3 Hilbert–Schmidt operators

Recall from Section 3.3 that giving examples of compact composition operators on the Hardy $H^2(D)$ or standard weight Bergman spaces $A_\alpha^2(D)$ which are not in some (or any) Schatten p–class is a delicate problem. In the realm of fast weight Bergman spaces, however, the matter is strikingly different. We will focus attention here on Hilbert–Schmidt composition operators. Our first goal is to show that for very fast weights, all compact C_φ are in fact Hilbert–Schmidt. Using the perfectly general statement (Exercise 3.3.1) that on any weighted Bergman space $A_G^2(D)$,

$$\|C_\varphi\|_{HS}^2 = \int_D \|K_{\varphi(z)}\|^2 G(|z|) \frac{dA(z)}{\pi} \qquad (5.3.1)$$

we have C_φ is Hilbert–Schmidt on $A_G^2(D)$ if and only if

$$\int_{r_0 < |z| < 1} \|K_{\varphi(z)}\|^2 G(|z|) \, dA(z) < \infty$$

for any fixed r_0, $0 < r_0 < 1$. Of course, $\|\cdot\|$ denotes the norm in $A_G^2(D)$. The next lemma is a straightforward estimate on $\|K_z\|$ when $G(r)$ is a positive, continuous and non-increasing weight function.

LEMMA 5.15
Let G be a positive, continuous and non-increasing weight function and fix b with $0 < b < 1$. For any z in the unit disk

$$\|K_z\|^2 \leq \frac{1}{\pi(1-b)^2(1-|z|)^2 G(1 - b(1-|z|))}$$

PROOF Let z be in the unit disk and denote by $D_e(z, \delta)$ the Euclidean disk centered at z of radius $\delta = (1-b)(1-|z|)$. Then

$$|K_z(z)| = \left| \frac{1}{\pi\delta^2} \int_{D_e(z,\delta)} K_z(w) \, dA(w) \right|$$

$$\leq \frac{1}{\sqrt{G(|z|+\delta)}} \int_{D_e(z,\delta)} |K_z(w)| \sqrt{G(|z|+\delta)}\, \frac{dA(w)}{\pi\delta^2}$$

Since G is non-increasing, $G(|z|+\delta) \leq G(|w|)$ for all w in $D_e(z,\delta)$ and

$$|K_z(z)| \leq \frac{1}{\sqrt{G(|z|+\delta)}} \int_{D_e(z,\delta)} |K_z(w)| \sqrt{G(|w|)}\, \frac{dA(w)}{\pi\delta^2}$$

Since $dA/(\pi\delta^2)$ is a probability measure on $D_e(z,\delta)$,

$$|K_z(z)| \leq \frac{1}{\sqrt{G(|z|+\delta)}} \left(\int_{D_e(z,\delta)} |K_z(w)|^2 G(|w|)\, \frac{dA(w)}{\pi\delta^2} \right)^{1/2}$$

$$\leq \frac{1}{\sqrt{\pi}\delta} \frac{1}{\sqrt{G(|z|+\delta)}} \left(\int_D |K_z(w)|^2 G(|w|)\, dA(w) \right)^{1/2}$$

Thus

$$\|K_z\|^4 = |K_z(z)|^2 \leq \frac{1}{\pi\delta^2} \frac{1}{G(|z|+\delta)} \|K_z\|^2$$

Since $G(|z|+\delta) = G(1 - b(1 - |z|))$, we have the desired result. ∎

In the next theorem, we consider non-increasing weights $G(r)$ that are sufficiently far into the realm of fast weights that

$$h(r) \equiv -\frac{\log G(r)}{\log^2(1-r)} \tag{5.3.2}$$

is non-decreasing near 1 and tends to infinity as r approaches 1. This assumption on $G(r)$ guarantees that $G(r)$ is regular, as well, so that Theorem 5.14 will apply (see Exercise 5.3.1).

THEOREM 5.16
Suppose that

$$G(r) = \exp\left(-h(r)\log^2(1-r)\right)$$

where $h(r)$ is non-decreasing near 1 and tends to infinity as r approaches 1. Then every compact composition operator on $A_G^2(D)$ is Hilbert–Schmidt.

PROOF   Suppose C_φ is a compact operator so that by Theorem 5.14 we have $|\varphi'(\zeta)| > 1$ for all ζ in D. This implies that there exist $r_0 < 1$ and $A > 1$ so that

$$\frac{1 - |\varphi(z)|}{1 - |z|} \geq A \text{ for } r_0 \leq |z| < 1$$

To show that C_φ is Hilbert–Schmidt it suffices to verify that

$$\int_{r_0 \leq |z| < 1} \|K_{\varphi(z)}\|^2 G(|z|)\, dA < \infty \tag{5.3.3}$$

We will actually show that the integrand in Expression (5.3.3) is bounded.

Choose $b < 1$ close enough to 1 so that $bA > 1$. For $|z| \geq r_0$ we have $1 - |\varphi(z)| \geq A(1 - |z|)$ and therefore using Lemma 5.15

$$\|K_{\varphi(z)}\|^2 = \|K_{|\varphi(z)|}\|^2 \leq \|K_{1-A(1-|z|)}\|^2$$

$$\leq \frac{1}{\pi(1-b)^2 A^2(1-|z|)^2 G(1-bA(1-|z|))}$$

Now letting $a = bA$, the hypothesis shows $G(1 - bA(1 - |z|))$ is

$$\exp\left(-h(1 - a(1 - |z|))\log^2(a(1 - |z|))\right) \geq \exp\left(-h(|z|)\log^2(a(1 - |z|))\right)$$

for $|z|$ sufficiently close to 1, since h is non-decreasing near 1. For such z, the integrand in Expression (5.3.3) is bounded above by

$$\frac{c(b)}{A^2(1-|z|)^2} \exp\left(-h(|z|)\left(\log^2(1 - |z|) - \log^2(a(1 - |z|))\right)\right)$$

where $c(b) = 1/(\pi(1-b)^2)$. Thus the integrand is bounded by a constant multiple of

$$\frac{1}{(1-|z|)^2} \exp\left(h(|z|)\left(2\log a \log(1 - |z|) + \log^2 a\right)\right)$$

which tends to 0 as $|z|$ approaches 1 since $a > 1$ and $h(|z|)$ converges to infinity.

∎

We conclude this section by considering those fast weights which decay more slowly than the weights of Theorem 5.16.

If $\varphi(z) = z$, then C_φ is the identity which is not compact on any $A_G^2(D)$ so, using Equation (5.3.1), we have

$$\int_0^1 \|K_r\|^2 G(r)r\,dr = \infty$$

For any standard weight, $G(r) = (1 - r)^\alpha$, we have $(1 - r)\|K_r\|^2 G(r)r \sim \frac{r}{1-r}$ which also just fails to be integrable. The next lemma generalizes this to arbitrary non-increasing weights $G(r)$.

LEMMA 5.17

If $0 < r < 1$ and K_r is the kernel function at r in $A_G^2(D)$, where G is any positive, continuous, non-increasing weight function, then

$$\int_0^1 (1 - r)\|K_r\|^2 G(r)r\,dr = \infty$$

PROOF   Since

$$\|K_z\|^2 = \frac{1}{2} \sum_{n=0}^{\infty} \frac{|z|^{2n}}{p_n}$$

where

$$p_n = \int_0^1 r^{2n+1} G(r) \, dr$$

we have, for each ρ in $(0, 1)$

$$\|K_{\rho z}\|^2 = \frac{1}{2} \sum_{n=0}^{\infty} \frac{\rho^{2n}}{p_n} |z|^{2n}$$

so that

$$\int_0^1 \|K_{\rho r}\|^2 G(r) r \, dr = \frac{1}{2} \sum_{n=0}^{\infty} \frac{\rho^{2n}}{p_n} \int_0^1 r^{2n+1} G(r) \, dr$$

$$= \frac{1}{2} \sum_{n=0}^{\infty} \rho^{2n} = \frac{1}{2(1 - \rho^2)}$$

Since G is non-increasing this gives

$$\int_0^1 \|K_{\rho r}\|^2 G(\rho r) r \, dr \geq \int_0^1 \|K_{\rho r}\|^2 G(r) r \, dr = \frac{1}{2(1 - \rho^2)}$$

By substituting $t = \rho r$ in the left-most integral this yields

$$\int_0^\rho \|K_t\|^2 G(t) t \, dt \geq \frac{\rho^2}{2(1 - \rho^2)}$$

An integration by parts shows

$$\int_0^\rho (1 - r) \|K_r\|^2 G(r) \, dr$$

$$= (1 - \rho) \int_0^\rho \|K_t\|^2 G(t) \, dt + \int_0^\rho \left(\int_0^r \|K_t\|^2 G(t) t \, dt \right) dr$$

$$\geq \int_0^\rho \left(\int_0^r \|K_t\|^2 G(t) t \, dt \right) dr$$

$$\geq \int_0^\rho \frac{r^2}{2(1 - r^2)} \, dr$$

which diverges as ρ approaches 1.   ∎

We will use this lemma to study Hilbert–Schmidt composition operators on $A_G^2(D)$ where the weight $G(r)$ satisfies, in contrast to the weights of Theorem 5.16,

$$\lim_{r \to 1} \frac{\log G(r)}{\log^2(1-r)} = 0$$

THEOREM 5.18

Suppose

$$G(r) = \exp\left(-h(r) \log^2(1-r)\right)$$

where $h(r)$ is non-increasing near 1 and tends to 0 as r approaches 1. If C_φ is a Hilbert–Schmidt operator on $A_G^2(D)$ then $|\varphi'(\zeta)| = \infty$ for almost all ζ in the unit circle.

Weights satisfying the hypothesis of this theorem need not be fast weights; however the main interest of the theorem is for this case.

Recall by Lemma 2.41 that if $|\varphi'(\zeta)| = d < \infty$ then φ maps each internally tangent disk $E(k, \zeta) = \{z \in D : |\zeta - z|^2 \le k(1 - |z|^2)\}$ into $E(dk, \varphi(\zeta))$. In particular $\varphi(r\zeta)$ will be in $E(d(1-r)/(1+r), \varphi(\zeta))$ so that

$$1 - |\varphi(r\zeta)| \le \frac{2d(1-r)}{1+r+d(1-r)}$$

provided r is sufficiently close to 1 that $2d(1-r)/\left(1+r+d(1-r)\right) < 1$, that is, $d < (1+r)/(1-r)$.

PROOF (of Theorem 5.18) Let C_φ be Hilbert–Schmidt on $A_G^2(D)$ and assume further that there exists $A < \infty$ and a set Q in the unit circle with $|Q| > 0$ so that $|\varphi'(\zeta)| \le A$ for all ζ in Q. Without loss of generality, we may assume $A > 1$. Set $r_0 = 1 - 1/A$ so that, in particular, $A < (1 + r_0)/(1 - r_0)$. The preceding remarks show that for $r_0 < r < 1$ and ζ in Q

$$1 - |\varphi(r\zeta)| \le \frac{2A(1-r)}{1+r+A(1-r)} = \frac{A(1-r)}{1+(A-1)(1-r)/2} < A(1-r)$$

since $A > 1$. Thus $|\varphi(r\zeta)| > 1 - A(1-r)$ whenever $r > r_0$. Since C_φ is Hilbert–Schmidt,

$$\infty > \|C_\varphi\|_{HS}^2 \ge \int_{D \backslash r_0 D} \|K_{\varphi(z)}\|^2 G(|z|) \frac{dA(z)}{\pi}$$

$$\ge \int_Q \int_{r_0}^1 \|K_{1-A(1-r)}\|^2 G(r) r \frac{dr \, d\theta}{\pi}$$

$$\ge C \int_{r_0}^1 \|K_{1-A(1-r)}\|^2 G(r) \, dr$$

where the constant C reflects both the measure of Q and the omission of the factor r/π in the integrand. Changing variables by $t = 1 - A(1 - r)$ we have

$$\infty > \frac{C}{A} \int_0^1 \|K_t\|^2 G(1 - \frac{1-t}{A})\, dt$$

$$\geq \frac{C}{A} \int_0^1 \|K_t\|^2 \frac{G(1 - \frac{1-t}{A})}{G(t)} G(t) t\, dt$$

By Lemma 5.17 this last integral must diverge if

$$\frac{G(1 - \frac{1-t}{A})}{G(t)} \geq 1 - t \tag{5.3.4}$$

for t near 1. Since $G(t) = \exp\left(-h(t)\log^2(1 - t)\right)$ the ratio in Inequality (5.3.4) is

$$\exp\left(-h(1 - \frac{1-t}{A})\log^2(\frac{1-t}{A}) + h(t)\log^2(1 - t)\right)$$

where, for t near 1,

$$h(1 - \frac{1-t}{A}) \leq h(t)$$

since h is non-increasing. Thus the left hand side of Inequality (5.3.4) is bounded below, for t near 1, by

$$\exp\left(-h(t)(\log^2\frac{1-t}{A} - \log^2(1 - t))\right)$$

$$= \exp\left(-h(t)(\log^2 A - 2\log(1 - t)\log A)\right)$$

$$= (1 - t)^{2\alpha h(t)} \exp\left(-\alpha^2 h(t)\right)$$

where $\alpha = \log A > 0$. Since $h(t)$ converges to 0 as t goes to 1 this is greater than $(1 - t)$ for t near 1 and we have our desired contradiction. ∎

As a consequence of Theorem 5.14 and 5.18 note that it is easy to give examples of compact C_φ which are not Hilbert–Schmidt on the space $A_G^2(D)$ where G is a fast regular weight satisfying the hypothesis of Theorem 5.18. The maps $\varphi(z) = z^n$, for $n = 2, 3, \ldots$ provide simple examples. This is in contrast to the much more difficult construction done in Section 3.3 of a composition operator that is compact on $H^2(D)$ but not Hilbert–Schmidt.

Exercises

5.3.1 For $h(r)$ as defined in Equation (5.3.2) show that if $h(r)$ is non-decreasing and tends to infinity as r goes to 1 then $G(r)$ is regular.

5.3.2 Suppose that $G(r) = \exp(-c\log^2(1-r))$ for some $c > 0$. Show that C_φ is Hilbert–Schmidt on $A_G^2(D)$ if $|\varphi'(\zeta)| > e^{1/(2c)}$ for all ζ.

5.3.3 Suppose that $G(r) = \exp(-c\log^2(1-r))$ for some $c > 0$. Show that if C_φ is Hilbert–Schmidt on $A_G^2(D)$ then $|\varphi'(\zeta)| > e^{1/(2c)}$ almost everywhere.

Notes

The main results of this section, Theorems 5.16 and 5.18, have analogues for the Schatten p–classes $\mathcal{S}_p$, $p > 0$ given by T. L. Kriete and B. D. MacCluer in [KrM92]. There it is shown that for fast weights G which decay sufficiently rapidly every compact C_φ on $A_G^2(D)$ will be in $\mathcal{S}_p$ for every $p > 0$, while for slower fast weights a necessary condition for C_φ to lie in any $\mathcal{S}_p$ class is $|\varphi'(\zeta)| = \infty$ almost everywhere. The kernel function approach given here for the Hilbert–Schmidt class is due to T. L. Kriete [Kri94], as are the results of Exercises 5.3.2 and 5.3.3.

6

Special Results for Several Variables

6.1 Compactness revisited

Our goal in this chapter is to continue the study begun in Section 3.5 of boundedness and compactness of C_φ on function spaces in B_N when $N > 1$. The Carleson measure criterion (Theorem 3.35) continues to underlie the results in this chapter, but our focus will be to obtain deeper consequences of this characterization than those in Section 3.5. We begin with a compactness question which is motivated by one of the earliest results on compactness in one variable: if $\varphi(D)$ is contained in a nontangential approach region in D, then C_φ is compact on $H^p(D)$ for all $p < \infty$ (Proposition 3.25). We will consider C_φ when $\varphi(B_N)$ is contained in a Koranyi approach region

$$\Gamma(\zeta, \alpha) = \{z \in B_N : |1 - \langle z, \zeta \rangle| < \frac{\alpha}{2}(1 - |z|^2)\}$$

for some $\alpha > 1$ and ζ in ∂D. Our first lemma exploits the compatibility of the Carleson sets $\mathcal{S}(\eta, t)$ with the Koranyi approach regions where, as in Section 2.2,

$$\mathcal{S}(\eta, t) = \{z \in \overline{B_N} : |1 - \langle z, \eta \rangle| < t\}$$

LEMMA 6.1
There exists $C < \infty$, depending only on α, so that whenever η in ∂B_N and $t > 0$ are such that

$$\mathcal{S}(\eta, t) \cap \Gamma(\zeta, \alpha) \neq \emptyset$$

then $|1 - \langle \zeta, \eta \rangle| \leq Ct$.

PROOF For z, w in the closed ball, let $d(z, w) = |1 - \langle z, w \rangle|$. Recall that $d(z, w)$ is a quasi-metric on $\overline{B_N}$ (Exercise 3.5.7). If w is in $\mathcal{S}(\eta, t) \cap \Gamma(\zeta, \alpha)$ then $d(w, \zeta) \leq (\alpha/2)(1 - |w|^2)$ and $d(w, \eta) \leq t$. So we have

$$d(\zeta, \eta) \leq 3\left(d(\zeta, w) + d(w, \eta)\right) \leq 3\left((\alpha/2)(1 - |w|^2) + t\right)$$

But since w is in $\mathcal{S}(\eta, t)$, $1 - |w| \leq t$. Thus

$$d(\zeta, \eta) \leq 3(\alpha t + t) = 3(\alpha + 1)t$$

which gives the conclusion with $C = 3(\alpha + 1)$. ∎

LEMMA 6.2
If $\varphi(B_N) \subset \Gamma(\zeta, \alpha)$ then on $H^p(B_N)$ we have

(1) C_φ is bounded if and only if $\sigma_N \varphi^{-1} \mathcal{S}(\zeta, t) \leq C t^N$ for some $C < \infty$ independent of t.

(2) C_φ is compact if and only if $\sigma_N \varphi^{-1} \mathcal{S}(\zeta, t) = o(t^N)$ as t tends to 0.

The point of the lemma is that if $\varphi(B_N) \subset \Gamma(\zeta, \alpha)$ the Carleson condition of Theorem 3.35 need only be checked at ζ.

PROOF   By Theorem 3.35, only the ($\Leftarrow$) direction of each assertion needs proof. Let η be an arbitrary point of ∂B_N and let $t > 0$. If $\mathcal{S}(\eta, t)$ and $\Gamma(\zeta, \alpha)$ are disjoint, then $\varphi^{-1} \mathcal{S}(\eta, t)$ is empty and $\sigma_N \varphi^{-1} \mathcal{S}(\eta, t) = 0$. On the other hand, if $\mathcal{S}(\eta, t) \cap \Gamma(\zeta, \alpha)$ is non-empty then Lemma 6.1 and the absorption property of the Carleson balls (Exercise 3.5.7) show that there exists $K < \infty$, independent of η, t so that

$$\mathcal{S}(\zeta, Kt) \supset \mathcal{S}(\eta, t)$$

and hence

$$\sigma_N \varphi^{-1} \mathcal{S}(\eta, t) \leq \sigma_N \varphi^{-1} \mathcal{S}(\zeta, Kt)$$
$$\leq C K^N t^N = C' t^N$$

Thus, if the Carleson criterion holds for Carleson sets based at ζ, it holds for all Carleson sets, giving the first part of the lemma. Similarly, if we know that $\sigma_N \varphi^{-1} \mathcal{S}(\zeta, t) = \epsilon(t) t^N$ where $\epsilon(t) \to 0$ as $t \to 0$ then the same argument shows

$$\sigma_N \varphi^{-1} \mathcal{S}(\eta, t) \leq K^N \epsilon(t) t^N$$

This gives the compactness assertion. ∎

In the next lemma we use the argument of Proposition 3.25 to show in a quantitative way why maps of the disk with image contained in a nontangential approach region give compact operators on $H^p(D)$. Specifically, the lemma relates the size of the pre-images of Carleson sets with the size of the angular opening of the approach region.

LEMMA 6.3

Suppose $\varphi : D \to D$ is analytic with $\varphi(D) \subset \{z : |1 - z| < (\alpha/2)(1 - |z|^2)\}$. Then there exists a finite constant C which depends only on $\varphi(0)$ and α so that

$$\sigma\varphi^{-1}S(1,t) \leq Ct^\beta$$

where

$$\beta = \frac{\pi}{2\cos^{-1}(1/\alpha)}$$

PROOF   From the proof of Proposition 3.25, we know that if

$$\varphi(D) \subset \{z : |1 - z| < (\alpha/2)(1 - |z|^2)\}$$

then we have

$$\frac{1}{1 - \varphi(z)} = F \circ \tau$$

where

$$F(z) = \left(\frac{1 + z}{1 - z}\right)^{2b/\pi}$$

with $b = \cos^{-1}(1/\alpha)$ and τ an analytic map of D into D. From this it follows that

$$|1 - \varphi(z)| < t \text{ if and only if } |F \circ \tau| > \frac{1}{t}$$

and

$$\{z : |F \circ \tau(z)| > \frac{1}{t}\} = \left\{z : \left|\frac{1 + \tau(z)}{1 - \tau(z)}\right| > \left(\frac{1}{t}\right)^{\pi/(2b)}\right\}$$

$$\subset \left\{z : \frac{2}{|1 - \tau(z)|} > \left(\frac{1}{t}\right)^{\pi/(2b)}\right\}$$

$$= \left\{z : |1 - \tau(z)| < 2t^{\pi/(2b)}\right\}$$

Thus

$$\sigma(\varphi^{-1}S(1,t)) \leq \sigma(\tau^{-1}S(1, 2t^{\pi/(2b)})) \leq Ct^\beta$$

for some constant C depending only on $\tau(0)$ (and hence ultimately on $\varphi(0)$ and α).   ∎

We use this series of lemmas in the proof of the next theorem which continues our study of maps into Koranyi regions.

THEOREM 6.4

For each $N > 1$, let $\alpha_N = \sec(\pi/(2N))$.

(1) If $\varphi(B_N) \subset \Gamma(\zeta, \alpha_N)$, then C_φ is bounded on $H^p(B_N)$, $p < \infty$.

(2) If $\varphi(B_N) \subset \Gamma(\zeta, \alpha)$ for some $\alpha < \alpha_N$, then C_φ is compact on $H^p(B_N)$, $p < \infty$.

(3) Both of the above results are sharp in the following sense: There exists φ with $\varphi(B_N) \subset \Gamma(\zeta, \alpha_N)$ and C_φ not compact on $H^p(B_N)$, and given $\gamma > \alpha_N$ there exists φ with $\varphi(B_N) \subset \Gamma(\zeta, \gamma)$ and C_φ not bounded on $H^p(B_N)$.

PROOF We verify (1) first. Suppose $\varphi(B_N) \subset \Gamma(\zeta, \alpha_N)$. We may take $\zeta = e_1$. By Lemma 6.2 we need only check that

$$\sigma_N(\varphi^{-1}\mathcal{S}(e_1, t)) \leq Ct^N$$

for some constant C independent of t. Set $A = \varphi^{-1}\mathcal{S}(e_1, t) \cap \partial B_N$. By slice integration (Formula 2.1.6 of Section 2.1) we have

$$\sigma_N(A) = \int_{\partial B_N} \chi_A \, d\sigma_N = \int_{\partial B_N} \int_0^{2\pi} \chi_A(e^{i\theta}\zeta) \frac{d\theta}{2\pi} \, d\sigma_N(\zeta)$$

For each ζ in ∂B_N define

$$\varphi^\zeta : D \to D$$

by $\varphi^\zeta(z) = \varphi_1(z\zeta)$. The hypothesis that $\varphi(B_N)$ is contained in $\Gamma(e_1, \alpha_N)$ guarantees that

$$|1 - \varphi^\zeta(z)| < \frac{\alpha_N}{2}(1 - |\varphi^\zeta(z)|^2)$$

so that $\varphi^\zeta(D)$ is contained in a nontangential approach region at 1 in the disk with "aperture" α_N. For each ζ, $\varphi^\zeta(0) = \varphi_1(0)$. By Lemma 6.3 we have

$$\sigma_1(\varphi^\zeta)^{-1}\mathcal{S}(1, t) \leq Ct^\beta$$

where

$$\beta = \frac{\pi}{2\cos^{-1}(1/\alpha_N)}$$

and C is independent of ζ and t. Notice that $\chi_A(e^{i\theta}\zeta) = 1$ if and only if $|1 - \varphi_1(e^{i\theta}\zeta)| < t$ if and only if $\varphi^\zeta(e^{i\theta})$ is in $\mathcal{S}(1, t)$. Thus, in the above slice integration formula,

$$\sigma_N(A) = \int_{\partial B_N} \int_0^{2\pi} \chi_A(e^{i\theta}\zeta) \frac{d\theta}{2\pi} \, d\sigma_N(\zeta) \leq Ct^\beta = Ct^N$$

where the last equality follows from the definition of α_N. Thus (1) follows from Lemma 6.2.

When $\alpha < \alpha_N$ and $\varphi(B_N) \subset \Gamma(e_1, \alpha)$ the same computations give

$$\sigma_N \varphi^{-1} \mathcal{S}(e_1, t) \le Ct^\beta$$

where now

$$\beta = \frac{\pi}{2 \sec^{-1}(\alpha)} > N$$

which gives (2), again by Lemma 6.2.

To show the results are sharp we work with maps Φ defined on B_N by

$$\Phi = (\psi \circ \varphi, 0')$$

where $\varphi : B_N \to D$ is an inner function with $\varphi(0) = 0$ and $\psi : D \to D$ is a biholomorphic map of D onto a nontangential approach region

$$\Gamma = \{z \in D : |1 - z| < \frac{\alpha}{2}(1 - |z|^2)\}$$

with $\psi(1) = 1$. We have $\Phi(B_N)$ contained in $\Gamma(e_1, \alpha)$. Since inner functions mapping 0 to 0 are measure preserving maps of ∂B_N into ∂D ([Ru80, p. 405])

$$\sigma_N \varphi^{-1}(E) = \sigma_1(E)$$

for every Borel set E in ∂D.

Roughly speaking, the map ψ must behave like $1 - (1 - z)^b$ near 1, where

$$b = \frac{2 \cos^{-1}(1/\alpha)}{\pi}$$

To make this precise, note that the boundary curves of the region Γ in D may be parametrized near 1 by

$$w_+(\theta) = 1 - (2 \cos \theta - \frac{2}{\alpha})e^{i\theta}$$

and $w_-(\theta) = \overline{w_+(\theta)}$ for $\cos^{-1}(1/\alpha) - \delta \le \theta \le \cos^{-1}(1/\alpha)$ with δ small. Since w'_+ and w'_- are continuous, a theorem of Warschawski [Pom92, p. 52] shows that if b is as defined above, then

$$\frac{1 - \psi(z)}{(1 - z)^b} \equiv g(z)$$

is continuous and non-zero in $\overline{D} \cap V$ where V is a small closed disk centered at 1, and

$$\psi^{-1} \mathcal{S}(1, t) \subset \overline{D} \cap V$$

for t sufficiently small. For such t

$$\{z : \Phi(z) \in \mathcal{S}(e_1, t)\} = \{z : |1 - \psi \circ \varphi(z)| < t\}$$
$$= \{z : |g(\varphi(z))||1 - \varphi(z)|^b < t\}$$

$$\supset \{z : |1 - \varphi(z)| < (t/M)^{1/b}\}$$
$$= \varphi^{-1}\mathcal{S}(1, (t/M)^{1/b})$$

where M is the maximum of $|g|$ on $\overline{D} \cap V$. Thus

$$\sigma_N \Phi^{-1}\mathcal{S}(e_1, t) \geq \sigma_N \varphi^{-1}\mathcal{S}(1, (t/M)^{1/b}) = \sigma_N \varphi^{-1}Q(1, (t/M)^{1/b})$$
$$= \sigma_1 Q(1, (t/M)^{1/b}) \sim Ct^{1/b}$$

Choosing $\alpha = \alpha_N = \sec(\pi/(2N))$ gives $1/b = N$ and thus C_Φ is not compact on $H^p(B_N)$.

If we choose $\alpha > \alpha_N$ then $1/b < N$ and the above calculation shows that

$$\sigma_N \Phi^{-1}\mathcal{S}(e_1, t) \neq O(t^N)$$

so that C_Φ is not bounded on $H^p(B_N)$. ∎

The same slice integration technique which is at the heart of the proof of Theorem 6.4 will be used in the next theorem, which gives a boundedness result not by restricting φ but rather by enlarging the range space for the operator C_φ.

THEOREM 6.5
If φ is any analytic map of B_N into itself, then C_φ maps $H^p(B_N)$ boundedly into the weighted Bergman space $A^p_{N-2}(B_N)$.

PROOF Carleson measure considerations show that we need only verify that for some finite constant C independent of ζ and t

$$\int_{\varphi^{-1}S(\zeta,t)} (1 - |z|^2)^{N-2} \, d\nu_N(z) \leq Ct^N$$

Since $H^p(B_N)$ is automorphism invariant we may assume that $\varphi(0) = 0$. For each ζ in the sphere, define a family of analytic self-maps of the disk by setting $\tau_\eta(z) = \langle \varphi(z\eta), \zeta \rangle$, where η is also in the sphere. Notice that $\tau_\eta(0) = 0$. Littlewood's Subordination Principle guarantees that each τ_η is a bounded composition operator with norm 1 on any standard weight Bergman space $A^p_\alpha(D)$, $\alpha > -1$. By Theorem 2.36 and Exercise 2.2.7 this translates into the estimate

$$\mu_\alpha \tau_\eta^{-1} S(e^{i\theta}, t) \leq Ct^{\alpha+2}$$

where μ_α is the weighted area measure on D given by $d\mu_\alpha = (1 - |z|^2)^\alpha \, dA$ where C does not depend on η or ζ since each map τ_η fixes 0.

Write A for $\varphi^{-1}S(\zeta, t)$, convert to polar coordinates (Exercise 2.1.13) and use slice integration to obtain

$$\int_{B_N} \chi_A(z)(1 - |z|^2)^{N-2} \, d\nu_N(z)$$

$$= \int_0^1 r^{2N-1} \int_{\partial B_N} \chi_A(r\eta)(1-r^2)^{N-2} \, d\sigma(\eta) \, dr$$

$$= \int_0^1 r^{2N-1}(1-r^2)^{N-2} \int_{\partial B_N} \int_0^{2\pi} \chi_A(re^{i\theta}\eta) \frac{d\theta}{2\pi} \, d\sigma(\eta) \, dr$$

$$\le c \int_{\partial B_N} \int_D \chi_A(z\eta)(1-|z|^2)^{N-2} \, dA(z) \, d\sigma(\eta)$$

Now $\chi_A(z\eta) = 1$ if and only if $\varphi(z\eta)$ is in $S(\zeta,t)$ which occurs if and only if $|1 - \langle \varphi(z\eta), \zeta \rangle| < t$ if and only if $\tau_\eta(z)$ is in $S(1,t)$, so the inner integral is bounded by some absolute constant times t^N and therefore

$$\int_{\varphi^{-1}S(\zeta,t)} (1-|z|^2)^{N-2} \, d\nu_N(z) \le Ct^N$$

where C does not depend on either ζ or t.   ∎

Exercises

6.1.1 Let $\varphi : B_N \to B_N$ be an arbitrary analytic map. Show $\sigma\varphi^{-1}S(\zeta,t) \le Ct$ for all ζ in ∂B_N and $t > 0$, where C depends only on $\varphi(0)$.

6.1.2 If $\varphi : B_N \to B_N$ has no finite angular derivative at any point of ∂B_N then C_φ mapping $H^p(B_N)$ into $A^p_{N-2}(B_N)$ is compact.

6.1.3 Consider $\varphi(z_1, z_2, \ldots, z_N) = (2z_1z_2, 0')$. Show that C_φ does not map $H^2(B_N)$ into $A^2_\alpha(B_N)$ for any $\alpha < (N-3)/2$. Compare this with the statement of Theorem 6.5. Do you get a better result by considering $\varphi(z_1, z_2, \ldots, z_N) = (c(\alpha)z^\alpha, 0')$ where α is a multi-index other than $(1, 1, 0')$ with at least two non-zero entries, and

$$c(\alpha) = \frac{|\alpha|^{|\alpha|/2}}{\prod_{\alpha_i \ne 0} \alpha_i^{\alpha_i/2}} ?$$

Notes

Theorem 6.4 is due to B. D. MacCluer [Mc85]. Theorem 6.5 is a special case of a result due to B. D. MacCluer and P. R. Mercer [McM93]. Both Theorem 6.4 and Theorem 6.5 have generalizations that hold for composition operators on smoothly bounded, strongly convex domains in C^N; see [McM93] for these results. Theorem 6.5 has also been generalized by J. A. Cima and P. R. Mercer [CiMe94] to composition operators between two weighted Bergman spaces in the ball (or more generally in smoothly bounded strongly convex domains in C^N). Their result shows, in particular, that C_φ will always map $A^p_\alpha(B_N)$ boundedly into $A^p_{\alpha+N-1}(B_N)$.

The result of Exercise 6.1.3 also appears in [McM93].

6.2 Wogen's theorem

In this section we look at a remarkable result that gives a necessary and sufficient condition for a smooth φ to induce a bounded composition operator on $H^p(B_N)$ for $0 < p < \infty$. By smooth we will mean that φ is of class C^3 on $\overline{B_N}$; *this will be a standing hypothesis on φ throughout this section*, and we will regard φ as being defined on $\overline{B_N}$. Our condition for boundedness of C_φ will involve a relationship between certain directional derivatives of the coordinate functions of φ at points of ∂B_N which map to ∂B_N. We begin with a few facts about some directional derivatives at a fixed point of ∂B_N.

LEMMA 6.6
For $\varphi : B_N \to B_N$ analytic and of class C^3 on $\overline{B_N}$ with fixed point at $e_1 = (1, 0')$

(1) $D_1\varphi_1(e_1) \geq (1 - |\varphi_1(0)|)/(1 + |\varphi_1(0)|) > 0$

(2) $D_k\varphi_1(e_1) = 0$ *for $k = 2, 3, \ldots, N$*

(3) $D_1\varphi_1(e_1) \geq |D_{kk}\varphi_1(e_1)|$ *for $k = 2, 3, \ldots, N$*

where $\varphi = (\varphi_1, \varphi_2, \ldots, \varphi_N)$.

PROOF If $\varphi_1(0) = 0$ then (1) follows immediately from the Schwarz Lemma (applied to the restriction of φ_1 to the complex line through e_1) and the Julia–Carathéodory Theorem (Theorem 2.81). If $\varphi_1(0) = a \neq 0$ consider $\psi : D \to D$ defined by $\psi(z) = \varphi_a \circ [\varphi_1]_{e_1}$ where $[\varphi_1]_{e_1}(w) = \varphi_1(we_1)$ and $\varphi_a(w) = (a - w)/(1 - \overline{a}w)$. Since $\psi(0) = 0$ and $|\psi(1)| = 1$ the Julia–Carathéodory Theorem implies that $|\psi'(1)| \geq 1$. This translates into $|\varphi_a'(1)D_1\varphi_1(e_1)| \geq 1$ or

$$|D_1\varphi_1(e_1)| \geq \frac{|1 - a|^2}{1 - |a|^2} \geq \frac{1 - |a|}{1 + |a|}$$

as desired.

For (2) and (3) fix k, $2 \leq k \leq N$ and consider φ on the circle $\Gamma(t) = (\cos t, 0, \ldots, \lambda \sin t, \ldots, 0)$ where λ is any constant of modulus 1 and $\lambda \sin t$ appears as the kth component of $\Gamma(t)$. Set $h_\lambda(t) = \text{Re}\,\varphi_1(\Gamma(t))$; this function has a maximum at $t = 0$, since $\varphi_1(e_1) = 1$. A calculation shows that $h_\lambda'(0) = \text{Re}\,\lambda D_k\varphi_1(e_1)$ for all λ with $|\lambda| = 1$. Since $h_\lambda'(0) = 0$ we have $D_k\varphi_1(e_1) = 0$, giving (2). Similarly, one computes

$$h_\lambda''(0) = -D_1\varphi_1(e_1) + \text{Re}\,\lambda^2 D_{kk}\varphi_1(e_1)$$

and this must be non-positive. So

$$\text{Re}\,\lambda^2 D_{kk}\varphi_1(e_1) \leq D_1\varphi_1(e_1)$$

for all λ with $|\lambda| = 1$ and (3) follows. ∎

From this point on, we confine our attention to the case $N = 2$ and write B for B_2 and σ for σ_2. This allows certain notational simplifications while still capturing all the essential ingredients of the arguments. At the end of the section, we will indicate how the results generalize for $N > 2$.

PROPOSITION 6.7

If $\varphi(e_1) = e_1$ and $D_1\varphi_1(e_1) = |D_{22}\varphi_1(e_1)|$ then C_φ is unbounded on $H^p(B)$ for $0 < p < \infty$.

PROOF First, we give a normalization argument to show it suffices to assume that $D_1\varphi_1(e_1) = D_{22}\varphi_1(e_1)$. For $|\lambda| = 1$ define the unitary map $U(z_1, z_2) = (z_1, \lambda z_2)$. Choose λ so that $\lambda^2 D_{22}\varphi_1(e_1) = |D_{22}\varphi_1(e_1)|$. If $\psi = \varphi \circ U$ then $\psi(e_1) = e_1$, $D_1\psi_1(e_1) = D_1\varphi_1(e_1)$, and $D_{22}\psi_1(e_1) = |D_{22}\varphi_1(e_1)|$. Thus ψ fixes e_1 and satisfies $D_1\psi_1(e_1) = D_{22}\psi_1(e_1)$. But, since C_U is an isometric isomorphism of $H^p(B)$, C_φ is unbounded if and only if $C_\psi = C_U C_\varphi$ is unbounded, so we may assume $D_1\varphi_1(e_1) = D_{22}\varphi_1(e_1)$.

By Theorem 3.35 it will suffice to show that $\sigma\varphi^{-1}S(e_1, \delta) \neq O(\delta^2)$. To do this we will produce sets $\mathcal{A}(\delta) \subset \partial B$ and a positive constant C so that for all sufficiently small $\delta > 0$,

(i) $\varphi(\mathcal{A}(\delta)) \subset S(e_1, C\delta)$

(ii) $\sigma(\mathcal{A}(\delta)) \sim \delta^2 \log \frac{1}{\delta}$

Parametrize a neighborhood of e_1 in ∂B by the one-to-one map $\Lambda : \{x \in R^3 : |x| < 1/2\} \to \partial B$ where

$$\Lambda(x_1, x_2, x_3) = (\sqrt{1 - |x|^2} + ix_1, x_2 + ix_3)$$

Note that for a set A in the image of Λ, $\sigma(A) \sim m_3(\Lambda^{-1}(A))$ where dm_3 is Lebesgue volume measure on R^3, since

$$d\sigma = \frac{dm_3}{\sqrt{1 - |x|^2}}$$

We will describe our desired sets $\mathcal{A}(\delta)$ as $\Lambda(\Omega(\delta))$ where (see Figure 6.1)

$$\Omega(\delta) = \{(x_1, x_2, x_3) : 0 < x_1 < \delta, \delta^{2/3} < x_3 < \delta^{1/2}, 0 < x_2 < \delta/x_3\}$$

Calculate

$$m_3(\Omega(\delta)) = \int_0^\delta dx_1 \int_{\delta^{2/3}}^{\delta^{1/2}} \frac{\delta}{x_3} dx_3$$

$$= \int_0^\delta \delta(\log \delta^{1/2} - \log \delta^{2/3}) dx_1$$

$$= c\delta^2 \log \frac{1}{\delta}$$

Thus $\sigma(\mathcal{A}(\delta)) \sim \delta^2 \log \frac{1}{\delta}$ which is (ii).

To verify (i) set $g(x) = \varphi_1(\Lambda(x)) - 1$ so that g is C^3 on $\{|x| < 1/2\} \subset R^3$ with $g(0) = 0$. From the definition of $\mathcal{S}(e_1, \delta)$ we will have $\varphi(\Lambda(\Omega(\delta)) \subset \mathcal{S}(e_1, C\delta)$ if $|g(x)| < C\delta$ for all x in $\Omega(\delta)$. Using the chain rule and the fact that $D_2\varphi_1(e_1) = 0$ we compute that

$$\frac{\partial g}{\partial x_2}(0) = 0 \text{ and } \frac{\partial g}{\partial x_3}(0) = 0$$

while the hypothesis that $D_1\varphi_1(e_1) = D_{22}\varphi_1(e_1)$ gives that

$$\frac{\partial^2 g}{\partial x_2^2}(0) = 0$$

We use these calculations to write the Taylor expansion of g about 0:

$$g(x) = \frac{\partial g}{\partial x_1}(0)x_1 + \frac{1}{2}\sum_{j,k=1}^{3}\frac{\partial^2 g}{\partial x_j \partial x_k}(0)x_j x_k + O(|x|^3) \qquad (6.2.1)$$

where $\frac{\partial^2 g}{\partial x_2^2}(0) = 0$. If $x = (x_1, x_2, x_3)$ is in $\Omega(\delta)$ we certainly have $|x_1| < \delta$, $|x_2| < \delta^{1/3}$, $|x_3| < \delta^{1/2}$, and $|x_2 x_3| < \delta$ and all terms on the right side of Equation (6.2.1) are $O(\delta)$ for x in $\Omega(\delta)$ and $\delta < 1$, giving (i). ∎

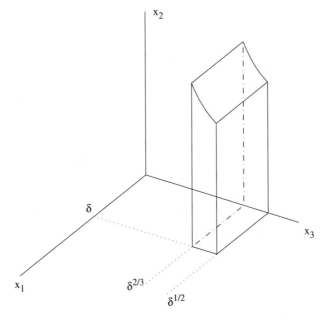

FIGURE 6.1

The sets $\Omega(\delta)$ in R^3.

Lemma 6.6 and Proposition 6.7 have non-normalized versions that do not assume $\varphi(e_1) = e_1$. To state these, we first set some notation. For η in ∂B_N let $\varphi_\eta(z) = \langle \varphi(z), \eta \rangle$ be the coordinate of φ in the η–direction. In the special case $\eta = e_1 = (1, 0)$ or $e_2 = (0, 1)$ we will continue to write φ_1 and φ_2 for these coordinate functions of φ. For ζ in ∂B_N with $\zeta = (\zeta_1, \zeta_2)$, let $\zeta^\perp$ denote $(-\overline{\zeta_2}, \overline{\zeta_1})$ so that $\langle \zeta, \zeta^\perp \rangle = 0$. Note that $e_1^\perp = e_2$. The directional derivative D_ζ in the ζ direction satisfies $D_\zeta \varphi_\eta(z) = \langle \varphi'(z)\zeta, \eta \rangle$. With this notation we have the following generalization of Lemma 6.6.

LEMMA 6.8
If $\varphi(\zeta) = \eta$ where $|\zeta| = |\eta| = 1$ then

(1) $D_\zeta \varphi_\eta(\zeta) \geq (1 - |\varphi(0)|)/(1 + |\varphi(0)|) > 0$

(2) $D_{\zeta^\perp} \varphi_\eta(\zeta) = 0$

(3) $D_\zeta \varphi_\eta(\zeta) \geq |D_{\zeta^\perp \zeta^\perp} \varphi_\eta(\zeta)|$

The proof is left as an exercise.

Proposition 6.7 generalizes similarly:

PROPOSITION 6.9
Suppose $\varphi : B \to B$ maps a point ζ in ∂B to η in ∂B with

$$D_\zeta \varphi_\eta(\zeta) = |D_{\zeta^\perp \zeta^\perp} \varphi_\eta(\zeta)|$$

Then C_φ is unbounded on $H^p(B)$ for $0 < p < \infty$.

The proof of this result is also left as an exercise.

More difficult is a converse to Proposition 6.9, which we turn to next. The theorem we will prove is as follows.

THEOREM 6.10
Suppose $D_\zeta \varphi_\eta(\zeta) > |D_{\zeta^\perp \zeta^\perp} \varphi_\eta(\zeta)|$ for all ζ in ∂B with $\eta = \varphi(\zeta)$ in ∂B. Then C_φ is bounded on $H^p(B)$.

Recall that our standing assumption is that φ is of class C^3 on $\overline{B}$. Denote by S_0 the set of all ζ in ∂B with $|\varphi(\zeta)| = 1$. Since C_φ is trivially bounded on $H^p(B)$ if S_0 is empty, we assume from this point on that S_0 is non-empty. Since S_0 is compact, for the hypothesis of Theorem 6.10 to hold we must have

$$\inf\{D_\zeta \varphi_\eta(\zeta) - |D_{\zeta^\perp \zeta^\perp} \varphi_\eta(\zeta)| : \zeta \in S_0, \varphi(\zeta) = \eta\} > 0$$

A key lemma in the proof of Theorem 6.10 will be an inequality that gives a local Lipschitz invertibility condition (relative to a non-isotropic metric) at certain points of ∂B.

For z and w in the closed unit ball, set $d(z, w) = |1 - \langle z, w \rangle|$ and recall that $d(z, w)^{1/2}$ satisfies the triangle inequality in $\overline{B}$ and is a metric on ∂B. Moreover, $d(z, w)$ is a quasi-metric on ∂B in the following sense: For z, w, and v in ∂B,

$$3 \left(d(z, w) + d(w, v) \right) \geq d(z, v)$$

(see Exercise 3.5.7). It is an easy exercise to see that there is an absolute constant c so that $|z - \zeta|^2 \leq cd(z, \zeta)$ for all ζ in ∂B and z in $\overline{B}$. We will write $d(z, K)$ for $\inf\{d(z, \zeta) : \zeta \in K\}$ where K is a subset of ∂B.

Next we introduce a variety of constants whose definitions will stay fixed throughout the remainder of this section.

- $A_0 = A_0(\varphi) = \inf\{D_\zeta \varphi_\eta(\zeta) - |D_{\zeta^\perp \zeta^\perp} \varphi_\eta(\zeta)| : \zeta \in S_0(\varphi), \varphi(\zeta) = \eta\}$.
- $M_1 = M_1(\varphi) = \max\{|D\varphi_\tau(w)| : \tau, w \in \partial B$ and D any directional derivative of order 1$\}$.
- $M_2 = M_2(\varphi) = \max\{|D\varphi_\tau(w)| : \tau, w \in \partial B$ and D any directional derivative of order at most 2$\}$.
- $K_1 = K_1(\varphi) = \sup\{|\varphi_\eta(z) - P_\eta^1(z, \zeta)|/d(z, \zeta) : \zeta, \eta \in \partial B, z \in \overline{B}\}$ where $P_\eta^1(z, \zeta)$ is the first order Taylor polynomial of φ_η at ζ.
- $K_2 = K_2(\varphi) = \sup\{|\varphi_\eta(z) - P_\eta^2(z, \zeta)|/(d(z, \zeta)^{3/2}) : \zeta, \eta \in \partial B, z \in \overline{B}\}$ where $P_\eta^2(z, \zeta)$ is the second order Taylor polynomial of φ_η at ζ.

Note that the finiteness of K_1 and K_2 follow from Taylor's theorem and the inequality $|z - \zeta|^2 \leq cd(z, \zeta)$, while M_1 and M_2 are finite by our smoothness hypothesis.

Initially, we also let δ_0 denote a positive constant chosen small enough so that

$$\frac{1}{2} M_2 \delta_0 + 2M_2 \sqrt{\delta_0} + K_2 \sqrt{\delta_0} < \frac{1}{16} A_0 \qquad (6.2.2)$$

A further restriction on δ_0 will be imposed later.

Our first result compares $d(z, \zeta)$ and $d(\varphi(z), \varphi(\zeta))$ when ζ is in S_0 and z is near ζ. *From this point on we assume the hypothesis of Theorem 6.10 holds; that is, that $A_0(\varphi) > 0$.*

LEMMA 6.11
There are positive constants A and B so that if ζ in S_0 and z in the closed ball satisfy $d(z, \zeta) < \delta_0$, then

$$Ad(z, \zeta) \leq d(\varphi(z), \varphi(\zeta)) \leq Bd(z, \zeta) \qquad (6.2.3)$$

PROOF The upper bound can be obtained from Lemma 3.40. This approach is outlined in Exercise 6.2.6. A second approach uses first order Taylor approximates: Suppose first that $\zeta = e_1$ and $\varphi(e_1) = \eta$ in ∂B. Setting $E_\eta^1(z, e_1) = \varphi_\eta(z) - P_\eta^1(z, e_1)$ we have

$$\varphi_\eta(z) = \varphi_\eta(e_1) + D_1 \varphi_\eta(e_1)(z_1 - 1) + D_2 \varphi_\eta(e_1)z_2 + E_\eta^1(z, e_1)$$

$$= 1 + D_1\varphi_\eta(e_1)(z_1 - 1) + E_\eta^1(z, e_1)$$

since $D_2\varphi_\eta(e_1) = 0$ by Lemma 6.8. Thus

$$d(\varphi(z), \eta) = |1 - \varphi_\eta(z)| \leq |D_1\varphi_\eta(e_1)(z_1 - 1)| + |E_\eta^1(z, e_1)|$$
$$\leq (M_1 + K_1)d(z, e_1)$$

which gives the upper bound in Inequality (6.2.3), with $B = M_1 + K_1$, in the case $\zeta = e_1$.

Next we consider the lower bound, still in the case that $\zeta = e_1$ and $\varphi(e_1) = \eta$. The second order Taylor polynomial $P_\eta^2(z, e_1)$ of $\varphi_\eta(z)$ at e_1 is

$$1 + D_1\varphi_\eta(e_1)(z_1 - 1)$$

$$+ \frac{1}{2}\left(D_{11}\varphi_\eta(e_1)(z_1 - 1)^2 + 2D_{12}\varphi_\eta(e_1)(z_1 - 1)z_2 + D_{22}\varphi_\eta(e_1)z_2^2\right)$$

since $\varphi_\eta(e_1) = 1$ and $D_2\varphi_\eta(e_1) = 0$ by Lemma 6.8. Also note that $D_1\varphi_\eta(e_1) > 0$ by Lemma 6.8. The fact that (z_1, z_2) is in the closed unit ball implies

$$|z_2|^2 \leq 1 - |z_1|^2 \leq 2(1 - |z_1|) \leq 2|z_1 - 1|$$

From this and the definitions of the constants M_2 and K_2, we have the following estimates

- $|\frac{1}{2}D_{11}\varphi_\eta(e_1)(z_1 - 1)^2| \leq \frac{1}{2}M_2|z_1 - 1|^2$
- $|D_{12}\varphi_\eta(e_1)(z_1 - 1)z_2| \leq M_2|z_1 - 1||z_2| \leq 2M_2|z_1 - 1|^{3/2}$
- $|\frac{1}{2}D_{22}\varphi_\eta(e_1)z_2^2| \leq |D_{22}\varphi_\eta(e_1)||z_1 - 1|$
- $|P_\eta^2(z, e_1) - \varphi_\eta(z)| \leq K_2|z_1 - 1|^{3/2}$

From these we obtain

$$d(\varphi(z), \varphi(e_1)) = d(\varphi(z), \eta) = |1 - \langle \varphi(z), \eta \rangle|$$

$$\geq ||1 - P_\eta^2(z, e_1)| - |P_\eta^2(z, e_1) - \varphi_\eta(z)||$$

$$\geq (D_1\varphi_\eta(e_1) - |D_{22}\varphi_\eta(e_1)|)\,|z_1 - 1|$$

$$- \left(\frac{1}{2}M_2|z_1 - 1|^2 + 2M_2|z_1 - 1|^{3/2} + K_2|z_1 - 1|^{3/2}\right)$$

$$\geq \left(A_0 - \frac{1}{2}M_2|z_1 - 1| - 2M_2|z_1 - 1|^{1/2} - K_2|z_1 - 1|^{1/2}\right)|z_1 - 1|$$

$$\geq (A_0 - \frac{1}{16}A_0)d(z, e_1)$$

provided $d(z, e_1) < \delta_0$. Thus a normalized version, for $\zeta = e_1$, of the result holds with, for simplicity, $A = \frac{1}{2}A_0$.

To remove this normalization, suppose ζ is an arbitrary point of S_0 with $\varphi(\zeta) = \eta$. Find a unitary map U taking e_1 to ζ and e_2 to $\zeta^\perp$ so that $(U\zeta)^\perp = U\zeta^\perp$. Consider $\psi = \varphi \circ U$.

By Exercise 6.2.3, $A_0(\varphi) = A_0(\psi)$, $M_2(\varphi) = M_2(\psi)$, and $K_2(\varphi) = K_2(\psi)$, so the previous calculations applied to ψ show that whenever $d(z, e_1) < \delta_0$, or equivalently $d(Uz, \zeta) < \delta_0$, then

$$d(\varphi(Uz), \varphi(\zeta)) = d(\psi(z), \psi(e_1)) > \frac{1}{2} A_0(\psi) d(z, e_1) = \frac{1}{2} A_0(\varphi) d(Uz, \zeta)$$

Furthermore, since $M_1(\psi) = M_1(\varphi)$ and $K_1(\psi) = K_1(\varphi)$, a similar argument shows $d(\varphi(Uz), \varphi(\zeta)) \leq B d(Uz, \zeta)$ where $B = M_1 + K_1$ and we are done. ∎

The definitions of the constants $A = A(\varphi) = \frac{1}{2} A_0$ and $B = B(\varphi) = M_1 + K_1$ that appear in Lemma 6.11 will be fixed through the rest of this section.

We emphasize that the previous lemma applies for ζ in S_0 only. We will need a version of this result which may be applied to certain points of ∂B which are not in S_0, specifically those points that are near S_0 and at which $d(\varphi(z), \eta)$ has a local minimum on ∂B for some η in ∂B with η near $\varphi(S_0)$. To this end we begin by imposing further requirements on the constant δ_0, which may be satisfied since $D\varphi_\eta(\zeta)$ is continuous in ζ and η, where D is any derivative of order at most 2. Thus we may assume that in addition to its original definition, δ_0 is sufficiently small so that if ζ, η are in ∂B with $d(\zeta, \zeta_0) < \delta_0$ and $d(\eta, \varphi(\zeta_0)) < \delta_0$ for some ζ_0 in S_0 then

$$\mathrm{Re}\, D_\zeta \varphi_\eta(\zeta) - |D_{\zeta^\perp \zeta^\perp} \varphi_\eta(\zeta)| \geq \frac{1}{2} A_0(\varphi) = A \tag{6.2.4}$$

$$|\mathrm{Im}\, D_\zeta \varphi_\eta(\zeta)| < A/8 \tag{6.2.5}$$

$$|\arg D_\zeta \varphi_\eta(\zeta)| < \pi/6 \tag{6.2.6}$$

where arg denotes the principal value of the argument. Notice that Inequality (6.2.6) follows, with room to spare, from

$$\mathrm{Re}\, D_\zeta \varphi_\eta(\zeta) \geq A \quad \text{while} \quad |\mathrm{Im}\, D_\zeta \varphi_\eta(\zeta)| < A/8$$

From this point on, we assume that δ_0 satisfies these requirements in addition to Inequality (6.2.2).

In Lemma 6.11, where $\varphi(\zeta) = \eta$ and η is in ∂B, we made essential use of the fact that $D_{\zeta^\perp} \varphi_\eta(\zeta) = 0$. To extend Lemma 6.11 to points ζ in ∂B but not in S_0 we define $R_\eta^\varphi = R_\eta$, for arbitrary ζ and η in the unit sphere and z in the closed ball, by

$$R_\eta^\varphi(z, \zeta) = R_\eta(z, \zeta) = \varphi_\eta(z) - \varphi_\eta(\zeta) - D_{\zeta^\perp} \varphi_\eta(\zeta)\langle z, \zeta^\perp \rangle$$

LEMMA 6.12

Suppose ζ and η are in ∂B with $d(\zeta, \zeta_0) < \delta_0$ and $d(\eta, \varphi(\zeta_0)) < \delta_0$ for some ζ_0 in S_0. Then

(1) $(A/2)d(z, \zeta) \leq |R_\eta(z, \zeta)| \leq Bd(z, \zeta)$ if $d(z, \zeta) < \delta_0$ and

(2) $|\arg R_\eta(z, \zeta)| \geq \pi/3$ if $R_\eta(z, \zeta) \neq 0$ and $d(z, \zeta) < \delta_0$.

PROOF   To simplify notation, we first prove the result if $\zeta = e_1$ and assume $d(e_1, \zeta_0) < \delta_0$ and $d(\eta, \varphi(\zeta_0)) < \delta_0$ for some ζ_0 in S_0. Write

$$
\begin{aligned}
R_\eta(z, e_1) &= \varphi_\eta(z) - \varphi_\eta(e_1) - D_2\varphi_\eta(e_1)z_2 \\
&= P_\eta^2(z, e_1) + E_\eta^2(z, e_1) - \varphi_\eta(e_1) - D_2\varphi_\eta(e_1)z_2 \\
&= \varphi_\eta(e_1) + D_1\varphi_\eta(e_1)(z_1 - 1) + D_2\varphi_\eta(e_1)z_2 \\
&\quad + \frac{1}{2}\left(D_{11}\varphi_\eta(e_1)(z_1 - 1)^2 + 2D_{12}\varphi_\eta(e_1)(z_1 - 1)z_2\right. \\
&\quad\quad \left. + D_{22}\varphi_\eta(e_1)z_2^2\right) \\
&\quad + E_\eta^2(z, e_1) - \varphi_\eta(e_1) - D_2\varphi_\eta(e_1)z_2
\end{aligned}
$$

where $E_\eta^2(z, e_1) = \varphi_\eta(z) - P_\eta^2(z, e_1)$ so that

$$
\begin{aligned}
|R_\eta(z, e_1)| \geq &\left| D_1\varphi_\eta(e_1)(z_1 - 1) + \frac{1}{2}D_{22}\varphi_\eta(e_1)z_2^2 \right| \\
&- \left| \frac{1}{2}D_{11}\varphi_\eta(e_1)(z_1 - 1)^2 + D_{12}\varphi_\eta(e_1)(z_1 - 1)z_2 + E_\eta^2(z, e_1) \right|
\end{aligned}
$$

The estimate proceeds as in the proof of Lemma 6.11:

$$
\left| D_1\varphi_\eta(e_1)(z_1 - 1) + \frac{1}{2}D_{22}\varphi_\eta(e_1)z_2^2 \right| \geq \left(|D_1\varphi_\eta(e_1)| - |D_{22}\varphi_\eta(e_1)| \right)|z_1 - 1|
$$

while if $d(z, e_1) < \delta_0$

$$
\left| \frac{1}{2}D_{11}\varphi_\eta(e_1)(z_1 - 1)^2 + D_{12}\varphi_\eta(e_1)(z_1 - 1)z_2 + E_\eta^2(z, e_1) \right|
$$

$$
\leq \frac{1}{2}M_2|z_1 - 1|^2 + 2M_2|z_1 - 1|^{3/2} + K_2|z_1 - 1|^{3/2}
$$

$$
\leq \left(\frac{1}{2}M_2|z_1 - 1| + 2M_2|z_1 - 1|^{1/2} + K_2|z_1 - 1|^{1/2} \right)|z_1 - 1|
$$

$$
\leq \frac{A_0}{16}|z_1 - 1| = \frac{A_0}{16}d(z, e_1)
$$

Thus for $d(z, e_1) < \delta_0$

$$
|R_\eta(z, e_1)| \geq \left(|D_1\varphi_\eta(e_1)| - |D_{22}\varphi_\eta(e_1)| \right)|z_1 - 1| - \frac{A_0}{16}|z_1 - 1|
$$

$$\geq \left(\operatorname{Re} D_1 \varphi_\eta(e_1) - |D_{22}\varphi_\eta(e_1)| \right) |z_1 - 1| - \frac{A_0}{16} |z_1 - 1|$$

$$\geq \frac{A_0}{4} d(z, e_1)$$

where we have used Inequality (6.2.4) in the last inequality.

The upper bound in (1) is easier. We have

$$R_\eta(z, e_1) = \varphi_\eta(z) - \varphi_\eta(e_1) - D_2\varphi_\eta(e_1)z_2$$

$$= \varphi_\eta(e_1) + D_1\varphi_\eta(e_1)(z_1 - 1) + D_2\varphi_\eta(e_1)z_2$$

$$+ E_\eta^1(z, e_1) - \varphi_\eta(e_1) - D_2\varphi_\eta(e_1)z_2$$

$$= D_1\varphi_\eta(e_1)(z_1 - 1) + E_\eta^1(z, e_1)$$

where $E_\eta^1(z, e_1) = \varphi_\eta(z) - P_\eta^1(z, e_1)$ satisfies $|E_\eta^1(z, e_1)| \leq K_1 d(z, e_1)$. Thus

$$|R_\eta(z, e_1)| \leq (|D_1\varphi_\eta(e_1)| + K_1)d(z, e_1)$$

$$\leq (M_1 + K_1)d(z, e_1) = Bd(z, e_1)$$

We remove the normalization $\zeta = e_1$ by noting that if ζ in ∂B is arbitrary, and if U is unitary with $Ue_1 = \zeta$ and $Ue_2 = \zeta^\perp$ then for $\psi = \varphi \circ U$, $R_\eta^\varphi(Uz, \zeta) = R_\eta^\psi(z, e_1)$. This observation, together with Exercise 6.2.3, gives (1) in the general case.

For (2) we consider first the case $\zeta = e_1$ and write

$$\operatorname{Re} R_\eta(z, e_1)$$

$$= \operatorname{Re} \left(D_1\varphi_\eta(e_1)(z_1 - 1) + \frac{1}{2} D_{22}\varphi_\eta(e_1)z_2^2 \right)$$

$$+ \operatorname{Re} \left(\frac{1}{2} D_{11}\varphi_\eta(e_1)(z_1 - 1)^2 + D_{12}\varphi_\eta(e_1)(z_1 - 1)z_2 + E_\eta^2(z, e_1) \right)$$

where $E_\eta^2(z, e_1) = \varphi_\eta(z) - P_\eta^2(z, e_1)$. Notice that

$$\operatorname{Re} \left(D_1\varphi_\eta(e_1)(z_1 - 1) + \frac{1}{2} D_{22}\varphi_\eta(e_1)z_2^2 \right)$$

$$\leq \operatorname{Re} D_1\varphi_\eta(e_1)\operatorname{Re}(z_1 - 1) - \operatorname{Im} D_1\varphi_\eta(e_1)\operatorname{Im} z_1 + \left| \frac{1}{2} D_{22}\varphi_\eta(e_1)z_2^2 \right|$$

$$\leq \operatorname{Re} D_1\varphi_\eta(e_1)\operatorname{Re}(z_1 - 1) - \operatorname{Im} D_1\varphi_\eta(e_1)\operatorname{Im} z_1$$

$$+ |D_{22}\varphi_\eta(e_1)|(1 - \operatorname{Re} z_1)$$

$$\leq \operatorname{Re} D_1\varphi_\eta(e_1)\operatorname{Re}(z_1 - 1) - \operatorname{Im} D_1\varphi_\eta(e_1)\operatorname{Im} z_1$$

$$+ \operatorname{Re} D_1\varphi_\eta(e_1)(1 - \operatorname{Re} z_1)$$

$$= -\operatorname{Im} D_1\varphi_\eta(e_1)\operatorname{Im} z_1$$

since $\operatorname{Re} D_1 \varphi_\eta(e_1) - |D_{22}\varphi_\eta(e_1)| > 0$ by Inequality (6.2.4). Thus

$\operatorname{Re} R_\eta(z, e_1)$

$$\leq -\operatorname{Im} D_1\varphi_\eta(e_1)\operatorname{Im} z_1$$

$$+\operatorname{Re}\left(\frac{1}{2}D_{11}\varphi_\eta(e_1)(z_1 - 1)^2 + D_{12}\varphi_\eta(e_1)(z_1 - 1)z_2 + E_\eta^2(z, e_1)\right)$$

$$\leq -\operatorname{Im} D_1\varphi_\eta(e_1)\operatorname{Im} z_1$$

$$+\left|\frac{1}{2}D_{11}\varphi_\eta(e_1)(z_1 - 1)^2 + D_{12}\varphi_\eta(e_1)(z_1 - 1)z_2 + E_\eta^2(z, e_1)\right|$$

But

$$\left|\frac{1}{2}D_{11}\varphi_\eta(e_1)(z_1 - 1)^2 + D_{12}\varphi_\eta(e_1)(z_1 - 1)z_2 + E_\eta^2(z, e_1)\right|$$

is bounded above by

$$\frac{1}{2}M_2|z_1 - 1|^2 + 2M_2|z_1 - 1|^{3/2} + K_2|z_1 - 1|^{3/2} \leq \frac{1}{16}A_0 d(z, e_1)$$

provided $d(z, e_1) < \delta_0$ while $|\operatorname{Im} D_1\varphi_\eta(e_1)| < A/8 = A_0/16$ so that

$$-\operatorname{Im} D_1\varphi_\eta(e_1)\operatorname{Im} z_1 \leq \frac{A}{8}|z_1 - 1| = \frac{A_0}{16}d(z, e_1)$$

Thus

$$\operatorname{Re} R_\eta(z, e_1) \leq \frac{A_0}{8}d(z, e_1) \ \text{ if } \ d(z, e_1) < \delta_0$$

If $R_\eta(z, e_1) \neq 0$ and $d(z, e_1) < \delta_0$

$$\frac{\operatorname{Re} R_\eta(z, e_1)}{|R_\eta(z, e_1)|} \leq \frac{(A_0/8)d(z, e_1)}{(A_0/4)d(z, e_1)} = \frac{1}{2}$$

so that $|\arg R_\eta(z, e_1)| \geq \pi/3$.

The non-normalized version follows directly as before, from $R_\eta^\varphi(Uz, \zeta) = R_\eta^\psi(z, e_1)$ where $\psi = \varphi \circ U$ for U unitary with $Ue_1 = \zeta$ and $Ue_2 = \zeta^\perp$. ∎

LEMMA 6.13
Suppose, for η in ∂B, that $d(\varphi(z), \eta) = |1 - \varphi_\eta(z)|$ has a local minimum on ∂B at some ζ in ∂B, where $d(\zeta, \zeta_0) < \delta_0$ and $d(\eta, \varphi(\zeta_0)) < \delta_0$ for some ζ_0 in S_0. Then

(1) $D_{\zeta^\perp}\varphi_\eta(\zeta) = 0$

(2) $1 - \varphi_\eta(\zeta) = aD_\zeta\varphi_\eta(\zeta)$ *for some* $a \geq 0$

(3) $d(\varphi(\zeta), \eta) + Bd(z, \zeta) \geq d(\varphi(z), \eta) \geq \frac{A}{4}d(z, \zeta)$ *for* $d(z, \zeta) < \delta_0$

PROOF   We verify the results when $\zeta = e_1$ and leave it to the reader to carry out the, by now familiar, argument to extend from this normalized case to the general result. If $\varphi_\eta(e_1) = 1$ then $\varphi(e_1) = \eta$ and the result is contained in Lemmas 6.8 and 6.11. Thus for the rest of the proof we may suppose that $g(e_1) \neq 0$ where $g(z) = 1 - \varphi_\eta(z)$, but that $|g|$ has a local minimum on ∂B at $\zeta = e_1$. Let δ_1 be sufficiently small so that $\delta_1 < \delta_0$ and $|g(e_1)| \leq |g(z)|$ for all z in $Q(e_1, \delta_1)$.

For an arbitrary C^1 curve $\gamma(t) = (\gamma_1(t), \gamma_2(t))$ lying in $Q(e_1, \delta_1)$ and satisfying $\gamma(0) = e_1$, let $\widetilde{\gamma} = g \circ \gamma$ so that $\widetilde{\gamma}$ is a C^1 curve in $\{w \in C : |w| \geq |g(e_1)|\}$ with $\widetilde{\gamma}(0) = g \circ \gamma(0) = g(e_1)$. The tangent vector to $\widetilde{\gamma}$ at $t = 0$ is perpendicular (in R^2) to $g(e_1)$ and therefore is parallel to $ig(e_1)$. By the chain rule

$$D_1\varphi_\eta(e_1)\gamma_1'(0) + D_2\varphi_\eta(e_1)\gamma_2'(0) = aig(e_1) \tag{6.2.7}$$

for some real number a. We interpret this with three special choices for $\gamma(t)$:

- If $\gamma(t) = (\cos t, \sin t)$ then Equation (6.2.7) implies $D_2\varphi_\eta(e_1) = a_1 ig(e_1)$ for some real a_1

- If $\gamma(t) = (\cos t, i\sin t)$ then Equation (6.2.7) implies $iD_2\varphi_\eta(e_1) = a_2 ig(e_1)$ for some real a_2

- If $\gamma(t) = (\cos t + i\sin t, 0)$ then Equation (6.2.7) implies $iD_1\varphi_\eta(e_1) = a_3 ig(e_1)$ for some real a_3

We conclude that

(i) $D_2\varphi_\eta(e_1) = 0$ and

(ii) $D_1\varphi_\eta(e_1) = a_3(1 - \varphi_\eta(e_1))$ for some real a_3

Note that $|\arg(1 - \langle \varphi(e_1), \eta \rangle)| = |\arg(1 - \varphi_\eta(e_1))| < \pi/2$ (principal value). Since by hypothesis $d(e_1, \zeta_0) < \delta_0$ and $d(\eta, \varphi(\zeta_0)) < \delta_0$ for some ζ_0 in S_0, we may apply Inequality (6.2.6) to conclude that $|\arg D_1\varphi_\eta(e_1)| < \pi/6$. Thus $D_1\varphi_\eta(e_1) = a_3(1 - \varphi_\eta(e_1))$ for some $a_3 > 0$ ($a_3 \neq 0$ since $\mathrm{Re}\, D_1\varphi_\eta(e_1) > 0$ by Inequality (6.2.4)) and we have (2).

Using (i) we may write

$$|1 - \varphi_\eta(z)| = |1 - \varphi_\eta(e_1) - R_\eta(z, e_1)|$$

where $|\arg(1 - \varphi_\eta(e_1))| = |\arg D_1\varphi_\eta(e_1)| < \pi/6$ by Inequality (6.2.6), and $|\arg R_\eta(z, e_1)| \geq \pi/3$ by Lemma 6.12, if $d(z, e_1) < \delta_0$. From this it follows (Exercise 6.2.5) that $|1 - \varphi_\eta(z)| \geq \frac{1}{2}|R_\eta(z, e_1)|$ for $d(z, e_1) < \delta_0$. For such z, Lemma 6.12 gives

$$d(\varphi(z), \eta) = |1 - \varphi_\eta(z)| \geq \frac{1}{2}|R_\eta(z, e_1)| \geq \frac{A}{4}d(z, \zeta)$$

and also

$$d(\varphi(z), \eta) \leq d(\varphi(e_1), \eta) + |R_\eta(z, e_1)| \leq d(\varphi(e_1), \eta) + Bd(z, e_1)$$

which completes the proof in the normalized case.

We are now ready to assemble these lemmas into a proof of the sufficiency theorem.

PROOF (of Theorem 6.10)   We prove the theorem by verifying the Carleson condition

$$\sigma\varphi^{-1}S(\eta,t) = O(t^2)$$

for η in ∂B and $t > 0$. Clearly it is enough to check this for all sufficiently small t. If we set $\Omega = \{z \in \partial B : d(z, S_0) < \delta_0/(6B_1)\}$ where $B_1 = \max\{1, B\}$ and let $m = \max\{|\varphi(z)| : z \in \partial B \setminus \Omega\}$, then $m < 1$ and for $t \leq 1 - m$

$$\varphi^{-1}S(\eta,t) \cap \partial B \subset \Omega \tag{6.2.8}$$

Let $t_0 = \min\{1 - m, \delta_0/6, A\delta_0/4\}$. We will verify the Carleson condition for all $t < t_0/2$.

Suppose η is in ∂B with $d(\eta, \varphi(S_0)) \geq \delta_0$. If z is in $\varphi^{-1}S(\eta, t_0) \cap \partial B$ then by the above remarks, z is in Ω and there exists ζ in S_0 with $d(z, \zeta) < \delta_0/(6B_1)$. By Lemma 6.11, $d(\varphi(z), \varphi(\zeta)) < \delta_0/6$ and

$$d(\varphi(\zeta), \eta) < 3\{d(\varphi(\zeta), \varphi(z)) + d(\varphi(z), \eta)\} < 3\{\delta_0/6 + t_0\} \leq \delta_0$$

which contradicts $d(\eta, \varphi(S_0)) \geq \delta_0$. Thus $\varphi^{-1}S(\eta, t_0) \cap \partial B = \emptyset$ if $d(\eta, \varphi(S_0)) \geq \delta_0$ and for the rest of the argument we may confine our attention to estimating $\sigma\varphi^{-1}S(\eta, t)$ when $t < t_0/2$ and η in ∂B satisfies $d(\eta, \varphi(S_0)) < \delta_0$.

Fix such an η and consider those components of $\varphi^{-1}S(\eta, t_0) \cap \partial B$ that also meet $\varphi^{-1}S(\eta, t_0/2)$. If O_j is such a component, then on the boundary of O_j, we have $|1 - \varphi_\eta(z)| = t_0$ and we may find ζ_j in O_j at which $d(\varphi(z), \eta) = |1 - \varphi_\eta(z)|$ attains a local minimum, necessarily less than $t_0/2$:

$$d(\varphi(\zeta_j), \eta) < t_0/2$$

Since ζ_j must be in Ω by Inequality (6.2.8) there exists ζ_j^0 in S_0 with $d(\zeta_j, \zeta_j^0) < \delta_0/(6B_1)$. Moreover, using Lemma 6.11

$$d(\eta, \varphi(\zeta_j^0)) \leq 3\left(d(\eta, \varphi(\zeta_j)) + d(\varphi(\zeta_j), \varphi(\zeta_j^0))\right) \leq 3\left(\frac{t_0}{2} + \frac{B\delta_0}{6B_1}\right) < \delta_0$$

Now Lemma 6.13 applies to conclude that for z in ∂B satisfying $d(z, \zeta_j) < \delta_0$ we have

$$d(\varphi(z), \eta) \geq \frac{A}{4}d(z, \zeta_j) \tag{6.2.9}$$

If z is in O_j, $d(\varphi(z), \eta) < t_0$, and we see from Inequality (6.2.9) that

$$Q(\zeta_j, \delta_0) \cap O_j \subset Q(\zeta_j, \frac{4t_0}{A})$$

But O_j is connected and $4t_0/A < \delta_0$ so $O_j \subset Q(\zeta_j, 4t_0/A)$. On the other hand, Lemma 6.13 also implies that

$$d(\varphi(\zeta_j), \eta) + Bd(z, \zeta_j) \geq d(\varphi(z), \eta)$$

whenever $d(z, \zeta_j) < \delta_0$. In particular, if $d(z, \zeta_j) < t_0/(2B)$ this shows that

$$d(\varphi(z), \eta) < B\frac{t_0}{2B} + \frac{t_0}{2}$$

and

$$Q(\zeta_j, \frac{t_0}{2B}) \subset O_j$$

As a consequence of this we see that there is an upper bound J_0 on the number of components of $\varphi^{-1}\mathcal{S}(\eta, t_0) \cap \partial B$ which meet $\varphi^{-1}\mathcal{S}(\eta, t_0/2)$; J_0 must satisfy $J_0 \sigma Q(\zeta_j, t_0/(2B)) \leq 1$.

Finally, if $t < t_0/2$ and z is in $\varphi^{-1}\mathcal{S}(\eta, t) \cap O_j$ with O_j as defined above, a component of $\varphi^{-1}\mathcal{S}(\eta, t_0) \cap \partial B$ which intersects $\varphi^{-1}\mathcal{S}(\eta, t_0/2)$, then $d(\varphi(z), \eta) < t$ so by Lemma 6.13 (which we may apply since $d(z, \zeta_j) < 4t_0/A \leq \delta_0$)

$$t > d(\varphi(z), \eta) \geq \frac{A}{4}d(z, \zeta_j)$$

Thus

$$\varphi^{-1}\mathcal{S}(\eta, t) \cap O_j \subset Q(\zeta_j, 4t/A)$$

By definition of O_j, every point of $\varphi^{-1}\mathcal{S}(\eta, t) \cap \partial B$ lies in one of the at most J_0 components O_j and

$$\sigma\varphi^{-1}\mathcal{S}(\eta, t) \leq J_0 \sigma Q(\zeta, 4t/A) \sim Ct^2$$

where C is a constant independent of η and t. ∎

Both Theorem 6.10 and Proposition 6.9 generalize to B_N for $N > 2$. If ζ, η, and τ are in ∂B_N with $\varphi(\zeta) = \eta$ and $\langle \zeta, \tau \rangle = 0$ then we always have $D_\zeta \varphi_\eta(\zeta) > 0$ and $D_\zeta \varphi_\eta(\zeta) \geq |D_{\tau\tau}\varphi_\eta(\zeta)|$. The following two results characterize the bounded composition operators on $H^p(B_N)$ with smooth symbol.

THEOREM 6.14
If $\varphi : B_N \to B_N$ is analytic and φ is of class C^3 on $\overline{B_N}$ and if there exist ζ, η, and τ in ∂B_N with $\varphi(\zeta) = \eta$ and $\langle \zeta, \tau \rangle = 0$ so that

$$D_\zeta \varphi_\eta(\zeta) = |D_{\tau\tau}\varphi_\eta(\zeta)|$$

then C_φ is unbounded on $H^p(B_N)$.

THEOREM 6.15
If $\varphi : B_N \to B_N$ is analytic and φ is of class C^3 on $\overline{B_N}$ and if

$$D_\zeta \varphi_\eta(\zeta) > |D_{\tau\tau}\varphi_\eta(\zeta)|$$

for all ζ in ∂B_N with $\varphi(\zeta) = \eta$ in ∂B_N and τ in ∂B_N with $\langle \zeta, \tau \rangle = 0$, then C_φ is bounded on $H^p(B_N)$.

Exercises

6.2.1 Show that there is an absolute constant c so that $|z - \zeta|^2 \leq cd(z, \zeta)$ for all ζ in ∂B and z in $\overline{B}$.

6.2.2 If φ is of class C^3 on $\overline{B}$ with

$$\sup \left\{ \frac{\sigma \varphi^{-1} S(\eta, t)}{t^2} : \eta \in \partial B, t > 0 \right\} = \infty$$

then there is an η_0 in ∂B such that

$$\sup \left\{ \frac{\sigma \varphi^{-1} S(\eta_0, t)}{t^2} : t > 0 \right\} = \infty$$

that is, if the Carleson criterion fails to hold, then it fails at a single point.

6.2.3 If $\psi = \varphi \circ U$ where U is unitary, then $A_0(\varphi) = A_0(\psi)$, $K_1(\varphi) = K_1(\psi)$, $K_2(\varphi) = K_2(\psi)$, $M_1(\varphi) = M_1(\psi)$, and $M_2(\varphi) = M_2(\psi)$. Note that from this it also follows that $A(\varphi) = A(\psi)$ and $B(\varphi) = B(\psi)$ where A and B are the constants of Lemma 6.11.

6.2.4 Prove Lemma 6.8 and Proposition 6.9.

6.2.5 Use the Law of Sines to show that in the proof of Lemma 6.13

$$|1 - \varphi_\eta(z)| \geq \frac{1}{2} |R_\eta(z, e_1)|$$

where $1 - \varphi_\eta(e_1) - R_\eta(z, e_1) = 1 - \varphi_\eta(z)$ and $|\arg(1 - \varphi_\eta(e_1))| < \pi/6$ while $|\arg R_\eta(z, e_1)| > \pi/3$.

6.2.6 Assume that $\varphi : B \to B$ is analytic and φ' is uniformly bounded. Note that by Lemma 3.40 there exist positive constants c and t_0 so that if $\varphi(\zeta) = \eta$ where $|\eta| = 1$ then

$$\varphi(Q(\zeta, t)) \subset S(\eta, ct)$$

for all $t < t_0$. The constant c depends on $\varphi(0)$ but not on t for $t < t_0$. Use this to show that there exists a finite constant $\widetilde{B}$ so that if $d(z, \zeta) < t_0$ where ζ is in S_0 then $d(\varphi(z), \varphi(\zeta)) \leq \widetilde{B}d(z, \zeta)$.

6.2.7 Suppose A is an $N \times N$ matrix with $\|A\| \leq 1$ and set $\varphi(z) = Az$. Show that C_φ is bounded on $H^p(B_N)$. Indeed, show that C_φ is bounded on any weighted Hardy space $H^2(\beta, B_N)$.

Notes

We have followed W. Wogen's [Wog88] original argument very closely; we have however corrected a missing hypothesis in Lemmas 6 and 7 of that paper.

6.3 Examples

Propositions 6.7 and 6.9 of the previous section do not, by themselves, make it any easier to *construct* examples of smooth $\varphi : B_N \rightarrow B_N$ for which C_φ will be unbounded on $H^p(B_N)$. In this section we develop some tools which will help construct some unbounded composition operators whose symbols, in contrast to the examples in Chapter 3, are highly non-degenerate maps of B_N into B_N. We will concentrate on examples in B_2 and we write B for B_2 throughout this section.

We begin with the disk automorphism $\varphi(z) = (z+a)/(1+az)$ for $0 < a < 1$ and its second order Taylor polynomial $P_2(z)$ at $z = 1$. A computation shows that

$$P_2(z) = \frac{a + 3a^2}{(1+a)^2} + \frac{1 + 2a - 3a^2}{(1+a)^2}z - \frac{a(1-a)}{(1+a)^2}z^2 \tag{6.3.1}$$

Notice that $P_2(1) = 1$ and $P_2(\partial D)$ has second order contact with ∂D at 1. Set $c_1 = a + 3a^2$, $c_2 = 1 + 2a - 3a^2$, and $c_3 = -a(1-a)$ so that $P_2(z) = (c_1 + c_2 z + c_3 z^2)/(1+a)^2$. The next lemma will allow us to estimate $|P_2(z)|$ on circles of radius r, $0 < r < 1$.

LEMMA 6.16
If $f(z) = c_1 + c_2 z + c_3 z^2$ with c_1, c_2, and c_3 as above, then for any r in $(0,1)$ and z with $|z| = r$ we have $|f(z)| \leq f(r)$; moreover the inequality is strict if $z \neq r$.

PROOF Write $z = x + iy$ where $|z| = r$. Then

$$|f(z)|^2 = c_1^2 + c_2^2 r^2 + c_3^2 r^4 + 2c_1 c_2 x + 2\left(c_1 c_3 x^2 - c_1 c_3 y^2 + c_2 c_3 x(x^2 + y^2)\right)$$
$$= c_1^2 + c_2^2 r^2 + c_3^2 r^4 + 2c_1 c_2 x + 2c_1 c_3 (x^2 - y^2) + 2c_2 c_3 x r^2$$

From the definitions of c_1, c_2 and c_3 we have $-4c_1 c_3 = c_1 c_2 + c_2 c_3$. Some algebra shows that $|f(z)|^2$ is equal to

$$c_1^2 + c_1 c_2 + c_2 c_3 + (c_2^2 - 2c_1 c_3)r^2 + c_3^2 r^4 - (c_1 c_2 + c_2 c_3)(1-x)^2 - 2c_2 c_3 x(1-r^2)$$

Since $-c_2 c_3$ and $c_1 c_2 + c_2 c_3$ are positive this is bounded above by

$$c_1^2 + c_1 c_2 + c_2 c_3 + (c_2^2 - 2c_1 c_3)r^2 + c_3^2 r^4 - (c_1 c_2 + c_2 c_3)(1-r)^2 - 2c_2 c_3 r(1-r^2)$$

which is $f(r)^2$ and equality holds only if $x = r$. ∎

As a consequence of this lemma notice that if z is in D with $|z| = r$ then

$$|P_2(z)| = \frac{1}{(1+a)^2}|f(z)| \leq \frac{1}{(1+a)^2}f(r) \leq 1$$

where the last inequality uses the fact that $0 < a < 1$. Thus $P_2(z)$ maps D into D. Moreover P_2 is one-to-one on $\overline{D}$ since if $c_1 + c_2 z + c_3 z^2 = c_1 + c_2 w + c_3 w^2$ we have $c_2|z - w| = |c_3||z - w||z + w|$ so that $c_2 = |c_3||z + w|$ when $z \neq w$. This implies $1 + 2a - 3a^2 = a(1 - a)|z + w| \leq 2a - 2a^2$ which contradicts $a < 1$.

Next we will construct a mapping of the form

$$\varphi(z_1, z_2) = (P_2(z_1) + k_1 z_2^2, k_2 z_2^2) \tag{6.3.2}$$

with $k_2 \neq 0$ and real so that $\varphi(B) \subset B$ and yet C_φ is unbounded on $H^p(B)$. For such a map $\varphi(z_1, z_2) = \varphi(w_1, w_2)$ if and only if $z_1 = w_1$ and $z_2^2 = w_2^2$, since P_2 is one-to-one on $\overline{D}$, so that φ will be at worst two-to-one on $\overline{B}$. Notice that $\varphi(e_1) = e_1$ and $D_1\varphi_1(e_1) = (1-a)/(1+a)$ while $D_{22}\varphi_1(e_1) = 2k_1$, so provided $\varphi(B) \subset B$, C_φ will be unbounded (by Proposition 6.7) if

$$k_1 = \frac{1 - a}{2 + 2a} = \left(\frac{a}{1 + a}\right)^2 \frac{c_2 + 2c_3}{2a^2}$$

Thus it remains only to see whether, with this choice of k_1, there is a choice of k_2 so that $\varphi(B) \subset B$. To this end we write $|\varphi(z_1, z_2)|^2 = |\varphi_1(z_1, z_2)|^2 + |\varphi_2(z_1, z_2)|^2$ and let $z = (z_1, z_2)$ be in ∂B with $|z_1| = r$ so that $|z_2|^2 = 1 - r^2$. We have

$$\varphi_1(z) = P_2(z_1) + k_1 z_2^2$$

$$= \frac{a^2}{(1 + a)^2} \left(\frac{c_1}{a^2} + \frac{c_2}{a^2} z_1 + \frac{c_3}{a^2} z_1^2 + \frac{c_2 + 2c_3}{2a^2} z_2^2\right)$$

Using Lemma 6.16 we have

$$|\varphi_1(z)| \leq \left(\frac{a}{1 + a}\right)^2 \left(\left|\frac{c_1}{a^2} + \frac{c_2}{a^2} z_1 + \frac{c_3}{a^2} z_1^2\right| + \left|\frac{c_2 + 2c_3}{2a^2} z_2^2\right|\right)$$

$$\leq \left(\frac{a}{1 + a}\right)^2 \left(\frac{c_1}{a^2} + \frac{c_2}{a^2} r + \frac{c_3}{a^2} r^2 + \left|\frac{c_2 + 2c_3}{2a^2}\right| (1 - r^2)\right)$$

Since $c_2 + 2c_3 > 0$ we have

$$|\varphi_1(z)| \leq \left(\frac{1}{1 + a}\right)^2 \left(c_1 + c_2 r + c_3 r^2 + \frac{c_2 + 2c_3}{2} - \frac{(c_2 + 2c_3)r^2}{2}\right)$$

$$= \frac{1}{(1 + a)^2} \left(c_1 + c_2 + c_3 - \frac{c_2}{2}(1 - r)^2\right)$$

$$= \frac{1}{(1 + a)^2} \left((1 + a)^2 - \frac{c_2}{2}(1 - r)^2\right) \leq 1$$

since $c_2 > 0$. Thus

$$|\varphi(z)|^2 \leq |\varphi_1(z)|^2 + |\varphi_2(z)|^2 \leq |\varphi_1(z)| + |\varphi_2(z)|^2$$

$$\leq \frac{1}{(1 + a)^2} \left((1 + a)^2 - \frac{c_2}{2}(1 - r)^2\right) + 4k_2^2(1 - r)^2$$

$$= 1 - (1 - r)^2 \left(\frac{c_2}{2(1 + a)^2} - 4k_2^2\right)$$

Any choice of k_2 satisfying

$$\frac{c_2}{2(1+a)^2} > 4k_2^2$$

will guarantee that $\varphi(B) \subset B$; for each a in $(0, 1)$ some such choice of k_2 will be possible since $c_2 = 1 + 2a - 3a^2 > 0$. As a specific example let $a = 1/2$ so that $c_1 = 5/4, c_2 = 5/4, c_3 = -1/4$ and $P_2(z) = (5 + 5z - z^2)/9$. Then

$$\varphi(z_1, z_2) = \left(\frac{5}{9} + \frac{5}{9}z_1 - \frac{1}{9}z_1^2 + \frac{1}{6}z_2^2, k_2 z_2^2\right)$$

where k_2 is chosen to satisfy $0 < k_2 < \sqrt{5}/(6\sqrt{2})$ will be an at worst two-to-one map of B into B, analytic in $\overline{B}$, for which C_φ is unbounded on $H^p(B)$.

Our next goal is to construct a *biholomorphic* map of B into B which gives rise to an unbounded composition operator. Again the Taylor polynomial

$$P_2(z) = \frac{a + 3a^2}{(1+a)^2} + \frac{1 + 2a - 3a^2}{(1+a)^2}z - \frac{a(1-a)}{(1+a)^2}z^2$$

will form the basis of our example. Let $f(z_1, z_2)$ be the first coordinate function of φ as defined in Equation (6.3.2), that is let

$$f(z_1, z_2) = P_2(z_1) + k_1 z_2^2$$

where $k_1 = (1-a)/(2+2a)$ with a in $(0, 1)$, so that as before we have $D_1 f(e_1) = D_{22}f(e_1)$. From f we construct a map ψ by

$$\psi(z_1, z_2) = \frac{1}{2}(1 + f(z_1, z_2), z_2(1 - f(z_1, z_2))) \qquad (6.3.3)$$

Note that the components of ψ are polynomials of degree at most 3. Several observations are essentially immediate:

- $\psi(e_1) = e_1$
- $D_1\psi_1(e_1) = D_{22}\psi_1(e_1)$
- ψ is one-to-one on $\overline{B}$

For the third of these observations note that $\psi(z_1, z_2) = \psi(w_1, w_2)$ for points (z_1, z_2) and (w_1, w_2) in $\overline{B}$ implies that $f(z_1, z_2) = f(w_1, w_2)$ and $z_2 = w_2$, since $f(z) = 1$ for z in $\overline{B}$ only if $z = e_1$. The assertion then follows since $f(z_1, z_2) = P_2(z_1) + k_1 z_2^2$ and P_2 is one-to-one on $\overline{D}$.

We claim that ψ maps B into B. To verify this we estimate $|\psi(z_1, z_2)|^2$ when $|z_1| = r$ and $|z_2|^2 = 1 - r^2$ for $0 \le r \le 1$. Then

$$4|\psi(z_1, z_2)|^2 = |1 + f(z_1, z_2)|^2 + |z_2|^2|1 - f(z_1, z_2)|^2$$
$$= 1 + 2\,\text{Re}\,f + |f|^2 + (1 - r^2)(1 - 2\,\text{Re}\,f + |f|^2)$$
$$= 2 + |f|^2 - r^2 + 2r^2\,\text{Re}\,f + (1 - r^2)|f|^2$$
$$\le 2 + |f|^2 - r^2 + 2r^2|f| + (1 - r^2)|f|^2$$

Recall our previous estimate on $|f(z_1, z_2)|$ when $|z_1| = r$ and $|z_2|^2 = 1 - r^2$:

$$|f| \le 1 - \frac{c_2}{2(1+a)^2}(1-r)^2$$

where $0 < a < 1$ and c_2 is the positive number $1 + 2a - 3a^2$. In particular

$$|f|^2 \le 1 - \frac{c_2}{(1+a)^2}(1-r)^2 + \frac{c_2^2}{4(1+a)^4}(1-r)^4$$

Using these estimates in the above calculation we see that on ∂B, $4|\psi(z_1, z_2)|^2$ is at most

$$4 - \frac{2c_2}{(1+a)^2}(1-r)^2 + \frac{c_2^2}{4(1+a)^4}(1-r)^4(2-r^2)$$

Thus $\psi(B) \subset B$ provided

$$\frac{2c_2}{(1+a)^2}(1-r)^2 \ge \frac{c_2^2}{4(1+a)^4}(1-r)^4(2-r^2)$$

for r in $[0, 1]$. This is immediate from the definition of c_2. Thus ψ is a map of B into B with polynomial coordinate functions, which is one-to-one on $\overline{B}$ and such that C_ψ is unbounded on $H^p(B)$, by Proposition 6.7.

Exercises

6.3.1 Let $P_2(z)$ be the Taylor polynomial from Equation (6.3.1) and set

$$\varphi(z_1, z_2) = \left(P_2(z_1) + \frac{a(1-a)}{(1+a)^2} z_2^2, \; \frac{1+2a-3a^2}{(1+a)^2} z_2 - \frac{2a(1-a)}{(1+a)^2} z_1 z_2 \right)$$

where $a \in (1/2, 1)$. Show that $\varphi(\overline{B} \setminus \{e_1\}) \subset B$ while $\varphi(e_1) = e_1$. Hints: Show that $(1+a)^4|\varphi(z)|^2$ is bounded above by

$$|c_1 + c_2 z_1 + c_3 z_1^2|^2 + 2|c_1 + c_2 z_1 + c_3 z_1^2||c_3 z_2^2| + |c_3 z_2^2|^2 + |z_2|^2|c_2 + 2c_3 z_1|^2$$

where $c_1 = a + 3a^2$, $c_2 = 1 + 2a - 3a^2$, and $c_3 = -a(1-a)$. Then let $|z_1| = r$ and $|z_2| = \sqrt{1-r^2}$ and use the calculations from Lemma 6.16 to show that $|c_1 + c_2 z_1 + c_3 z_1^2|^2$ is equal to

$$c_1^2 + c_1 c_2 + c_2 c_3 + (c_2^2 - 2c_1 c_3)r^2 + c_3^2 r^4 - (c_1 c_2 + c_2 c_3)(1-x_1)^2 - 2c_2 c_3 x_1(1-r^2)$$

where $x_1 = \operatorname{Re} z_1$. From this obtain an upper estimate on $(1+a)^4|\varphi(z)|^2$ of

$$(1+a)^4 - 2c_2 c_3(1-r)(1+r)(r-x_1) - (c_1 c_2 + c_2 c_3)(1-x_1)^2$$

which is at most $(1+a)^4$ provided $a > 1/2$, since

$$-2c_2 c_3(1-r)(1+r)(r-x_1) \le 4c_2|c_3|(1-x_1)^2$$

6.3.2 Let φ be defined as in Exercise 6.3.1. For z in $\overline{B}$ show that

$$(1+a)^2 \operatorname{Re} \varphi_1(z_1, z_2) > c_1 - c_2 + 2c_3$$

and hence that $\operatorname{Re} \varphi_1(z_1, z_2) > 0$ provided $(3 + \sqrt{41})/16 < a < 1$.

6.3.3 Let φ be defined as in Exercise 6.3.1. Show that φ is one-to-one on $\overline{B}$. Hints: The hypothesis $\varphi_2(z_1, z_2) = \varphi_2(w_1, w_2)$ implies that

$$|1 + 2a - 3a^2 - 2a(1 - a)z_1||z_2 - w_2| = |2a(1 - a)w_2||z_1 - w_1|$$

From this obtain the estimate

$$|z_2 - w_2| \leq \frac{2a}{1 + a}|z_1 - w_1|$$

The hypothesis $\varphi_1(z_1, z_2) = \varphi_1(w_1, w_2)$ implies that

$$|z_1 - w_1||1 + 2a - 3a^2 - a(1 - a)(z_1 + w_1)| = a(1 - a)|w_2 - z_2||w_2 + z_2|$$

and hence that

$$|z_1 - w_1| \leq \frac{2a}{1 + a}|w_2 - z_2|$$

6.3.4 Let $\psi(z_1, z_2) = \frac{1}{2}(1 + z_1^2 + z_2^2, z_2(1 - z_1^2 - z_2^2))$. Show that $\psi(B) \subset B$ and C_ψ is unbounded on $H^p(B)$.

6.3.5 Set $\tau = \psi \circ \varphi$ where ψ is as in Exercise 6.3.4 and φ as in Exercise 6.3.1. By these exercises, τ maps B into B.

    (a) Show there exists choices of a, $a < 1$, so that τ is one-to-one on $\overline{B}$. Hint: Show that ψ is univalent on $\{z \in B : \text{Re } z_1 > 0\}$.

    (b) Check that $\tau(e_1) = e_1$ and $D_1\tau_1(e_1) = D_{22}\tau_1(e_1)$. Conclude that C_τ is a biholomorphic map of B into B, with polynomial coordinate functions, and yet C_τ is unbounded on $H^p(B)$.

6.3.6 Let $\Phi = \psi \circ \varphi$ where

$$\varphi(z_1, z_2) = \frac{1}{2}(1 + z_1, z_2)$$

and

$$\psi(z_1, z_2) = \frac{1}{2}(1 + z_1^2 + z_2^2, z_2(1 - z_1^2 - z_2^2))$$

We know that both φ and ψ map B into B so that Φ does also. Since clearly $S_0(\varphi) = \{e_1\}$ and $\psi(e_1) = e_1$ we have $S_0(\Phi) = \{e_1\}$.

    (a) Show that C_Φ is bounded on $H^p(B_2)$.

    (b) Show that Φ is univalent on B_2 yet

$$\Omega(z) = \frac{\|\Phi'(z)\|^2}{|J_\Phi(z)|^2}$$

    is not bounded in B_2.

Thus the sufficiency condition of Theorem 3.41 is not necessary.

Notes

The examples from Equation (6.3.2) are due to J. A. Cima and W. R. Wogen. With a specific choice made for a they appear in [CiW87]. This paper predates Wogen's theorem from Section 6.2, so the original verification that the associated composition operators are unbounded used an ad hoc Carleson measure argument. Cima and Wogen's paper [CiW87] also contains an example of a biholomorphic map, with polynomial coordinate functions, which induces an unbounded composition operator. This example is outlined in Exercises 6.3.1, 6.3.2, 6.3.3, 6.3.4 and 6.3.5. The much simpler example $\psi(z_1, z_2)$ from Equation (6.3.3) was recently discovered by W. R. Wogen (private communication) and we are grateful to Professor Wogen for allowing us to present it here.

The result of Exercise 6.3.6 is also due to J. A. Cima and W. R. Wogen [CiW87].

7

Spectral Properties

7.1 Introduction

In this chapter, we investigate the spectra of composition operators. We have seen in other situations that the structure of composition operators depends on the location and nature of the fixed points of the underlying map. Spectra of composition operators show, perhaps even more clearly, that this is the case. Even for fairly simple functions φ, the structures of the spectra of the associated composition operators exhibit surprising diversity and complexity. As an introduction to our study of spectra, in this section we will give, without proof, examples of spectra of composition operators on $H^2(D)$ whose symbols are linear fractional transformations mapping the disk into itself. That is, we will assume that

$$\varphi(z) = \frac{\alpha z + \beta}{\gamma z + \delta}$$

for some complex numbers α, β, γ, and δ such that φ maps the unit disk D into itself. In later sections of this chapter, we will find relationships between characteristics of the symbol and properties of spectra that include each of the examples given.

In the table below, we list some typical maps, their most important properties, and the spectra of the associated composition operators. The Denjoy–Wolff point (or the fixed point of φ in D) will be denoted a. Given that the linear fractional model for analytic functions on the disk includes four cases, we should probably expect ten examples in the table: four cases for the automorphisms, four for non-automorphisms with no compactness, and two with compactness. However, there are no automorphisms with Denjoy–Wolff point on the unit circle, derivative 1 there, whose model for iteration is the plane. Also, the non-automorphism cases in which the Denjoy–Wolff point is on the circle with a halfplane model have the same sort of spectra for the derivative less than 1 or equal to 1.

The perspective we take in the remaining sections is that spectrum of a composition operator should be closely related to the spectrum of a linear fractional composition operator whose model is the same. While understanding of the spectra of composition operators is far from complete and a few exceptions to this point of view are known, this perspective is relatively successful in explaining the spectra known at this time. This table is taken from [Co88].

Example	*Properties*	*Spectrum C_φ on $H^2(D)$*
$\varphi(z) = \zeta z$ where $\vert \zeta \vert = 1$	$a = 0$ elliptic, inner	closure$\{\zeta^k : k = 0, 1, \dots\}$
$\varphi(z) = \frac{3z+1}{z+3}$	$a = 1,\ \varphi'(1) = \frac{1}{2}$ hyperbolic, inner	$\{\lambda : \frac{1}{\sqrt{2}} \le \vert \lambda \vert \le \sqrt{2}\}$
$\varphi(z) = \frac{(1+i)z-1}{z+i-1}$	$a = 1,\ \varphi'(1) = 1$ parabolic, inner	$\{\lambda : \vert \lambda \vert = 1\}$
$\varphi(z) = sz + 1 - s$ where $0 < s < 1$	$a = 1,\ \varphi'(1) = s$	$\{\lambda : \vert \lambda \vert \le \frac{1}{\sqrt{s}}\}$
$\varphi(z) = \frac{(2-t)z+t}{-tz+2+t}$ where $\mathrm{Re}(t) > 0$	$a = 1,\ \varphi'(1) = 1$ $\varphi''(1) = t$	$\{e^{\beta t} : \beta \le 0\} \cup \{0\}$ (a spiral)
$\varphi(z) = \frac{rz}{1-(1-r)z}$ where $0 < r < 1$	$a = 0,\ \varphi'(0) = r$ $\varphi(1) = 1$	$\{\lambda : \vert \lambda \vert \le \sqrt{r}\} \cup \{1\}$
$\varphi(z) = -\frac{1}{2}z$	$a = 0,\ \varphi'(0) = -\frac{1}{2}$ C_φ compact	$\{(-\frac{1}{2})^k : k = 0, 1, \dots\} \cup \{0\}$
$\varphi(z) = -\frac{1}{2}z + \frac{1}{2}$	$a = \frac{1}{3},\ \varphi'(\frac{1}{3}) = -\frac{1}{2}$ C_φ^2 compact, C_φ not	$\{(-\frac{1}{2})^k : k = 0, 1, \dots\} \cup \{0\}$

7.2   Invertible operators on the classical spaces on the disk

Among the special cases in which good spectral information is available are the invertible composition operators on the classical Hardy and weighted Bergman spaces in the disk. As expected, a description of the spectrum involves the fixed point character of the symbol. For invertible $C_\varphi \ne I$, φ will be an automorphism and the fixed point set of φ relative to $\overline{D}$ will fall into one of three categories:

- two fixed points on the unit circle
- one fixed point on the unit circle
- one fixed point in the open unit disk

We describe $\sigma(C_\varphi)$ in each of these cases. A number of the results we will obtain for the case of two boundary fixed points will have analogs under the much more general assumption that $\varphi : D \to D$ has Denjoy–Wolff point a in ∂D with $\varphi'(a) < 1$, by virtue of the good information contained in the linear fractional model in this case. This more general situation will be considered in Section 7.5.

Our methods for the automorphisms will apply equally well to all the classical spaces arising as weighted Hardy spaces in the disk where the weight sequence is $\beta(n)^2 = (n+1)^{1-\gamma}$ for $\gamma \geq 1$. Recall that this includes $H^2(D)$ (by the choice $\gamma = 1$) and all the standard weight Bergman spaces.

From Exercise 2.3.5, we see that if φ is an automorphism of the disk, there is an automorphism ψ so that $\psi^{-1} \circ \varphi \circ \psi$ is $(z+r)/(1+rz)$ for $0 < r < 1$ if φ has two fixed points on the circle, either $((1+i)z-1)/(z+i-1)$ or $((1-i)z-1)/(z-i-1)$ if φ has one fixed point on the circle, and λz for $|\lambda| = 1$ if φ has a fixed point in the unit disk. The spaces we are considering are automorphism invariant, so on these spaces C_φ is similar to the operator $C_{\psi^{-1} \circ \varphi \circ \psi} = C_\psi C_\varphi C_\psi^{-1}$. Since similar operators have the same spectra, we may use these special automorphisms in the proofs of the results.

The similarity to operators whose symbol is a special automorphism makes the computation of the spectrum especially easy in the case of elliptic automorphisms. It is easy to see that the result is valid in any automorphism invariant functional Banach space that contains the polynomials and for which $\|C_U\| = 1$ whenever U is a rotation.

THEOREM 7.1

Let φ, not the identity map, be an automorphism of D with fixed point a in D. If $H^2(\beta)$ is an automorphism invariant weighted Hardy space, then $\sigma(C_\varphi)$ is the closure of the positive powers of $\varphi'(a)$. This closure is a finite subgroup of the circle if $\varphi'(a)^n = 1$ for some n and is the unit circle otherwise.

PROOF If φ is an elliptic automorphism, then C_φ is similar to $C_{\lambda z}$ where $\lambda = \varphi'(a)$ and $|\lambda| = 1$. Since $C_{\lambda z}(z^k) = \lambda^k z^k$ and $\{z^k/\beta(k)\}$ is an orthonormal basis for $H^2(\beta)$, the matrix for $C_{\lambda z}$ with respect to this basis is diagonal with diagonal entries λ^k for $k = 0, 1, 2, \ldots$. The spectrum of C_φ is therefore the closure of the set of diagonal entries, which is the finite set λ^k for $k = 1, 2, \ldots, n$ if $\lambda^n = 1$ for some n or is the unit circle otherwise. $\blacksquare$

For automorphisms that do not have a fixed point in the disk, finding the spectrum is somewhat more difficult. The next result estimates the spectral radius of C_φ, denoted $\rho(C_\varphi)$, for a parabolic or hyperbolic automorphism φ of the disk.

LEMMA 7.2

If φ is a parabolic or hyperbolic automorphism and $\beta(n)^2 = (n+1)^{1-\gamma}$ for $\gamma \geq 1$, then the spectral radius of C_φ on $H^2(\beta)$ satisfies $\rho(C_\varphi) \leq \varphi'(a)^{-\gamma/2}$ where a is the Denjoy–Wolff point of φ.

PROOF   Suppose $\varphi(z) = \lambda(u - z)/(1 - \overline{u}z)$, for $|\lambda| = 1$ and $|u| < 1$, is a parabolic or hyperbolic automorphism.

When $\gamma = 1$, that is, on the Hardy space $H^2(D)$, the result is contained in Theorem 3.9.

When $\gamma > 1$, for the purposes of the spectral radius formula, we may compute norms using the equivalent weighted Bergman space description of $H^2(\beta)$ (Exercise 2.1.5): f is in $H^2(\beta)$ if and only if

$$\int_D |f(z)|^2 (1 - |z|^2)^{\gamma-2}\, dA(z) < \infty$$

Using the identity (Equation (2.3.1))

$$1 - |\varphi^{-1}(w)|^2 = \frac{(1 - |w|^2)(1 - |u|^2)}{|1 - \overline{\lambda}uw|^2}$$

and a change of variable, we have

$$\int_D |f(\varphi(z))|^2 (1 - |z|^2)^{\gamma-2}\, dA(z)$$

$$= \int_{\varphi^{-1}(D)} |f(w)|^2 (1 - |\varphi^{-1}(w)|^2)^{\gamma-2}\, dA(\varphi^{-1}(w))$$

$$= \int_D |f(w)|^2 \frac{(1 - |w|^2)^{\gamma-2}(1 - |u|^2)^{\gamma-2}}{|1 - \overline{\lambda}uw|^{2(\gamma-2)}} |(\varphi^{-1})'(w)|^2\, dA(w)$$

$$= \int_D |f(w)|^2 \frac{(1 - |w|^2)^{\gamma-2}(1 - |u|^2)^{\gamma}}{|1 - \overline{\lambda}uw|^{2(\gamma-2)}|1 - \overline{\lambda}uw|^4}\, dA(w)$$

$$\leq \frac{(1 - |u|^2)^{\gamma}}{(1 - |u|)^{2\gamma}} \int_D |f(w)|^2 (1 - |w|^2)^{\gamma-2}\, dA(w)$$

$$= \left(\frac{1 + |u|}{1 - |u|}\right)^{\gamma} \int_D |f(w)|^2 (1 - |w|^2)^{\gamma-2}\, dA(w)$$

so the equivalence of norms gives a constant M so that

$$\|C_\varphi\| \leq M \left(\frac{1 + |\varphi(0)|}{1 - |\varphi(0)|}\right)^{\gamma/2}$$

A direct calculation for parabolic automorphisms (see Exercise 7.2.2) shows that $\varphi'(a) = \lim_{z \to a}(1 - |\varphi(z)|)/(1 - |z|)$ where z approaches a along an oricycle and by Theorem 2.44 the same equality holds for hyperbolic automorphisms where z approaches a nontangentially. Thus,

$$\rho(C_\varphi) = \lim_{n \to \infty} \|C_{\varphi_n}\|^{1/n} \leq \lim_{n \to \infty} M^{1/n} \left(\frac{1 + |\varphi_n(0)|}{1 - |\varphi_n(0)|}\right)^{\gamma/(2n)}$$

$$= \lim_{n \to \infty} (1 - |\varphi_n(0)|)^{-\gamma/(2n)}$$

$$= \lim_{n\to\infty} \left(\prod_{k=0}^{n-1} \frac{1 - |\varphi_k(0)|}{1 - |\varphi_{k+1}(0)|} \right)^{\gamma/(2n)}$$

$$= \lim_{n\to\infty} \left(\frac{1 - |\varphi_n(0)|}{1 - |\varphi_{n+1}(0)|} \right)^{\gamma/2} = \varphi'(a)^{-\gamma/2}$$

where in the last line we use the fact that the iterates $\varphi_n(0)$ approach a nontangentially when φ is a hyperbolic automorphism, and approach a along an oricycle when φ is a parabolic automorphism. ∎

Before finding the spectrum for a hyperbolic automorphism, we need a calculation that determines for what real values of s the function

$$\left(\frac{1+z}{1-z} \right)^s$$

belongs to the spaces $H^2(\beta)$ we are considering.

LEMMA 7.3
For $\gamma \geq 1$ and $\beta(n)^2 = (n+1)^{1-\gamma}$, the function

$$\left(\frac{1+z}{1-z} \right)^s$$

belongs to $H^2(\beta)$ if and only if $|s| < \gamma/2$.

PROOF   Since $(1+z)^s$ and $(1-z)^s$ are bounded whenever s is positive, and therefore multipliers of all the spaces in question, it suffices to determine for which positive s the functions

$$\left(\frac{1}{1-z} \right)^s$$

and

$$\left(\frac{1}{1+z} \right)^s$$

are in $H^2(\beta)$. Since

$$\frac{1}{(1-z)^s} = \sum_{n=0}^{\infty} \frac{\Gamma(n+s)}{\Gamma(n+1)\Gamma(s)} z^n$$

this lies in the desired space when

$$\sum_{n=0}^{\infty} \frac{\Gamma(n+s)^2}{\Gamma(n+1)^2} (n+1)^{1-\gamma} < \infty$$

For n large, Stirling's formula

$$\lim_{n\to\infty} \frac{\sqrt{2n\pi}(n/e)^n}{n!} = 1$$

shows that the terms in this sum are of order $n^{2s-\gamma-1}$ so the sum converges, for positive s, if and only if $s < \gamma/2$. The same calculation works for the function

$$\frac{1}{(1+z)^s}$$

which gives the desired result.   ∎

The key observation in the proof of the next theorem is that the functions of Lemma 7.3 are eigenfunctions for C_φ for the automorphisms $\varphi = (z+r)/(1+rz)$, where $0 < r < 1$, which are hyperbolic and have Denjoy–Wolff point 1.

THEOREM 7.4
If φ is a hyperbolic disk automorphism with Denjoy–Wolff point a and $\beta(n)^2 = (n+1)^{1-\gamma}$ for $\gamma \geq 1$, then each point of the annulus

$$\varphi'(a)^{\gamma/2} < |\lambda| < \varphi'(a)^{-\gamma/2}$$

is an eigenvalue of C_φ on $H^2(\beta)$ of infinite multiplicity. Moreover, the spectrum and the essential spectrum of C_φ are the annulus

$$\varphi'(a)^{\gamma/2} \leq |\lambda| \leq \varphi'(a)^{-\gamma/2}$$

PROOF   It is enough to prove the theorem for the automorphism $\varphi(z) = (z+r)/(1+rz)$ which has Denjoy–Wolff point 1 and $\varphi'(1)^{-1} = (1+r)/(1-r)$ for $0 < r < 1$. Fix t real and s in $(-\gamma/2, \gamma/2)$. Let

$$f(z) = \exp\left((s+it)\log\frac{1+z}{1-z} \right)$$

where $\log$ is the principal branch of the natural logarithm function. That is, f is a branch of the function

$$\left(\frac{1+z}{1-z}\right)^s \left(\frac{1+z}{1-z}\right)^{it}$$

Lemma 7.3 shows that $((1+z)/(1-z))^s$ is in $H^2(\beta)$. This and the fact that $((1+z)/(1-z))^{it}$ is bounded and hence is a multiplier of the weighted Hardy spaces under consideration shows that f is in $H^2(\beta)$. Moreover

$$f \circ \varphi(z) = \left(\frac{1+r}{1-r}\right)^{s+it} f(z)$$

As s varies over $(-\gamma/2, \gamma/2)$ and t varies over the real line, this gives each point of the open annulus as an eigenvalue infinitely often.

Lemma 7.2 showed that the spectrum of C_φ is contained in the disk $\{\lambda :$ $|\lambda| \leq \varphi'(1)^{-\gamma/2}\}$. Moreover, applying Lemma 7.2 to $C_\varphi{}^{-1} = C_{\varphi^{-1}}$, we see the spectrum of $C_\varphi{}^{-1}$ is also contained in $\{\lambda : |\lambda| \leq \varphi'(1)^{-\gamma/2}\}$ because φ^{-1} has Denjoy–Wolff point -1 and $(\varphi^{-1})'(-1) = 1/\varphi'(-1) = \varphi'(1)$. This means that the spectrum of C_φ is contained in the set $\{\lambda : |\lambda| \geq \varphi'(1)^{\gamma/2}\}$. Since the spectrum is closed and the set of eigenvalues is dense in this annulus, the closed annulus must be the spectrum.

Finally, since the essential spectrum is a closed set contained in the spectrum that contains all the eigenvalues of infinite multiplicity, the essential spectrum of C_φ must be the spectrum in this case. ∎

We will see in Exercise 7.2.3 that if φ is a hyperbolic automorphism of the unit disk, then C_φ is similar to $e^{i\theta}C_\varphi$ for all real θ, when C_φ acts on a weighted Hardy space $H^2(\beta)$ for which the bounded analytic functions are all multipliers. Under these assumptions the spectrum of C_φ will be circularly symmetric: λ in $\sigma(C_\varphi)$ implies $e^{i\theta}\lambda$ is in $\sigma(C_\varphi)$ for all real θ. This behavior is typical of weighted shifts; several authors have established strong connections between these composition operators and weighted shifts. In particular, the weight sequences for these shifts explain the radii of the annulus that is the spectrum.

We next turn our attention to those automorphisms that fix a single point of the boundary of D, and we prove an analogous theorem in this case.

THEOREM 7.5

If φ is a parabolic automorphism of D, then each point of the unit circle is an eigenvalue of infinite multiplicity for C_φ acting on $H^2(\beta)$ with $\beta(n)^2 = (n+1)^{1-\gamma}$ and $\gamma \geq 1$. Moreover, in this case, the spectrum and essential spectrum of C_φ are the unit circle.

PROOF   Note first that if a is the Denjoy–Wolff point of φ, then $\varphi'(a) = 1$ so by Lemma 7.2 that the spectrum of C_φ is contained in the disk $\{\lambda : |\lambda| \leq 1\}$. Since $C_\varphi{}^{-1} = C_{\varphi^{-1}}$ also has spectrum contained in the closed unit disk, the spectrum of C_φ is contained in the unit circle.

There is no loss of generality in assuming that either $\varphi(z) = ((1+i)z-1)/(z+i-1)$ or $\varphi(z) = ((1-i)z-1)/(z-i-1)$ as C_φ is similar to a composition operator with such a symbol. In either case let

$$f(z) = \exp(s(z+1)/(z-1))$$

and note that f is a bounded analytic function for s non-negative, and thus f is in $H^2(\beta)$ for β as in the hypothesis. Moreover, in the first case, $f(\varphi(z)) = e^{-2is}f(z)$ so f is an eigenvector with eigenvalue e^{-2is} for $s \geq 0$ and in the other case, $f(\varphi(z)) = e^{2is}f(z)$ so f is an eigenvector with eigenvalue e^{2is} for $s \geq 0$. Since every point of the unit circle has infinitely many representations of each of these forms, in each case, every point of the unit circle is an eigenvalue of infinite multiplicity. This shows the spectrum and the essential spectrum of C_φ are each equal to the unit circle. ∎

In fact, the proof shows that when φ is a parabolic automorphism and C_φ is acting on any weighted Hardy space $H^2(\beta)$ which contains the bounded analytic functions, every point of the unit circle is an eigenvalue of C_φ of infinite multiplicity.

Exercises

7.2.1 Justify the assertions made in the table of Section 7.1 about composition operators whose symbols are automorphisms of the disk.

7.2.2 Suppose φ is a parabolic automorphism with Denjoy–Wolff point a. Show that
$$\lim_{z \to a} \frac{1 - |\varphi(z)|}{1 - |z|} = \varphi'(a)$$
if z approaches a along the oricycle $|a - z|^2 = k(1 - |z|^2)$.

7.2.3 Suppose $0 < s < 1$ and $\varphi(z) = ((1 + s)z + (1 - s))/((1 - s)z + (1 + s))$, the automorphism of the disk with $\varphi(\pm 1) = \pm 1$ and $\varphi'(1) = s$.

(a) Given λ with $|\lambda| = 1$, find infinitely many functions f analytic on the unit disk so that both f and $1/f$ are bounded and $f(\varphi(z)) = \lambda f(z)$.

(b) Suppose that $H^2(\beta)$ is a weighted Hardy space for which the bounded analytic functions are all multipliers (such as $\beta(n)^2 = (n + 1)^{1-\gamma}$ for $\gamma \geq 1$ as in Theorem 7.4). Show that $M_f^{-1} C_\varphi M_f = \lambda C_\varphi$ on $H^2(\beta)$, that is, C_φ is similar to all rotates of itself.

7.2.4 Suppose $0 < s < 1$ and $\varphi(z) = ((1 + s)z + (1 - s))/((1 - s)z + (1 + s))$ and let $z_n = \varphi_n(0)$ for all integers n.

(a) Show that the set $\{z_n\}_{-\infty}^\infty$ is a Blaschke sequence in the disk. Let B be the Blaschke product with this zero set and $B'(0) > 0$.

(b) Let $f_n = (1 - z_n^2)^{1/2}/(1 - z_n z)$, that is, f_n is a multiple of the reproducing kernel on H^2 for z_n so that $\|f_n\| = 1$. Show that the closed span of $\{f_n\}_{-\infty}^\infty$ is $(BH^2)^\perp$.

(c) Let $\{e_n\}_{-\infty}^\infty$ be the standard orthonormal basis for $\ell^2(\mathbf{Z})$ and let $A : \ell^2 \mapsto (BH^2)^\perp$ be given by $A(e_n) = f_n$. Show that A is bounded and has bounded inverse.

(d) Using the above show that the restriction of C_φ^* to $(BH^2)^\perp$ is similar to a weighted shift. Find the spectrum of this weighted shift.

Notes

E. Nordgren [No68] was the first to consider spectra of invertible composition operators and the proofs presented here are extensions of his proofs which were given for $H^2(D)$. Exercises 7.2.1 and 7.2.3 are special cases of Theorems 4.3 and 4.5 of [Co83]. Exercise 7.2.4 is contained in P. Hurst's paper [Hur94] but is closely related to the shift analogies of [Co83] and especially to the model for C_φ due to E. Nordgren, P. Rosenthal, and F. Wintrobe [NoRW87].

7.3   Invertible operators on the classical spaces on the ball

In this section we consider the spectra of invertible composition operators in the ball B_N for $N > 1$. We give a complete description for the spaces $H^2(\beta) = H^2(\beta, B_N)$ when $\beta(n) = (n+1)^{1-\gamma}$ with $\gamma \geq 1$, that is, on the Hardy space $H^2(B_N)$ and the standard weight Bergman spaces, and partial results in some other spaces. Based on the work in the last section, it is reasonable to expect these descriptions of the spectra to depend on the fixed point character of the symbol φ in $\text{Aut}(B_N)$. Since φ is actually analytic in a neighborhood of $\overline{B_N}$, the Brouwer fixed point theorem guarantees that φ will have at least one fixed point in $\overline{B_N}$. If φ has no fixed points in the open ball we know by Exercise 2.6.3 that φ has at most two fixed points in ∂B_N. The Heisenberg translations of B_N and the non-isotropic dilations of B_N (see Equation (2.6.3) and Exercise 2.6.5) provide examples where the fixed point set is exactly 1 or exactly 2 boundary fixed points respectively. For the purposes of our description of the spectra we distinguish three cases, just as in the case $N = 1$:

- φ fixes two points of ∂B_N and no points in B_N.
- φ fixes one point of ∂B_N and no points in B_N.
- φ fixes at least one point in B_N.

Roughly speaking our plan for these three cases is as follows: the case of two fixed points on the sphere will be handled by an argument which allows us to reduce to the analogous situation in the disk, while the case of an interior fixed point will be a straightforward adaptation of the argument from Theorem 7.1 in the last section. The remaining case, while in spirit reminiscent of the one boundary fixed point case in the disk, will require some of the ideas which were developed in Section 2.6 to prove the Denjoy–Wolff theorem for these automorphisms.

We begin with the case that φ has interior fixed point(s). As was the situation with this case in the disk, the proof depends on the automorphism invariance of the space and the invariance of the norm under unitary maps of the ball.

THEOREM 7.6

Suppose φ is an automorphism of B_N fixing at least one point of B_N and suppose that $H^2(\beta)$ is an automorphism invariant weighted Hardy space on B_N. Then the spectrum of C_φ on $H^2(\beta)$ is the closure of all possible products of the eigenvalues of $\varphi'(z_0)$, where z_0 is any interior fixed point. This closure will either be the unit circle, or a finite subgroup of the circle.

PROOF   If φ fixes z_0 there is τ in $\mathrm{Aut}(B_N)$ so that

$$U \equiv \tau \circ \varphi \circ \tau^{-1}(z_1, \ldots, z_N) = (e^{i\theta_1} z_1, \ldots, e^{i\theta_N} z_N)$$

(τ is the composition of a unitary map with an automorphism that sends z_0 to 0). Note that the eigenvalues of $\varphi'(z_0)$ and the eigenvalues of U are the same, namely $\{e^{i\theta_1}, \ldots, e^{i\theta_N}\}$. Since $H^2(\beta)$ is automorphism invariant, the spectrum of C_φ and C_U are the same.

If

$$\lambda = \prod (e^{i\theta_j})^{n_j}$$

where $n_j \geq 0$ are integers, then setting

$$f(z_1, \ldots, z_N) = \prod z_j^{n_j}$$

gives $f \circ U = \lambda f$ and f is in $H^2(\beta)$ regardless of what β is. Thus λ is an eigenvalue of C_φ and $\sigma(C_\varphi)$ contains the closure of the set of all products of eigenvalues of $\varphi'(z_0)$. Denote this closure by P.

Since $\|C_U\| = \|C_U^{-1}\| = 1$, the spectrum of C_φ is contained in the unit circle. If $P = \partial D$ we are done. If not, then P is a finite subset of the unit circle that is closed under multiplication, so it must be a subgroup of the circle, and P contains all M^{th} roots of 1 for some integer M. In particular, $e^{i\theta_j M} = 1$ for $1 \leq j \leq N$. Thus U_M is the identity where U_M is the Mth iterate, and C_U^M is the identity. If λ is any point in $\sigma(C_U)$, $\lambda^M = 1$ and λ must therefore be in P.   ∎

The result for a normalized version of the two boundary fixed point case gives the spectrum as an annulus.

THEOREM 7.7
Suppose φ in $\mathrm{Aut}(B_N)$ fixes the points e_1 and $-e_1$ only, with Denjoy–Wolff point e_1. Then the spectrum and essential spectrum of C_φ on $H^2(\beta)$ in B_N where $\beta(n)^2 = (n+1)^{1-\gamma}, \gamma \geq 1$ is the closed annulus centered at 0 with radii $K^{(N+\gamma-1)/2}$ and $K^{-(N+\gamma-1)/2}$ where $K = D_1\varphi_1(-e_1)$ and points in the interior of the annulus are eigenvalues of C_φ of infinite multiplicity.

The proof of this theorem will require a lemma about automorphisms that map a slice in the ball into itself.

LEMMA 7.8
If $\varphi = (\varphi_1, \varphi_2, \ldots, \varphi_N)$ in $\mathrm{Aut}(B_N)$ maps $[e_1] \cap B_N = \{(\lambda, 0') : |\lambda| < 1\}$ into itself, then φ_1 depends on z_1 only.

PROOF   We need to show that if (z_1, z') and (z_1, w') are two points of B_N with the same first coordinate, then $\varphi_1(z_1, z') = \varphi_1(z_1, w')$. Let $\widetilde{\varphi} : D \to D$ be given by $\widetilde{\varphi}(\lambda) = \varphi_1(\lambda, 0')$. Note that $\widetilde{\varphi}$ is an automorphism of D (we are using Theorem 2.74). If $\widetilde{\varphi}^{-1} = e^{i\theta}(u - z)/(1 - \overline{u}z)$ we can extend $\widetilde{\varphi}^{-1}$ to an automorphism ψ of B_N by defining

$$\psi(z_1, z') = \left(e^{i\theta} \frac{u - z_1}{1 - \overline{u}z_1}, \frac{-sz'}{1 - \overline{u}z_1} \right)$$

where $s = (1 - |u|^2)^{1/2}$. Notice that the first coordinate function of ψ depends only on z_1 and that $\psi\varphi$ is the identity on $\{(\lambda, 0') : |\lambda| < 1\}$ in B_N. Thus $\psi\varphi$ is a unitary map whose matrix (with respect to the standard basis) is

$$\begin{pmatrix} 1 & 0 & \cdots & 0 \\ 0 & & & \\ \vdots & & V & \\ 0 & & & \end{pmatrix}$$

where V is unitary on C^{N-1}. Now consider $\varphi = \psi^{-1}(\psi\varphi)$ applied to our points (z_1, z') and (z_1, w'). The images of these points under $\psi\varphi$ have the same first coordinate. But ψ^{-1} takes points with the same first coordinate to points with the same first coordinate. This shows $\varphi_1(z_1, z') = \varphi_1(z_1, w')$ as desired. ∎

A generalization of this result that will be needed later is given in Exercise 7.3.2.

PROOF (of Theorem 7.7)   We first show that every point in the interior of this annulus is an eigenvalue of infinite multiplicity for C_φ. Note that φ restricted to the complex line $[e_1] \cap B_N$ is an automorphism of the disk fixing 1 and -1, since φ must map this affine set to an affine set in B_N containing e_1 and $-e_1$. By Lemma 7.8 we have

$$\varphi_1(z_1, z') = \frac{z_1 + r}{1 + rz_1}$$

for some r with $0 < r < 1$. Denote by $\widetilde{\varphi}_1$ the disk automorphism that maps w to $(w + r)/(1 + rw)$ and set

$$\widetilde{\beta}(k)^2 = \frac{(N-1)!}{(N-1+k)\cdots(k+1)}\beta(k)^2 = \frac{(N-1)!}{(N-1+k)\cdots(k+1)}(k+1)^{1-\gamma}$$

Recall that by Proposition 2.21 the extension operator that maps $H^2(\widetilde{\beta}, D)$ into $H^2(\beta, B_N)$ defined by $Ef(z_1, z') = f(z_1)$ is an isometry. Moreover if we set

$$\widehat{\beta}(k)^2 = \frac{(N-1)!}{(k+1)^{N-1}}(k+1)^{1-\gamma} = (N-1)!(k+1)^{2-\gamma-N}$$

then $H^2(\widehat{\beta}, D) \subset H^2(\widetilde{\beta}, D)$ since $\widehat{\beta}(k)^2 \geq \widetilde{\beta}(k)^2$. By Theorem 7.4 we know that the point spectrum of $C_{\widetilde{\varphi}_1}$ on $H^2(\widehat{\beta}, D)$ includes the annulus

$$\{\lambda : K^{-(N+\gamma-1)/2} < |\lambda| < K^{(N+\gamma-1)/2}\}$$

where $K = (1+r)/(1-r) = D_1\varphi_1(-e_1)$. Thus if λ is in this open annulus we may find infinitely many linearly independent functions f in $H^2(\hat\beta, D) \subset H^2(\tilde\beta, D)$ so that $f \circ \tilde\varphi_1 = \lambda f$. Since f is in $H^2(\tilde\beta, D)$, Ef is in $H^2(\beta, B_N)$ and

$$Ef \circ \varphi(z_1, z') = Ef(\varphi_1(z_1, z'), \dots, \varphi_N(z_1, z')) = f \circ \varphi_1(z_1, z')$$
$$= f\left(\frac{z_1 + r}{1 + rz_1}\right) = f \circ \tilde\varphi_1(z_1) = \lambda f(z_1) = \lambda Ef(z_1, z')$$

This verifies that λ is an eigenvalue of infinite multiplicity for C_φ on $H^2(\beta)$.

For the remainder of the proof we first compute the spectral radius of C_φ on $H^2(\beta)$ exactly as in Lemma 7.2:

$$\rho(C_\varphi) = \lim_{n\to\infty} \|C_{\varphi_n}\|^{1/n}$$

From Exercise 3.5.4 it follows that on $H^2(B_N)$, which is $H^2(\beta)$ with $\beta(k) \equiv 1$, we have

$$\|C_\varphi\| \leq M \left(\frac{1 + |\varphi(0)|}{1 - |\varphi(0)|}\right)^{N/2}$$

for some constant M. On the other weighted Hardy spaces under consideration ($H^2(\beta)$ with $\beta(k) = (k+1)^{1-\gamma}$ for $\gamma > 1$) it suffices, for the purposes of the spectral radius calculation, to estimate $\|C_\varphi\|$ on the equivalent weighted Bergman space $\{f : \|f\|^2 = \int_{B_N} |f|^2 (1 - |z|^2)^{\gamma-2} \, d\nu_N < \infty\}$. Here the calculations are just like in Lemma 7.2, using the fundamental identity (Equation (2.5.3)) and the fact that, for φ an automorphism,

$$|J_\varphi(z)|^2 = \left[\frac{1 - |\varphi^{-1}(0)|^2}{|1 - \langle z, \varphi^{-1}(0)\rangle|^2}\right]^{N+1}$$

(the details of this computation can be found in [Ru80, p. 28]) to obtain that $\|C_\varphi\|$ on the weighted Bergman space is not greater than some constant multiple of

$$\left(\frac{1 + |\varphi(0)|}{1 - |\varphi(0)|}\right)^{(\gamma+N-1)/2}$$

Thus for all $\gamma \geq 1$

$$\rho(C_\varphi) \leq \liminf_{n\to\infty} \left(\frac{1 + |\varphi_n(0)|}{1 - |\varphi_n(0)|}\right)^{(\gamma+N-1)/(2n)}$$
$$= \liminf_{n\to\infty} (1 - |\varphi_n(0)|)^{-(\gamma+N-1)/(2n)}$$
$$\leq \limsup_{n\to\infty} \left(\frac{1 - |\varphi_n(0)|}{1 - |\varphi_{n+1}(0)|}\right)^{(\gamma+N-1)/2}$$

Since $\varphi_n(0) \to e_1$ by the Denjoy–Wolff theorem,

$$\limsup_{n\to\infty} \frac{1 - |\varphi_n(0)|}{1 - |\varphi_{n+1}(0)|} \leq \frac{1}{D_1\varphi_1(e_1)} = \frac{1+r}{1-r} = D_1\varphi_1(-e_1)$$

and thus $\rho(C_\varphi) \leq K^{(\gamma+N-1)/2}$. A completely analogous calculation shows that $\rho(C_\varphi{}^{-1}) = \rho(C_{\varphi^{-1}}) \leq K^{(\gamma+N-1)/2}$. This shows that the spectrum is contained in the annulus in the statement of the theorem. Since the essential spectrum is contained in the spectrum and contains all the eigenvalues of infinite multiplicity, this completes the proof. ∎

For the case of automorphisms with exactly one fixed point in the boundary of the ball there is no obvious mechanism for reducing to a one variable situation. In the disk, the automorphisms that fix 1 only correspond to translations in the upper halfplane; this is exactly what makes the proof of Theorem 7.5 work. Recall from Section 2.5 that the ball B_N for $N > 1$ is biholomorphic to the Siegel upper half space $\Omega = \{(w_1, \ldots, w_N) = (w_1, w') : \operatorname{Im} w_1 > |w'|^2\}$ via the Cayley transform

$$\Phi(z) = i\frac{e_1 + z}{1 - z_1}$$

and for each $b = (b_1, b')$ in $\partial\Omega$, we have a Heisenberg translation h_b in $\operatorname{Aut}(\Omega)$ defined by $h_b(w_1, w') = (w_1 + b_1 + 2i\langle w', b'\rangle, w' + b')$ which fixes ∞ only; the corresponding map $\Phi^{-1} \circ h_b \circ \Phi$ is an automorphism of B_N fixing e_1 only, and is called a Heisenberg translation of B_N. By Theorem 2.90 the automorphisms of B_N that fix e_1 only are either Heisenberg translations or they fix, *as a set*, some non-empty, proper, affine subset of B_N. This fact allows us to treat the non-Heisenberg translation automorphisms in an inductive fashion. We first dispense with those φ which are Heisenberg translations.

LEMMA 7.9

Suppose (b_1, b') is in $\partial\Omega$ with associated Heisenberg translation $h_{(b_1,b')}$ and let $H_{(b_1,b')} = \Phi^{-1} \circ h_{(b_1,b')} \circ \Phi$ be the corresponding automorphism of B_N. Then every λ in the unit circle is an eigenvalue of infinite multiplicity for C_φ on $H^\infty(B_N)$.

PROOF We first suppose that $b' \neq 0$ and claim that the composition operator with symbol $H_{(b_1,b')}$ is similar to the operator with symbol $H_{(t\operatorname{Re}b_1+i,r')}$ where $\sum_2^N |b_j|^2 = 1/t$ and $r' = (r_2, \ldots, r_N)$ with $r_j = \sqrt{t}|b_j|, j = 2, \ldots, N$. To see this it suffices to find an automorphism τ of Ω with $\tau \circ h_{(b_1,b')} \circ \tau^{-1} = h_{(t\operatorname{Re}b_1+i,r')}$. The desired map τ is constructed in two steps: Let U in $\operatorname{Aut}(\Omega)$ be defined by $U(w_1, w_2, \ldots, w_N) = (w_1, e^{-i\theta_2}w_2, \ldots, e^{-i\theta_N}w_N)$ where $e^{-i\theta_j}b_j = s_j \geq 0$ for $j = 2, \ldots, N$. Then with $1/t = \sum_2^N s_j^2 = \sum_2^N |b_j|^2$ let $\delta_{1/t}(w_1, w') = ((w_1)/t, w'/\sqrt{t})$. We have $\delta_{1/t}$ in $\operatorname{Aut}(\Omega)$ with inverse $\delta_t(w_1, w') = (tw_1, \sqrt{t}w')$. A calculation shows that $\tau = \delta_t \circ U$ has the desired property.

We are left to consider $(b_1, b') = (a+i, r')$ where a is real and $r' = (r_2, \ldots, r_N)$ with $r_j \geq 0, j = 2, \ldots, N$ and (necessarily) $\sum_2^N r_j^2 = 1$. For λ in the unit circle, write $\lambda = e^{i\gamma(a-c)}$ for some $c, \gamma > 0$. The function

$$f(w_1, w_2, \ldots, w_N) = \exp\{i\gamma(w_1 - i\sum_2^N w_j^2 - c\sum_2^N r_j w_j)\}$$

is in $H^\infty(\Omega)$. To see this note

$$|f(w)| = \exp\left(-\gamma \text{Im}\,(w_1 - i\sum_2^N w_j^2 - c\sum_2^N r_j w_j)\right)$$

Writing $w_j = x_j + iy_j$ for $j = 2, \ldots, N$ and using the fact that $\sum_2^N x_j^2 + y_j^2$ is less than $\text{Im}\,w_1$ a calculation gives

$$\text{Im}\,(w_1 - i\sum_2^N w_j^2 - c\sum_2^N r_j w_j) \geq -c^2/8$$

and hence $|f(w)| \leq e^{\gamma c^2/8}$. Moreover

$$f \circ h_{(a+i,r')}(w_1, w') = f(w_1 + a + i + 2i\langle w', r'\rangle, w' + r')$$

$$= \exp\left(i\gamma(w_1 - i\sum_2^N w_j^2 - c\sum_2^N r_j w_j + a - c)\right)$$

$$= e^{i\gamma(a-c)}f(w_1, w') = \lambda f(w_1, w')$$

where we have used $\sum_2^N r_j^2 = 1$. This proves the theorem except for the special case $b' = 0, b_1$ real which we leave to the reader as Exercise 7.3.3.   ∎

THEOREM 7.10
Suppose φ is an automorphism of B_N fixing e_1 only. Then every λ in the unit circle is an eigenvalue of infinite multiplicity for C_φ on any automorphism invariant weighted Hardy space $H^2(\beta)$ containing $H^\infty(B_N)$.

PROOF   The proof is by induction on the dimension N, with the $N = 1$ case following from the proof of Theorem 7.5. We assume then that $N > 1$ and that the result holds for all $k < N$. Let φ be as in the statement of the theorem. If φ is a Heisenberg translation of B_N then the previous lemma gives the desired conclusion. If φ is not a Heisenberg translation, then Theorem 2.90 shows that φ fixes as a set some non-empty proper affine subset $\mathcal{A}$ of B_N; we may find τ in $\text{Aut}(B_N)$ so that $\tau(e_1) = e_1$ and

$$\tau(\mathcal{A}) = \{(z_1, \ldots, z_N) \in B_N : z_i = 0 \text{ for } i > r\}$$

where $1 \leq r < N$. If ρ is the restriction of $\tau \circ \varphi \circ \tau^{-1}$ to B_r, then ρ is an automorphism of B_r fixing e_1' of B_r only. By induction we may find (for any λ

in the unit circle) infinitely many linearly independent functions f in $H^\infty(B_r)$ with $f \circ \rho = \lambda f$. Each such f extends to a bounded analytic function F on B_N by $F(z', z'') = f(z')$ where $z' = (z_1, \ldots, z_r)$ and $z'' = (z_{r+1}, \ldots, z_N)$. Since Exercise 7.3.2 shows that the first r coordinate functions of $\tau \circ \varphi \circ \tau^{-1}$ depend only on $z_1, \ldots, z_r$ we have

$$F \circ \tau \varphi \tau^{-1}(z', z'') = f((\tau \circ \varphi \circ \tau^{-1})_1(z', z''), \ldots, (\tau \circ \varphi \circ \tau^{-1})_r(z', z''))$$
$$= f((\tau \circ \varphi \circ \tau^{-1})_1(z', 0''), \ldots, (\tau \circ \varphi \circ \tau^{-1})_r(z', 0''))$$
$$= f \circ \rho(z') = \lambda f(z') = \lambda F(z', z'')$$

∎

To complete our analysis of $\sigma(C_\varphi)$ when φ fixes only a single point of the boundary of B_N we need a spectral radius computation.

PROPOSITION 7.11

Suppose φ is an automorphism of B_N fixing e_1 only. Then the spectral radius of C_φ on $H^2(\beta), \beta(k) = (k+1)^{1-\gamma}, \gamma \geq 1$ is 1.

PROOF Set

$$\liminf_{n \to \infty} \frac{1 - |\varphi_{n+1}(0)|}{1 - |\varphi_n(0)|} = L$$

We claim that $L \geq 1$. To see this recall that if φ is an automorphism fixing e_1 only then

(i) $\varphi_n(0) \to e_1$

(ii) For all z in B_N

$$\frac{|1 - \varphi^1(z)|^2}{1 - |\varphi(z)|^2} = \frac{|1 - z_1|^2}{1 - |z|^2}$$

where φ^1 denotes the first coordinate function of φ.

where (i) follows from Theorem 2.93 and (ii) from Lemma 2.92. Since by Julia's Lemma

$$\frac{|1 - \varphi^1(z)|^2}{1 - |\varphi(z)|^2} \leq L \frac{|1 - z_1|^2}{1 - |z|^2}$$

(ii) shows that $L \geq 1$.

Our standard spectral radius computation, exactly as in the proof of Theorem 7.7, using

$$\|C_{\varphi_n}\| \leq M \left(\frac{1 + |\varphi_n(0)|}{1 - |\varphi_n(0)|} \right)^{(\gamma + N - 1)/2}$$

where φ is an automorphism shows that

$$\rho(C_\varphi) \leq \limsup_{n\to\infty} \left(\frac{1 - |\varphi_n(0)|}{1 - |\varphi_{n+1}(0)|} \right)^{(\gamma+N-1)/2} \leq 1$$

Since 1 is an eigenvalue of C_φ, $\rho(C_\varphi) = 1$ (notice this implies $L = 1$) and we are done. ∎

THEOREM 7.12
If φ is an automorphism of B_N fixing ζ in ∂B_N only, then on $H^2(\beta)$, for $\beta(k) = (k+1)^{1-\gamma}$ with $\gamma \geq 1$, the spectrum of C_φ is the unit circle.

PROOF We may in the usual way normalize so that $\zeta = e_1$ since the spaces in question are invariant under unitary transformations of the ball. Theorem 7.10 shows that the spectrum contains the unit circle. Since $\rho(C_\varphi) = 1$ by the previous proposition, the spectrum is contained in the closed unit disk. But $\rho(C_\varphi^{-1}) = 1$ also, since φ^{-1} is an automorphism fixing e_1 only. The spectral mapping theorem now completes the proof. ∎

Exercises

7.3.1 Use Exercise 2.6.5 to show that if φ in $\mathrm{Aut}(B_N)$ fixes e_1 and $-e_1$ only, with Denjoy–Wolff point e_1, then

$$\varphi_1(z) = \frac{z_1 + r}{1 + rz_1}$$

for some r with $0 < r < 1$. This gives an alternate argument for the first part of the proof of Theorem 7.7.

7.3.2 Show that if φ in $\mathrm{Aut}(B_N)$ maps $B_r = \{(z_1, \ldots, z_N) \subset B_N : z_i = 0 \text{ for } i > r\}$ into itself, then the first r coordinate functions for φ depend only on the variables $z_1, \ldots, z_r$.

7.3.3 Complete the proof of Lemma 7.9 for the case $h(w_1, w') = (w_1 + b_1, w')$ where b_1 is real and non-zero.

7.3.4 What is the spectrum of C_φ on $H^p(B_N)$ for $p \geq 1$, $p \neq 2$ in the case that φ is an automorphism fixing two points of ∂B_N?

Notes

The spectrum of C_φ acting on $H^p(B_N)$, $N > 1$ and $p \geq 1$, when φ is an automorphism of B_N was identified by B. D. MacCluer in [Mc84b]. The results given here are minor modifications of the arguments in that paper.

7.4 Spectra of compact composition operators

The general principle that the structure of the composition operator C_φ is closely related to the fixed point behavior of the function φ is well illustrated by compact composition operators. Determining the spectrum of a compact operator is equivalent to finding the eigenvalues of the operator. We will prove here a theorem of considerable generality, which will show that in essentially all of the spaces of interest to us these eigenvalues are determined by the derivative of φ at the Denjoy–Wolff fixed point of φ. This statement will need some clarification in the case of spaces defined on B_N, $N > 1$ as then "the" Denjoy–Wolff point may not be well-defined.

Our first result generalizes Theorem 2.17 on the behavior of the normalized kernel functions in weighted Hardy spaces in the disk.

PROPOSITION 7.13

Suppose $H^2(\beta)$ is a weighted Hardy space in the disk. If

$$\sum_{n=0}^{\infty} \frac{n^{2k}}{\beta(n)^2} = \infty$$

for some non-negative integer k, then letting $K_w^{(k)}$ denote the kernel function for evaluation of the kth derivative at w, we have

$$\frac{K_w^{(k)}}{\|K_w^{(k)}\|} \to 0$$

weakly as $|w| \to 1$.

PROOF The $k = 0$ version of this proposition was proved in Theorem 2.17, so we may consider $k \geq 1$. By Theorem 2.16

$$K_w^{(k)}(z) = \sum_{n=k}^{\infty} \frac{n!}{(n-k)!} \frac{\overline{w}^{n-k} z^n}{\beta(n)^2}$$

so

$$\|K_w^{(k)}\|^2 = \sum_{n=k}^{\infty} \left| \frac{n!}{(n-k)!} \frac{\overline{w}^{n-k}}{\beta(n)^2} \right|^2 \beta(n)^2$$

$$= \frac{1}{|w|^{2k}} \sum_{n=k}^{\infty} \left(\frac{n!}{(n-k)!} \right)^2 \frac{|w|^{2n}}{\beta(n)^2}$$

Since $n^k/2^k \leq n!/(n-k)! \leq n^k$ for $n \geq 2k$, the last sum tends to ∞ as $|w| \to 1$

precisely when

$$\sum_{n=0}^{\infty} \frac{n^{2k}}{\beta(n)^2} = \infty$$

which is our hypothesis.

Now consider any weak limit point f of

$$\frac{K_w^{(k)}}{\|K_w^{(k)}\|}$$

for a sequence of w's tending to the boundary of the disk. If p is any polynomial we have

$$|\langle p, f \rangle| = \lim_{w \to \partial D} |\langle p, \frac{K_w^{(k)}}{\|K_w^{(k)}\|} \rangle| = \lim_{w \to \partial D} \frac{|p^{(k)}(w)|}{\|K_w^{(k)}\|} = 0$$

Since the polynomials are dense in $H^2(\beta)$ we must have $f = 0$. ∎

The same expression for $\|K_w^{(j)}\|$ used in the proof of this result shows that if

$$\sum_{n=0}^{\infty} \frac{n^{2j}}{\beta(n)^2} < \infty$$

for some j, then $\|K_w^{(j)}\|$ is *bounded* as w approaches the unit circle, a fact that we will need shortly.

The next result will be key to describing the spectrum of C_φ when C_φ is compact.

THEOREM 7.14

Let $H^2(\beta)$ be a weighted Hardy space on the disk. Suppose

$$\sum_{n=0}^{\infty} \frac{n^{2k}}{\beta(n)^2} = \infty$$

for some non-negative integer k. If C_φ is compact on $H^2(\beta)$, then φ has a unique fixed point in the open unit disk.

PROOF   Let k be the least non-negative integer for which the sum in the statement of the lemma diverges. If $k = 0$ the argument is familiar: Assume φ has no fixed point in D and let ζ denote its Denjoy–Wolff point in ∂D, so that the angular derivative at ζ is not greater than 1. By Julia's Lemma (Theorem 2.41), φ maps the disks

$$\{z \in D : |z - \zeta|^2 \le c(1 - |z|^2)\}$$

into themselves, for each $c > 0$. In particular, $|\varphi(r\zeta)| \geq r$ for all r in $(0, 1)$. We have, by Proposition 3.13

$$\|C_\varphi\|_e \geq \limsup_{r \to 1} \frac{\|K_{\varphi(r\zeta)}\|}{\|K_r\|} \geq 1$$

so C_φ cannot be compact, which is our desired contradiction. Uniqueness of the interior fixed point is obvious, since φ is not the identity.

For the case $k \geq 1$ we must work a little harder. We begin by identifying $C_\varphi^*(K_w^{(k)})$. To do this notice that for f in $H^2(\beta)$

$$\langle f, C_\varphi^* K_w^{(k)} \rangle = f'(\varphi(w))\varphi'(w) \quad \text{when } k = 1$$

and for $k > 1$

$$\begin{aligned}
\langle f, C_\varphi^* K_w^{(k)} \rangle &= D^k(f \circ \varphi)(w) \\
&= f^{(k)}(\varphi(w))(\varphi'(w))^k + f'(\varphi(w))\varphi^{(k)}(w) + \text{ lower order terms}
\end{aligned}$$

where $f^{(k)}$ and $\varphi^{(k)}$ indicate the kth derivative and "lower order terms" means terms involving products of derivatives of f at $\varphi(w)$ of order less than k and derivatives of φ at w of order less than k. Thus if $k = 1$

$$C_\varphi^*(K_w^{(1)}) = \overline{\varphi'(w)} K_{\varphi(w)}^{(1)}$$

and if $k > 1$

$$C_\varphi^*(K_w^{(k)}) = \overline{[\varphi'(w)]}^k K_{\varphi(w)}^{(k)} + \overline{\varphi^{(k)}(w)} K_{\varphi(w)}^{(1)} + \text{ lower order terms}$$

where the "lower order terms" involve kernels for derivatives of order less than k at $\varphi(w)$ with coefficients involving products of derivatives of φ at w of order less than k. If φ has no fixed point in D and therefore has Denjoy–Wolff point at some ζ in ∂D consider

$$\left\| C_\varphi^* \frac{K_{r\zeta}^{(k)}}{\|K_{r\zeta}^{(k)}\|} \right\| = \begin{cases} \dfrac{\left\| \overline{\varphi'(r\zeta)} K_{\varphi(r\zeta)}^{(1)} \right\|}{\|K_{r\zeta}^{(1)}\|} & \text{if } k = 1 \\[4ex] \left\| \dfrac{\overline{\varphi'(r\zeta)}^k K_{\varphi(r\zeta)}^{(k)}}{\|K_{r\zeta}^{(k)}\|} + \dfrac{\overline{\varphi^{(k)}(r\zeta)} K_{\varphi(r\zeta)}^{(1)}}{\|K_{r\zeta}^{(k)}\|} + \cdots \right\| & \text{if } k > 1 \end{cases}$$

$$(7.4.1)$$

When $k > 1$

$$\frac{\varphi^{(k)}(r\zeta)}{\|K_{r\zeta}^{(k)}\|} = \langle \varphi, \frac{K_{r\zeta}^{(k)}}{\|K_{r\zeta}^{(k)}\|} \rangle \to 0$$

as r tends to 1 by Proposition 7.13, since φ is in $H^2(\beta)$. As $\|K_w^{(1)}\|$ is bounded as

w tends to ∂D, we have for the case $k > 1$

$$\left\| \frac{\overline{\varphi^{(k)}(r\zeta)} K^{(1)}_{\varphi(r\zeta)}}{\|K^{(k)}_{r\zeta}\|} \right\| \to 0$$

as r tends to 1. Moreover since for all $j < k$, $\|K^{(j)}_w\|$ is bounded as w tends to ∂D, as is $|\varphi^{(j)}(w)| = |\langle \varphi, K^{(j)}_w \rangle| \leq \|\varphi\| \|K^{(j)}_w\|$ while $\|K^{(k)}_w\| \to \infty$ as $|w| \to 1$, so that, in the case $k > 1$, the terms in Equation (7.4.1) that are not explicitly shown all tend to 0 as r tends to 1.

Putting it all together we see that if C^*_φ is compact and $k > 0$ is the least integer for which $\sum n^{2k}/\beta(n)^2 = \infty$ we must have

$$\frac{|\varphi'(r\zeta)|^k \|K^{(k)}_{\varphi(r\zeta)}\|}{\|K^{(k)}_{r\zeta}\|} \to 0$$

as $r \to 1$. Since ζ is the Denjoy–Wolff point of φ, we must have $\varphi'(r\zeta) \to \varphi'(\zeta) \neq 0$ and therefore

$$\frac{\|K^{(k)}_{\varphi(r\zeta)}\|}{\|K^{(k)}_{r\zeta}\|} \to 0$$

which is a contradiction since $|\varphi(r\zeta)| \geq r$. Thus φ must have fixed point in D; and the uniqueness of this fixed point is obvious. ∎

The proof shows that the important features of the weighted Hardy spaces that were used were that the norms of the kernel functions only depend on the modulus of the point and that the norms for the kernel functions for the appropriate order derivatives tend to infinity as one approaches the boundary. The more difficult part of the proof of this last result is only needed in the case that $\sum \beta(n)^{-2} < \infty$ and $H^2(\beta)$ is not automorphism invariant, since otherwise Theorem 4.5 would apply to show that C_φ compact implies $\|\varphi\|_\infty < 1$ and φ then clearly has a unique fixed point in D.

This theorem can be used to show that some weighted Hardy spaces contain only smooth functions.

COROLLARY 7.15
Let $\{\beta(n)\}$ be an increasing sequence satisfying

$$\sum_{n=0}^{\infty} \frac{\beta([\alpha n])}{\beta(n)} < \infty$$

for some α, $1/2 < \alpha < 1$, where $[x]$ denotes the greatest integer less than or equal to x. If $H^2(\beta)$ is the associated weighted Hardy space on the disk, then each function in $H^2(\beta)$ is in $C^\infty(\overline{D})$.

PROOF In Example 4.6, we showed that under the hypothesis above, the composition operator C_φ with symbol $\varphi(z) = (z+1)/2$ is compact on $H^2(\beta)$. Since $a = 1$ is the only fixed point of φ in the closed disk, Theorem 7.14 implies

$$\sum_{n=0}^{\infty} \frac{n^{2k}}{\beta(n)^2} < \infty$$

for every positive integer k. Now suppose $f(z) = \sum a_n z^n$ is a function in $H^2(\beta)$. Then by the Cauchy–Schwarz inequality

$$\left(\sum n^k |a_n|\right)^2 \le \left(\sum \frac{n^{2k}}{\beta(n)^2}\right)\left(\sum |a_n|^2 \beta(n)^2\right)$$

$$= \left(\sum \frac{n^{2k}}{\beta(n)^2}\right) \|f\|^2 < \infty$$

Since $|f^{(k)}(z)| \le \sum n^k |a_n|$ for all points of the closed unit disk, we see that f is in $C^\infty(\overline{D})$. ∎

The part of the proof of Theorem 7.14 that works with kernel functions for evaluation of derivatives does not generalize to several variables, chiefly because the expression for the adjoint of C_φ applied to kernel functions for derivatives of order at least one involve, as coefficients, derivatives of all the components of φ, and these may not all be well behaved as the unit sphere is approached. Nevertheless, the several variable generalization of Theorem 7.14 is still true, as we shall see next.

THEOREM 7.16
If C_φ is compact on a weighted Hardy space $H^2(\beta)$ in the ball B_N, $N > 1$ where

$$\sum_{n=0}^{\infty} \frac{n^{2k}}{\beta(n)^2} = \infty$$

for some non-negative integer k, then φ has exactly one fixed point in the open ball.

PROOF We suppose first that φ has no interior fixed point, so that φ must have Denjoy–Wolff point in ∂B_N (Theorem 2.83). We may assume without loss of generality that the Denjoy–Wolff point is $e_1 = (1, 0')$. We then know that each ellipsoid

$$E(c, e_1) = \{|1 - z_1|^2 < c(1 - |z|^2)\}$$

is mapped into itself by φ. Define an analytic map ψ of the disk into itself by $\psi(\lambda) = \varphi_1(\lambda e_1)$. Note that $\psi(1) = 1$ and ψ has angular derivative not greater than 1 at 1. The latter fact follows, for example, from considering $\varphi(z)$ for

$z = (r, 0')$ where $0 < r < 1$. For such $z, \varphi(z)$ is in the closure of the ellipsoid $E(c, e_1)$ with $c = (1 - r)/(1 + r)$. This says

$$\frac{|1 - \varphi_1(z)|^2}{1 - |\varphi_1(z)|^2} \leq \frac{|1 - \varphi_1(z)|^2}{1 - |\varphi(z)|^2} \leq \frac{1 - r}{1 + r} \text{ where } z = (r, 0')$$

so that

$$\frac{|1 - \psi(r)|^2}{1 - |\psi(r)|^2} \leq \frac{1 - r}{1 + r}$$

An elementary geometric argument will then show that $|\psi(r)| \geq r$ and therefore

$$\liminf_{z \to 1} \frac{1 - |\psi(z)|}{1 - |z|} \leq 1$$

The next step is to show that C_ψ is compact on the weighted Hardy space in the disk defined by the weight sequence

$$\widetilde{\beta}(s)^2 = \frac{(N - 1)!}{(N - 1 + s) \cdots (s + 1)} \beta(s)^2$$

We verify compactness by using the criteria of Exercise 3.2.1. Suppose that $\{f_n\}$ is a bounded sequence in $H^2(\widetilde{\beta}, D)$ and that f_n converges to 0 uniformly on compact subsets of D. Set $E f_n = F_n$ where E is the extension operator defined by $E f_n(z_1, z') = f_n(z_1)$ for (z_1, z') in B_N. Note that F_n tends to 0 uniformly on compact subsets of B_N and by Proposition 2.21, $\|E f_n\|_\beta = \|f_n\|_{\widetilde{\beta}}$ (where $\| \cdot \|_\beta$ and $\| \cdot \|_{\widetilde{\beta}}$ denote the norms in $H^2(\beta, B_N)$ and $H^2(\widetilde{\beta}, D)$ respectively) so that $\{F_n\}$ is a bounded sequence in $H^2(\beta, B_N)$. Since C_φ is compact we must have

$$\|F_n \circ \varphi\|_\beta \to 0$$

But

$$\|f_n \circ \psi\|_{\widetilde{\beta}} = \|R(F_n \circ \varphi)\|_{\widetilde{\beta}} \leq \|F_n \circ \varphi\|_\beta$$

where R is the restriction operator defined on analytic functions in B_N by $RG(\lambda) = G(\lambda e_1)$, again using Proposition 2.21. We conclude that $\|f_n \circ \psi\|_{\widetilde{\beta}}$ tends to 0 as n goes to infinity, which shows that C_ψ is compact on $H^2(\widetilde{\beta}, D)$.

Its easy to see that if

$$\sum_{n=0}^{\infty} \frac{n^{2k}}{\beta(n)^2} \text{ diverges for some non-negative integer } k$$

then also

$$\sum_{n=0}^{\infty} \frac{n^{2k'}}{\widetilde{\beta}(n)^2} \text{ diverges for some non-negative integer } k'$$

So Theorem 7.14 for compact operators on weighted Hardy spaces in the disk applies to ψ to give that ψ has a unique fixed point in D, contradicting the fact that $\psi(1) = 1$ with $\psi'(1) \leq 1$.

We conclude that our original map φ must have a fixed point in the open ball. To see that this fixed point is unique, recall that if φ has more than one fixed point in B_N then it must in fact fix every point of some affine subset A of B_N. For each a in A, $C_\varphi^*(K_a) = K_a$ so 1 is an eigenvalue of C_φ^* with infinite multiplicity, contradicting the hypothesis that C_φ is compact. ∎

Remark. The uniqueness part of the above argument applies in very great generality to arbitrary compact C_φ. Indeed, suppose C_φ is compact on some functional Banach space $\mathcal{Y}$ on B_N, where $\mathcal{Y}$ contains the polynomials. If φ has a non-trivial affine set of fixed points in B_N then for each fixed point a we have $C_\varphi^*(K_a) = K_a$, where K_a in the dual of $\mathcal{Y}$ is the linear functional of evaluation at a. Since $\mathcal{Y}$ contains the polynomials, these functionals are linearly independent, and 1 is an eigenvalue of infinite multiplicity for C_φ^*, contradicting the assumption that C_φ is compact. Thus for compact C_φ on any functional Banach space in B_N we may speak of "the" Denjoy–Wolff point of φ, since if it is an interior point it must be unique. (Denjoy–Wolff points on ∂B_N are always unique — see Theorem 2.83).

To proceed with our goal of identifying the spectrum of C_φ when C_φ is compact we need the following lemma to reduce the eigenvalue problem to a finite dimensional problem.

LEMMA 7.17
Suppose $\mathcal{H}$ is a Hilbert space with $\mathcal{H} = \mathcal{K} \oplus \mathcal{L}$ where $\mathcal{K}$ is finite dimensional and C is a bounded operator on $\mathcal{H}$ that leaves $\mathcal{K}$ or $\mathcal{L}$ invariant. If the operator C has the matrix representation

$$C = \begin{pmatrix} X & Y \\ 0 & Z \end{pmatrix} \quad or \quad C = \begin{pmatrix} X & 0 \\ Y & Z \end{pmatrix}$$

with respect to this decomposition, then $\sigma(C) = \sigma(X) \cup \sigma(Z)$.

PROOF Clearly, it is enough to show that C is invertible if and only if both X and Z are invertible. Moreover, by taking adjoints, we see that it is enough to consider only the case

$$C = \begin{pmatrix} X & Y \\ 0 & Z \end{pmatrix}$$

Suppose C is invertible and suppose C^{-1} has block matrix

$$C^{-1} = \begin{pmatrix} P & Q \\ R & S \end{pmatrix}$$

Then

$$\begin{pmatrix} P & Q \\ R & S \end{pmatrix} \begin{pmatrix} X & Y \\ 0 & Z \end{pmatrix} = \begin{pmatrix} I & 0 \\ 0 & I \end{pmatrix}$$

implies $PX = I$. Since P and X act on the finite dimensional space $\mathcal{K}$, this means X is invertible. Furthermore, the invertibility of X and the equality $RX = 0$ imply $R = 0$, so the lower right corner gives $I = RY + SZ = SZ$. Finally, multiplying in the opposite order gives $I = ZS$, so Z is invertible.

Conversely, if X and Z are invertible, it is easy to see that

$$\begin{pmatrix} X & Y \\ 0 & Z \end{pmatrix}^{-1} = \begin{pmatrix} X^{-1} & -X^{-1}YZ^{-1} \\ 0 & Z^{-1} \end{pmatrix}$$

∎

DEFINITION 7.18 *If $\mathcal{H}$ is a space of functions analytic on the domain Ω and a is in the closure of Ω, then a will be called a **point of strong evaluation for $\mathcal{H}$** if evaluation of derivatives of all orders at a are bounded linear functionals on $\mathcal{H}$ and if whenever f is in $\mathcal{H}$ and $f(a)$ and all derivatives of f at a are 0, then f must be 0.*

Our spectral theorem will apply to any compact C_φ where φ has Denjoy–Wolff point which is a point of strong evaluation. Theorems 7.14 and 7.16 show that for most of the familiar weighted Hardy spaces, the Denjoy–Wolff point is in the domain of analyticity for functions in the space, and hence it is a point of strong evaluation. Another example will be considered after we prove the theorem. The theorem will say that the eigenvalues of C_φ are 1 and 0 and all powers of $\varphi'(a)$ in the one variable case and all products of the eigenvalues of $\varphi'(a)$ in the several variable case.

There are technical difficulties in the several variable case that we need to consider before we will be ready to give the proof. First, we recall a few facts about the tensor product of two matrices. If A is an $m \times n$ matrix and B is of size $p \times q$ then $A \otimes B$ is the $mp \times nq$ matrix defined by the block matrix $A \otimes B = [a_{ij}B]$. It's easy to check directly (see Exercise 7.4.6) that if $Au = \lambda u$ and $Bv = \mu v$ then

$$(A \otimes B)(u \otimes v) = \lambda\mu(u \otimes v)$$

that is, that the eigenvectors of $A \otimes B$ are the tensor products of the eigenvectors of A and B and the eigenvalues are the pairwise products of the eigenvalues of A and B.

If $[n_1, n_2, \ldots, n_m]$ is an m–tuple of the integers $1, \ldots, N$, let $\kappa_a^{[n_1,\ldots,n_m]}$ denote the kernel for evaluation of the corresponding partial derivative at a, that is,

$$\langle f, \kappa_a^{[n_1,\ldots,n_m]} \rangle = \frac{\partial^m f}{\partial z_{n_1} \cdots \partial z_{n_m}}(a)$$

for all f in $H^2(\beta)$. If $\varphi(a) = a$ and a is a point of strong evaluation for $H^2(\beta)$, then the evaluation kernels for all derivatives at a exist and the closure of K_a and their span is $H^2(\beta)$. For any positive integer m, let $\mathcal{K}_m$ be the subspace spanned by K_a and the derivative evaluation kernels at a for total order up to and including m. For example, $\mathcal{K}_0$ is the subspace spanned by K_a and $\mathcal{K}_1$ is the subspace spanned by K_a and the evaluation kernels for the first partials at a, $\kappa_a^{[1]}, \ldots, \kappa_a^{[N]}$. We arrange the spanning set so that lower total order evaluation kernels come before higher order ones, and lexicographically within each group with the same total order. For example, for $N = 2$ the spanning set for $\mathcal{K}_2$ would be

$$K_a, \kappa_a^{[1]}, \kappa_a^{[2]}, \kappa_a^{[1,1]}, \kappa_a^{[1,2]}, \kappa_a^{[2,1]}, \kappa_a^{[2,2]}$$

(The alert reader will have noticed that by the equality of mixed partials, $\kappa_a^{[1,2]} = \kappa_a^{[2,1]}$, but we will ignore this annoying detail for the moment.)

We want to find a block matrix for C_φ^* with respect to this spanning set, where the blocks correspond to the order of the partial derivatives involved. For the kernel for evaluation of functions at a we have $C_\varphi^*(K_a) = K_{\varphi(a)} = 1K_a$ so $\mathcal{K}_0$ is an invariant subspace for C_φ^* and the matrix for the restriction of C_φ^* to $\mathcal{K}_0$ is (1). Continuing to the kernels for the first partial derivatives at a, applying the chain rule and remembering that $\varphi(a) = a$, we find that

$$\langle f, C_\varphi^* \kappa_a^{[j]} \rangle = \langle C_\varphi f, \kappa_a^{[j]} \rangle = \langle f \circ \varphi, \kappa_a^{[j]} \rangle$$

$$= \sum_{k=1}^{N} (D_k f)(\varphi(a))(D_j \varphi_k)(a) = \sum_{k=1}^{N} (D_k f)(a)(D_j \varphi_k)(a)$$

$$= \langle f, \sum_{k=1}^{N} \overline{(D_j \varphi_k)(a)} \kappa_a^{[k]} \rangle \tag{7.4.2}$$

This calculation shows that $\mathcal{K}_1$ is invariant for C_φ^* and the matrix for the restriction of C_φ^* to $\mathcal{K}_1$ with respect to the basis $K_a, \kappa_a^{[1]}, \ldots, \kappa_a^{[N]}$ is the block matrix

$$\begin{pmatrix} 1 & 0 \\ 0 & \overline{\varphi'(a)} \end{pmatrix}$$

In the same way,

$$\langle f, C_\varphi^* \kappa_a^{[j_1,j_2]} \rangle = \langle f \circ \varphi, \kappa_a^{[j_1,j_2]} \rangle$$

$$= \sum_{k_1,k_2} (D_{k_1 k_2} f)(\varphi(a))(D_{j_1} \varphi_{k_1})(a)(D_{j_2} \varphi_{k_2})(a)$$

$$+ \text{ terms with first partials of } f$$

$$= \langle f, \sum_{k_1,k_2} \overline{(D_{j_1} \varphi_{k_1})(a)(D_{j_2} \varphi_{k_2})(a)} \kappa_a^{[k_1,k_2]} \rangle$$

$$+ \langle f, \text{ terms with kernels for first partials at } a \rangle$$

This calculation shows that $\mathcal{K}_2$ is invariant for C_φ^* and that, formally, the matrix for the restriction of C_φ^* to $\mathcal{K}_2$ with respect to $K_a, \kappa_a^{[1]}, \ldots, \kappa_a^{[N]}, \kappa_a^{[1,1]}, \kappa_a^{[1,2]}, \ldots,$ $\kappa_a^{[N,N]}$ is the block matrix

$$
\begin{pmatrix}
1 & 0 & 0 \\
0 & \overline{\varphi'(a)} & * \\
0 & 0 & \overline{\varphi'(a) \otimes \varphi'(a)}
\end{pmatrix}
$$

where we have grouped the kernels for partials of the same total order. More generally, we can see from the chain and the product rules for differentiation, that the subspaces $\mathcal{K}_m$ are invariant for C_φ^* and the matrix for the restriction of C_φ^* to $\mathcal{K}_m$ with respect to our set of kernels is a block upper triangular matrix whose diagonal entries are the complex conjugates of $1, \varphi'(a), \varphi'(a) \otimes \varphi'(a), \ldots,$ $\varphi'(a) \otimes \cdots \otimes \varphi'(a)$.

But is it not nonsense to write a "matrix for T" with respect to a spanning set that might not be a basis? It turns out that it is not!

DEFINITION 7.19 *If T is a linear transformation of the finite dimensional vector space $\mathcal{V}$ into itself and $v_1, v_2, \ldots, v_n$ is a spanning set for $\mathcal{V}$, we say $A = \left(a_{ij}\right)_{i,j=1}^{n}$ is a **redundant matrix** for T if*

$$
Tv_j = \sum_{i=1}^{n} a_{ij} v_i
$$

for each $j = 1, \ldots, n$.

With this terminology, we have shown that there is a redundant matrix for the restriction of C_φ^* to $\mathcal{K}_m$ with respect to the spanning set $K_a, \kappa_a^{[1]}, \ldots, \kappa_a^{[N]}, \kappa_a^{[1,1]},$ $\kappa_a^{[1,2]}, \ldots, \kappa_a^{[N,\ldots,N]}$ that is a block upper triangular matrix whose diagonal entries are the complex conjugates of $1, \varphi'(a), \varphi'(a) \otimes \varphi'(a), \ldots, \varphi'(a) \otimes \cdots \otimes \varphi'(a)$. Because the eigenvalues of a tensor product are the products of the eigenvalues of the matrices in the product, the eigenvalues of this redundant matrix for the restriction of C_φ^* to $\mathcal{K}_m$ are 1 and the conjugates of all possible products of m or fewer eigenvalues of $\varphi'(a)$. We want to relate the eigenvalues of the redundant matrix to the eigenvalues of the restriction of C_φ^* to $\mathcal{K}_m$.

For $v_1, v_2, \ldots, v_n$ a spanning set for $\mathcal{V}$, the **representing operator** is the linear transformation $R : C^n \mapsto \mathcal{V}$ defined by

$$
R\begin{pmatrix} d_1 \\ d_2 \\ \vdots \\ d_n \end{pmatrix} = \sum_{j=1}^{n} d_j v_j
$$

Since $v_1, v_2, \ldots, v_n$ spans $\mathcal{V}$, the rank of R is the dimension of $\mathcal{V}$ so R has a right inverse S, that is, $RS = I_\mathcal{V}$ where $I_\mathcal{V}$ is the identity on $\mathcal{V}$. If A is a redundant matrix for T with respect to the spanning set $v_1, v_2, \ldots, v_n$, then

$$T\left(\sum_{j=1}^{n} d_j v_j\right) = \sum_{j=1}^{n} d_j T(v_j) = \sum_{j=1}^{n} d_j \left(\sum_{i=1}^{n} a_{ij} v_i\right)$$

$$= \sum_{i=1}^{n} \left(\sum_{j=1}^{n} a_{ij} d_j\right) v_i$$

This equality says that if $d = (d_1, \ldots, d_n)$ is a vector in C^n, then

$$TRd = RAd$$

In particular, this equation justifies the use of redundant matrices to represent transformations — it shows that if a vector v is represented in any way as a linear combination of the spanning set, $v = Rd$, then Tv is represented using the redundant matrix, $Tv = R(Ad)$.

We can now show that the spectrum of T is a subset of the spectrum of A and give a condition for an eigenvalue of A to be an eigenvalue of T. Writing I_n for the identity on C^n, suppose $A - \lambda I_n$ is invertible, that is, λ is not in the spectrum of A, and suppose B satisfies $(A - \lambda I_n)B = I_n$. Then

$$(T - \lambda I_V)RBS = (TR - \lambda R)BS = (RA - \lambda R)BS$$

$$= R(A - \lambda I_n)BS = RI_n S = I_V$$

Since V is a finite dimensional space, RBS is also a left inverse of $(T - \lambda I_V)$ and λ is not in the spectrum of T. In other words, $\sigma(T) \subset \sigma(A)$.

Conversely, if d is an eigenvector for A with eigenvalue λ, that is, $Ad = \lambda d$, then $T(Rd) = R(Ad) = \lambda(Rd)$ so either Rd is an eigenvector for T or $Rd = 0$.

Recall that the eigenvalues of the redundant matrix for the restriction of C_φ^* to $\mathcal{K}_m$ are 1 and the conjugates of all possible products of m or fewer eigenvalues of $\varphi'(a)$. The containment above shows the restriction of C_φ^* to $\mathcal{K}_m$ has no other eigenvalues and in order to show that each of these is actually an eigenvalue of the restriction, we need to see that for each eigenvalue of the redundant matrix, there is an eigenvector d with $Rd \neq 0$.

We will prove this by induction on the number of terms in the tensor product which corresponds to the order of the partial derivatives involved in the kernel functions. For clarity, we will let R_j denote the representing operator for the j–fold tensor product, that is, that takes a vector in C^{N^j} into the corresponding sum of kernels for evaluation of partial derivatives of order j. For example, if B_2 is the underlying space,

$$R_2((c_{11}, c_{12}, c_{21}, c_{22})) = c_{11}\kappa_a^{[1,1]} + c_{12}\kappa_a^{[1,2]} + c_{21}\kappa_a^{[2,1]} + c_{22}\kappa_a^{[2,2]}$$

$$= c_{11}\kappa_a^{[1,1]} + (c_{12} + c_{21})\kappa_a^{[1,2]} + c_{22}\kappa_a^{[2,2]} \qquad (7.4.3)$$

Since $\kappa_a^{[1]}, \ldots, \kappa_a^{[N]}$ are linearly independent, R_1 is an isomorphism and the result is true if there is only one term in the product. Now suppose the result is true for products of s or fewer terms and suppose λ is an eigenvalue of the the $(s+1)$–fold product $\overline{\varphi'(a)} \otimes \cdots \otimes \overline{\varphi'(a)}$. There is an eigenvector for this eigenvalue consisting of the tensor product of an eigenvector u of $\overline{\varphi'(a)}$ and an eigenvector v of the s–fold product $\overline{\varphi'(a)} \otimes \cdots \otimes \overline{\varphi'(a)}$ such that, by the induction hypothesis, $R_1(u)$ and $R_s(v)$ are non-zero. We want to see that $R_{s+1}(u \otimes v)$ is non-zero. Rewrite $R_s(v)$ and $R_{s+1}(u \otimes v)$ in the more usual way as a linear combination of the K_a^α where, for $\alpha = (\alpha_1, \alpha_2, \ldots, \alpha_N)$, K_a^α denotes the kernel for evaluation of the partial derivative at a α_1 times with respect to z_1, α_2 times with respect to z_2, etc., ordered first by total order and then for fixed total order by K_a^α precedes K_a^β if $\alpha_j = \beta_j$ for $j < j_0$ and $\alpha_{j_0} > \beta_{j_0}$. For example, continuing the calculation of Equation (7.4.3) above,

$$R_2((c_{11}, c_{12}, c_{21}, c_{22})) = c_{11}\kappa_a^{[1,1]} + (c_{12} + c_{21})\kappa_a^{[1,2]} + c_{22}\kappa_a^{[2,2]}$$
$$= c_{11}K_a^{(2,0)} + (c_{12} + c_{21})K_a^{(1,1)} + c_{22}K_a^{(0,2)}$$

Suppose $R_1 u = \sum a_j K_a^{(j)}$ and $R_s v = \sum b_\alpha K_a^\alpha$ are the ordered sums where $a_j = 0$ for $j < j_0$ and $a_{j_0} \neq 0$ and $b_\alpha = 0$ for $\alpha < \beta$ and $b_\beta \neq 0$. The term of the tensor product of u and v coming from j and α corresponds to following the partial derivative represented by α by the partial derivative with respect to z_j. Because no partial derivative with respect to z_j for $j > j_0$ coming from u or α for $\alpha > \beta$ coming from v can form the partial associated with $\gamma = (\beta_1, \ldots, \beta_{j_0-1}, \beta_{j_0} + 1, \beta_{j_0+1}, \ldots, \beta_N)$, the coefficient of K_a^γ in $R_{s+1}(u \otimes v)$ is $a_{j_0}b_\beta \neq 0$, so $R_{s+1}(u \otimes v) \neq 0$. Thus, we conclude that the eigenvalues of C_φ^* restricted to $\mathcal{K}_m$ are the products of m or fewer eigenvalues of $\overline{\varphi'(a)}$. This gives the following theorem.

THEOREM 7.20

Suppose φ has Denjoy–Wolff point a and a is a point of strong evaluation for a weighted Hardy space $H^2(\beta)$ on B_N, $N \geq 1$. If C_φ is a compact operator on $H^2(\beta)$, then the spectrum of C_φ is the set consisting of 0, 1, and all possible products of the eigenvalues of $\varphi'(a)$.

PROOF   Since C_φ^* is compact, 0 is in the spectrum; we will show that the eigenvalues of C_φ^* are the conjugates of the remaining numbers listed.

Now for each m, let $\mathcal{L}_m$ be the orthogonal complement of $\mathcal{K}_m$ in $H^2(\beta)$. The block matrix for C_φ^* is then

$$C_\varphi^* = \begin{pmatrix} X_m & Y_m \\ 0 & Z_m \end{pmatrix}$$

The compactness of C_φ^* and the fact that the span of the $\mathcal{K}_m$ is dense in $H^2(\beta)$ implies $\|Z_m\|$ tends to zero as m tends to infinity. In particular, for $\epsilon > 0$, taking m

large enough that $\|Z_m\| < \epsilon$, and applying Lemma 7.17, we see that the spectrum of C_φ^* is the union of the spectra of X_m and Z_m. From the preceding discussion, the spectrum of X_m is 1 together with all products of at most m eigenvalues of $\overline{\varphi'(a)}$. Since the spectrum of Z_m must be contained in the disk of radius ϵ, and ϵ is arbitrary, we see that the spectrum of C_φ^* is 0, 1, together with the set of all products of eigenvalues of $\overline{\varphi'(a)}$. ∎

Theorem 7.20 almost completely answers the question of the spectrum of a compact composition operator on a weighted Hardy space. If the space is large enough so that $\sum n^{2k}\beta(n)^{-2} = \infty$ for some k then the theorem applies. As a non-trivial application of the theorem in the case that $\sum n^{2k}\beta(n)^{-2} < \infty$ for all k we consider the spaces $H^2(\beta, D)$ for the choice $\beta(n) = \exp(n^a)$, with $1/2 \le a < 1$. Recall that in Chapter 4 we showed that $\varphi(z) = (1+z)/2$ gives a compact operator on these spaces. Note that φ has Denjoy–Wolff point 1. Though we will not give the details here, these spaces are non-trivial in the sense of containing functions which are not continuable across any point of ∂D, yet each is a quasi-analytic class on $\overline{D}$, so that 1 is a point of strong evaluation [Cal62, p. 331]. This means Theorem 7.20 applies and identifies $\sigma(C_\varphi)$ as $\{0, 1\} \cup \{1/2^k : k = 1, 2, \ldots\}$.

Exercises

7.4.1 Justify the assertions made in the table of Section 7.1 about composition operators whose symbols are $\varphi(z) = -z/2$ and $\varphi(z) = (1 - z)/2$.

7.4.2 Suppose $\varphi(0) = 0$ and C_φ is compact on $H^2(D)$. Find a recursive formula for the coefficients of the eigenfunctions for C_φ and prove that if f is an eigenfunction of C_φ, then f^k is in H^2 for every positive integer k.

7.4.3 Show that each eigenspace for a compact composition operator on the disk has dimension 1, but give an example of a compact composition operator on a ball with an eigenspace of dimension bigger than 1.

7.4.4 Give an example of a compact composition operator on $H^2(D)$ that has more than one fixed point in the closed disk.

7.4.5 Show that if $H^2(\beta)$ is a weighted Hardy space on the unit disk for which $\sum \beta(k)^{-2}$ converges and C_φ is a compact composition operator on $H^2(\beta)$, then φ has exactly one fixed point in the closed disk.

7.4.6 Show that if u and v are eigenvectors of A and B with eigenvalues λ and μ respectively, then

$$(A \otimes B)(u \otimes v) = \lambda\mu u \otimes v$$

and that all the eigenvalues of $A \otimes B$ arise in this way.

Notes

J. Caughran and H. Schwartz [CaS75] began the study of spectra of compact composition operators by proving Theorems 7.14 and 7.20 for the classical Hardy space. B. Mac-Cluer [Mc84a] proved these theorems for the Hardy spaces in the ball, and N. Zorboska [Zo89b] partially extended them to some weighted Hardy spaces. Zorboska [Zo89b] discovered the property in Exercise 7.4.5. The examples spaces $H^2(\beta, D)$ for the choice $\beta(n) = \exp(n^a)$, with $1/2 \leq a < 1$ were noted in J. H. Shapiro's paper [Sho87b, p. 54]. Theorem 7.20 is an extension of all the results on spectra of compact composition operators of which we are aware.

7.5   Spectra: boundary fixed point, $\varphi'(a) < 1$

In this section we consider the spectrum of a composition operator on $H^2(D)$ whose symbol has Denjoy–Wolff point a on the unit circle with $\varphi'(a) < 1$. Some of the results hold and will be presented in this generality but others need a smoothness hypothesis. Rather than striving always for the minimal hypotheses, we will assume enough to allow us to concentrate on the operator theoretic implications without worrying too much about the function theory. For example, if φ is an analytic map of the disk into the disk that is analytic in a neighborhood of the closed disk, then it has only a finite number of fixed points in the closed disk and the angular derivatives at the fixed points on the boundary are finite. Several of the results are presented in this setting because the finitely many fixed points and the good boundary behavior near them facilitate the arguments.

The first observation is that these composition operators are circularly symmetric; the proof illustrates a connection between the Toeplitz operators and composition operators.

THEOREM 7.21
If φ is an analytic mapping of the unit disk to itself with Denjoy–Wolff point a on the unit circle and $\varphi'(a) < 1$, then for real θ the operator C_φ on $H^2(D)$ is similar to the operator $e^{i\theta} C_\varphi$. In particular, if λ is in the spectrum of C_φ then so also is then $e^{i\theta} \lambda$ for all real θ.

PROOF   For φ as in the hypotheses, let $s = \varphi'(a)$ and let θ be a real number. The map φ has a halfplane/dilation linear fractional model, that is, there is a map σ of the disk into the right halfplane such that $\sigma(\varphi(z)) = s\sigma(z)$ for all z in the disk. Let $F(w) = \exp(i\theta \log(w)/\log(s))$ for w in the right halfplane, where log denotes the principle branch of the logarithm. Simple calculations show $F(sw) = e^{i\theta} F(w)$ and $e^{\pi\theta/(2\log(s))} < |F(w)| < e^{-\pi\theta/(2\log(s))}$ if $\theta \geq 0$ while $e^{-\pi\theta/(2\log(s))} < |F(w)| < e^{\pi\theta/(2\log(s))}$ if θ is negative. Writing $f = F \circ \sigma$ we see that $f \circ \varphi = e^{i\theta} f$ and both f and $1/f$ are in $H^\infty(D)$. This means the analytic

Toeplitz operator T_f given by $T_f(h)(z) = f(z)h(z)$ is bounded and invertible. For h in $H^2(D)$,

$$((T_f)^{-1}C_\varphi T_f)(h) = (T_f)^{-1}((f \circ \varphi)(h \circ \varphi)) = e^{i\theta}(T_f)^{-1}T_f C_\varphi h = e^{i\theta}(C_\varphi h)$$

so C_φ is similar to $e^{i\theta}C_\varphi$.

The second conclusion follows because similar operators have the same spectrum, but the spectral mapping theorem implies the spectrum of $e^{i\theta}C_\varphi$ is $e^{i\theta}$ times the spectrum of C_φ. ∎

The proofs of the main results of this section are based on a weighted shift analogy. The following definition introduces terminology we use to set up the analogy. The K and M in the definition can be finite integers or $\pm\infty$.

DEFINITION 7.22   *We say the sequence of points $\{z_k\}_{k=K}^M$ in D is an **iteration sequence for** φ if $\varphi(z_k) = z_{k+1}$ for $K < k < M$.*

We will see that when the points of an iteration sequence are far apart, the kernel functions for an iteration sequence act like an orthogonal basis and the adjoint of a composition operator acts like a weighted shift on this basis. In particular, the span of these kernel functions is an invariant subspace for C_φ^* and spectral information about C_φ can be obtained from spectral information about the weighted shift.

If $\{z_k\}_{k=-\infty}^0$ is an iteration sequence of distinct points for φ, not an elliptic automorphism of the disk onto itself, then $\lim_{k \to -\infty} |z_k| = 1$. Indeed, suppose b were a limit point of this sequence and $|b| < 1$. If a is the Denjoy–Wolff point of φ and $b = a$, then the pseudohyperbolic distance from b to z_{k_1} is greater than the pseudohyperbolic distance from b to $z_{k_2} = \varphi_n(z_{k_1})$ for $n = k_2 - k_1 > 0$, but this is incompatible with the assumption that a subsequence $\{z_{k_j}\}$ of the iteration sequence converges to b with k_j tending to $-\infty$. On the other hand, if $b \neq a$, then there is $\epsilon > 0$ so that $D_\epsilon = \{z : |z - b| \leq \epsilon\}$ is contained in D and does not contain a. In this case, the iterates of D_ϵ converge to a and there is n so that $\varphi_n(D_\epsilon) \cap D_\epsilon = \emptyset$, which also contradicts the assumption that a subsequence $\{z_{k_j}\}$ of the iteration sequence converges to b with k_j tending to $-\infty$.

If φ is not an elliptic automorphism of the disk onto itself and z_0 is a point of the disk, then $z_k = \varphi_k(z_0)$ defines an iteration sequence of distinct points for $k \geq 0$ or else there is a least M so that $\varphi_M z_0 = a$, the Denjoy–Wolff point of φ in D, and there is an iteration sequence of distinct points defined for $0 \leq k \leq M$. Moreover, either there is no point w of D with $\varphi(w) = z_0$ or we can find z_{-1} so that $\varphi(z_{-1}) = z_0$. So every point of the disk is in at least one iteration sequence $\{z_k\}_{k=K}^M$ of distinct points for which either $M = \infty$ and $\lim_{k \to \infty} z_k = a$ or M is finite and $\varphi_M(z_0) = a$ and either $K = -\infty$ and $\lim_{k \to -\infty} |z_k| = 1$ or K is finite and there is no point w in the disk with $\varphi(w) = z_K$.

The case $\{z_k\}_{k=-\infty}^0$ in which $\lim_{k \to -\infty} z_k = b$ where b is a fixed point of φ on the circle is especially interesting. If b is not the Denjoy–Wolff point, then we have seen $\varphi'(b) > 1$ (Theorem 2.48). If φ is analytic in a neighborhood of b, then there is $\epsilon > 0$ so that φ is univalent on $D_\epsilon = \{z : |z - b| \leq \epsilon\}$ and φ^{-1} maps D_ϵ into itself. A modification of Theorem 2.53 shows that there is a analytic map ψ of D_ϵ into the plane so that $\psi(b) = 0$ and

$$\psi(\varphi^{-1}(z)) = \frac{1}{\varphi'(b)}\psi(z)$$

for z in D_ϵ. Since φ^{-1} is univalent near b, ψ is also. Possibly multiplying ψ by a constant of modulus 1, we may assume that ψ maps the radial segment $\{rb : 1 - \epsilon < r < 1\}$ to a curve tangent to the positive real axis at 0. Now the conformality of ψ shows that if $\delta > 0$ is small enough that the interval $[0, \delta)$ is in $\psi(D_\epsilon)$, the image of $[0, \delta)$ under ψ^{-1} is a curve tangent to the radius $\{rb : 0 < r < 1\}$ in the disk. Now for w so that $0 < w < \delta$, defining $z_{-k} = \psi^{-1}(\varphi'(b)^{-k}w)$ for $k = 0, 1, \ldots$ gives an iteration sequence $\{z_j\}_{j=-\infty}^{0}$ so that z_j converges to b nontangentially as j approaches negative infinity. In particular, there are uncountably many such iteration sequences.

THEOREM 7.23

Suppose φ is an analytic map of the unit disk to itself, not an automorphism of the disk, and suppose C_φ is the corresponding composition operator on $H^2(D)$. If b is a fixed point of φ on the unit circle with $\varphi'(b) > 1$ and φ is analytic in neighborhood of b, then for each $\rho < \varphi'(b)^{-1/2}$, the circle of radius ρ centered at the origin intersects the spectrum of C_φ.

PROOF Suppose φ is as in the hypotheses. The discussion preceding the theorem shows that the analyticity of φ near the boundary fixed point b implies there are uncountably many iteration sequences $\{z_j\}_{j=-\infty}^{0}$ with $\lim_{j\to-\infty} z_j = b$ nontangentially. Given such an iteration sequence, let $k_j(z) = (1 - |z_j|^2)^{1/2}/(1 - \overline{z_j}z)$ denote the normalized reproducing kernel for the point z_j.

Suppose $\rho < \rho_1 < \varphi'(b)^{-1/2}$ and $|\lambda| = \rho$. Let

$$h_\lambda = \sum_{j=-\infty}^{-1} \lambda^{-j-1}\left(\frac{1 - |z_0|^2}{1 - |z_j|^2}\right)^{1/2} k_j$$

Since

$$\lim_{j\to-\infty} \frac{1 - |z_{j+1}|^2}{1 - |z_j|^2} = \varphi'(b)$$

there is a constant c so that

$$|\lambda|^{-j-1}\left(\frac{1 - |z_0|^2}{1 - |z_j|^2}\right)^{1/2} = |\lambda|^{-j-1}\left(\prod_{l=j}^{-1} \frac{1 - |z_{l+1}|^2}{1 - |z_l|^2}\right)^{1/2}$$

$$\leq c \left(\frac{|\lambda|}{\rho_1} \right)^{-j} = c \left(\frac{\rho}{\rho_1} \right)^{-j}$$

Since $\|k_j\| = 1$, the series for h_λ converges absolutely which implies it converges in $H^2(D)$, and $(C_\varphi^* - \lambda)h_\lambda$ is

$$\sum_{j=-\infty}^{-1} \lambda^{-j-1} \left(\frac{1 - |z_0|^2}{1 - |z_{j+1}|^2} \right)^{1/2} k_{j+1} - \sum_{j=-\infty}^{-1} \lambda^{-j} \left(\frac{1 - |z_0|^2}{1 - |z_j|^2} \right)^{1/2} k_j$$

$$= k_0$$

Moreover, if $\sigma(C_\varphi)$ does not intersect the circle $|\lambda| = \rho$, then $(C_\varphi^* - \rho e^{i\theta})^{-1}$ exists for each real θ and we define Q by

$$Q = \frac{1}{2\pi} \int_0^{2\pi} (C_\varphi^* - \rho e^{i\theta})^{-1} \, d\theta$$

Now

$$Qk_0 = \frac{1}{2\pi} \int_0^{2\pi} (C_\varphi^* - \rho e^{i\theta})^{-1} k_0 \, d\theta = \frac{1}{2\pi} \int_0^{2\pi} h_{\rho e^{i\theta}} \, d\theta$$

$$= \sum_{j=-\infty}^{-1} \frac{1}{2\pi} \int_0^{2\pi} \rho^{-j-1} e^{i(-j-1)\theta} \left(\frac{1 - |z_0|^2}{1 - |z_j|^2} \right)^{1/2} k_j \, d\theta$$

$$= \left(\frac{1 - |z_0|^2}{1 - |z_{-1}|^2} \right)^{1/2} k_{-1}$$

Thus, writing $K_w(z) = 1/(1 - \overline{w}z)$, the kernel for evaluation at w, we have $QK_{z_0} = K_{z_{-1}}$, which means $C_\varphi^* Q K_{z_0} = K_{z_0}$. Since Q is a rational function of C_φ^*, they commute and we have $QC_\varphi^* K_{z_0} = K_{z_0}$ also.

For each such iteration sequence $\{z_k\}_{k=-\infty}^0$, the argument above showed that $C_\varphi^* Q K_{z_0} = Q C_\varphi^* K_{z_0} = K_{z_0}$. Since there are uncountably many such z_0, their kernel functions span $H^2(D)$ and we get $C_\varphi^* Q = Q C_\varphi^* = I$. But it was assumed that φ is not an automorphism, so this is a contradiction. This means our assumption that $\sigma(C_\varphi)$ does not intersect the circle $|\lambda| = \rho$ must be false, which is the conclusion of the theorem. ∎

The function theory of Section 2.4 showed that Schroeder's functional equation, except for the multiple 0, always has an infinite dimensional solution space. Schroeder's functional equation is the eigenvalue equation for the composition operator, but in order for a solution of the functional equation to be an eigenfunction for the operator C_φ, the function must be in $H^2(D)$. The work of the following lemmas is therefore in the growth estimates that determine when the solutions of Schroeder's functional equation are in the space. Exercise 7.5.4 shows that these results are close to best possible. We first find an annulus of eigenvalues for C_φ.

LEMMA 7.24

Suppose φ is an analytic map of the unit disk to itself that has Denjoy–Wolff point a on the unit circle with $\varphi'(a) < 1$. If

$$\varphi'(a)^{1/2} < |\lambda| < \varphi'(a)^{-1/2}$$

then λ is an eigenvalue of C_φ acting on $H^2(D)$ of infinite multiplicity.

PROOF Let $s = \varphi'(a)$. Suppose λ is a positive number satisfying $s^{1/2} < \lambda < s^{-1/2}$. The proof of Theorem 7.21 shows that the conclusion will follow if we show λ is an eigenvalue: for $|\mu| = \lambda$, the similarity transformation associated with $e^{i\theta}$ carries an eigenfunction for the eigenvalue λ to linearly independent eigenfunctions for the eigenvalue $\mu = \lambda e^{i\theta}$ for each of the values of $\theta = \arg \mu$.

The map φ has a halfplane/dilation model in which σ is a map of the disk into the right halfplane satisfying $\sigma \circ \varphi = s\sigma$. Let $x = \log \lambda / \log s$, where log is the principal branch of the logarithm function, so that $(\sigma^x) \circ \varphi = (s\sigma)^x = \lambda \sigma^x$. In other words, we can conclude σ^x is an eigenfunction of C_φ for the eigenvalue λ if we can show it is in $H^2(D)$.

The analytic function $\Psi(z) = (1 - \sigma(z))/(1 + \sigma(z))$ maps the disk into itself. For $-1/2 < x < 1/2$ the function $F(z) = \big((1 - z)/(1 + z)\big)^x$ is in $H^2(D)$, and therefore so also is $\sigma^x = F \circ \Psi$. In other words, we see easily that λ is an eigenvalue when $s^{1/2} < \lambda < s^{-1/2}$. (Note that this corresponds to Theorem 7.4 for hyperbolic automorphisms and this justification is really giving it as a corollary of that result.) ∎

We are now ready to shrink the inner radius of the annulus of eigenvalues for smooth functions. By Exercise 7.5.1, if φ is analytic in a neighborhood of the closed unit disk and φ is not inner, then $\{e^{i\theta} : |\varphi(e^{i\theta})| = 1\}$ is at most a finite set of points. In the hypothesis of the next result we assume that fixed points of φ are the only points of the circle that get mapped to the circle. This eliminates inner functions, but, as we will see when we identify the spectrum, the result can be used to get information about smooth functions that are not inner.

LEMMA 7.25

Suppose φ is analytic in a neighborhood of the closed unit disk, maps the unit disk to itself, and has Denjoy–Wolff point a on the unit circle with $\varphi'(a) < 1$. Suppose also that $\{e^{i\theta} : |\varphi(e^{i\theta})| = 1\} = \{a, b_1, b_2, \ldots, b_k\}$, where $\varphi(b_j) = b_j$ for $j = 1, \ldots, k$. If

$$\max\{\varphi'(b_j)^{-1/2} : j = 1, 2, \ldots, k\} < |\lambda| < \varphi'(a)^{-1/2}$$

then λ is an eigenvalue of C_φ on $H^2(D)$ of infinite multiplicity.

PROOF Lemma 7.24 gives the result for λ with $\varphi'(a)^{1/2} < |\lambda| < \varphi'(a)^{-1/2}$. Let $r_0 = \min\{\varphi'(b_j) : j = 1, 2, \ldots, k\}$ and let $s = \varphi'(a)$. Suppose λ and λ_0 are a positive numbers satisfying $r_0^{-1/2} < \lambda_0 < \lambda < 1$. As in the previous lemma, the proof of Theorem 7.21 shows that the conclusion will follow if we show λ is an eigenvalue.

As in the proof of Lemma 7.24 we have $\sigma \circ \varphi = s\sigma$ where σ is a map of the disk into the right halfplane and we wish to show the eigenfunction σ^x, for $x = \log \lambda / \log s$, is in $H^2(D)$.

Since $|\varphi(z)| < 1$ for z in the closed disk, except for $z = a \equiv b_0$ or $z = b_j$, where $j = 1, \ldots, k$, for every $\epsilon > 0$, the set

$$\{\varphi(z) : |z| \leq 1 \text{ and } |z - b_j| \geq \epsilon \text{ for } j = 0, 1, 2, \ldots, k\}$$

is a compact subset of the open unit disk. It follows from the construction of the model for iteration that there is a constant M_ϵ so that $|\sigma(z)| \leq M_\epsilon$ for $|z - b_j| \geq \epsilon$. Since $x > 0$ for the values of λ under consideration, to show σ^x is in $H^2(D)$ we need only control the growth of σ near b_j for $j = 1, \ldots, k$. Since $\varphi'(b_j)^{-1/2} < \lambda_0 < 1$, we have

$$\frac{|\log s|}{-2 \log \lambda_0} > \frac{|\log s|}{\log \varphi'(b_j)}$$

so by Theorem 2.64, there are M_j and ϵ_j so that

$$|\sigma(z)| \leq M_j |z - b_j|^{-|\log s|/|2 \log \lambda_0|}$$

for $|z - b_j| < \epsilon_j$. Putting these inequalities together, we see that there is M so that for z in D,

$$|\sigma(z)| \leq M \prod_{j=1}^{k} |z - b_j|^{-|\log s|/|2 \log \lambda_0|}$$

and for $r_0^{-1/2} < \lambda_0 < \lambda < 1$. This implies

$$|\sigma(z)^{2x}| \leq M^{2x} \prod_{j=1}^{k} \left(|z - b_j|^{-|\log s|/|2 \log \lambda_0|} \right)^{2|\log \lambda / \log s|}$$

$$= M^{2x} \prod_{j=1}^{k} |z - b_j|^{-|\log \lambda / \log \lambda_0|}$$

Noting that $0 < \lambda_0 < \lambda < 1$ implies $|\log \lambda / \log \lambda_0| < 1$, the product can be seen to be integrable on the unit circle, so σ^x is in $H^2(D)$.

Thus each λ, $r_0^{-1/2} < \lambda < s^{-1/2}$, is an eigenvalue of C_φ on $H^2(D)$. As in the proof of Lemma 7.24, for every real θ, $\lambda e^{i\theta}$ is also an eigenvalue and the corresponding eigenfunctions are linearly independent. This completes the proof.
∎

We are now ready to put our results together to find the spectrum of C_φ for non-inner maps of the type we are considering.

THEOREM 7.26

If φ, not an inner function, is analytic in a neighborhood of the closed unit disk, maps the unit disk to itself, and has Denjoy–Wolff point a on the unit circle with $\varphi'(a) < 1$, then for C_φ acting on the Hardy space $H^2(D)$,

$$\sigma(C_\varphi) = \{\lambda : |\lambda| \leq \varphi'(a)^{-1/2}\}$$

PROOF Since φ is analytic in a neighborhood of closed disk and not a finite Blaschke product, $\{e^{i\theta} : |\varphi(e^{i\theta})| = 1\}$ is finite, by Exercise 7.5.1. Furthermore, by Exercise 7.5.2, this implies that for some positive integer n, the set $\{e^{i\theta} : |\varphi_n(e^{i\theta})| = 1\}$ consists entirely of fixed points $\{a, b_1, \ldots, b_k\}$ of φ_n. Note that $\varphi'_n(a) = \varphi'(a)^n < 1$, so a is the Denjoy–Wolff point of φ_n also. Theorem 7.23 implies that $\sigma(C_{\varphi_n})$ intersects the circle of radius r for $0 < r < r_0^{-1/2}$ for $r_0 = \min\{\varphi'_n(b_j)\}$. On the other hand, Lemma 7.25 implies $\sigma(C_{\varphi_n})$ includes the circle of radius r with $r_0^{-1/2} < r < (\varphi'(a)^n)^{-1/2}$. Since $(C_\varphi)^n = C_{\varphi_n}$, the spectral mapping theorem implies $\sigma(C_\varphi)$ intersects the circle of radius r for $0 < r < \varphi'(a)^{-1/2}$. Now, Theorem 7.21 implies the spectrum of C_φ includes the disk $\{\lambda : |\lambda| \leq \varphi'(a)^{-1/2}\}$. Since the spectral radius of C_φ is $\varphi'(a)^{-1/2}$ by Theorem 3.9, the proof is complete. ∎

Circular symmetry of an operator, such as that noted in Theorem 7.21, is associated with weighted shifts (see, for example, [Ge77] or [AHHK84] for specific theorems). We have seen that (adjoints of) composition operators have analogues of shifts embedded in them. Much of the study of the spectrum of a composition operator is closely connected with and can be motivated by this analogy or alternatively, by its failure in particular cases. To pursue the analogy in more detail in the case under consideration in this section, we must first identify the shifted basis. The starting point is the observation that $C_\varphi^*(K_w) = K_{\varphi(w)}$.

Recall that an ***interpolating sequence*** $\{z_j\}$ in the disk is one for which, given any bounded sequence $\{c_j\}$ of complex numbers, there is a bounded analytic function f so that $f(z_j) = c_j$. Interpolating sequences are Blaschke sequences that are spread out enough for there to be a positive number δ so that for each z_k in the sequence, if B is the Blaschke product that has zeros z_j for $j \neq k$, then $|B(z_k)| > \delta$. We have seen (Theorem 2.65) that if φ has a halfplane linear fractional model, then a sequence $\{z_j\}$ for which $\varphi(z_j) = z_{j+1}$ is an interpolating sequence. In the case we are considering, $\lim_{j \to \infty} z_j = a$ and, if the sequence is indexed by all the integers, $\lim_{j \to -\infty} z_j = b$ where b is a different fixed point of φ. Moreover, $\lim_{j \to \infty}(1 - |z_{j+1}|)/(1 - |z_j|) = \varphi'(a)$ and, if applicable, $\lim_{j \to -\infty}(1 - |z_{j+1}|)/(1 - |z_j|) = \varphi'(b)$, provided the $z'_j s$ are approaching b nontangentially. .

The following shows that the normalized reproducing kernels corresponding to an iteration sequence form an almost orthonormal set. More precisely, we will show that they are equivalent in the Banach space sense to an orthonormal set. Operators for which these vectors form a natural coordinate basis are similar to operators that act formally the same on an orthonormal basis. This result is not the most general, but is enough for our work.

THEOREM 7.27

Let φ be a map of the unit disk into itself that is analytic in a neighborhood of the closed disk and suppose the Denjoy–Wolff point a of φ is on the unit circle and satisfies $\varphi'(a) < 1$. Suppose $\{z_j\}$ is a sequence in the unit disk for which $\varphi(z_j) = z_{j+1}$ for all j. Denoting by B the Blaschke product with zeros $\{z_j\}$ and by $\{e_j\}$ the usual orthonormal basis for ℓ^2, if the operator S is given by

$$S\left(\frac{\sqrt{1 - |z_j|^2}}{1 - \overline{z_j}z}\right) = e_j$$

then S is a bounded map of $(BH^2)^\perp$ onto ℓ^2 and S has bounded inverse.

PROOF   Let $\{z_j\}$ be an iteration sequence as in the hypotheses and suppose B is the Blaschke product with these zeros. Without loss of generality, we may assume that either $j = 0$ is the least index occurring in the sequence or j can be any integer. Let $k_j(z) = \sqrt{1 - |z_j|^2}/(1 - \overline{z_j}z)$ denote the normalized reproducing kernel for the point z_j. Since z_j is a zero of B, we have $\langle Bf, k_j \rangle = \sqrt{1 - |z_j|^2}B(z_j)f(z_j) = 0$ for every f in $H^2(D)$, and k_j is in $(BH^2)^\perp$. On the other hand, the closed span of the k_j is $(BH^2)^\perp$ because if g is in $H^2(D)$ and $\langle g, k_j \rangle = 0$ for each j, then $g(z_j) = 0$ for each j which means $g = Bf$ for some f in $H^2(D)$. Our goal is to show that the k_j form an almost orthonormal basis for $(BH^2)^\perp$.

If the sequence is indexed by all the integers and b is the fixed point satisfying $\lim_{j \to -\infty} z_j = b$, we have $\lim_{j \to -\infty}(1 - |z_j|)/(1 - |z_{j+1}|) = 1/\varphi'(b)$, since the smoothness hypothesis on φ guarantees that z_j approaches b nontangentially as j goes to negative infinity. By Exercise 2.3.15, $1/\varphi'(b)$ is at most $\varphi'(a)$. In particular, there is a positive integer J and a number ρ for which

$$\frac{1 - |z_{j+1}|}{1 - |z_j|} < \rho < 1$$

for all $j > J$ and, if applicable,

$$\frac{1 - |z_j|}{1 - |z_{j+1}|} < \rho < 1$$

for all $j < -J$.

For m a positive integer, we break the sequence $\{z_j\}$ into m pieces according to the equivalence class of j modulo m. Note that when m is large enough, after perhaps choosing a slightly larger ρ than in the hypothesis,

$$\frac{1 - |z_{m(j+1)+r}|}{1 - |z_{mj+r}|} < \rho^m$$

for $r = 0, 1, \ldots, m - 1$ and $j \geq 0$ and similarly for $j < 0$. We will choose m later to make ρ^m small enough, in addition to this normalization.

For $r = 0, 1, \ldots, m - 1$, let B_r be the Blaschke product with zeros z_{jm+p} for $p = r, \ldots, m - 1$ and let $B_m = 1$. (Note that $B_0 = B$.) For $r = 0, 1, \ldots, m - 1$ and j an integer (or non-negative integer, as appropriate), let $e_{r,j}$ be the vector in $(BH^2)^\perp$ given by

$$e_{r,j}(z) = k_{mj+r}(z) B_{r+1}(z) \prod_{l=j+1}^{\infty} \frac{|z_{ml+r}|}{z_{ml+r}} \frac{z_{ml+r} - z}{1 - \overline{z_{ml+r}} z}$$

The vectors $\{e_{r,j}\}$ form an orthonormal basis for $(BH^2)^\perp$! If we write $(r_1, j_1) < (r_2, j_2)$ when either $r_1 = r_2$ and $j_1 < j_2$ or $r_1 < r_2$, then for $(r_1, j_1) < (r_2, j_2)$, we find

$$\langle e_{r_1,j_1}, e_{r_2,j_2} \rangle = \langle k_{mj_1+r_1} \prod_{(r_1,j_1)<(p,l)\leq(r_2,j_2)} \frac{|z_{ml+p}|}{z_{ml+p}} \frac{z_{ml+p} - z}{1 - \overline{z_{ml+p}} z}, k_{mj_2+r_2} \rangle$$

$$= (1 - |z_{mj_2+r_2}|^2)^{1/2} k_{mj_1+r_1}(z_{mj_2+r_2}).$$

$$\prod_{(r_1,j_1)<(p,l)\leq(r_2,j_2)} \frac{|z_{ml+p}|}{z_{ml+p}} \frac{z_{ml+p} - z_{mj_2+r_2}}{1 - \overline{z_{ml+p}} z_{mj_2+r_2}}$$

which is zero because the last term in the product is zero. Also,

$$\langle e_{r,j}, e_{r,j} \rangle = \langle k_{mj+r}, k_{mj+r} \rangle = 1$$

Define the operator A on $(BH^2)^\perp$ by $A(e_{r,j}) = k_{mj+r}$. Ordering the basis $\{e_{r,j}\}$ by the lexicographic ordering on the subscripts as above, the matrix for A with respect to this basis has entries

$$\langle Ae_{r_1,j_1}, e_{r_2,j_2} \rangle = \langle k_{mj_1+r_1}, e_{r_2,j_2} \rangle = \sqrt{1 - |z_{mj_1+r_1}|^2} \, \overline{e_{r_2,j_2}(z_{mj_1+r_1})}$$

This matrix is lower triangular, since $e_{r_2,j_2}(z_{mj_1+r_1}) = 0$ when $(r_1, j_1) > (r_2, j_2)$. We will regard the matrix as an $m \times m$ block matrix with the blocks corresponding to the vectors $\{e_{r,j}\}$ for fixed r, that is, blocked according to the decomposition $B_{r+1}H^2 \ominus B_r H^2$ for $r = 0, 1, \ldots, m - 1$.

The matrix entries for $(r_1, j_1) < (r_2, j_2)$ satisfy

$$|\langle Ae_{r_1,j_1}, e_{r_2,j_2} \rangle| \leq \sqrt{1 - |z_{mj_1+r_1}|^2} |k_{mj_2+r_2}(z_{mj_1+r_1})|$$

$$= \frac{\sqrt{(1 - |z_{mj_1+r_1}|^2)(1 - |z_{mj_2+r_2}|^2)}}{|1 - \overline{z_{mj_2+r_2}} z_{mj_1+r_1}|}$$

Now for $0 < |w| < 1$ and $|z| < 1$,

$$\frac{\sqrt{(1 - |z|^2)(1 - |w|^2)}}{|1 - \overline{w}z|} \leq \frac{\sqrt{(1 - |z|^2)(1 - |w|^2)}}{1 - |w||z|}$$

$$\leq \frac{\sqrt{(1 - |z|^2)(1 - |w|^2)}}{|w| - |w||z|} = \frac{\sqrt{(1 + |z|)(1 + |w|)}}{|w|} \sqrt{\frac{1 - |w|}{1 - |z|}}$$

Thus, since there are only finitely many points of the sequence in the disk of radius $1/2$, there is a constant C so that if $0 \leq j_1 \leq j_2$,

$$|\langle Ae_{r_1,j_1}, e_{r_2,j_2} \rangle| \leq C \sqrt{\frac{1 - |z_{mj_2+r_2}|}{1 - |z_{mj_1+r_1}|}} \leq C\rho^{m|j_2-j_1|/2}$$

Similarly, if $0 \leq j_2 \leq j_1$, then

$$|\langle Ae_{r_1,j_1}, e_{r_2,j_2} \rangle| \leq C \sqrt{\frac{1 - |z_{mj_1+r_1}|}{1 - |z_{mj_2+r_2}|}} \leq C\rho^{m|j_2-j_1|/2}$$

Analogously, if j_1 and j_2 are both negative, we find

$$|\langle Ae_{r_1,j_1}, e_{r_2,j_2} \rangle| \leq C\rho^{m|j_2-j_1|/2}$$

On the other hand, if j_1 and j_2 have opposite signs, since $\lim_{j \to \infty} z_j = a$ and $\lim_{j \to -\infty} z_j = b$,

$$\frac{\sqrt{(1 - |z_{mj_1+r_1}|^2)(1 - |z_{mj_2+r_2}|^2)}}{|1 - \overline{z_{mj_2+r_2}} z_{mj_1+r_1}|} \approx \frac{\sqrt{(1 - |z_{mj_1+r_1}|^2)(1 - |z_{mj_2+r_2}|^2)}}{|1 - \overline{b}a|}$$

and we have in this case, as well, that

$$|\langle Ae_{r_1,j_1}, e_{r_2,j_2} \rangle| \leq C\sqrt{(1 - |z_{mj_1+r_1}|)(1 - |z_{mj_2+r_2}|)} \leq C\rho^{m|j_2-j_1|/2}$$

That is, each block of the matrix for A is dominated by the Toeplitz matrix with entries $C\rho^{m|j_2-j_1|/2}$ that defines a bounded operator. By Exercise 7.5.3, this means each block of A is a bounded operator and since A has only finitely many blocks, A is bounded.

To investigate the invertibility of A, we recall that if A has (finite) block lower triangular form and each block on the main diagonal is invertible, then A is invertible (see Lemma 7.17). Each block on the diagonal is lower triangular and the diagonal entries of the block satisfy

$$|\langle Ae_{r,j}, e_{r,j} \rangle| = |\langle k_{mj+r}, e_{r,j} \rangle| = \sqrt{1 - |z_{mj+r}|^2}\, |e_{r,j}(z_{mj+r})|$$

$$= |B_{r+1}(z_{mj+r})| \prod_{l=j+1}^{\infty} \left| \frac{z_{ml+r} - z_{mj+r}}{1 - \overline{z_{ml+r}} z_{mj+r}} \right|$$

The fact that the iteration sequence is an interpolating sequence implies that the values of the Blaschke products in the expressions for the diagonal entries are bounded below by δ independent of m, r, and j. If we let D_r be the diagonal matrix with entries $\langle Ae_{r,j}, e_{r,j} \rangle$ and A_r the block on the diagonal of A corresponding to the equivalence class $[r]$, then A_r is invertible if and only if $D_r^{-1} A_r$ is invertible.

The diagonal entries of $D_r^{-1} A_r$ are all 1's and, by the estimates above, the entries below the diagonal satisfy

$$|\langle D_r^{-1} A_r e_{r,j_1}, e_{r,j_2} \rangle| \leq C\rho^{m(j_1 - j_2)/2}/\delta$$

This means $D_r^{-1} A_r - I$ is dominated by the strictly lower triangular Toeplitz matrix with entries $C\rho^{m(j_1 - j_2)/2}/\delta$ which has norm $C\rho^{m/2}/(\delta(1 - \rho^{m/2}))$. Choose m so that this number is less than 1; then for each r, $\|D_r^{-1} A_r - I\| < 1$ which means $D_r^{-1} A_r$ is invertible. Thus, each A_r is invertible and A is invertible.

This establishes the equivalence of the normalized reproducing kernels with an orthonormal basis for $(BH^2)^\perp$. Since all infinite dimensional separable Hilbert spaces are unitarily equivalent, combining A with a unitary map of $(BH^2)^\perp$ onto ℓ^2 gives the desired operator S. ∎

If $\{z_j\}$ is an iteration sequence for φ, the operator C_φ^* shifts the basis k_j:

$$C_\varphi^*(k_j) = C_\varphi^* \left(\sqrt{1 - |z_j|^2} K_{z_j} \right) = \sqrt{1 - |z_j|^2} K_{z_{j+1}} = \sqrt{\frac{1 - |z_j|^2}{1 - |z_{j+1}|^2}} k_{j+1}$$

Notice that if $\lim_{j \to \infty} z_j = a$ and $\lim_{j \to -\infty} z_j = b$, then the spectrum of the weighted shift with weight sequence $(1 - |z_j|^2)^{1/2}(1 - |z_{j+1}|^2)^{-1/2}$ is

$$\{\lambda : \varphi'(b)^{-1/2} \leq |\lambda| \leq \varphi'(a)^{1/2}\}$$

Thus, the weighted shifts explain the outer part of the spectrum, but the central part of the spectrum does not seem to be connected to these weighted shifts. We record the most obvious of these observations.

COROLLARY 7.28

If φ is a map of the disk into the disk that is analytic in a neighborhood of the closed disk and has Denjoy–Wolff point a on the unit circle with $\varphi'(a) < 1$, then there are subspaces M of $H^2(D)$ such that M is invariant subspace for C_φ^* and the restriction of C_φ^* to M is similar to a weighted shift.

The ideas of Theorem 7.27 can be extended to an iteration sequence $\{z_j\}$ converging, as j tends to negative infinity, to a fixed point b whose angular derivative is bigger than 1 as well. Moreover, although it is not generally true that the union of two interpolating sequences is again an interpolating sequence, if the one sequence converges to b on the circle and the other converges to $b' \neq b$, the union is again an interpolating sequence. We may then put the corresponding normalized reproducing kernel functions together into a single almost orthonormal basis. This construction, together with the observation of Exercise 7.5.6, suggests the following theorem that exhibits more structure to the spectrum of C_φ, but we will not give the details of the proof. Note that the hypothesis $z_0 = z_0'$ implies φ is not univalent on the disk and that b and b' are fixed points of φ with $\varphi'(b), \varphi'(b') > \varphi'(a)^{-1}$ where a is the Denjoy–Wolff point of φ.

THEOREM 7.29

Let φ be a map of the disk into the disk that is analytic in a neighborhood of the closed disk and suppose $\{z_j\}_{j=-\infty}^0$ and $\{z_j'\}_{j=-\infty}^0$ are iteration sequences with $\lim_{j \to -\infty} z_j = b$ and $\lim_{j \to -\infty} z_j' = b'$ with $b \neq b'$. If $\varphi'(b') \leq \varphi'(b)$ and $|\lambda| < \varphi'(b)^{-1/2}$, then λ is an eigenvalue of C_φ^* on $H^2(D)$.

Exercises

7.5.1 Suppose φ is analytic in a neighborhood of the closed unit disk and φ is not inner. Apply the identity principle to $\varphi(\overline{z}^{-1})$ to show that $|\varphi(z)| = 1$ for at most a finite number of points of the unit circle.

7.5.2 Show that if φ is analytic in a neighborhood of the closed unit disk and is not an inner function then there is a positive integer n so that $\{e^{i\theta} : |\varphi_n(e^{i\theta})| = 1\}$ is either empty or consists entirely of fixed points of φ_n.

7.5.3 (a) Suppose $A = (a_{jk})$ and $B = (b_{jk})$ are matrices with $|a_{jk}| \leq b_{jk}$. Show that if B defines a bounded operator on ℓ^2 (with respect to the usual basis), then so does A and $\|A\| \leq \|B\|$.

   (b) Show that if $T = (t_{jk})$ is the Toeplitz matrix with $t_{jk} = r^{|j-k|}$ for $0 < r < 1$, then T is a bounded operator on ℓ^2. What is the norm of T?

7.5.4 Suppose φ is analytic in a neighborhood of the closed unit disk, maps the unit disk to itself and has Denjoy–Wolff point a on the unit circle with $\varphi'(a) < 1$. Prove that if b is another point of the unit circle with $\varphi(b) = b$ and $|\lambda| < (\varphi'(b))^{-1/2}$, then λ is not an eigenvalue of C_φ.

7.5.5 For φ an analytic mapping of the unit disk to itself with Denjoy–Wolff point a on the unit circle and $\varphi'(a) < 1$, show that every eigenvalue of C_φ^* has infinite multiplicity.

7.5.6 Let $\{e_n, e_n', d_n\}_{n=0}^\infty$ be an orthonormal basis for the Hilbert space $\mathcal{H}$ and let $0 < w_n, w_n'$ be given. Let W be the operator on $\mathcal{H}$ given by $W e_n = w_n e_{n-1}$ and $W e_n' = w_n' e_{n-1}'$ for $n \geq 1$, $W e_0 = W e_0' = d_0$, and $W d_n = d_{n+1}$ for $n \geq 0$. (Such an operator W, called a ***branched weighted shift***, can be expressed in terms of a weighted composition operator where the underlying map of the set of integers into itself is onto and one-to-one except at a single point.) Show that if $\lim_{n \to \infty} w_n = \rho$ and $\lim_{n \to \infty} w_n' = \rho'$ where $\rho \leq \rho'$, then $|\lambda| < \rho$ implies λ is an eigenvalue of W.

Notes

Theorem 7.26 was stated by H. Kamowitz [Kam75] but there is an error in the proof. Many of the results of this section, including the correction of the error in Kamowitz's paper, appear in somewhat greater generality in Cowen's paper [Co83]. For example, under some less restrictive smoothness hypotheses, for non-automorphisms φ with Denjoy–Wolff point on the boundary and $\varphi'(a) < 1$, the spectrum of C_φ on $H^2(D)$ is the disk of Theorem 7.26.

It follows from work of H. S. Shapiro and A. L. Shields [ShS61] that Theorem 7.27 is a special case of a more general result, namely, that normalized kernel functions corresponding to an interpolating sequence form an almost orthonormal set in $H^2(D)$. The proof given here depends on the fact that for iteration sequences, the convergence to the boundary is geometric and this simplifies the proof considerably.

Exercise 7.5.6 is due to J. Carlson ([Caj85] and [Caj90b]) where the motivation is to study situations as in Theorem 7.29. The results of Exercise 7.5.1 and Exercise 7.5.2 are in Lemmas 1.3, 1.4 and 1.5 of [Kam75].

7.6 Spectra: interior fixed point

In this section, we consider the spectra for non-compact composition operators whose symbols have a fixed point in the ball. We will see, at least in the case φ is univalent and not an automorphism or unitary on a slice, that the spectrum contains a disk in addition to the eigenvalues. Note that the function theory of the eigenfunctions is the same as in the discussion of compact composition operators: the analytic functions satisfying $f(\varphi(z)) = \lambda f(z)$ exist for the special values of λ as in the previous section, but they need not be in the space. Thus, we shall see that in these cases, the spectrum includes some of these eigenvalues and a disk centered at the origin. For the sake of clarity, the proofs will be presented only in the special case of composition operators on $H^2(D)$ whose symbols are univalent maps of the disk into itself. The proofs for other known cases are closely related but must overcome more technical hurdles. Some of the generalizations for other cases are stated without proof. The main result of the section is the following theorem.

THEOREM 7.30

Let φ, not an automorphism, be univalent on the disk with $\varphi(D) \subset D$ and $\varphi(a) = a$ for some point a with $|a| < 1$. If C_φ is the associated composition operator on $H^2(D)$, then

$$\sigma(C_\varphi) = \{\lambda : |\lambda| \leq \rho\} \cup \{\varphi'(a)^k : k = 1, 2, \ldots\} \cup \{1\}$$

where ρ is the essential spectral radius of C_φ.

The proofs of this and other results of this section are based on the weighted shift analogy developed in the last section. Again, the kernel functions for an iteration sequence act like an orthogonal basis and the adjoint of a composition operator acts like a weighted shift on this basis, but in this case, because the iterates of any point are converging to a point of the disk, these kernel functions are becoming less and less like an orthogonal basis. Therefore, we restrict our attention to an invariant subspace on which the kernel functions get small. The easy estimate for the essential norm given in Proposition 3.13 is sharp for univalent maps and allows a computation of the essential spectral radius as well.

THEOREM 7.31
If φ is a univalent map of D into itself, then the essential norm of C_φ on $H^2(D)$ satisfies

$$\|C_\varphi\|_e = \limsup_{|w| \to 1} \frac{\|K_{\varphi(w)}\|}{\|K_w\|}$$

and the essential spectral radius, ρ, satisfies

$$\rho = \lim_{k \to \infty} \left(\limsup_{|w| \to 1} \frac{\|K_{\varphi_k(w)}\|}{\|K_w\|} \right)^{1/k}$$

where φ_k denotes the kth iterate of φ.

PROOF Since φ is assumed to be univalent, $N_\varphi(u) = -\log|\varphi^{-1}(u)|$ for u in $\varphi(D)$. Since $\lim_{x \to 1}(1 - x^2)/\log x = -2$, using Theorem 3.20 and letting $\varphi(w) = u$, we get

$$\|C_\varphi\|_e^2 = \limsup_{|u| \to 1} \frac{N_\varphi(u)}{-\log|u|} = \limsup_{|w| \to 1} \frac{\|K_{\varphi(w)}\|^2}{\|K_w\|^2}$$

Now the spectral radius formula in the Calkin algebra (the Banach algebra of bounded operators modulo the compact operators) shows the essential spectral radius of C_φ is

$$\lim_{n \to \infty} \left(\|C_\varphi{}^n\|_e \right)^{1/n} = \lim_{n \to \infty} \left(\|C_{\varphi_n}\|_e \right)^{1/n}$$

which is the formula in the theorem.   ∎

If $\varphi(a) = a$ and ψ is an automorphism of the disk interchanging 0 and a, then C_φ and $C_{\psi^{-1} \circ \varphi \circ \psi}$ are similar and $\psi^{-1} \circ \varphi \circ \psi(0) = 0$. Without loss of generality, then, we may assume $\varphi(0) = 0$ to prove Theorem 7.30.

Letting $\mathcal{H}_m = z^m H^2(D)$ for non-negative integers m, $\varphi(0) = 0$ implies $\mathcal{H}_m$ is invariant for C_φ. Moreover, since $H^2(D)$ is a Hilbert space of analytic functions, so is the subspace $\mathcal{H}_m$ and K_w^m, the kernel for evaluation of functions in $\mathcal{H}_m$ at the point w of the disk, is

$$K_w^m(z) = \sum_{j=m}^{\infty} (\overline{w}z)^j = \frac{(\overline{w}z)^m}{1 - \overline{w}z}$$

In particular, $\|K_w^m\|^2 = |w|^{2m}/(1 - |w|^2)$.

Since $\mathcal{H}_m^{\perp}$ is span$(1, z, \ldots, z^{m-1})$, the orthogonal complement of $\mathcal{H}_m$ is finite dimensional and Lemma 7.17 shows that we can compute the spectrum of C_φ by putting together the spectra of C_φ on $\mathcal{H}_m$ and the compression of C_φ on $\mathcal{H}_m^{\perp}$. The following proposition is a generalization of Theorem 7.20 on the spectrum of compact composition operators. In several variables, just as in Section 7.4, an analogous argument works with $\mathcal{H}_m$ being the subspace spanned by the monomials of total degree greater than or equal to m. In this case, $\varphi'(0)$ is an $N \times N$ matrix and the eigenvalues of the matrix X in Lemma 7.17 are 1 and all products of $m - 1$ or fewer eigenvalues of $\varphi'(0)$.

PROPOSITION 7.32

If φ is map of the disk into itself with $\varphi(a) = a$ for some a, $|a| < 1$, then the spectrum of C_φ on $H^2(D)$ includes 1 and $\varphi'(a)^n$ for $n = 1, 2, \ldots$.

PROOF We may assume that $a = 0$ as noted above so that $\varphi(z) = \varphi'(0)z + c_2 z^2 + \cdots$. The matrix for C_φ with respect to the usual orthonormal basis $\{z^k\}$ for $H^2(D)$ is lower triangular and the diagonal entries are 1, $\varphi'(0)$, $\varphi'(0)^2$, ... Taking $\mathcal{L}$ to be $\mathcal{H}_m$ in Lemma 7.17, the matrix X is an $m \times m$ lower triangular matrix with diagonal entries 1, $\varphi'(0)$, $\varphi'(0)^2$, ..., $\varphi'(0)^{m-1}$, so these numbers are the spectrum of X. The conclusion of Lemma 7.17 implies these numbers are in the spectrum of C_φ. Since m was arbitrary, the conclusion follows. ∎

The most important technical aspect of the proof is the control of the sizes of the kernel functions that will be part of the shifted basis. In classical weighted shifts, the shifted basis is an orthogonal basis so the eigenvectors, or approximate eigenvectors, are expressed as infinite series whose norms are easily computed. In our case, the kernel functions in the series are not orthogonal and the norms of the series are more delicate to compute. Since the points at which these kernels evaluate are iterates under φ, this means that we need to understand how far various iterates are from the origin (which is the limit of $\varphi_n(z)$ for each z). The next lemma gives a rate at which points leave the boundary of the disk.

LEMMA 7.33

If φ is an analytic map, not an automorphism, of the unit disk into itself and $\varphi(0) = 0$, then for $0 < r < 1$, there is $A > 1$ so that

$$\frac{1 - |\varphi(z)|}{1 - |z|} > A$$

for all z with $|z| \geq r$.

PROOF   The Schwarz lemma and the fact that φ is not an automorphism implies that $|\varphi(z)| < |z|$ for every non-zero z in D, so

$$\frac{1 - |\varphi(z)|}{1 - |z|} > 1 \tag{7.6.1}$$

for non-zero z in D. On the other hand, Proposition 2.46 says there is ζ_0 on the unit circle so that

$$\inf_{\zeta \in \partial D} |\varphi'(\zeta)| = \liminf_{|z| \to 1} \frac{1 - |\varphi(z)|}{1 - |z|} = |\varphi'(\zeta_0)|$$

and Inequality (7.6.1) says $|\varphi'(\zeta_0)| \geq 1$. Now if $|\varphi'(\zeta_0)| = 1$, Julia's Lemma (Lemma 2.41) would imply for any $k > 0$, φ maps the disk

$$\frac{|\zeta_0 - z|^2}{1 - |z|^2} \leq k$$

into the image of this disk under a rotation. Since $(1 - k)\zeta_0/(1 + k)$ is the point of this disk closest to the origin, it is easy to see that this implies that, for $0 < k < 1$,

$$\left| \varphi \left(\frac{1 - k}{1 + k} \zeta_0 \right) \right| \geq 1 - \frac{2k}{1 + k} = \frac{1 - k}{1 + k}$$

contradicting $|\varphi(z)| < |z|$. Thus, $|\varphi'(\zeta_0)| > 1$ and there is r between 0 and 1 so that $(1 - |\varphi(z)|)/(1 - |z|) > (|\varphi'(\zeta_0)| + 1)/2 \equiv A$ whenever $r \leq |z| \leq 1$.   ∎

Recall that a sequence $\{z_k\}$ in the disk is called an ***interpolating sequence*** if the map $h \to (h(z_k))$ is a map of $H^\infty(D)$ onto ℓ^∞. The closed graph theorem then guarantees that for any sequence of values, there is a corresponding h whose norm is comparable to the norm of the sequence. Moreover, the constant relating the norms depends only on the relative distance of the $\{z_k\}$ from each other not on the exact locations of the points. For example, if there is a number $a < 1$ so that

$$\frac{1 - |z_k|}{1 - |z_{k+1}|} < a < 1$$

then the sequence $\{|z_k|\}$ is interpolating for $H^\infty(D)$ [Hof62, p. 203] and the interpolating constant depends only on a. For A as in the previous lemma, taking $a = 1/A$, we get the following result.

LEMMA 7.34
Let φ and r be as in the previous lemma. There exists $M < \infty$ such that if $\{z_k\}_{-K}^\infty$ is an iteration sequence with $|z_n| \geq r$ for some integer n with $n \geq 0$ and if $\{w_k\}_{-K}^n$ is arbitrary, then there is h in $H^\infty(D)$ such that

$$h(z_k) = w_k \quad \text{for} \quad -K \leq k \leq n$$

and

$$\|h\|_\infty \leq M \sup\{|w_k| : -K \leq k \leq n\}$$

The rate at which iterates move toward the fixed point is also under control.

LEMMA 7.35
If φ is an analytic map, not an automorphism, of the unit disk into itself with $\varphi(0) = 0$ and $\{z_k\}$ is any iteration sequence, then there exists $c < 1$ such that

$$\frac{|z_{k+1}|}{|z_k|} \leq c$$

whenever $|z_k| \leq .5$.

PROOF   Since $\varphi(0) = 0$ and φ is not an automorphism of the disk, the Schwarz Lemma guarantees that $|\varphi'(0)| < 1$ and $|\varphi(z)| < |z|$ when $|z| < 1$. In particular, the function $|\varphi(z)|/|z|$, defined to be $\varphi'(0)$ for $z = 0$, is continuous and less than 1 on the open disk. It follows that on the compact set $|z| \leq .5$, the maximum of the function is less than 1 and the conclusion follows.   ∎

We are now ready to prove Theorem 7.30.

PROOF (of Theorem 7.30)   If $\rho = 0$, there is nothing to prove because φ is not an automorphism and is therefore not invertible. Thus, we assume $\rho > 0$ and $0 < |\lambda| < \rho$. We will show that, for m large enough, $(C_m - \bar\lambda I)^*$ is not bounded below, where C_m is the restriction of C_φ to $\mathcal{H}_m$. Lemma 7.17 will show that $\bar\lambda$ is in the spectrum of C_φ, and the fact that the spectrum is closed gives the conclusion of the theorem.

If $\{z_k\}_{-K}^\infty$ is an iteration sequence with $|z_0| > .5$, let n be the integer

$$n = \max\{k : |z_k| > .25\}$$

Since $\{|z_k|\}$ is a decreasing sequence converging to 0, it follows that $n \geq 0$ and $|z_k| < .25$ for $k > n$ and $|z_k| \geq .25$ for $k \leq n$.) If $|z_n| > .5$, then $|z_{n+1}| < .25$ implies $|z_{n+1}| \leq .5|z_n|$. Lemma 7.35, together with this observation, shows there is a number c with $.5 \leq c < 1$ so that $|z_{k+1}| \leq c|z_k|$ whenever $k \geq n$. Applying this inequality repeatedly, we find, for $k \geq n$,

$$|z_k| = |z_n| \left(\frac{|z_{n+1}|}{|z_n|}\right) \cdots \left(\frac{|z_k|}{|z_{k-1}|}\right) \leq |z_n| c^{k-n} \tag{7.6.2}$$

Let M be the interpolation constant for Lemma 7.34 with $r = .25$. (Of course, $M \geq 1$.) Choose a positive integer m large enough that

$$\frac{c^m}{|\lambda|} < \frac{1}{7M} \tag{7.6.3}$$

We will show $C_m^* - \lambda I$ is not bounded below on $\mathcal{H}_m$.

If $\{z_k\}_{-K}^{\infty}$ is an iteration sequence for φ with $|z_0| > .5$, then the series

$$\sum_{k=-K}^{\infty} \lambda^{-k} K_{z_k}^m$$

converges absolutely. Indeed, for $k > n$, since $|z_k| < .25$, we see that

$$\|K_{z_k}^m\| = \frac{|z_k|^m}{\sqrt{1 - |z_k|^2}} \leq 2|z_k|^m$$

and it follows that

$$\sum_{k=n+1}^{\infty} |\lambda|^{-k} \|K_{z_k}^m\| \leq 2 \sum_{k=n+1}^{\infty} \frac{|z_k|^m}{|\lambda|^k} \leq 2 \frac{|z_n|^m}{|\lambda|^n} \sum_{k=n+1}^{\infty} \left(\frac{c^m}{|\lambda|}\right)^{k-n} < \infty$$

We also need a lower bound for $\|\sum_{k=-K}^{\infty} \lambda^{-k} K_{z_k}^m\|$. By Lemma 7.34, we can find a function f in $H^{\infty}(D)$ so that $\|f\|_{\infty} \leq M$ and for $k \leq n$,

$$|f(z_k)| = 1 \quad \text{and} \quad \frac{z_k^m f(z_k)}{\lambda^k (1 - z_0 \overline{z_k})} > 0 \tag{7.6.4}$$

Now by Eq. (7.6.4)

$$\langle \sum_{-K}^{\infty} \lambda^{-k} K_{z_k}^m, \frac{z^m f}{1 - \overline{z_0} z} \rangle$$

$$= \sum_{-K}^{n-1} \frac{|z_k|^m}{|\lambda|^k |1 - \overline{z_0} z_k|} + \frac{|z_n|^m}{|\lambda|^n |1 - \overline{z_0} z_n|} + \sum_{k=n+1}^{\infty} \frac{\overline{z_k}^m \overline{f(z_k)}}{\lambda^k (1 - z_0 \overline{z_k})}$$

By Inequality (7.6.3) and Lemma 7.35 and the fact that $|1 - z_0 \overline{z_k}| \geq .75$

$$\left| \sum_{k=n+1}^{\infty} \frac{\overline{z_k}^m \overline{f(z_k)}}{\lambda^k (1 - z_0 \overline{z_k})} \right| \leq \sum_{k=n+1}^{\infty} \frac{|z_k|^m |f(z_k)|}{|\lambda|^k |1 - z_0 \overline{z_k}|} \leq \frac{4M|z_n|^m}{3|\lambda|^n} \sum_{k=n+1}^{\infty} \left(\frac{c^m}{|\lambda|}\right)^{k-n}$$

$$= \frac{4M|z_n|^m}{3|\lambda|^n} \frac{\frac{c^m}{|\lambda|}}{1 - \frac{c^m}{|\lambda|}} \leq \frac{4M|z_n|^m}{3|\lambda|^n} \frac{\frac{1}{7M}}{1 - \frac{1}{7M}} \leq \frac{4M|z_n|^m}{3|\lambda|^n} \frac{1}{6M} = \frac{2|z_n|^m}{9|\lambda|^n}$$

This means

$$\left| \langle \sum_{-K}^{\infty} \lambda^{-k} K_{z_k}^m, \frac{z^m f}{1 - \overline{z_0} z} \rangle \right|$$

$$\geq \sum_{-K}^{n-1} \frac{|z_k|^m}{|\lambda|^k |1 - \overline{z_0} z_k|} + \frac{|z_n|^m}{|\lambda|^n |1 - \overline{z_0} z_n|} - \left| \sum_{k=n+1}^{\infty} \frac{\overline{z_k}^m \overline{f(z_k)}}{\lambda^k (1 - z_0 \overline{z_k})} \right|$$

$$= \sum_{-K}^{n-1} \frac{|z_k|^m}{|\lambda|^k |1 - \overline{z_0} z_k|} + \frac{1}{2} \frac{|z_n|^m}{|\lambda|^n |1 - \overline{z_0} z_n|}$$

$$+ \left(\frac{1}{2} \frac{|z_n|^m}{|\lambda|^n |1 - \overline{z_0} z_n|} - \left| \sum_{k=n+1}^{\infty} \frac{\overline{z_k}^m \overline{f(z_k)}}{\lambda^k (1 - z_0 \overline{z_k})} \right| \right)$$

$$\geq \sum_{-K}^{n-1} \frac{|z_k|^m}{|\lambda|^k |1 - \overline{z_0} z_k|} + \frac{1}{2} \frac{|z_n|^m}{|\lambda|^n |1 - \overline{z_0} z_n|} + \frac{|z_n|^m}{|\lambda|^n} \left(\frac{1}{4} - \frac{2}{9} \right)$$

$$\geq \sum_{-K}^{n-1} \frac{|z_k|^m}{|\lambda|^k |1 - \overline{z_0} z_k|} + \frac{1}{2} \frac{|z_n|^m}{|\lambda|^n |1 - \overline{z_0} z_n|} \geq \frac{1}{2} \frac{|z_0|^m}{1 - |z_0|^2}$$

We have used $|1 - \overline{z_0} z_n| \leq 2$ and the final inequality is obtained by ignoring terms with $k \neq 0$. Since

$$\left\| \frac{z^m f}{1 - \overline{z_0} z} \right\| \leq \frac{M}{\sqrt{1 - |z_0|^2}}$$

the Cauchy–Schwarz inequality now yields the desired lower bound:

$$\left\| \sum_{-K}^{\infty} \lambda^{-k} K_{z_k}^m \right\| \geq \frac{\left| \langle \sum_{-K}^{\infty} \lambda^{-k} K_{z_k}^m, \frac{z^m f}{1 - \overline{z_0} z} \rangle \right|}{\left\| \frac{z^m f}{1 - \overline{z_0} z} \right\|} \geq \frac{|z_0|^m}{2M \sqrt{1 - |z_0|^2}} = \frac{1}{2M} \| K_{z_0}^m \|$$

Note that since $\mathcal{H}_m$ is invariant for C_φ, for any g in $\mathcal{H}_m$,

$$\langle g, C_\varphi^* K_w^m \rangle = \langle C_\varphi g, K_w^m \rangle = \langle g \circ \varphi, K_w^m \rangle = g(\varphi(w)) = \langle g, K_{\varphi(w)}^m \rangle$$

It follows that

$$\left(C_m^* - \lambda I \right) \left(\sum_{-K}^{\infty} \lambda^{-k} K_{z_k}^m \right) = -\lambda^{K+1} K_{z_{-K}}^m$$

and

$$\frac{\left\| \left(C_m^* - \lambda I \right) \left(\sum_{-K}^{\infty} \lambda^{-k} K_{z_k}^m \right) \right\|}{\left\| \sum_{-K}^{\infty} \lambda^{-k} K_{z_k}^m \right\|} \leq 2M |\lambda|^{K+1} \frac{\| K_{z_{-K}}^m \|}{\| K_{z_0}^m \|}$$

Finally we make our choice of iteration sequence $\{z_k\}_{-K}^{\infty}$. Since

$$\rho = \lim_{k \to \infty} \left(\limsup_{|w| \to 1} \frac{\| K_{\varphi_k(w)} \|}{\| K_w \|} \right)^{1/k}$$

and we are assuming $\rho > 0$ we must have

$$\limsup_{|w| \to 1} \| K_{\varphi_k(w)} \| = \infty$$

for every k. Since m is fixed and $\|K_u^m\| \leq \|K_u\| \leq \|K_u^m\| + m$ for all u in D, for the w near ∂D important in the computation of ρ

$$\frac{\|K_{\varphi_k(w)}^m\|}{\|K_w^m\|} \approx \frac{\|K_{\varphi_k(w)}\|}{\|K_w\|}$$

so the essential spectral radius may be computed

$$\rho = \lim_{k \to \infty} \left(\limsup_{|w| \to 1} \frac{\|K_{\varphi_k(w)}^m\|}{\|K_w^m\|} \right)^{1/k}$$

Pick ρ' with $|\lambda| < \rho' < \rho$. For any positive integer K, there are points w near ∂D so that $|\varphi^{(K)}(w)| > .5$ and

$$\frac{\|K_{\varphi_K(w)}^m\|}{\|K_w^m\|} > (\rho')^K$$

Let $z_{-K} = w$ and $z_{k+1} = \varphi(z_k)$ for $k \geq -K$. Thus the iteration sequence $\{z_k\}_{-K}^\infty$ has $|z_0| > .5$ and our calculations give, for this iteration sequence

$$\frac{\|(C_m^* - \lambda I)\left(\sum_{-K}^\infty \lambda^{-k} K_{z_k}^m\right)\|}{\|\sum_{-K}^\infty \lambda^{-k} K_{z_k}^m\|} \leq 2M|\lambda|^{K+1} \frac{\|K_{z_{-K}}^m\|}{\|K_{z_0}^m\|} \leq 2M|\lambda| \left(\frac{|\lambda|}{\rho'} \right)^K$$

Since $|\lambda| < \rho'$ we may choose K sufficiently large so that this is as small as desired. Thus $C_m^* - \lambda I$ is not bounded below and $\bar{\lambda}$ is in $\sigma(C_\varphi)$ as desired. ∎

Kamowitz [Kam75] was the first to investigate spectra of composition operators whose symbol is not an inner function and has a fixed point in the disk. The hypothesis of his theorem is analyticity in a neighborhood of the closed disk. From the perspective of the proof of Theorem 7.30, this hypothesis allows him to calculate the essential spectral radius in terms of the derivative of the function φ at its fixed points on the unit circle and to find iteration sequences on which the estimates in the proof are still valid.

Specifically, if φ is analytic in a neighborhood of the closed disk and is not an inner function, Exercise 7.5.2 guarantees that there is an integer n so that $S_n = \{w : |w| = 1 \text{ and } |\varphi_n(w)| = 1\}$ is either empty or consists only of the (finitely many) fixed points of φ_n on the circle. If φ has a fixed point a in the open disk, then, for such an n, the derivatives of φ_n at each of the fixed points on the circle are bigger than 1 and the essential spectral radius of C_φ^n is

$$\max\{\varphi_n'(w)^{-1/2} : w \in S_n\}$$

Applying the spectral mapping theorem for the Calkin algebra, we see the essential spectral radius of C_φ is

$$\rho = \max\{\varphi_n'(w)^{-1/2n} : w \in S_n\}$$

As we see in Kamowitz's result below, with this extra hypothesis, the same conclusion as in Theorem 7.30 is true even without univalence.

THEOREM 7.36
Let φ, not an inner function, be analytic in a neighborhood of the closed disk with $\varphi(D) \subset D$ and $\varphi(a) = a$ for some point a with $|a| < 1$. If C_φ is the associated composition operator on $H^2(D)$, then

$$\sigma(C_\varphi) = \{\lambda : |\lambda| \le \rho\} \cup \{\varphi'(a)^k : k = 1, 2, \ldots\} \cup \{1\}$$

where ρ is the essential spectral radius of C_φ.

The fact that the inner functions are excluded from the theorems presented thus far is no indication that they are more difficult, indeed, they are the easiest of the functions that have a fixed point in the disk. We will see in Section 7.8 that if φ is an inner function with a fixed point in the disk but not an automorphism, then C_φ is similar to a unilateral shift of infinite multiplicity and the spectrum and the essential spectrum of C_φ are the closed unit disk.

Just as the automorphisms of the disk are a special case because they give invertible composition operators, there are similar complications in higher dimensions. Indeed, difficulties arise in higher dimensions for some maps that are not automorphisms, for example, if $\varphi(0) = 0$ and the restriction of φ to some slice of the ball is unitary, and we will not consider such mappings. Recall that φ *is **unitary on a slice in** B_N* if there exist η and ζ in ∂B_N such that $\varphi(\lambda \eta) = \lambda \zeta$ for every scalar λ in D.

In addition, for $N > 1$, further restrictions on φ are necessary to ensure that C_φ is a bounded operator on the Hardy space $H^2(B_N)$. A convenient sufficient condition for boundedness of C_φ on these spaces involves the quantity

$$\Omega(z) = \frac{\|\varphi'(z)\|^2}{|J_\varphi(z)|^2}$$

where $\|\varphi'(z)\|$ denotes the norm of the derivative of φ and $J_\varphi(z)$ is the complex Jacobian of φ. If φ is univalent, Theorem 3.41 shows that boundedness of $\Omega(z)$ in B_N is sufficient to ensure that C_φ is bounded on $H^2(B_N)$. The following several variable version of Theorem 7.30 is known.

THEOREM 7.37
Suppose φ is an analytic map of B_N into B_N that is univalent, $\varphi(0) = 0$, and φ is not unitary on any slice of B_N. If $\Omega(z)$ is bounded in B_N, then the spectrum of C_φ as an operator on $H^2(B_N)$ includes the disk

$$\left\{\lambda : |\lambda| \le \frac{\rho^N}{\|\Omega\|_\infty^{N/2}}\right\}$$

where ρ is the essential spectral radius of C_φ on $H^2(B_N)$.

For $N = 1$, we have $\Omega(z) \equiv 1$ so the disk in this theorem specializes to the disk in the one variable result, Theorem 7.30.

Exercises

7.6.1 Given B an inner function on the disk, let $\mathcal{H} = BH^2$. For w in the disk with $B(w) \neq 0$, find the kernel K_w^B for evaluation of functions in $\mathcal{H}$. That is, find K_w^B in $\mathcal{H}$ so that for f in $\mathcal{H}$, $\langle f, K_w^B \rangle = f(w)$. Can you solve the analogous problem for arbitrary subspaces of a Hilbert space of analytic functions?

7.6.2 Justify the assertions made in the table of Section 7.1 about the composition operator whose symbol is $\varphi(z) = rz/(1 - (1 - r)z)$.

7.6.3 For $\varphi(z) = -(z^3 + z)/2$, find the spectrum of C_φ on $H^2(D)$.

7.6.4 Let $\varphi(z) = (z^8 + z)/2$. Find the spectrum of C_φ on $H^2(D)$ and find the Riesz projection associated with the disk in the spectrum.

7.6.5 Suppose A is an $N \times N$ matrix with $\|A\| \leq 1$. Find the spectrum of C_φ on $H^2(B_N)$ where φ is the map $\varphi(z) = Az$.

Notes

E. Nordgren [No68] showed that composition operators with inner symbol and fixed point in the disk are similar to isometries and found the Wold decomposition of the isometries. In particular, this gave the spectrum in this case. H. Kamowitz [Kam75] proved Theorem 7.36 and illuminated the way for investigation of the spectra of non-compact composition operators. Kamowitz's approach was directly through C_φ, rather than through C_φ^* as we have done. The proof is accomplished by showing that C_φ does not map $H^2(D)$ onto itself. Theorems 7.30 and 7.37 and generalizations to other settings appear in [CoM94]. The proof of Theorem 7.30 given here is a simplification, for the special hypotheses, of the proof appearing in [CoM94].

There is still much to be done on spectra in this case. The parts of the spectrum are not classified and the non-univalent case needs work. J. Carlson's work [Caj85] and [Caj90b] on branched weighted shifts would seem relevant in treating the case of non-univalent symbols. It would appear that, on the disk, the spectrum of C_φ is connected except for the eigenvalues $\varphi'(a)^n$, but this has not been proved. Is there an explanation for the circular symmetry of the spectrum, except for the eigenvalues $\varphi'(a)^n$? For example, is there a subspace for which the restriction of C_φ is similar to rotates of itself?

7.7 Spectra: boundary fixed point, $\varphi'(a) = 1$

When $\varphi'(a) = 1$, the situation is much more difficult: indeed, this hypothesis includes φ in both the halfplane/translation case and the plane/translation case. These cases in the model are not readily distinguishable, yet the composition operators exhibit quite different properties. One feature distinguishing the cases, that any iteration sequence is an interpolating sequence if φ is in the halfplane/translation case (Exercise 2.4.18), can be used with an analogue of Theorem 7.27 to show that in this case, there are invariant subspaces on which C_φ^* is similar to a weighted shift. It appears that there is no such direct connection with shifts when φ is in the plane/translation case. In general, the best we can say about the spectrum in this case follows from the spectral radius calculation of Theorem 3.9, that $\sigma(C_\varphi) \subset \overline{D}$. However, even for non-invertible C_φ, the spectrum can be a proper subset of the closed unit disk.

If $\varphi'(a) = 1$ and φ is in the halfplane/translation case of the model for iteration, some of the analogues of the results of Section 7.5 are true, but not all of them. For example, we have the following easy analog of Theorem 7.21 on the similarity of C_φ to multiples of C_φ.

THEOREM 7.38

Suppose φ is an analytic mapping of the unit disk to itself with Denjoy–Wolff point a on the unit circle and $\varphi'(a) = 1$ and suppose σ is a map of the disk to the upper halfplane such that $\sigma \circ \varphi(z) = \sigma(z) + 1$ for z in D. For each positive real number θ, the function f_θ defined by $f_\theta(z) = \exp(i\theta\sigma(z))$ satisfies $|f_\theta(z)| < 1$ for all z in D and $f_\theta \circ \varphi = e^{i\theta} f_\theta$.

PROOF Since $\sigma(D)$ is a subset of the upper halfplane, $i\theta\sigma(z)$ is in the left halfplane and $f_\theta(z)$ satisfies $|f_\theta(z)| < 1$. Moreover,

$$f_\theta(\varphi(z)) = \exp(i\theta\sigma(\varphi(z))) = \exp(i\theta(\sigma(z) + 1)) = e^{i\theta} f_\theta(z)$$

∎

Clearly, a similar theorem is true if the model for φ is translation by -1 on the upper halfplane; just replace $f_\theta \circ \varphi = e^{i\theta} f_\theta$ by $f_\theta \circ \varphi = e^{-i\theta} f_\theta$.

COROLLARY 7.39

If φ is an analytic mapping of the unit disk with a halfplane/translation model for iteration, then the spectrum and essential spectrum of C_φ on $H^2(D)$ contain the unit circle. Moreover, if λ is an eigenvalue of C_φ, then $e^{i\theta}\lambda$ is also an eigenvalue of C_φ for each positive number θ.

PROOF If g is an eigenvector for C_φ with eigenvalue λ, since f_θ is in $H^\infty(D)$,

$f_\theta g$ is in $H^2(D)$ and

$$C_\varphi(f_\theta g)(z) = f_\theta(\varphi(z))g(\varphi(z)) = e^{i\theta}f_\theta(z)\lambda g(z)$$

so $f_\theta g$ is an eigenvector for C_φ with eigenvalue $e^{i\theta}\lambda$.

Since the constant function 1 is an eigenvector for C_φ with eigenvalue 1, each point of the unit circle is a eigenvalue of infinite multiplicity for C_φ and the unit circle is contained in the spectrum and the essential spectrum of C_φ. ∎

However, the functions f_θ do not induce a similarity between C_φ and $e^{i\theta}C_\varphi$ as in the case $\varphi'(a) < 1$ because f_θ is never invertible in $H^\infty(D)$. Indeed, since the set $\sigma(V)$ of Theorem 2.53 is a fundamental set for translation in the upper halfplane, for each positive number M, there is z in D so that $\text{Im}\,\sigma(z) > M$, so for each $\epsilon > 0$, there is z in D so that $|f_\lambda(z)| < \epsilon$.

Results pertaining to the fixed points of φ besides the Denjoy–Wolff point, such as Theorems 7.23 and 7.29, apply to C_φ for φ in the halfplane/translation case as well. For φ not an automorphism but in the halfplane/translation case, the picture that emerges is that the spectrum of C_φ is a subset of the closed unit disk that contains the unit circle and a disk (perhaps degenerate) centered at the origin. However, no examples are known in which the spectrum is a proper subset of the unit disk for this case and the spectrum of C_φ for $|a| = 1$ and $\varphi'(a) < 1$ (given by Theorem 7.26) suggests that the spectrum should be the unit disk.

However, the situation is radically different for φ in the plane/translation case. The most obvious reason this should be so is that the weighted shift analogy that works in the halfplane/dilation case depends on the near orthogonality of the kernel functions for the iterates of a point but in the plane/translation case, these kernels are not close to orthogonal. The evidence that this is more than just a technical inconvenience is that the spectra for C_φ when φ is in the plane/translation case need not show any circular symmetry. This case is poorly understood; we present a class of examples that permit calculation of spectra but do not suggest plausible general techniques.

DEFINITION 7.40   *A set $\{\varphi_t : t \geq 0\}$ will be called a **one-parameter semigroup of analytic functions on** D if φ_t is an analytic map of the disk into itself for each t with $\varphi_0(z) \equiv z$, $\varphi_t \circ \varphi_s = \varphi_{s+t}$ and $(z, t) \mapsto \varphi_t(z)$ is jointly continuous. A semigroup indexed by a complex parameter t will be called a **holomorphic semigroup of analytic functions** if there is $\tau > 0$ such that the semigroup properties hold for $|\arg t| < \tau$ and on this sector the map $t \mapsto \varphi_t(z)$ is analytic for each z in D.*

An easy example of a one-parameter semigroup of analytic functions is $\{r^t z : t \geq 0\}$ where $|r| \leq 1$ and other simple examples can be constructed from other linear fractional maps of the disk into itself (see Exercise 7.7.1). For $0 < r < 1$, the one-parameter semigroup above can be extended to a holomorphic semigroup $\{r^t z : \text{Re}\,t > 0\}$. The model of Theorem 2.53 provides a good way to understand semigroups of analytic functions. Since automorphisms of the plane or the halfplane are always part of a semigroup, the intertwining $\Phi \circ \sigma = \sigma \circ \varphi$, which implies $\Phi_n \circ \sigma = \sigma \circ \varphi_n$ for positive integers n, can be formally extended to get $\Phi_t \circ \sigma = \sigma \circ \varphi_t$. It is not difficult to show that every function in a semigroup of analytic functions is univalent (Exercise 7.7.2), so the intertwining map σ is univalent and we see that we should have $\varphi_t = \sigma^{-1} \circ \Phi_t \circ \sigma$. This relation can be used to construct a semigroup φ_t as long as the set $\sigma(D)$ is invariant under Φ_t for $t \geq 0$.

If φ_t is a semigroup of analytic functions, $\{C_{\varphi_t}\}$ is a strongly continuous semigroup of operators which can be studied with the tools developed in that theory [HiP57]. (A set of bounded operators $\{T_t\}$ is called a ***strongly continuous semigroup of operators*** if $T_s T_t = T_{s+t}$, $T_0 = I$, and $t \mapsto T_t x$ is continuous for each x in the space.) The examples of spectra of composition operators that we will exhibit arise as part of a semigroup in this way.

Given θ with $0 < \theta \leq \pi$, let G be the sector $G = \{\zeta : |\arg \zeta| < \theta\}$, and let σ be the map

$$\sigma(z) = \left(\frac{1+z}{1-z}\right)^{2\theta/\pi}$$

which is a conformal map of D onto G. Let $\tau(G) = G$ if $\theta \leq \pi/2$ and $\tau(G) = \{t : |\arg t| < \pi - \theta\}$ if $\theta > \pi/2$ so that, in either case, $\zeta + t$ is in G whenever ζ is in G and t is in $\tau(G)$. Now, for t in $\tau(G)$, define φ_t by $\varphi_t(z) = \sigma^{-1}(\sigma(z) + t)$ so that φ_t is analytic in D with $\varphi_t(D) \subset D$. Replacing σ by σ/t, we see the model for φ_t is also halfplane/translation and the function φ_t also has Denjoy–Wolff point 1 with $\varphi_t'(1) = 1$. We will write C_t for the composition operator C_{φ_t}. In the case $\theta = \pi/2$ (so G is a halfplane)

$$\varphi_t(z) = \frac{t + (2-t)z}{(2+t) - tz}$$

for $\operatorname{Re} t > 0$. In the case $\theta = \pi/4$, we find

$$\varphi_t(z) = \frac{t^2 + 2t\sqrt{1-z^2} + (2-t^2)z}{(2+t^2) + 2t\sqrt{1-z^2} - t^2 z}$$

for $|\arg t| < \pi/4$.

We will prove a spectral containment theorem using the theories of holomorphic semigroups and Banach algebras that will allow us to identify the spectrum of C_t in some cases.

THEOREM 7.41
Let θ, G, $\tau(G)$, and C_t be as above. Then on $H^2(D)$

$$\sigma(C_t) \subset \{e^{-\beta t} : |\arg \beta| \leq |\pi/2 - \theta|\} \cup \{0\}$$

for all t in $\tau(G)$.

PROOF The set $\{C_t : t \in \tau(G)\}$ is a holomorphic semigroup of operators. Indeed, since operator valued functions are analytic in the norm topology if and only if they are analytic in the weak-operator topology (Theorem 3.10.1 of [HiP57, p. 93]), it is sufficient to check that the map $t \mapsto \langle C_t(f), K_z \rangle$ is holomorphic in $\tau(G)$ for each f in $H^2(D)$ and each z in D. But, $\langle C_t(f), K_z \rangle = f(\sigma^{-1}(\sigma(z)+t))$ which is holomorphic in t because σ^{-1} and f are holomorphic. In particular, $t \mapsto C_t$ is a continuous and holomorphic function of t in the norm topology for t in $\tau(G)$.

Let $\mathcal{A}$ be the norm closed algebra of operators generated by $\{I\} \cup \{C_t : t \in \tau(G)\}$. Thus $\mathcal{A}$ is a commutative Banach algebra with identity and the Gelfand theory applies: the spectrum of C_t as an element of $\mathcal{A}$, denoted $\sigma_{\mathcal{A}}(C_t)$, is the set

$$\{\Lambda(C_t) : \Lambda \text{ is a multiplicative linear functional on } \mathcal{A}\}$$

For Λ a multiplicative linear functional on $\mathcal{A}$, let $\lambda(t) = \Lambda(C_t)$ for t in $\tau(G)$. Since Λ is multiplicative, the Gelfand theory implies $\|\Lambda\| = 1$ and since C_t is a norm-holomorphic semigroup, we see that $\lambda(t)$ is a holomorphic function on $\tau(G)$ such that

$$\lambda(t_1 + t_2) = \Lambda(C_{t_1+t_2}) = \Lambda(C_{t_1})\Lambda(C_{t_2}) = \lambda(t_1)\lambda(t_2)$$

This means, either $\lambda(t) \equiv 0$ or $\lambda(t) = e^{-\beta t}$ for some complex number β. In addition, we see that for every t in $\tau(G)$, using Theorem 3.9 and the fact that $\|\Lambda\| = 1$,

$$|e^{-\beta t}| = \lim_{n\to\infty} |e^{-\beta nt}|^{1/n} = \lim_{n\to\infty} |\Lambda(C_t^n)|^{1/n}$$

$$\leq \lim_{n\to\infty} \|C_t^n\|^{1/n} = \varphi_t'(1)^{-1/2} = 1.$$

The definition of $\tau(G)$ and this inequality imply $|\arg \beta| \leq |\pi/2 - \theta|$.

Putting all our information together, we have

$$\sigma(C_t) \subset \sigma_{\mathcal{A}}(C_t) = \{\Lambda(C_t) : \Lambda \text{ is a multiplicative linear functional on } \mathcal{A}\}$$

$$\subset \{e^{-\beta t} : |\arg \beta| \leq |\pi/2 - \theta|\} \cup \{0\}$$

as was to be proved.   ∎

COROLLARY 7.42

Suppose $0 < \theta \leq \pi/2$ and G, $\tau(G)$, and C_t are as above. Then on $H^2(D)$

$$\sigma(C_t) = \{e^{-\beta t} : |\arg \beta| \leq \pi/2 - \theta\} \cup \{0\}$$

for all t in $\tau(G)$.

PROOF When $|\arg \beta| \leq \pi/2 - \theta$, the real part of $\beta\sigma(z)$ is positive, so $f(z) = \exp(-\beta\sigma(z))$ is in $H^\infty(D)$. Since $f(\varphi_t(z)) = \exp(-\beta\sigma(z) - \beta t) = e^{-\beta t} f(z)$,

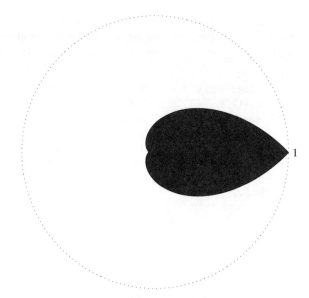

FIGURE 7.1

Spectrum of $C_\heartsuit$ in the unit disk.

we see that $e^{-\beta t}$ is an eigenvalue of C_t, so $\sigma(C_t) \supset \{e^{-\beta t} : |\arg \beta| \leq \pi/2 - \theta\}$. Theorem 7.41 and the fact that $\sigma(C_t)$ is closed yield the conclusion.   ∎

When $\theta = \pi/2$, each of these sets is a logarithmic spiral from 1 to 0. In particular, taking $t = 2$ and $\theta = \pi/2$, we find $\sigma(C_\varphi)$ is the closed interval $[0, 1]$ for $\varphi(z) = (2 - z)^{-1}$.

When $\theta = \pi/4$ and $t = 1$, that is,

$$\heartsuit(z) = \frac{1 + z + 2\sqrt{1 - z^2}}{3 - z + 2\sqrt{1 - z^2}}$$

we find $\sigma(C_\heartsuit)$ is the heart-shaped region $\{e^{-\beta} : |\arg \beta| \leq \pi/4\} \cup \{0\}$ (see Figure 7.1).

When $\pi/2 < \theta \leq \pi$, Theorem 7.41 gives a hint as to what the spectrum might be, but there are no obvious eigenvalues, so we are unable to go further.

The techniques used above depend heavily on the special semigroup properties of the defining maps under iteration. While maps that are part of a holomorphic semigroup are quite rare, the situation might not be completely hopeless: it can be shown from the model of Theorem 2.53 that if φ is analytic in the disk with $\varphi(D) \subset D$ and $\varphi'(a) \neq 0$, then for each z in the disk, there is a non-negative number t_z so that $\varphi_t(z)$ is defined for all $t > t_z$ and the dependence is holomorphic in z and real analytic in t. It remains to be seen whether these partial semigroups can help in finding $\sigma(C_\varphi)$.

Exercises

7.7.1 (a) Find a one-parameter semigroup of analytic functions that contains the automorphism $(2z + 1)/(2 + z)$.

(b) Find a one-parameter semigroup of analytic functions that contains the automorphism $((1 + i)z - 1)/(z + i - 1)$.

(c) Extend the result of (b) to find a holomorphic semigroup of analytic functions that contains the automorphism $((1 + i)z - 1)/(z + i - 1)$.

(d) Find a one-parameter semigroup of analytic functions that contains $.5 + .5z$.

7.7.2 Show that every function in a semigroup of analytic functions is univalent.

7.7.3 Give an example of a univalent map of the unit disk into itself that is not in a semigroup of analytic functions.

Notes

The results of this section are, for the most part, found in C. Cowen's paper [Co83]. If φ is analytic on the closed disk and $\varphi'(a) = 1$, then H. Kamowitz [Kam75, p. 142] proved that the spectrum of C_φ is contained in the closed unit disk and asks if the inclusion can ever be proper. Theorem 7.41 shows that this answer is sometimes yes. Nevertheless, Kamowitz's question is still interesting when φ is restricted to the halfplane/translation case.

A. Siskakis ([Sis85] – [Sis94b]) has made an extensive study of semigroups of composition operators, their infinitesimal generators, and their spectra.

The idea for the proof of Theorem 7.41 (which is Theorem 6.1 of [Co83]) was suggested by R. Kaufman.

7.8 Spectra: inner functions

In this section we consider the spectra of composition operators whose symbol is an inner function, that is, $|\varphi(e^{i\theta})| = 1$ almost everywhere on the circle, but not an automorphism. (Automorphisms were considered in Section 7.2.) In this context, inner functions are best thought of as many-to-one maps of the disk (almost) onto the disk. For example, z^3 is a three-to-one map of the disk onto itself and the singular inner function $\exp((z + 1)/(z - 1))$ is an infinite-to-one map of the disk onto the disk except 0.

The situation is most easily understood when φ has a fixed point in the disk, where $\varphi(z) = z^3$ is a typical example.

THEOREM 7.43

If φ is an inner function with a fixed point in the disk, not an automorphism, then C_φ acting on $H^2(D)$ is similar to an isometry whose unitary part is the identity *on a one dimensional space and whose purely isometric part is a unilateral shift of infinite multiplicity. Moreover,*

$$\sigma(C_\varphi) = \sigma_e(C_\varphi) = \{\lambda : |\lambda| \leq 1\}$$

PROOF If $\varphi(a) = a$ for $|a| < 1$, let $\psi(z) = (a - z)/(1 - \bar{a}z)$ which is an automorphism of the disk with $\psi(a) = 0$, $\psi(0) = a$, and $\psi^{-1} = \psi$. Then $\psi \circ \varphi \circ \psi$ is an inner function that has fixed point 0 in the disk and $C_{\psi \circ \varphi \circ \psi} = C_\psi^{-1} C_\varphi C_\psi$ is similar to C_φ. That is, without loss of generality, we may assume φ is an inner function with $\varphi(0) = 0$.

Theorem 3.8 shows that if $\varphi(0) = 0$, then $\|C_\varphi f\| = \|f\|$ for every f in $H^2(D)$, in other words, C_φ is an isometry. We need to find the Wold decomposition of the isometry C_φ. Since $C_\varphi 1 = 1$ and $C_\varphi^* 1 = C_\varphi^* K_0 = K_{\varphi(0)} = K_0 = 1$, we see that the subspace spanned by 1 is a reducing subspace for C_φ and that the restriction of C_φ to this subspace is the identity. If $f(z) = zg(z)$ is a function in $[1]^\perp = zH^2(D)$, then $\|C_\varphi^n f\| = \|f\| = \|g\| = \|C_\varphi^n g\|$ for all n. In particular, if h is in $\cap_n C_\varphi^n(zH^2(D))$, then for each n, there is $f = zg$ in $zH^2(D)$ so that $h = C_\varphi^n f$ and $\|h\| = \|f\| = \|g\|$. On the other hand, for w in the disk,

$$|h(w)| = |(C_\varphi^n f)(w)| = |\varphi_n(w)||(C_\varphi^n g)(w)|$$
$$\leq |\varphi_n(w)|\|C_\varphi^n g\|\|K_w\| = |\varphi_n(w)|\|h\|\|K_w\|$$

Since φ is not an automorphism and its Denjoy–Wolff point is 0, $|\varphi_n(w)|$ converges to 0 and $h(w) = 0$. Thus $\cap_n C_\varphi^n(zH^2(D)) = \{0\}$ which means the unitary part of C_φ is the restriction of C_φ to $[1]$ and the purely isometric part is the restriction to $zH^2(D)$.

Since C_φ is an isometry, the range of C_φ is closed and the multiplicity of the purely isometric part is simply the dimension of $C_\varphi(H^2(D))^\perp$. Our hypothesis is that φ is not an automorphism, so by Theorem 3.29, $C_\varphi(H^2(D))^\perp$ is infinite dimensional. (A direct proof of this is outlined in Exercise 7.8.1.)

Finally, since the purely isometric part of C_φ is a unilateral shift of infinite multiplicity, the spectrum and essential of C_φ are the closed unit disk. ∎

To study the operators arising from inner functions with Denjoy–Wolff point a on the unit circle, we will consider the point spectra of C_φ and C_φ^*. We know from Lemma 7.24 that in the case $|a| = 1$ with $\varphi'(a) < 1$ that the point spectrum of C_φ includes the annulus $\{\lambda : \varphi'(a)^{1/2} < |\lambda| < \varphi'(a)^{-1/2}\}$ and if φ is in the halfplane/translation case, that the point spectrum of C_φ includes the unit circle.

THEOREM 7.44
Let φ be an inner function, not an automorphism, with Denjoy–Wolff point a in the unit circle and let C_φ act on $H^2(D)$. If $|\lambda| < \varphi'(a)^{1/2}$ then λ is an eigenvalue of C_φ^* of infinite multiplicity.

PROOF We will show that if $|\lambda| = r < \varphi'(a)^{1/2}$, then $C_\varphi - \bar{\lambda}I$ is left invertible but not Fredholm. This means $C_\varphi^* - \lambda I$ is right invertible but not Fredholm, that is, that λ is an eigenvalue of C_φ^* of infinite multiplicity.

In the proof of Theorem 3.9, for $|a| = 1$, we saw that

$$\varphi'(a)^{1/2} = \lim_{n \to \infty} \left(\frac{1 - |\varphi_n(0)|}{1 + |\varphi_n(0)|} \right)^{1/2n}$$

Since $r < \varphi'(a)^{1/2}$, we may choose n large enough that

$$r^n < \left(\frac{1 - |\varphi_n(0)|}{1 + |\varphi_n(0)|} \right)^{1/2}$$

Since φ_n is an inner function, Theorem 3.8 implies

$$\left(\frac{1 - |\varphi_n(0)|}{1 + |\varphi_n(0)|} \right)^{1/2} \|f\| \le \|C_{\varphi_n} f\| = \|C_\varphi^n f\|$$

In particular, C_φ^n is one-to-one and the range $\mathcal{R}$ of C_φ^n is closed. Let P be the orthogonal projection of $H^2(D)$ onto $\mathcal{R}$ and let A be the inverse of C_φ^n as a map of $H^2(D)$ onto $\mathcal{R}$, that is, $AC_\varphi^n = I$ and $C_\varphi^n A = I_\mathcal{R}$. The estimate above shows

$$\|AP\| \le \|A\| \le \left(\frac{1 + |\varphi_n(0)|}{1 - |\varphi_n(0)|} \right)^{1/2}$$

so the series

$$L_\theta = \sum_{k=0}^{\infty} r^{nk} e^{ink\theta} (AP)^{k+1}$$

converges absolutely for each real θ.

Since $APC_\varphi^n = AC_\varphi^n = I$, it is easily checked that L_θ is a left inverse of $C_\varphi^n - r^n e^{in\theta}$. Because $\ker(C_\varphi^n) = \{0\}$ and C_φ^n has closed range, but C_φ^n is not invertible, Theorem 3.29 implies the kernel of C_φ^{*n}, which is $\mathcal{R}^\perp$, is infinite dimensional. Now if v is in $\mathcal{R}^\perp$, then $Pv = 0$, hence $L_\theta v = 0$. The fact that L_θ is a left inverse means that it is a left essential inverse so if $C_\varphi^n - r^n e^{in\theta}$ were invertible modulo the compacts, L_θ would be the inverse. Since L_θ has infinite dimensional kernel, it cannot be invertible modulo the compacts and $C_\varphi^n - r^n e^{in\theta}$ is not Fredholm. This means the essential spectrum of C_φ^n, $\sigma_e(C_\varphi^n)$, includes the circle $\{\mu : |\mu| = r^n\}$.

The identity

$$\prod_{k=1}^{n} \left(C_\varphi - r e^{i(\theta + 2\pi k/n)} \right) = C_\varphi^n - r^n e^{in\theta}$$

shows the operator $\widetilde{L_\theta}$ defined by

$$\widetilde{L_\theta} = L_\theta \prod_{k=1}^{n-1} \left(C_\varphi - r e^{i(\theta + 2\pi k/n)} \right)$$

is a left inverse for $C_\varphi - re^{i\theta}$. Since $\sigma_e(C_\varphi^n)$ includes the circle of radius r^n, the spectral mapping theorem for the Calkin algebra implies $\sigma_e(C_\varphi)$ intersects the circle of radius r. If $\sigma_e(C_\varphi)$ did not include this circle, there would be λ_0 on the boundary of $\sigma_e(C_\varphi)$ with $|\lambda_0| = r$. But this would imply, as λ approaches λ_0 in the essential resolvent of C_φ with $|\lambda| = r$ that the norm of the essential inverse of $C_\varphi - \lambda I$ approaches infinity. For the appropriate θ, since $\widetilde{L_\theta}$ is a left inverse of $C_\varphi - \lambda I$, it must be the essential inverse of $C_\varphi - \lambda I$ but

$$\|\widetilde{L_\theta}\|_e \leq \|\widetilde{L_\theta}\| \leq \|A\|(1 - r^n\|A\|)^{-1}(\|C_\varphi\| + r)^{n-1}$$

for all real θ. That is, the boundary of $\sigma_e(C_\varphi)$ does not intersect the circle of radius r but $\sigma_e(C_\varphi)$ does intersect the circle, so we conclude that $\sigma_e(C_\varphi)$ includes this circle.

We have shown that $C_\varphi - \lambda I$ is left invertible but not Fredholm for all λ with $|\lambda| = r$ for $r < \varphi'(a)^{1/2}$, as desired.   ∎

THEOREM 7.45

Let φ be an inner function, not an automorphism, with Denjoy–Wolff point a on the unit circle. Then on $H^2(D)$

$$\sigma(C_\varphi) = \sigma_e(C_\varphi) = \{\lambda : |\lambda| \leq \varphi'(a)^{-1/2}\}$$

PROOF   Theorem 7.44 shows $\{\lambda : |\lambda| < \varphi'(a)^{1/2}\}$ is a set of eigenvalues of C_φ^* of infinite multiplicity and Lemma 7.24 shows $\{\lambda : \varphi'(a)^{1/2} < |\lambda| < \varphi'(a)^{-1/2}\}$ is a set of eigenvalues of C_φ of infinite multiplicity so these sets are contained in the essential spectrum. Since Theorem 3.9 gives the spectral radius as $\varphi'(a)^{-1/2}$, the compactness of the spectrum and essential spectrum shows the conclusion holds. ∎

Exercises

    7.8.1   Suppose ψ is a non-trivial inner function and φ is the inner function $\varphi(z) = z\psi(z)$. Let $k(z)$ be a function orthogonal to $\psi H^2(D)$. Show that $zk\varphi^n$ is orthogonal to the range $\mathcal{R}$ of C_φ for each non-negative integer n. Thus, $\mathcal{R}^\perp$ is infinite dimensional.

Notes

Theorem 7.43 is found in E. A. Nordgren's seminal paper [No68]. The special case of Theorem 7.45 in which φ is a finite Blaschke product and $\varphi'(a) < 1$ where the Denjoy–Wolff point a is on the circle is covered by H. J. Kamowitz's general result on spectra in [Kam75] for $|a| = 1$ and $\varphi'(a) < 1$. Theorem 7.45 with this proof is found in Cowen's paper [Co83].

8

Normality

8.1 Normal and hyponormal composition operators

Among operators on Hilbert space, the normal operators, because of the spectral theorem, are as close as any class to being completely understood. In 1950, Halmos [Hal50] defined the broader classes of operators that we now call subnormal and hyponormal and in the intervening four decades a rich theory has developed for these operators as well. Because so much is known about these classes, it is appropriate to try to identify the composition operators that are closely related to them. In our investigation, we will use facts about normal, subnormal, and hyponormal operators without proof; an excellent reference for background on these ideas is Conway's book *The Theory of Subnormal Operators* [Con91].

Recall that an operator A is called **normal** if $A^*A = AA^*$ and that it is called **hyponormal** if $A^*A \geq AA^*$ where $\geq$ denotes the usual ordering on selfadjoint operators. The latter condition is easily seen to be equivalent to $\|Ax\| \geq \|A^*x\|$ for all vectors x. It is also easy to see that the hyponormality of A implies the hyponormality of the translates $A - \lambda I$. Combining these observations, we see that if x is an eigenvector for A, then x is also an eigenvector for A^* which means that the subspace spanned by x is a reducing subspace.

The locations of the fixed points of the symbols of composition operators limit the possibilities for phenomena related to normality. Normality is closely connected with the details of the norm structure and therefore is a very rigid notion. The constraints placed on the locations of the fixed points seem to result from the influence of the location of the fixed point on the symmetry in the operator.

Because the notions of normality and its relatives are intimately connected with the norm structure, care needs to be exercised in passing between equivalent norms. Arguments based on the size of the nullspaces of the operator and its adjoint are fairly robust and will work for a variety of equivalent norms and show that the operators are far from being normal or hyponormal. An argument showing that the norm of the operator is inconsistent with normality or hyponormality appears to be less robust and would require further examination before making inferences about the operator on spaces with an equivalent norm. In several arguments in this chapter, the generating function k for the weighted Hardy space (see Equation (2.1.4) and the associated discussion) plays an important role; in part it should be thought of as encoding the norm structure in the space. We begin with an easy application of the formula $K_w(z) = k(\langle z, w \rangle)$ to composition operators whose symbol is a linear map.

LEMMA 8.1
If A is an $N \times N$ matrix with $\|A\| \leq 1$ and $\varphi(z) = Az$, then C_φ is bounded on the weighted Hardy space $H^2(\beta, B_N)$ and $C_\varphi^* = C_\psi$ where $\psi(z) = A^* z$.

PROOF The boundedness of C_φ is Exercise 6.2.7. For z and w in the ball,

$$
\begin{aligned}
(C_\varphi^* K_w)(z) = K_{\varphi(w)}(z) &= k(\langle z, Aw \rangle) \\
&= k(\langle A^* z, w \rangle) = K_w(A^* z) = (C_\psi K_w)(z)
\end{aligned}
$$

In other words, on reproducing kernel functions, $C_\varphi^* = C_\psi$. Since the span of the kernel functions is dense in $H^2(\beta, B_N)$ and C_φ^* and C_ψ are continuous, we have $C_\varphi^* = C_\psi$. ∎

THEOREM 8.2
Suppose φ is an analytic mapping of the unit ball into itself and $H^2(\beta, B_N)$ is a weighted Hardy space on the ball on which C_φ is bounded. If C_φ is hyponormal, then $\varphi(0) = 0$. Moreover, C_φ is normal if and only if $\varphi(z) = Az$ for some normal $N \times N$ matrix A with $\|A\| \leq 1$.

PROOF Suppose C_φ is hyponormal. Since $C_\varphi(1) = 1 \circ \varphi = 1$, the vector 1 is an eigenvector for C_φ. The hyponormality of C_φ implies $1 = K_0$ (assuming the normalization $\beta(0) = 1$) is also an eigenvector of C_φ^*. But $C_\varphi^*(K_0) = K_{\varphi(0)}$, so we infer $K_{\varphi(0)} = K_0$, that is, $\varphi(0) = 0$.

Suppose C_φ is normal. Since normal operators are hyponormal, $\varphi(0) = 0$ and the Taylor series of φ is $\varphi(z) = Az + O(|z|^2)$ for some $N \times N$ matrix A. Let $\mathcal{K}_1$ be the subspace of $H^2(\beta, B_N)$ spanned by $\{1, z_1, \ldots, z_N\}$. Note that the kernel for evaluation of the partial derivative at 0 with respect to z_j is a multiple of z_j. As in the calculation of Equation (7.4.2) in the proof of Theorem 7.20 (with $a = 0$), the fact that $\varphi(0) = 0$ implies the subspace $\mathcal{K}_1$ is invariant for C_φ^*. Since $\mathcal{K}_1$ is finite dimensional, Exercise 8.1.1 implies $\mathcal{K}_1$ is an invariant subspace for both C_φ^* and C_φ and that the restrictions are normal. Recalling that $\varphi(0) = 0$, we see, for $j = 1, \ldots, N$, that $C_\varphi(z_j)$ is a linear combination of $z_1, \ldots, z_N$, that is, that $\varphi(z) = Az$. We are given that $\varphi(B_N) \subset B_N$, so $\|A\| \leq 1$.

Letting $\psi(z) = A^* z$, Lemma 8.1 shows $C_\varphi^* = C_\psi$. Since C_φ is normal,

$$ C_{\psi \circ \varphi} = C_\varphi C_\psi = C_\psi C_\varphi = C_{\varphi \circ \psi} $$

Thus, for every z in the ball, $A^*Az = \psi(\varphi(z)) = \varphi(\psi(z)) = AA^*z$ and A is normal.

Conversely, suppose A is a normal $N \times N$ matrix with $\|A\| \le 1$. Then $\varphi(z) = Az$ gives a bounded operator C_φ on the weighted Hardy space $H^2(\beta, B_N)$ and $C_\varphi^* = C_\psi$ where $\psi(z) = A^*z$. Since A is normal,

$$C_\varphi C_\psi = C_{\psi \circ \varphi} = C_{\varphi \circ \psi} = C_\psi C_\varphi$$

which shows C_φ is normal. ∎

This theorem shows that normal composition operators on weighted Hardy spaces are all trivial. Characterizations of selfadjoint and unitary composition operators are left to the exercises. (See Exercise 8.1.3).

Because most studies of phenomena related to normality have taken place in the one variable context and on the classical Hardy space case, we will primarily restrict our attention to this case. Some of what is here can be generalized to the several variable context, but much is subtle and the generalizations are unclear. The lemma below continues to explore the relationship between the location of fixed points and the rigidity needed for hyponormality.

LEMMA 8.3

If φ is a map of the unit disk into itself with Denjoy–Wolff point a and either $0 < |a| < 1$, or $|a| = 1$ and $\varphi'(a) = 1$, then neither C_φ nor C_φ^* is hyponormal on $H^2(D)$.

PROOF The spectral radius of a hyponormal operator is equal to its norm [Con91, p. 47]. Of course, the norms and spectral radii of C_φ and C_φ^* are the same and we have seen that, under the hypotheses given, the spectral radius of C_φ is 1 (Theorem 3.9), but the norm is greater than 1. ∎

Several examples of φ for which C_φ or C_φ^* is hyponormal will be given in the next section when we look at subnormality. The next two results show that hyponormality of the adjoint of a composition operator is a fairly restrictive condition.

THEOREM 8.4

If φ is a map of the unit disk into itself with Denjoy–Wolff point a and C_φ^* is hyponormal on $H^2(D)$, then either $|a| = 1$ and $\varphi'(a) < 1$, or C_φ is normal.

PROOF By Lemma 8.3, we need only examine the case of $\varphi(0) = 0$. By Equation (7.4.2) for $N = 1$, $C_\varphi^* z = \overline{\varphi'(0)} z$ which gives

$$|\varphi'(0)| = \|C_\varphi^* z\|_2 \ge \|C_\varphi z\|_2 = \|\varphi\|_2 \ge |\varphi'(0)|$$

This means $\|\varphi\|_2 = |\varphi'(0)|$, so $\varphi(z) = \varphi'(0)z$ and C_φ is normal. ∎

THEOREM 8.5

Suppose φ is a (non-constant) analytic mapping of the unit disk and $H^2(\beta, D)$ is a weighted Hardy space on the disk such that C_φ is bounded. If C_φ^* is hyponormal, then φ is a univalent map of the disk into itself.

PROOF For non-constant φ, by the open mapping theorem for analytic functions, the kernel of C_φ is (0). If C_φ^* is hyponormal, then the inequality $\|C_\varphi f\| \leq \|C_\varphi^* f\|$ characterizing hyponormality implies the kernel of C_φ^* is contained in the kernel of C_φ, so it must be (0) also. That is, the range of C_φ is dense. In particular, there is a sequence of polynomials p_n so that $C_\varphi(p_n) = p_n(\varphi)$ converges to z in $\mathcal{H}$. For this sequence of polynomials, $p_n(\varphi(w))$ converges to w for every point w of D which implies φ is univalent. ∎

Lemma 3.26, which only requires that the dimension of the kernel of C_φ^* be finite, could be invoked to give an alternate argument for the last part of the proof of this theorem. The advantage to the above proof is that it shows that for spaces in which the functions have some kind of boundary values, the univalence persists to the boundary as well. For example in $H^2(D)$, we can conclude that there is a set of measure zero on the unit circle so that off this set, the boundary function is univalent as well.

The extension of this result to the several variables case illustrates some of the subtlety that occurs in general. The critical difference, for the proof above, in the one and several variables case is that the kernel of C_φ need not be zero in several variables. The following theorem for the several variable case includes the result for one variable (with the extra hypothesis) since the set $\mathcal{M} \cap B_N$ of the theorem is the unit disk if φ is not the zero function.

THEOREM 8.6

Suppose φ is an analytic mapping of the unit ball into itself such that 0 is in $\varphi(B_N)$ and $H^2(\beta, B_N)$ is a weighted Hardy space on which C_φ is bounded. Let $\mathcal{M}$ be the subspace of C^N spanned by $\varphi(B_N)$. If C_φ^* is hyponormal, then φ is a univalent map of $\mathcal{M} \cap B_N$ into itself.

PROOF Let k be the generating function for the space $H^2(\beta, B_N)$; that is, k is analytic on the disk and $K_w(z) = k(\langle z, w \rangle)$ (Equation (2.1.4) and Exercise 2.1.17). The kernel of C_φ is the set of functions in $H^2(\beta, B_N)$ that vanish on the range of φ. Now suppose that C_φ^* is hyponormal and $\varphi(w_1) = \varphi(w_2)$. Then, $K_{w_1} - K_{w_2}$ is in the kernel of C_φ^*, hence, by the hyponormality, in the kernel of C_φ. That is, $K_{w_1}(\varphi(z)) - K_{w_2}(\varphi(z)) = 0$ for all z in B_N, or, in other words, $k(\langle \varphi(z), w_1 \rangle) = k(\langle \varphi(z), w_2 \rangle)$ for all z in B_N. Now Equation (2.1.4) implies $k'(0) \neq 0$ so there is $\epsilon > 0$ so that k is univalent on $\{\zeta \in D : |\zeta| < \epsilon\}$. Let z_0 be a point of B_N with $\varphi(z_0) = 0$ and let $\delta > 0$ be such that if $|z - z_0| < \delta$, then $|\varphi(z)| < \epsilon$. Thus, for $|z - z_0| < \delta$, both $|\langle \varphi(z), w_1 \rangle| < \epsilon$ and $|\langle \varphi(z), w_2 \rangle| < \epsilon$, so the equality $k(\langle \varphi(z), w_1 \rangle) = k(\langle \varphi(z), w_2 \rangle)$ implies $\langle \varphi(z), w_1 \rangle = \langle \varphi(z), w_2 \rangle$. In other words, for z close to z_0, we have $\langle \varphi(z), w_1 - w_2 \rangle = 0$. Since the function $z \mapsto \langle \varphi(z), w_1 - w_2 \rangle$ is analytic and is zero in neighborhood of z_0, it is identically zero in the ball.

To summarize, if $\varphi(w_1) = \varphi(w_2)$, then $\langle \varphi(z), w_1 - w_2 \rangle = 0$ for all z in B_N. To put it a different way, let $\mathcal{M}$ be the subspace of C^N spanned by $\varphi(B_N)$ and let $\mathcal{N}$ be the subspace spanned by

$$\{w_1 - w_2 : w_1, w_2 \in B_N \text{ and } \varphi(w_1) = \varphi(w_2)\}$$

If C_φ^* is hyponormal, then reasoning above shows $\mathcal{M} \subset \mathcal{N}^\perp$. Furthermore, the restriction of φ to $\mathcal{M} \cap B_N$ is univalent; indeed, if w_1 and w_2 are in $\mathcal{M} \cap B_N$ and $\varphi(w_1) = \varphi(w_2)$, then $w_1 - w_2$ is in both $\mathcal{N}$ and $\mathcal{M} \subset \mathcal{N}^\perp$ so $w_1 - w_2 = 0$. ∎

It is easily seen that if we assume k is univalent, which is the case for the usual Hardy and Bergman space in the disk and the usual Hardy space on the two-dimensional ball, then the above proof gives the conclusion without the hypothesis that 0 is in $\varphi(B_N)$ giving another generalization of Theorem 8.5.

Exercises

8.1.1 If C is an operator on a Hilbert space $\mathcal{H}$ and $\mathcal{M}$ is a closed subspace of $\mathcal{H}$, we say $\mathcal{M}$ is an **invariant subspace** of C if Cx is in $\mathcal{M}$ whenever x is in $\mathcal{M}$. We say the subspace $\mathcal{M}$ is a **reducing subspace** for C if $\mathcal{M}$ is an invariant subspace for C and C^*.

   (a) Suppose C is an operator and $\mathcal{M}$ is an invariant subspace for C. Prove that $\mathcal{M}$ is a reducing subspace if and only if $\mathcal{M}^\perp$ is also an invariant subspace for C.

   (b) Suppose C is an operator and $\mathcal{M}$ is an invariant subspace for C. Prove that the restriction of C to $\mathcal{M}$ is normal if and only if $\mathcal{M}$ is a reducing subspace for C.

   (c) Suppose C is a normal operator on $\mathcal{H}$ and $\mathcal{M}$ is an invariant subspace for C such that $\mathcal{M}^\perp$ is finite dimensional. Prove that $\mathcal{M}$ is a reducing subspace for C.

8.1.2 In general, an operator C is normal if and only if $\|Cv\| = \|C^*v\|$ for every vector v. Show that the composition operator C_φ on $H^2(D)$ is normal if and only if $\|C_\varphi 1\| = \|C_\varphi^* 1\|$ and $\|C_\varphi z\| = \|C_\varphi^* z\|$.

8.1.3 Which composition operators on a weighted Hardy space are unitary? selfadjoint?

8.1.4 Let $\varphi(z) = z^2$ and let C_φ be the corresponding composition operator on $H^2(D)$. For w in the unit disk, compute $C_\varphi^*(C_\varphi(K_w))$ and thereby show that it is, in general, different from $C_\varphi(C_\varphi^*(K_w))$.

8.1.5 Let $\varphi(z) = (2+z)/(4-z)$. (This map satisfies $\varphi(1) = 1$, $s = \varphi'(1) = 2/3$, and $r = -1/2$ in the normalization of Theorem 8.7.) Prove that C_φ^* is *not* hyponormal on $H^2(D)$.

Notes

H. J. Schwartz [Scz69] showed that C_φ is normal on $H^2(D)$ if and only if $\varphi(z) = Az$ with $|A| \leq 1$. The observation of Exercise 8.1.2 is due to D. Szajda and P. Bourdon (unpublished). Statements for $H^2(D)$ of several of the theorems of this section appear in C. Cowen and T. Kriete [CoK88]. The short proof of Theorem 8.4 presented here is due to P. S. Bourdon and D. Szajda. The observation of boundary univalence in the statement of the $H^2(D)$ version of Theorem 8.5 appears in [CoK88] with thanks to P. S. Bourdon. H. Sadraoui [Sad92] gave a necessary condition for C_φ^* to be hyponormal when φ is a linear fractional map of the unit disk into itself, but it does not appear to be sharp. The φ of Exercise 8.1.5 does not satisfy Sadraoui's condition. It is not known which linear fractional maps give rise to C_φ^* that are hyponormal, indeed, little is known about hyponormality of C_φ or C_φ^* in general. The results of this section for several variables appear to be new.

8.2 Subnormality of adjoints

An operator S is said to be ***subnormal*** on a Hilbert space $\mathcal{H}$ if there is a Hilbert space $\mathcal{K}$ containing $\mathcal{H}$ (as a closed subspace) and a normal operator N on $\mathcal{K}$ that has $\mathcal{H}$ as an invariant subspace and the restriction of N to $\mathcal{H}$ is S. It is easy to see that subnormal operators are hyponormal, so the results of the previous section apply.

If φ is an inner function with $\varphi(0) = 0$, then C_φ on the classical Hardy space is an isometry, so is subnormal. Also, we will see (Example 9.3) that if $\varphi = sz/(1 - (1 - s)z)$ for $0 < s < 1$, then C_φ is unitarily equivalent to $1 \oplus sC_{sz+(1-s)}^*$. We will see below that the latter operator is subnormal, so that in this case, C_φ is subnormal. The isometries, the (trivial) normal composition operators, and operators C_φ with φ linear fractional such that C_φ is unitarily equivalent to $1 \oplus C_\psi^*$ and C_ψ^* subnormal, as above, are the only composition operators known to be subnormal. On the other hand, there is no reason at this time to believe this is a complete list. Subnormality of adjoints of composition operators is somewhat better understood.

Rather than trying to give a comprehensive exposition of what is known about subnormality, we will try to give enough to indicate the nature of the results. The proofs depend on characterizations of subnormality that link it to moment conditions. In the background, the important object studied is the natural semigroup of iterates of C_φ. Although C_φ need not be *a priori* embeddable in a continuous semigroup, the model for iteration shows that the vestiges of a semigroup exist; the problem of subnormality is approached through it. The main theorem of the section is the following.

THEOREM 8.7

For $0 < s < 1$ and $0 \leq r \leq 1$, if

$$\varphi(z) = \frac{(r+s)z + (1-s)}{r(1-s)z + (1+sr)}$$

then C_φ^* is subnormal.

It is not difficult to show that a linear fractional map defined as in the theorem satisfies $\varphi(1) = 1$ and $\varphi'(1) = s$. Moreover, the other fixed point of φ is $-1/r$ and φ maps the unit disk into itself if and only if $|r| \leq 1$ (although $r = -1$ corresponds to the constant map $\varphi(z) \equiv 1$). For $r = 0$, the linear fractional map is $\varphi(z) = sz + 1 - s$ and C_φ^* is closely associated with the Cesàro operator (see Exercise 8.2.6). For $r = 1$, the resulting φ is an automorphism of the disk and the subnormality of C_φ^* can be established directly [Co88, p. 156] by using an adjoint calculation (see Section 9.1) and the characterization of subnormality in terms of positive operator-valued measures due to Embry [Em73].

For each such r, these linear fractional maps form a semigroup under iteration: if φ_t is the map obtained by replacing s by s^t, then $\varphi_{t_1+t_2} = \varphi_{t_1} \circ \varphi_{t_2}$. The semigroup structure is easier to recognize and to exploit if we change the focus of attention from the disk to the right halfplane. Let $\gamma(w) = (1+r-w)/(1+r+rw)$. For $0 \leq r < 1$, the map γ takes a disk in the right halfplane that is tangent to the imaginary axis onto the unit disk and satisfies

$$\varphi_t(z) = \gamma(s^t\gamma^{-1}(z)) \tag{8.2.1}$$

and when $r = 1$, γ is a map of the whole right halfplane onto the disk satisfying Equation (8.2.1). Since $r \geq 0$, $\gamma^{-1}(D)$ always contains the interval $(0, 1]$ in the right halfplane. The point of departure for the proof of the theorem is A. Lambert's version of a criterion for subnormality due to M. Embry. We state it as applied to C_φ^*.

THEOREM 8.8   (Embry [Em73], Lambert [La76])

C_φ^* *is subnormal if and only if for each f in a dense subset of $H^2(D)$, there is a finite positive measure ν_f on $[0, \|C_\varphi\|]$ such that*

$$\langle C_\varphi^{\,n} C_\varphi^{*n} f, f \rangle = \int x^{2n} d\nu_f(x) \qquad \text{for } n = 0, 1, 2, \dots$$

We will apply the theorem with the f's taken to be finite linear combinations of some kernel functions, where we use the function γ to express the parameter of the kernel functions. Let f be defined by

$$f = \sum_{j=1}^{m} \xi_j K_{\gamma(a_j)} \tag{8.2.2}$$

where $a_1, \ldots, a_m$ are points on the real axis with $0 < a_j \leq 1$ and $\xi_1, \ldots, \xi_m$ are complex numbers. Then for $n = 0, 1, 2, \ldots,$

$$\langle C_\varphi{}^n C_\varphi^{*n} f, f \rangle = \|C_{\varphi_n}^* f\|^2$$

$$= \sum_{i,j=1}^{m} \xi_i \overline{\xi_j} K_{\varphi_n(\gamma(a_i))}(\varphi_n(\gamma(a_j)))$$

$$= \sum_{i,j=1}^{m} \xi_i \overline{\xi_j} K_{\gamma(s^n a_i)}(\gamma(s^n a_j)) \qquad (8.2.3)$$

But for $0 < a, b \leq 1$, and x positive and small enough, there are continuous functions $\{h_n\}_{n=0}^{\infty}$ and an expansion

$$K_{\gamma(xa)}(\gamma(xb)) = \frac{1}{1 - \overline{\gamma(xa)}\gamma(xb)} = \frac{1}{x} \sum_{p=0}^{\infty} h_p(a, b) x^p \qquad (8.2.4)$$

(In this case, computing the functions h_p is tedious, but not difficult; we will get an expression for them below.) To see that Equation (8.2.4) is plausible, if the complex conjugate is left out of the expression, the resulting function is a function of x that is analytic in the neighborhood of the closed unit disk except for a simple pole at $x = 0$, so the expansion will be valid for $0 < x \leq 1$. Putting this into the above Equation (8.2.3), we get

$$\langle C_\varphi{}^n C_\varphi^{*n} f, f \rangle = \sum_{i,j=1}^{m} \xi_i \overline{\xi_j} K_{\gamma(s^n a_i)}(\gamma(s^n a_j))$$

$$= \sum_{i,j=1}^{m} \xi_i \overline{\xi_j} \frac{1}{s^n} \sum_{p=0}^{\infty} h_p(a_i, a_j) s^{np}$$

$$= \sum_{p=0}^{\infty} s^{n(p-1)} \sum_{i,j=1}^{m} \xi_i \overline{\xi_j} h_p(a_i, a_j).$$

This last expression is precisely

$$\int x^{2n} d\nu_f(x)$$

where ν_f is the purely atomic complex measure on $[0, s^{-1/2}]$ which is carried by $\{s^{\frac{p-1}{2}}\}_{p=0}^{\infty}$ with

$$\nu_f(\{s^{\frac{p-1}{2}}\}) = \sum_{i,j=1}^{m} \xi_i \overline{\xi_j} h_p(a_i, a_j) \qquad (8.2.5)$$

Integrals against x^{2n} determine any measure with compact support in $[0, \infty)$, so if the positive measure of Embry-Lambert exists, it must coincide with the complex measure ν_f given by Equation (8.2.5). In other words, C_φ^* is subnormal exactly when ν_f is a positive measure for each f that can be expressed as in Equation (8.2.2).

DEFINITION 8.9   *Let X be a set. A complex function h defined on $X \times X$ is positive definite provided*

$$\sum_{j,k=1}^{n} \xi_j \overline{\xi_k} h(a_j, a_k) > 0$$

for all n and for all non-zero vectors $(\xi_1, \ldots, \xi_n)$ in C^n and $\{a_j\}_{j=1}^n \subset X$.

This definition says, of course, that for every choice of n and set $\{a_j\}_{j=1}^n$, the $n \times n$ matrix whose entries are $h(a_j, a_k)$ is positive definite.

In other words, for φ of the form we are considering, we can see C_φ^* is subnormal if we can show the functions h_p are positive definite on $(0, 1] \times (0, 1]$. For homogeneous functions (see Exercise 8.2.1), a theorem of Bochner [RiN, p. 385] shows that positive definiteness is equivalent to the existence of a positive measure ν so that, for u real,

$$h(e^{-u/2}, e^{u/2}) = \int e^{-iuy} d\nu(y), \qquad \text{for } u \text{ in } R$$

where in this formula h is extended to $(0, \infty) \times (0, \infty)$ by homogeneity. Calculation of the h_p will show that h_p is homogeneous of degree $p - 1$ (or see Exercise 8.2.2) and that

$$h_0(e^{-u/2}, e^{u/2}) = \frac{1}{e^{u/2} + e^{-u/2}}$$

$$h_1(e^{-u/2}, e^{u/2}) = (1 + r)^{-1} \left(r + \frac{1 - r}{(e^{u/2} + e^{-u/2})^2} \right)$$

and

$$h_p(e^{-u/2}, e^{u/2}) = (1 + r)^{-p} \left[\frac{(1 - r)^p}{(e^{u/2} + e^{-u/2})^{p+1}} + \frac{r(1 - r)^{p-2}}{(e^{u/2} + e^{-u/2})^{p-1}} \right]$$

for $p \geq 2$. Using standard tables of Fourier transforms such as [Er53] and [Er54, p. 30], we find associated measures

$$d\nu_0(y) = \frac{1}{2\pi} |\Gamma(\tfrac{1}{2} + iy)|^2 dy$$

$$d\nu_1(y) = \frac{r}{1 + r} d\delta_0(y) + \frac{1}{2\pi} \left(\frac{1 - r}{1 + r} \right) |\Gamma(1 + iy)|^2 dy$$

where δ_0 is a unit point mass at 0, and

$$dv_p(y) =$$

$$\frac{(1-r)^p}{2\pi p!(1+r)^{p+2}} \left(\frac{(1-r)^2}{n(n+1)} (\frac{(p+1)^2}{4} + y^2) + r \right) |\Gamma(\frac{p+1}{2} + iy)|^2 dy$$

for $p \geq 2$. Since $0 \leq r \leq 1$, each of these measures is non-negative.

To summarize, this shows that for each f that is a finite linear combination of the kernel functions for evaluation at points $\gamma((0,1])$, the measure v_f that occurs in the integral for $\langle C_\varphi{}^n C_\varphi^{*n} f, f \rangle$ is a positive measure. Since such f's are dense in $H^2(D)$, C_φ^* is subnormal, by the Embry–Lambert criterion. This completes the outline of the proof of Theorem 8.7.

The Equations (8.2.4) and

$$h_p(e^{-u/2}, e^{u/2}) = \int e^{-iuy} dv_p(y) \tag{8.2.6}$$

allow us to write down a normal extension of C_φ^* via an interesting isomorphism. We define a measure μ on the halfplane $\{x + iy : x \geq -1/2\}$ by

$$d\mu(x + iy) = \sum_{p=0}^{\infty} d\delta_{\frac{p-1}{2}}(x) \, dv_p(y)$$

where $\delta_{\frac{p-1}{2}}$ is a unit point mass at $(p-1)/2$. In other words, μ is carried on the union of the vertical lines $\{x + iy : x = (p-1)/2\}$ for $p = 0, 1, 2, \ldots$ and v_p is the part of μ on the pth line.

For $0 < a, b \leq 1$ we calculate the inner product of the functions a^w and b^w in $L^2(\mu)$:

$$\int a^w \overline{b^w} \, d\mu(w) = \int a^{x+iy} b^{x-iy} \, d\mu(x + iy)$$

$$= \sum_{p=0}^{\infty} \int a^{\frac{p-1}{2}+iy} b^{\frac{p-1}{2}-iy} \, dv_p(y)$$

$$= \sum_{p=0}^{\infty} (ab)^{\frac{p-1}{2}} \int \left(\frac{b}{a}\right)^{-iy} dv_p(y) \tag{8.2.7}$$

Now write $b/a = e^u$. Then using Equation (8.2.6) we have

$$\int \left(\frac{b}{a}\right)^{-iy} dv_p(y) = h_p(e^{-u/2}, e^{u/2}) = h_p\left(\sqrt{\frac{a}{b}}, \sqrt{\frac{b}{a}}\right)$$

Thus the pth term in Equation (8.2.7) is exactly $h_p(a, b)$, where we are also using the homogeneity of h_p. We can invoke Equation (8.2.4) to find

$$\int a^w \overline{b^w}\, d\mu(w) = \sum_{p=0}^{\infty} h_p(a,b) = \langle K_{\gamma(a)}, K_{\gamma(b)} \rangle \qquad (8.2.8)$$

Note that taking $a = b = 1$ shows that μ is a finite measure of total mass $\|K_{\gamma(1)}\|^2$. Note also that the functions a^w are all in $L^\infty(\mu)$.

Now let $\mathcal{H}$ denote the subspace of $L^2(\mu)$ spanned by $\{a^w : 0 < a \le 1\}$. Let us define a map V on the kernel functions $K_{\gamma(a)}$ by $V(K_{\gamma(a)}) = a^w$. According to Equation (8.2.8), V preserves inner products, and so has a unique extension to a unitary operator V from $H^2(D)$ to $\mathcal{H}$. Moreover, for s fixed, $0 < s < 1$, we can define a normal operator N on $L^2(\mu)$ by $(Ng)(w) = s^w g(w)$ for g in $L^2(\mu)$. Since $N(a^w) = (sa)^w$, $\mathcal{H}$ is invariant for N and the restriction of N to $\mathcal{H}$ is a subnormal operator T.

THEOREM 8.10
C_φ^* *is unitarily equivalent to* T *via* V, *so that* N *can be considered to be a normal extension of* C_φ^*.

PROOF   It is enough to check that VC_φ^* and TV agree on the spanning set $\{K_{\gamma(a)} : 0 < a \le 1\}$ for $H^2(D)$.

$$VC_\varphi^* K_{\gamma(a)} = VK_{\varphi(\gamma(a))} = VK_{\gamma(sa)}$$
$$= (sa)^w = s^w a^w = TVK_{\gamma(a)}$$

∎

Under some extra hypotheses on smoothness, the converse of Theorem 8.7 can also be proved. For example, if φ is analytic in a neighborhood of the closed disk with $\varphi(1) = 1$, $s = \varphi'(1) < 1$, and C_φ^* is subnormal, then there is r, $0 \le r \le 1$, so that

$$\varphi(z) = \frac{(r+s)z + (1-s)}{r(1-s)z + (1+sr)}$$

The strategy for the proof of such a theorem is to use the smoothness of φ near the Denjoy–Wolff point to get an expansion for $\langle C_\varphi{}^n C_\varphi^{*}{}^n K_{z_1}, K_{z_2} \rangle$ for real z_1 and z_2 near 1 as above. This is possible because the model for iteration of analytic functions on the disk (Theorem 2.53) gives a partially defined semigroup and a function like γ in this case as well. The resulting series expansion gives functions h_p as above and the subnormality of C_φ^* implies the positive-definiteness of the h_p. An analysis of the positive-definiteness of the h_p puts restrictions on γ, ultimately showing that γ, hence φ, is a linear fractional map with parameters as in the conclusion. It seems likely that the smoothness conditions are not needed, that is, that the converse is true always, but the techniques used in the existing proofs seem to require some smoothness.

Similar analysis can be attempted on other weighted Hardy spaces. The same linear fractional maps as in Theorem 8.7 also give rise to composition operators whose adjoints are subnormal on the Bergman space $A^2(D)$. On the other hand, the corresponding analysis of h_p in the Bergman space case does not seem to imply γ is linear fractional. That is, it appears possible that there are some maps φ besides linear fractional maps for which C_φ^* is subnormal on the Bergman space, but, to date, no example of φ for which C_φ^* is subnormal is known besides the linear fractional ones.

Exercises

8.2.1 Let h be a continuous complex-valued function on $(0, 1] \times (0, 1]$ that is homogeneous of degree m, that is,

$$h(ta, tb) = t^m h(a, b)$$

whenever both sides make sense.

   (a) Show that h has a unique extension to $(0, \infty) \times (0, \infty)$ that is also homogeneous of degree m.

   (b) Show that h, so extended, is positive definite on $(0, 1] \times (0, 1]$ if and only if it is positive definite on $(0, \infty) \times (0, \infty)$.

   (c) Considering h as extended, let

$$g(u) = h(e^{-u/2}, e^{u/2})$$

   Show that h is positive definite if and only if g is positive definite in the sense that

$$\sum_{j,k=1}^{n} \xi_j \overline{\xi_k} g(u_j - u_k) \geq 0 \qquad (8.2.9)$$

   for all n, for all vectors ξ in C^n, and all $\{u_j\}$ in R.

   Note According to Bochner's Theorem, a continuous function g on R is positive definite in the sense of Equation (8.2.9) if and only if $g(u) = \int e^{-iuy} \, d\nu(y)$ for some finite, positive Borel measure ν on R.

8.2.2 Use the version of Equation (8.2.4),

$$K_{\gamma(xa)}(\gamma(xb)) = \frac{1}{1 - \gamma(xa)\gamma(xb)} = \frac{1}{x} \sum_{p=0}^{\infty} h_p(a, b) x^p$$

   valid for a, b, and x in $(0, 1]$, and the definition of γ to show directly that each h_p is homogeneous of degree $p - 1$ without computing a formula for h_p.

8.2.3 Let S denote the unilateral shift on ℓ^2. Show that $A = 2S + S^*$ is hyponormal but that A^2 is not. Show that B subnormal implies B^2 is subnormal and deduce that A is a hyponormal operator that is not subnormal.

8.2.4 It has been proved [Co92] that if φ is a map of the disk into itself such that C_φ^* is subnormal on the Dirichlet space, then C_φ^* is subnormal on $H^2(D)$. Use this fact and the converse to Theorem 8.7 to find the functions φ that are analytic in a neighborhood of the closed unit disk such that C_φ^* is subnormal on the Dirichlet space.

8.2.5 Find the spectrum of the normal operator N in Theorem 8.10.

8.2.6 In Theorem 8.7, consider the case $r = 0, 0 < s < 1$ and write φ_s for the mapping $\varphi_s(z) = sz + (1 - s) = \gamma(s\gamma^{-1}(z))$, where $\gamma(\zeta) = 1 - \zeta$.

(a) Show that there exists a unique bounded operator B on $H^2(D)$ with

$$\langle Bf, g \rangle = \int_0^1 \langle C_{\varphi_s} f, g \rangle \, ds$$

for all f, g in $H^2(D)$, and, moreover, $\|B\| \leq 2$.

(b) Show that for any f in $H^2(D)$

$$Bf(z) = \frac{1}{z - 1} \int_1^z f(\zeta) \, d\zeta$$

where the integral is over the line segment from 1 to z.

(c) Compute the point spectrum of B and conclude that $\|B\| = 2$.

(d) The Cesàro operator A on $H^2(D)$ is defined by

$$A\left(\sum_{n=0}^{\infty} a_n z^n\right) = \sum_{n=0}^{\infty} b_n z^n$$

where $b_n = \frac{1}{n+1} \sum_{k=0}^{n} a_k$. It is known to be bounded. Show that $A = B^*$. Hint: Compute the matrices for A and B with respect to the standard basis for $H^2(D)$.

(e) Note that the measure μ, the subspace $\mathcal{H}$ of $L^2(\mu)$ and the unitary operator V, all associated to φ_s as in Theorem 8.10, are independent of s. Show that for every f in $\mathcal{H}$

$$(VAV^* f)(w) = \frac{1}{w + 1} f(w)$$

μ–almost everywhere. Conclude the VAV^* is subnormal, and hence that A itself is subnormal, since subnormality is a unitary invariant.

(f) In part (e) you found a normal extension for VAV^*. Find its spectrum.

Notes

E. Nordgren [No68] observed that if φ is an inner function with $\varphi(0) = 0$, then C_φ is an isometry, hence subnormal. Theorems 8.7 and 8.10 are due to C. Cowen and T. Kriete and their paper [CoK88] also contains the converse of Theorem 8.7 and details of the computation of the minimal normal extension. The two special cases of Theorem 8.7, $r = 0$ and $r = 1$ had been proved earlier. The operator C_φ^* for $\varphi(z) = sz + (1 - s)$ was discovered by J. Deddens [Ded72] to be an analytic function of the Cesàro operator, which had been proved by T. Kriete and D. Trutt [KrT71] to be subnormal, so Deddens deduced the subnormality of C_φ^* in this case. (The invariant subspace structure of the Cesàro operator, hence of C_φ^*, was studied in Kriete and Trutt's paper [KrT74].) In C. Cowen's paper [Co84b], the operator C_φ^* was proved directly to be subnormal in this case and the subnormality of the Cesàro operator deduced from that. In E. Nordgren, P. Rosenthal, and F. Wintrobe's paper [NoRW87] and C. Cowen's paper [Co88], the subnormality of C_φ^* for the automorphism arising from the case $r = 1$ of Theorem 8.7 was established by different proofs.

9

Miscellanea

9.1 Adjoints of some composition operators

Even in $H^2(D)$, there are few composition operators whose norms are known exactly, although the estimates of earlier sections are sufficient to determine the norm within a factor of $\sqrt{2}$. This lack of information is due to a more fundamental lack of knowledge: the inability to compute adjoints of composition operators. Of course, we have the formula for the adjoint of a composition operator on kernel functions (and we have made good use of it), but most functions in the space do not have a convenient expression in terms of the kernel functions. Two exceptions to this lack of information are when φ is inner or φ is affine.

If φ is an inner function in $H^\infty(D)$ so that it may be considered to be a mapping of the unit circle onto itself, then a satisfactory formula can be obtained for C_φ^* by changing variables in the integral giving the inner product and interpreting the result as a weighted expectation operator. The formula is more or less simple depending on the multiplicity of φ. We will not take this point of view explicitly, but it is in the background.

Suppose first that Φ is an inner function with $\Phi(0) = 0$. Then $\{\Phi^n\}_{n=0}^\infty$ is an orthonormal set and C_Φ is an isometry:

$$\left\| C_\Phi \left(\sum a_n z^n \right) \right\| = \left\| \sum a_n \Phi^n \right\| = \left\| \sum |a_n|^2 \right\|^{1/2} = \left\| \sum a_n z^n \right\|$$

In particular, the range of C_Φ is closed and the powers of Φ are an orthonormal basis for it. The projection Q of $H^2(D)$ onto the range of C_Φ has the simple formula $Qf = \sum \langle f, \Phi^n \rangle \Phi^n$ or, writing it as an integral,

$$(Qf)(z) = \sum \int f(e^{i\theta}) \overline{\Phi^n(e^{i\theta})} \frac{d\theta}{2\pi} \Phi^n(z) = \int \frac{f(e^{i\theta})}{1 - \overline{\Phi(e^{i\theta})}\Phi(z)} \frac{d\theta}{2\pi}$$

Since C_Φ is an isometry, its adjoint is the projection onto the range followed by its inverse, that is, $C_\Phi^* f = \sum \langle f, \Phi^n \rangle z^n$ or

$$(C_\Phi^* f)(z) = \int \frac{f(e^{i\theta})}{1 - \overline{\Phi(e^{i\theta})} z} \frac{d\theta}{2\pi}$$

If φ is a general analytic function mapping D into itself, no satisfactory formula for C_φ^* on $H^2(D)$ is known. Here we obtain a simple formula for C_φ^* when φ is a linear fractional transformation. The adjoint is a product of Toeplitz operators and a composition operator.

LEMMA 9.1
If $\varphi(z) = (az + b)(cz + d)^{-1}$ is a linear fractional transformation mapping D into itself, where $ad - bc = 1$, then $\sigma(z) = (\bar{a}z - \bar{c})(-\bar{b}z + \bar{d})^{-1}$ maps D into itself.

PROOF   Linear fractional transformations may be regarded as one-to-one mappings of the Riemann sphere onto itself. Let $\check{D}$ denote the open set

$$\check{D} = \{z : |z| > 1\} \cup \{\infty\}$$

Now, φ maps D into itself, so $\gamma(z) = \overline{\varphi(\bar{z})}$ also maps D into itself. It follows that $\gamma^{-1}(z)$ maps $\check{D}$ into itself. An easy calculation shows that

$$\sigma(z) = \frac{1}{\gamma^{-1}(\frac{1}{z})}$$

which implies $\sigma(z)$ maps D into D.   ∎

Recall that for g in $L^\infty(\partial D)$, the Toeplitz operator T_g is the operator on H^2 given by $T_g(f) = Pfg$ for f in H^2 and P the orthogonal projection of L^2 onto H^2. (For general properties of Toeplitz operators, see [Do72, Chapter 7].)

THEOREM 9.2
Let $\varphi(z) = (az + b)(cz + d)^{-1}$ be a linear fractional transformation mapping D into itself, where $ad - bc = 1$. Then $\sigma(z) = (\bar{a}z - \bar{c})(-\bar{b}z + \bar{d})^{-1}$ maps D into itself, $g(z) = (-\bar{b}z + \bar{d})^{-1}$ and $h(z) = cz + d$ are in H^∞, and

$$C_\varphi^* = T_g C_\sigma T_h^*$$

PROOF   The function h is clearly in H^∞. By Lemma 9.1, σ maps D into itself and since the denominators of σ and g are the same, g is in H^∞. This means the formula makes sense.

Now, for w in D, let $K_w(z) = (1 - \bar{w}z)^{-1}$. This function is the reproducing kernel at w, that is, $\langle f, K_w \rangle = f(w)$ for f in $H^2(D)$. It is easily proved that $T_h^* K_w = \overline{h(w)} K_w$ and $C_\varphi^* K_w = K_{\varphi(w)}$. Calculation gives

$$T_g C_\sigma T_h^* (K_w)(z) = \overline{h(w)} T_g C_\sigma (K_w)(z)$$

$$= \overline{(cw + d)} \left(\frac{1}{-\bar{b}z + \bar{d}} \right) \left(\frac{1}{1 - \bar{w}\frac{\bar{a}z - \bar{c}}{-\bar{b}z + \bar{d}}} \right)$$

$$= \frac{\overline{cw} + \bar{d}}{-\bar{b}z + \bar{d} - \overline{wa}z + \overline{wc}}$$

$$= \frac{1}{1 - \overline{\varphi(w)}z} = K_{\varphi(w)}(z) = C_\varphi^*(K_w)(z)$$

Since the K_w span a dense set of H^2, the desired equality holds. ∎

EXAMPLE 9.3   A Unitary Equivalence.

For $t > 1$, let $\varphi(z) = z(t^2 - (t^2 - 1)z)^{-1}$. We want to understand the operator C_φ on $H^2(D)$. Easy calculations show that $\varphi(0) = 0$, $\varphi(1) = 1$, and $\varphi(D)$ is the disk of radius $t^2(2t^2 - 1)^{-1}$ inside the unit circle and tangent to it at 1, so C_φ is a bounded operator. Since $\varphi(0) = 0$, Theorem 3.1 says $\|C_\varphi\| = 1$. Since 1 is a fixed point of φ and $\varphi'(1) = t^2$, the chain rule gives $\varphi_n'(1) = t^{2n}$. Moreover, 1 is the only point of the circle that φ maps to the circle, so Exercise 3.2.5 shows $\|C_{\varphi_n}\|_e = t^{-n}$ and the essential spectral radius of C_φ is

$$\rho = \lim_{n \to \infty} \left(\|C_{\varphi_n}\|_e \right)^{1/n} = t^{-1}$$

Finally, since $\varphi'(0) = t^{-2} < \rho$, Theorem 7.30 shows that the spectrum of C_φ is $\{1\} \cup \{\lambda : |\lambda| \le t^{-1}\}$.

We have $C_\psi 1 = 1$ for every ψ mapping the disk to itself and since $\varphi(0) = 0$, we also have $C_\varphi^* 1 = 1$ which means the subspace spanned by 1 is a reducing subspace. To understand C_φ, it is sufficient to understand it on $zH^2(D) = [1]^\perp$ and it turns out that to do so, it will be helpful to find C_φ^*.

Normalizing the expression for φ as

$$\varphi(z) = \frac{z}{t^2 - (t^2 - 1)z} = \frac{t^{-1}z + 0}{(t^{-1} - t)z + t}$$

the functions needed in Theorem 9.2 for the calculation of C_φ^* are

$$\sigma(z) = \frac{t^{-1}z - (t^{-1} - t)}{-0z + t} = t^{-2}z + (1 - t^{-2})$$

$$g(z) = (-0z + t)^{-1} = t^{-1} \quad \text{and} \quad h(z) = (t^{-1} - t)z + t$$

and we find

$$C_\varphi^* = T_g C_\sigma T_h^* = C_\sigma T_{1 - (1 - t^{-2})\bar{z}}$$

Let F be in $H^2(D)$ so that zF is in $H^2(D)$ and $\|zF\| = \|F\|$.

$$
\begin{aligned}
C_\varphi^* zF &= C_\sigma T_{1-(1-t^{-2})\bar{z}}(zF) = C_\sigma \left(zF - (1 - t^{-2})F \right) \\
&= \left(\sigma - (1 - t^{-2}) \right) F \circ (\sigma) = t^{-2} zF \circ \sigma \\
&= z(t^{-2} F \circ \sigma)
\end{aligned}
$$

That is, if U is the unitary operator mapping $H^2(D)$ onto $zH^2(D)$ given by $UF = zF$ then

$$
C_\varphi^* U(F) = C_\varphi^* (zF) = z(t^{-2} C_\sigma F) = t^{-2} U C_\sigma F
$$

In other words, the restriction of C_φ^* to $zH^2(D)$ is $U(t^{-2}C_\sigma)U^*$. Taking adjoints, this says C_φ is unitarily equivalent to

$$
\begin{pmatrix} 1 & 0 \\ 0 & t^{-2} C_\sigma^* \end{pmatrix}
$$

where the blocks are with respect to the decomposition $H^2(D) = [1] \oplus zH^2(D)$.

In particular, since we have seen that C_σ^* is subnormal (take $r = 0$ and $s = t^{-2}$ in Theorem 8.7), C_φ is also subnormal.   ⬜

A norm calculation

The best general estimate of the norm of C_φ on $H^2(D)$ (Corollary 3.7) is

$$
\frac{1}{\sqrt{1 - |\varphi(0)|^2}} \leq \|C_\varphi\| \leq \frac{1 + |\varphi(0)|}{\sqrt{1 - |\varphi(0)|^2}}
$$

and both inequalities can be achieved by linear fractional transformations (see Exercise 3.1.2 and Theorem 3.8).

As we will see, it is possible to use the adjoint calculation above to find the norm of the composition operators with affine symbol. It will become clear that the norms of composition operators depend in a rather complex way on the parameters of the symbol.

THEOREM 9.4
If $\varphi(z) = sz + t$ *for* $|t| < 1$ *and* $|s| + |t| \leq 1$, *then*

$$
\|C_\varphi\| = \sqrt{\frac{2}{1 + |s|^2 - |t|^2 + \sqrt{(1 - |s|^2 + |t|^2)^2 - 4|t|^2}}}
$$

PROOF   In the trivial cases $t = 0$ and $s = 0$, the formula gives the correct norms, $\|C_\varphi\| = 1$ and $\|C_\varphi\| = (1 - |t|^2)^{-\frac{1}{2}}$ respectively. We therefore assume s and t are non-zero.

The function φ has not been presented in a way that we may directly apply the adjoint calculation; choosing $a^2 = s$ and $b = t/a$, a normalized expression for φ

is $\varphi(z) = (az+b)(0z+a^{-1})^{-1}$. In the notation of Theorem 9.2, $C_\varphi^* = T_g C_\sigma T_h^*$ where

$$g(z) = \left(-\bar{b}z + \bar{a}^{-1}\right)^{-1}$$

$$\sigma(z) = \frac{\bar{a}z}{-\bar{b}z + \bar{a}^{-1}}$$

and $h(z) = a^{-1}$

Thus,

$$C_\varphi^* C_\varphi = \bar{a}^{-1} T_g C_\sigma C_\varphi = T_{(1-\overline{ab}z)^{-1}} C_\sigma C_\varphi = T_f C_\psi$$

where $f(z) = (1 - \bar{t}z)^{-1}$ and

$$\psi(z) = \varphi(\sigma(z)) = \frac{(|s|^2 - |t|^2)z + t}{-\bar{t}z + 1}$$

Now

$$\|C_\varphi\|^2 = \|C_\varphi^* C_\varphi\| = \lim_{n \to \infty} \|(C_\varphi^* C_\varphi)^n\|^{\frac{1}{n}}$$

$$= \lim_{n \to \infty} \|T_f T_{f \circ \psi} \cdots T_{f \circ \psi_{n-1}} C_\psi^n\|^{\frac{1}{n}}$$

$$\leq \lim_{n \to \infty} \left(\|f\|_\infty \|f \circ \psi\|_\infty \cdots \|f \circ \psi_{n-1}\|_\infty\right)^{\frac{1}{n}} \lim_{n \to \infty} \|C_\psi^n\|^{\frac{1}{n}}$$

(Here ψ_k denotes the kth iterate of ψ.) The last quantity in this expression is just the spectral radius of C_ψ which was calculated in Theorem 3.9. If $|s| + |t| = 1$, then $\psi(t/|t|) = t/|t|$ and $\psi'(t/|t|) = 1$, so the spectral radius of C_ψ is 1. If $|s| + |t| < 1$, then ψ maps the closed disk into the open disk and C_ψ is compact and has spectral radius 1. Thus, the last quantity in this expression is always 1.

In the case $|s| + |t| = 1$, since $0 < |t| < 1$, we find

$$|\psi(-t/|t|)| = \left|3 - \frac{4}{1+|t|}\right| < 1$$

This information, together with the fact from the above paragraph that the Denjoy–Wolff point of ψ is $t/|t|$, implies that ψ maps the closed unit disk onto a proper subdisk internally tangent to the unit circle at $t/|t|$. In particular, this means that $\lim_{n \to \infty} \psi_n(z) = t/|t|$, uniformly, in the closed unit disk. It follows that $\lim_{n \to \infty} f \circ \psi_n = (1 - |t|)^{-1} = |s|^{-1}$ so

$$\lim_{n \to \infty} \left(\|f\|_\infty \|f \circ \psi\|_\infty \cdots \|f \circ \psi_n\|_\infty\right)^{\frac{1}{n}} = |s|^{-1}$$

The above inequality now implies that $\|C_\varphi\| \leq |s|^{-\frac{1}{2}}$.

On the other hand, taking $w = rt|s|/(s|t|)$ for $0 < r < 1$, we find

$$\|C_\varphi\|^2 \geq \lim_{r \to 1} \frac{\|C_\varphi^* K_w\|^2}{\|K_w\|^2} = \lim_{r \to 1} \frac{\|K_{\varphi(w)}\|^2}{\|K_w\|^2}$$

$$= \lim_{r \to 1} \frac{1 - r^2}{1 - (r|s| + |t|)^2} = \frac{1}{|s|}$$

Thus, $\|C_\varphi\| = |s|^{-\frac{1}{2}}$ when $|s| + |t| = 1$, which agrees with the conclusion for this case.

In the case $|s| + |t| < 1$, then $\psi(z) = (pz+t)(-\bar{t}z+1)^{-1}$, where $p = |s|^2 - |t|^2$, and the fixed point of ψ in D is the solution with smaller modulus of

$$\bar{t}z^2 + (p - 1)z + t = 0$$

The solutions of this quadratic equation are

$$z = \frac{1 - p \pm \sqrt{(p-1)^2 - 4|t|^2}}{2\bar{t}}$$

Noting that $-1 < p < 1$ and $(p - 1)^2 \geq \left(1 - (1 - |t|)^2 + |t|^2\right)^2 = 4|t|^2$, we see that the numerator is a positive number in either case, so the Denjoy–Wolff point is

$$\hat{w} = \frac{1 - p - \sqrt{(p-1)^2 - 4|t|^2}}{2\bar{t}}$$

As before, $\lim_{n \to \infty} f \circ \psi_n = f(\hat{w})$, and $\|C_\varphi\|^2 \leq |f(\hat{w})|$. On the other hand, since

$$C_\varphi^* C_\varphi(K_{\hat{w}}) = (C_\varphi^* C_\varphi)^*(K_{\hat{w}}) = (T_f C_\psi)^*(K_{\hat{w}})$$

$$= C_\psi^* T_f^*(K_{\hat{w}}) = \overline{f(\hat{w})} K_{\psi(\hat{w})} = \overline{f(\hat{w})} K_{\hat{w}}$$

we see that $\|C_\varphi\|^2 = \|C_\varphi^* C_\varphi\| \geq |f(\hat{w})|$. Therefore, in this case, $\|C_\varphi\| = \sqrt{|f(\hat{w})|}$ which is the conclusion of the theorem. ∎

Except when $s > 0$ and $|t| = 1 - s$, the operator $C_\varphi{}^2$ is compact and the spectral radius is 1. Except when $t = 0$, the norm of C_φ is greater than 1, so in general, we see the spectral radius is less than the norm.

Exercises

9.1.1 Show that $C_{z^2}^* \left(\sum a_k z^k \right) = \sum a_{2k} z^k$. Abusing notation, we may regard this as $\left(C_{z^2}^* f \right)(z) = \left(f(\sqrt{z}) + f(-\sqrt{z}) \right)/2$ which exhibits $C_{z^2}^*$ in terms of an averaging operator and the inverse of z^2.

9.1.2 Show that if $\varphi(z) = z(z-c)/(1-\bar{c}z)$, then

$$\left\{ \varphi^k, \; \frac{z\sqrt{1-|c|^2}}{1-\bar{c}z}\varphi^k : k = 0, 1, \ldots \right\}$$

is an orthonormal basis for $H^2(D)$. Find a formula for $C_\varphi^*(f)$ given the Fourier coefficients for f in this basis.

9.1.3 Prove a theorem similar to Theorem 9.2 for the Bergman space of the disk.

9.1.4 Let $u \neq 0$ and v satisfy $|v| \geq 1 + |u|$ and let $\varphi(z) = z(ux + v)^{-1}$. Show that if C_φ is hyponormal, then $v > 1$ and $|u| = v - 1$ which implies C_φ is subnormal.

Notes

The results on the adjoints of C_φ for φ inner might be considered folklore, but the majority of the results in this section and Exercise 9.1.4 are from C. C. Cowen's paper [Co88]. Exercise 9.1.3 and related extensions of work from this section are contained in P. R. Hurst's thesis [Hur95].

9.2   Equivalence of composition operators

Throughout this work, we have frequently used automorphisms to reduce a problem to a more convenient form. This procedure is justified by the observation that if φ is a map of the disk to the disk and $\psi(z) = e^{-i\theta}\varphi(e^{i\theta}z)$ then C_φ and C_ψ are unitarily equivalent on the spaces we have considered or if η is an automorphism of the disk and $\psi(z) = \eta^{-1}(\varphi(\eta(z)))$ then C_φ and C_ψ are similar, if the space is automorphism invariant. If we are investigating properties of operators that are invariant under unitary equivalence or similarity, then we may make the substitutions with no loss of generality. In this section we explore the extent to which the converse of these trivial equivalences hold; that is, suppose C_φ and C_ψ are similar or unitarily equivalent, is it the case that the functions φ and ψ are themselves related by an automorphism of the disk?

Some care must be used in phrasing the question. If there is a bounded invertible operator S so that $C_\psi = SC_\varphi S^{-1}$, it is too much to ask that S be an invertible composition operator. Indeed, if $\psi(z) = \eta^{-1}(\varphi(\eta(z)))$ so that $C_\psi = C_\eta C_\varphi C_\eta^{-1}$, then we also have $C_\psi = SC_\varphi S^{-1}$ for $S = 5C_\eta$, which is not a composition operator. That is, we must ask, given $C_\psi = SC_\varphi S^{-1}$, whether the equivalence *can be* given by an invertible composition operator.

There are easy examples of equivalence of composition operators in which there is no corresponding equivalence of the symbols. Theorem 7.43 shows that if φ is an inner function, not an automorphism, with $\varphi(0) = 0$ then C_φ is an isometry whose unitary part is the identity operator on a one dimensional space and whose purely isometric part is a unilateral shift of infinite multiplicity. Clearly, if φ and ψ are inner functions with $\varphi(0) = \psi(0) = 0$, φ and ψ not automorphisms, then C_φ and C_ψ are unitarily equivalent, but it is not necessarily true that $\psi(z) = \eta^{-1}(\varphi(\eta(z)))$. For example, $\varphi(z) = z^2$ and $\psi(z) = z^3$ have this property. Another example is given in Exercise 9.2.2. Nevertheless, the spirit of the main results in this section is that the theorem that equivalence of the operators implies equivalence of the symbols is "nearly true"!

We have proved several results about spectra and norms that can be used to show that composition operators are not similar or unitarily equivalent by using the fact that similarity preserves spectra and unitary equivalence preserves norms. For example, if φ is a hyperbolic, parabolic, or elliptic automorphism of the disk and C_φ is similar to C_ψ, then ψ is also a hyperbolic, parabolic, or elliptic automorphism of the disk (see Exercises 9.2.1 and 9.2.2).

The following theorem is an example of the rigidity to be expected in unitary equivalence. The example of $\varphi(z) = z^2$ and $\psi(z) = z^3$ given above shows one of the limitations of the conclusion. Nevertheless, as the corollary following the theorem shows, the conclusion does have force.

THEOREM 9.5

If φ and ψ are maps of the disk into itself and U is a unitary operator on $H^2(D)$ that takes constants to constants such that $C_\varphi = U^* C_\psi U$, then there is a real number θ such that $\varphi(z)$ and $e^{i\theta}\psi(e^{-i\theta}z)$ agree on the sequence $\{\varphi_n(0)\}_{n=0}^\infty$.

PROOF Of course, $C_\varphi = U^* C_\psi U$ implies $C_\varphi^* = U^* C_\psi^* U$ and $C_{\varphi_n}^* = U^* C_{\psi_n}^* U$ for all positive integers n.

Since U is an isometry, the hypothesis that U takes constants to constants implies $U(1) = \gamma$ with $|\gamma| = 1$. Now the kernel function K_0 is the constant 1, so

$$U K_{\varphi_n(0)} = U C_{\varphi_n}^* K_0 = C_{\psi_n}^* U K_0 = \gamma C_{\psi_n}^* K_0 = \gamma K_{\psi_n(0)}$$

For $n = 1$, equating norms gives

$$(1 - |\varphi(0)|^2)^{-1/2} = \|K_{\varphi(0)}\| = \|U K_{\varphi(0)}\| = \|\gamma K_{\psi(0)}\| = (1 - |\psi(0)|^2)^{-1/2}$$

from which it follows that $\varphi(0) = e^{i\theta}\psi(0)$ for some real number θ. In particular, if $\varphi(0) = 0$, then we also have $\psi(0) = 0$ and both sequences $\varphi_n(0)$ and $\psi_n(0)$ are identically 0, so the conclusion holds.

If $\varphi(0) \neq 0$,

$$\langle K_{\varphi(0)}, K_{\varphi_n(0)} \rangle = \langle U K_{\varphi(0)}, U K_{\varphi_n(0)} \rangle = \langle \gamma K_{\psi(0)}, \gamma K_{\psi_n(0)} \rangle$$
$$= \langle K_{\psi(0)}, K_{\psi_n(0)} \rangle$$

More explicitly, this says

$$(1 - \overline{\varphi(0)}\varphi_n(0))^{-1} = (1 - \overline{\psi(0)}\psi_n(0))^{-1}$$

which means

$$\varphi_n(0) = \frac{\overline{\psi(0)}}{\varphi(0)}\psi_n(0) = e^{i\theta}\psi_n(0)$$

for all positive integers n.

Finally,

$$\varphi(\varphi_n(0)) = \varphi_{n+1}(0) = e^{i\theta}\psi_{n+1}(0) = e^{i\theta}\psi(\psi_n(0)) = e^{i\theta}\psi(e^{-i\theta}\varphi_n(0))$$

so the functions $\varphi(z)$ and $e^{i\theta}\psi(e^{-i\theta}z)$ agree on the sequence $\{\varphi_n(0)\}_{n=0}^{\infty}$ as desired. ∎

A unitary equivalence between operators maps eigenspaces onto eigenspaces: If U is unitary with $UA = BU$ and $Av = \lambda v$, then $BUv = UAv = U(\lambda v) = \lambda Uv$, so if v is an eigenvector for A with eigenvalue λ, then Uv is an eigenvector of B with the eigenvalue λ. Since $U^{-1} = U^*$ satisfies $U^*B = AU^*$, we see that if w is an eigenvector of B with eigenvalue λ, then U^*w is an eigenvector of A with eigenvalue λ. With this in mind, we can apply Theorem 9.5 to the case in which φ has a fixed point in the disk.

COROLLARY 9.6

Suppose φ, not an automorphism, is a map of the disk into itself with Denjoy–Wolff point a with $|a| < 1$ and suppose $\varphi_n(0) \neq a$ for all positive integers n. Then C_φ and C_ψ are unitarily equivalent as operators on $H^2(D)$ if and only if $\varphi(z) = e^{i\theta}\psi(e^{-i\theta}z)$ for some real number θ.

PROOF The sufficiency of the condition is clear, so we assume $C_\varphi = U^*C_\psi U$ for some unitary operator U on $H^2(D)$.

Since φ has a fixed point in the unit disk, Theorem 2.63 shows that 1 is an eigenvalue of C_φ and $C_\varphi f = f$ if and only if f is constant. On the other hand, the constants are also eigenvectors of C_ψ with eigenvalue 1. Since the eigenspace of C_φ for the eigenvalue 1 is one dimensional, the eigenspace of C_ψ corresponding to the eigenvalue 1 is also one dimensional and this means that it also consists of the constants. Since U maps the eigenspaces of C_φ onto the eigenspaces of C_ψ, we see that U maps constant functions to constant functions and the hypotheses of Theorem 9.5 are satisfied.

Thus, there is a real number θ such that $\varphi(z) = e^{i\theta}\psi(e^{-i\theta}z)$ for $z = \varphi_n(0)$, for all positive integers n. Now by hypothesis, the Denjoy–Wolff point is not $\varphi_n(0)$ for any positive integer n. Since a is the only fixed point of φ in the disk and φ is not an automorphism, the sequence $\{\varphi_n(0)\}$ consists of infinitely many distinct points. Moreover, since the sequence $\{\varphi_n(0)\}$ converges to a in the disk, the identity principle implies the functions $\varphi(z)$ and $e^{i\theta}\psi(e^{-i\theta}z)$ are equal for all z in the disk. ∎

The relative paucity of isometric isomorphisms on the Hardy spaces besides the Hardy Hilbert space allows us to draw the strong conclusion of Corollary 9.6 for all composition operators.

THEOREM 9.7
If φ and ψ are analytic maps of the disk into itself such that $C_\varphi = W^{-1}C_\psi W$ for W an isometric isomorphism on $H^p(D)$ with $1 \le p < \infty$ and $p \ne 2$, then there is a real number θ so that $\varphi(z) = e^{i\theta}\psi(e^{-i\theta}z)$ on the disk.

PROOF Again, the sufficiency of the condition is clear, so we assume $C_\varphi = W^{-1}C_\psi W$ for W an isometric isomorphism on $H^p(D)$. Forelli's Theorem [Fo64, p. 726] says the isometric isomorphism W is

$$(Wf)(z) = \lambda \frac{(1-|w|^2)^{1/p}}{(1-\overline{w}z)^{2/p}} f(\eta(z))$$

where $|\lambda| = 1$, $|w| < 1$, and η is an automorphism $\eta(z) = e^{i\theta}(z-w)/(1-\overline{w}z)$ for some θ real.

On the one hand,

$$WC_\varphi(1) = W(1) = \lambda\frac{(1-|w|^2)^{1/p}}{(1-\overline{w}z)^{2/p}}$$

On the other hand,

$$WC_\varphi(1) = C_\psi W(1) = \lambda\frac{(1-|w|^2)^{1/p}}{(1-\overline{w}\psi(z))^{2/p}}$$

Taking absolute values, we see that these equalities imply

$$|1-\overline{w}z| = |1-\overline{w}\psi(z)|$$

for all z in the disk. It follows that $w = 0$ or $\psi(z) \equiv z$. In the latter case, $C_\psi = I$ which means $C_\varphi = I$ and $\varphi(z) \equiv z$ also.

If $w = 0$ then $W = \lambda C_\eta$ and $\eta(z) = e^{i\theta}z$ which means

$$C_\varphi = W^{-1}C_\psi W = C_{\eta^{-1}}C_\psi C_\eta = C_{\eta \circ \psi \circ \eta^{-1}}$$

and $\varphi(z) = e^{i\theta}\psi(e^{-i\theta}z)$ as desired. ∎

We conclude the section with the statement of a result on power compact composition operators.

THEOREM 9.8
Suppose C_φ^n is compact on $H^2(D)$ for some positive integer n and $\varphi'(a) \ne 0$ where a is the Denjoy–Wolff point of φ. Then C_φ is similar to C_ψ if and only if $\varphi = \eta \circ \psi \circ \eta^{-1}$. Furthermore, C_φ is unitarily equivalent to C_ψ if and only if $\varphi(z) = e^{i\theta}\psi(e^{-i\theta}z)$ for some real number θ.

The structure of the direct proofs of the two results of the theorem are similar, although the proof of the similarity result is more difficult than the proof of the unitary equivalence. In both cases, a basis is chosen for $H^2(D)$ related to the kernels for derivative evaluations at the Denjoy–Wolff point. With respect to this basis, the matrices for C_φ^* and C_ψ^* are triangular. An invertible (or unitary) operator S is assumed to satisfy $SC_\varphi^* = C_\psi^* S$ and estimates of the matrix entries of S are used to show that the functions φ and ψ are related as in the conclusion. The special case in which some of the Taylor coefficients for φ at a are zero causes considerable combinatorial difficulty. It is to be hoped that a more transparent argument for these results can be found.

The theorems of this section suggest that it is difficult for two composition operators to be similar or unitarily equivalent unless their symbols are related by an automorphism. It would be desirable to have results that would characterize the exceptional cases, if this is indeed the case.

Exercises

9.2.1 Suppose φ is an automorphism of the unit disk and C_φ and C_ψ are similar as operators on $H^2(D)$. Prove that ψ is an automorphism of the same type and prove that if φ is hyperbolic, there is an automorphism η so that $\psi(z) = \eta^{-1}(\varphi(\eta(z)))$.

9.2.2 Suppose λ and μ are distinct primitive n^{th} roots of unity for some positive integer n. Show that $C_{\lambda z}$ and $C_{\mu z}$ are unitarily equivalent as operators on $H^2(D)$ but that there is no automorphism η of the disk for which $\lambda z = \eta^{-1}(\mu\eta(z))$.

9.2.3 Suppose φ and ψ are maps of the disk into itself with Denjoy–Wolff points a and b respectively and suppose C_φ and C_ψ are similar as operators on $H^2(D)$. Prove that $|a| < 1$ if and only if $|b| < 1$.

9.2.4 Prove that if $0 < r, s < 1$, then C_{rz^2} and C_{sz^3} are not similar on $H^2(D)$.

Notes

The results and exercises of this section are taken from the work of R. K. Campbell–Wright [Car89], [Car91], [Car93], and [Car94]. Specifically, Theorem 9.5, Corollary 9.6, Theorem 9.7, Exercise 9.2.2 and Exercise 9.2.3 are in [Car91] and Theorem 9.8 is a combination of results in [Car93] and [Car94]. Exercise 9.2.4 is a special case of Theorem 2.6 of [Car91] which shows that if $\|\varphi\|_\infty < 1$, $\|\psi\|_\infty < 1$, $\varphi(0) = \psi(0) = 0$, and C_φ and C_ψ are similar on $H^p(D)$, then φ and ψ vanish to the same order at 0.

9.3   Topological structure

Let $\mathcal{C}(H^2)$ denote the collection of all composition operators C_φ on $H^2(D)$, equipped with the operator norm topology. Work of E. Berkson [Ber81] showed that certain highly non-compact composition operators are isolated in $\mathcal{C}(H^2)$; that is that the component of $\mathcal{C}(H^2)$ containing C_φ is the singleton $\{C_\varphi\}$. In this section we will look at Berkson's theorem as well as subsequent work on isolation and the component structure of $\mathcal{C}(H^2)$ in general.

The first result is an easy one that holds not just in $\mathcal{C}(H^2)$ but more generally in $\mathcal{C}(\mathcal{D}_\alpha)$ for $\alpha \geq 1$, again with the operator norm topology. Recall that this one parameter family of spaces $\mathcal{D}_\alpha$ includes $H^2(D)$ ($\alpha = 1$) and the standard weight Bergman spaces in the disk ($\alpha > 1$). Since some of the results in this section will be strictly $H^2(D)$ results while others will be set in $\mathcal{D}_\alpha$ we will use the notation $\|\cdot\|$ for the norm of a function in $H^2(D)$, and $\|\cdot\|_\alpha$ for the norm in $\mathcal{D}_\alpha$, when needed for clarity. For a bounded operator T acting on some Hilbert space $\mathcal{H}$, $\|T\|$ will always denote the operator norm.

PROPOSITION 9.9

The compact composition operators on $\mathcal{D}_\alpha$ form an arcwise connected set in $\mathcal{C}(\mathcal{D}_\alpha)$ for all $\alpha \geq 1$.

PROOF   Given C_φ compact on $\mathcal{D}_\alpha$ we first construct a continuous map of $[0, 1]$ into $\mathcal{C}(\mathcal{D}_\alpha)$ taking 1 to C_φ and 0 to C_{φ_0}, the composition operator whose symbol is the constant map $\varphi(0)$; that is, $C_{\varphi_0}(f)$ is the constant function with value $f(\varphi(0))$. Let $t \to C_{\varphi_t}$ where $\varphi_t(z) = \varphi(tz)$. Notice that for each t in $[0, 1]$, C_{φ_t} is compact on $\mathcal{D}_\alpha$, by hypothesis when $t = 1$ and because $\|\varphi_t\|_\infty < 1$ when $0 \leq t < 1$.

For $0 \leq t \leq 1$ let C_{tI} be the composition operator whose symbol is the dilation map $z \mapsto tz$. The following are true:

(1) $C_{tI} C_\varphi = C_{\varphi_t}$

(2) $\{C_{tI}\}$ is an equicontinuous family of operators in $\mathcal{C}(\mathcal{D}_\alpha)$, since $\|C_{tI}f - C_{tI}g\|_\alpha = \|C_{tI}(f - g)\|_\alpha \leq \|f - g\|_\alpha$ for all t in $[0, 1]$.

(3)   If s is fixed in $[0, 1]$ and t converges to s (for t also in $[0, 1]$) then $C_{tI}f$ converges to $C_{sI}f$ in $\mathcal{D}_\alpha$, for each f in $\mathcal{D}_\alpha$, since $C_{tI}f$ is just the dilation f_t.

Now $\{f \circ \varphi : \|f\|_\alpha \leq 1\}$ is a relatively compact set in $\mathcal{D}_\alpha$, so the pointwise convergence in (3) is actually uniform on this set; that is, given $\epsilon > 0$ there exists $\delta > 0$ so that if $|t - s| < \delta$, then

$$\|C_{tI}(f \circ \varphi) - C_{sI}(f \circ \varphi)\|_\alpha < \epsilon$$

for all f with $\|f\|_\alpha \leq 1$. By virtue of (1), this is equivalent to $\|C_{\varphi_t} - C_{\varphi_s}\| < \epsilon$, which shows that $t \mapsto C_{\varphi_t}$ is a continuous arc joining C_φ and C_{φ_0}.

To finish, suppose that ψ is another mapping with C_ψ compact. We wish to construct a continuous arc joining C_{φ_0} to C_{ψ_0}. Let $L(t) = (1-t)\varphi(0) + t\psi(0)$ be the line segment in D from $\varphi(0)$ to $\psi(0)$ and then map $[0,1]$ into $C(\mathcal{D}_\alpha)$ by $t \to C_{L(t)}$ where $C_{L(t)}f(z) = f(L(t))$. We leave it as an easy exercise to verify that this mapping is continuous. $\blacksquare$

Later we will return to the question of whether the compact composition operators form a (path) component in $C(\mathcal{D}_\alpha)$.

The next result has as a consequence Berkson's theorem that an analytic map of the disk into itself whose radial limits have modulus one on a set of positive measure must induce an isolated composition operator in $C(H^2)$. Even more is true: C_φ must be isolated in the weaker topology induced by the essential norm $\| \cdot \|_e$. Notice that by Theorem 3.22 and Exercise 3.2.7, no such result can hold in any of the standard weight Bergman spaces $\mathcal{D}_\alpha$ for $\alpha > 1$ since on these spaces one can construct compact C_φ with φ inner. Denote by $E(\varphi)$ the set of all ζ in ∂D with $|\varphi(\zeta)| = 1$. We give a lower bound on the distance from C_φ in $C(H^2)$ to any other C_ψ in terms of $|E(\varphi)|$, the normalized Lebesgue measure of $E(\varphi)$.

PROPOSITION 9.10
Suppose $\varphi \neq \psi$. Then on $H^2(D)$

$$\|C_\varphi - C_\psi\|_e^2 \geq |E(\varphi)| + |E(\psi)|$$

PROOF We obtain this lower estimate by considering $C_\varphi - C_\psi$ acting on normalized reproducing kernel functions

$$k_{r\zeta}(z) = \frac{\sqrt{1-r^2}}{1 - r\bar{\zeta}z}$$

where $0 \leq r < 1$ and ζ in ∂D. Since $k_{r\zeta}$ converges to 0 weakly in $H^2(D)$ as r tends to 1 (by Theorem 2.17), we have $\lim_{r\to 1} \|Qk_{r\zeta}\| = 0$ for every compact operator Q on $H^2(D)$. Thus $\|C_\varphi - C_\psi - Q\| \geq \|(C_\varphi - C_\psi)k_{r\zeta}\| - \|Qk_{r\zeta}\|$ and therefore

$$\|C_\varphi - C_\psi - Q\|^2 \geq \limsup_{r\to 1} \sup_{\zeta \in \partial D} \|(C_\varphi - C_\psi)k_{r\zeta}\|^2$$

$$\geq \limsup_{r\to 1} \int_{\partial D} \|(C_\varphi - C_\psi)k_{r\zeta}\|^2 \, d\sigma(\zeta)$$

where σ is normalized Lebesgue measure on the circle. Since this holds for arbitrary compact Q, $\|C_\varphi - C_\psi\|_e^2$ is bounded below by

$$\limsup_{r\to 1} \int_{\partial D} \left(\int_{\partial D} |k_{r\zeta} \circ \varphi(\eta) - k_{r\zeta} \circ \psi(\eta)|^2 \, d\sigma(\eta) \right) d\sigma(\zeta) \qquad (9.3.1)$$

Now the integrand in Equation (9.3.1) is

$$|k_{r\zeta} \circ \varphi(\eta) - k_{r\zeta} \circ \psi(\eta)|^2$$

$$= |k_{r\zeta} \circ \varphi(\eta)|^2 + |k_{r\zeta} \circ \psi(\eta)|^2$$
$$-k_{r\zeta}(\varphi(\eta))\overline{k_{r\zeta}(\psi(\eta))} - k_{r\zeta}(\psi(\eta))\overline{k_{r\zeta}(\varphi(\eta))}$$

$$= \frac{1-r^2}{|1-r\overline{\zeta}\varphi(\eta)|^2} + \frac{1-r^2}{|1-r\overline{\zeta}\psi(\eta)|^2}$$

$$- \frac{1-r^2}{(1-r\overline{\zeta}\varphi(\eta))(1-r\zeta\overline{\psi(\eta)})} - \frac{1-r^2}{(1-r\overline{\zeta}\psi(\eta))(1-r\zeta\overline{\varphi(\eta)})}$$

Interchanging the order of integration shows that $\|C_\varphi - C_\psi\|_e^2$ is not less than

$$\limsup_{r\to 1} \int_{\partial D} \int_{\partial D} I(\zeta, \eta)\, d\sigma(\zeta)\, d\sigma(\eta)$$

where

$$I(\zeta, \eta) = \frac{1-r^2}{|1-r\overline{\zeta}\varphi(\eta)|^2} + \frac{1-r^2}{|1-r\overline{\zeta}\psi(\eta)|^2}$$

$$- \frac{1-r^2}{(1-r\overline{\zeta}\varphi(\eta))(1-r\zeta\overline{\psi(\eta)})} - \frac{1-r^2}{(1-r\overline{\zeta}\psi(\eta))(1-r\zeta\overline{\varphi(\eta)})}$$

Now

$$\int_{\partial D} \frac{1-r^2}{|1-r\overline{\zeta}\varphi(\eta)|^2}\, d\sigma(\zeta) = (1-r^2)\langle K_{r\varphi(\eta)}, K_{r\varphi(\eta)}\rangle = \frac{1-r^2}{1-|r\varphi(\eta)|^2}$$

where K_w is the kernel function in $H^2(D)$ at w. Similarly

$$\int_{\partial D} \frac{1-r^2}{|1-r\overline{\zeta}\psi(\eta)|^2}\, d\sigma(\zeta) = \frac{1-r^2}{1-|r\psi(\eta)|^2}$$

while

$$\int_{\partial D} \frac{1-r^2}{(1-r\overline{\zeta}\varphi(\eta))(1-r\zeta\overline{\psi(\eta)})}\, d\sigma(\zeta) = (1-r^2)\langle K_{r\psi(\eta)}, K_{r\varphi(\eta)}\rangle$$

$$= \frac{1-r^2}{1-r^2\varphi(\eta)\overline{\psi(\eta)}}$$

and

$$\int_{\partial D} \frac{1-r^2}{(1-r\zeta\overline{\varphi(\eta)})(1-r\overline{\zeta}\psi(\eta))}\, d\sigma(\zeta) = \frac{1-r^2}{1-r^2\overline{\varphi(\eta)}\psi(\eta)}$$

At this point we know that $\|C_\varphi - C_\psi\|_e^2$ is bounded below by

$$\limsup_{r \to 1} \left(\int_{\partial D} \frac{1 - r^2}{1 - |r\varphi(\eta)|^2} + \frac{1 - r^2}{1 - |r\psi(\eta)|^2} \right.$$

$$\left. - \frac{1 - r^2}{1 - r^2\overline{\psi(\eta)}\varphi(\eta)} - \frac{1 - r^2}{1 - r^2\overline{\varphi(\eta)}\psi(\eta)} \, d\sigma(\eta) \right)$$

All four terms in the integrand are bounded for $0 \le r < 1$ and have pointwise limits a.e. as $r \to 1$:

$$\lim_{r \to 1} \frac{1 - r^2}{1 - |r\varphi(\eta)|^2} = \begin{cases} 1 & \text{if } |\varphi(\eta)| = 1 \\ 0 & \text{otherwise} \end{cases}$$

$$\lim_{r \to 1} \frac{1 - r^2}{1 - |r\psi(\eta)|^2} = \begin{cases} 1 & \text{if } |\psi(\eta)| = 1 \\ 0 & \text{otherwise} \end{cases}$$

$$\lim_{r \to 1} \frac{1 - r^2}{1 - r^2\overline{\psi(\eta)}\varphi(\eta)} = \lim_{r \to 1} \frac{1 - r^2}{1 - r^2\psi(\eta)\overline{\varphi(\eta)}} = \begin{cases} 1 & \text{if } \overline{\psi(\eta)}\varphi(\eta) = 1 \\ 0 & \text{otherwise} \end{cases}$$

But $\{\eta : \overline{\psi(\eta)}\varphi(\eta) = 1\} \subset \{\eta : \varphi(\eta) = \psi(\eta)\}$ and this is a set of measure 0 if $\varphi \ne \psi$. Thus $\|C_\varphi - C_\psi\|_e^2 \ge \sigma\{\eta : |\varphi(\eta)| = 1\} + \sigma\{\eta : |\psi(\eta)| = 1\}$ which is the desired result. ∎

So on the one hand if C_φ is compact, it cannot be isolated in $\mathcal{C}(H^2)$, while on the other hand, the last result shows that if C_φ is highly non-compact by virtue of the fact that φ has radial limits of modulus 1 on a set of positive measure, then C_φ must be isolated in $\mathcal{C}(H^2)$. It is natural then to ask if every non-compact C_φ will be isolated. We will answer this in the negative first with a specific example and then with a general non-isolation theorem. Our first step will be to obtain a useful expression for the $H^2(D)$ norm of the difference of two reproducing kernel functions.

LEMMA 9.11
For u and v in D and $K_u(z) = (1 - \overline{u}z)^{-1}$ we have

$$\|K_u - K_v\|^2 = \frac{|u - v|^2}{|1 - \overline{u}v|^2} \left(\frac{1}{1 - |u|^2} + \frac{1}{1 - |v|^2} - 1 \right)$$

PROOF

$$\|K_u - K_v\|^2$$

$$= \frac{1}{1 - |u|^2} + \frac{1}{1 - |v|^2} - 2\text{Re} \frac{1}{1 - \overline{u}v}$$

$$= \frac{(2 - |u|^2 - |v|^2)(|1 - \overline{u}v|^2) - 2(1 - \mathrm{Re}\, u\overline{v})(1 - |u|^2)(1 - |v|^2)}{(1 - |u|^2)(1 - |v|^2)|1 - \overline{u}v|^2}$$

$$= \frac{|u - v|^2}{|1 - \overline{u}v|^2}\left(\frac{1}{1 - |u|^2} + \frac{1}{1 - |v|^2} - 1\right)$$

∎

LEMMA 9.12

If $|\varphi(\zeta)|$ and $|\psi(\zeta)|$ are less than 1 a.e. on ∂D then on $H^2(D)$

$$\|C_\varphi - C_\psi\|^2 \leq \int_{\partial D} \frac{|\varphi(\zeta) - \psi(\zeta)|^2}{|1 - \overline{\varphi}(\zeta)\psi(\zeta)|^2}\left(\frac{1}{1 - |\varphi(\zeta)|^2} + \frac{1}{1 - |\psi(\zeta)|^2} - 1\right)d\sigma(\zeta)$$

PROOF

$$\|C_\varphi - C_\psi\|^2 = \sup\left\{\|C_\varphi f - C_\psi f\|^2 : \|f\| \leq 1\right\} \tag{9.3.2}$$

$$= \sup\left\{\int_{\partial D} |C_\varphi f(\zeta) - C_\psi f(\zeta)|^2\, d\sigma(\zeta) : \|f\| \leq 1\right\}$$

Since $\varphi(\zeta)$ and $\psi(\zeta)$ are in D for almost all ζ in ∂D, we have, for almost all ζ,

$$|C_\varphi f(\zeta) - C_\psi f(\zeta)|$$

$$= |f(\varphi(\zeta)) - f(\psi(\zeta))|$$

$$= |\langle f, K_{\varphi(\zeta)} - K_{\psi(\zeta)}\rangle|$$

$$\leq \|f\|\|K_{\varphi(\zeta)} - K_{\psi(\zeta)}\|$$

$$\leq \|f\|\frac{|\varphi(\zeta) - \psi(\zeta)|}{|1 - \overline{\varphi}(\zeta)\psi(\zeta)|}\left(\frac{1}{1 - |\varphi(\zeta)|^2} + \frac{1}{1 - |\psi(\zeta)|^2} - 1\right)^{1/2}$$

Substitution of this estimate into Equation (9.3.2) yields the desired result.   ∎

The result is also trivially true if either φ or ψ has radial limits of modulus 1 on a set of positive measure, as in this case the integral must diverge if $\varphi \neq \psi$, since $|\{\zeta : \varphi(\zeta) = \psi(\zeta)\}| = 0$.

EXAMPLE 9.13

Consider the family of maps $\varphi_t(z) = 1 + \frac{1}{2}(z - 1) + t(z - 1)^3$ where t lies in an interval $[-\epsilon, \epsilon]$ to be specified. When $t = 0$ we will simply write $\varphi(z)$ for $\varphi_0(z) = (1 + z)/2$; this is, of course, the canonical example of a mapping that satisfies $|\varphi(\zeta)| < 1$ almost everywhere on the unit circle and yet induces a non-compact composition operator on $H^2(D)$. Since for every z in the closed disk,

$\varphi(z)$ lies in the internally tangent disk $\{z : |1 - z|^2 \leq 1 - |z|^2\}$ with center at $1/2$ and radius $1/2$, we have

$$|1 - \varphi(z)|^2 \leq 1 - |\varphi(z)|^2$$

for all z in $\overline{D}$. Thus

$$1 - |\varphi(z)| = \frac{1 - |\varphi(z)|^2}{1 + |\varphi(z)|} \geq \frac{1 - |\varphi(z)|^2}{2} \geq \frac{|1 - \varphi(z)|^2}{2} = \frac{1}{8}|z - 1|^2$$

while if $|t| < \frac{1}{32}$

$$1 - |\varphi_t(z)| \geq 1 - |\varphi(z)| - |t||z - 1|^3 \geq \frac{1}{8}|z - 1|^2 - |t||z - 1|^3 \geq \frac{1}{16}|z - 1|^2$$

In particular, for such t we have $\varphi_t(\overline{D} \setminus \{1\}) \subset D$; of course $\varphi_t(1) = 1$. Also notice that if $\max\{|t|, |s|\} < \frac{1}{128}$ then

$$|1 - \overline{\varphi}_s \varphi_t| = |1 - (\overline{\varphi} + s\overline{(z - 1)}^3)(\varphi + t(z - 1)^3)|$$

$$= \left|1 - |\varphi|^2 - s\varphi(z)\overline{(z - 1)}^3 - t\overline{\varphi(z)}(z - 1)^3 - st|z - 1|^6\right|$$

$$\geq 1 - |\varphi|^2 - (|s| + |t|)|z - 1|^3 - |st||z - 1|^6$$

$$\geq \frac{1}{16}|z - 1|^2$$

Set $\epsilon = \frac{1}{128}$. Now computing in $H^2(D)$ using Lemma 9.12 gives for t, s in $[-\epsilon, \epsilon]$

$$\|C_{\varphi_t} - C_{\varphi_s}\|^2$$

$$\leq \int_{\partial D} \frac{|\varphi_t(\zeta) - \varphi_s(\zeta)|^2}{|1 - \overline{\varphi}_t(\zeta)\varphi_s(\zeta)|^2} \left(\frac{1}{1 - |\varphi_t(\zeta)|^2} + \frac{1}{1 - |\varphi_s(\zeta)|^2}\right) d\sigma(\zeta)$$

$$\leq C \int_{\partial D} \frac{|t - s|^2 |\zeta - 1|^6}{|\zeta - 1|^6} d\sigma(\zeta) = C|t - s|^2$$

for some absolute constant C. This shows that for $t \in [-\epsilon, \epsilon]$, the parametrization $t \mapsto C_{\varphi_t}$ gives a continuous arc in $\mathcal{C}(H^2)$ containing the non-compact operator C_φ. ☐

The techniques used in this example can be generalized to show that whenever $\varphi(D)$ has "finite order of contact" with ∂D, then C_φ will lie in an arc in $\mathcal{C}(H^2)$. Instead of pursuing this approach further, we will see how to obtain the same result as a consequence of the next theorem which shows that only extreme points of the unit ball in $H^\infty(D)$ can induce isolated composition operators. The extreme points of the ball in $H^\infty(D)$ are those $\varphi : D \to D$ satisfying

$$\int_{\partial D} \log(1 - |\varphi(\zeta)|) \, d\sigma(\zeta) = -\infty$$

Geometrically this implies that φ is not a proper convex combination of two distinct maps in the ball of $H^\infty(D)$ ([Dur70, §7.6]), though it is the integral condition, rather than this geometric interpretation, that will concern us here. If φ is not extreme, so that $1 - |\varphi|$ is a bounded non-negative function with $\log(1 - |\varphi|)$ in $L^1(\partial D)$, we may find a bounded outer function $\omega(z)$ in D satisfying $|\omega| = (1 - |\varphi|)^{3/2}$ almost everywhere on the unit circle:

$$\omega(z) = \exp\left(\frac{1}{2\pi}\int_0^{2\pi} \frac{e^{it}+z}{e^{it}-z}\log(1-|\varphi(e^{it})|)^{3/2}\,dt\right)$$

THEOREM 9.14

If

$$\int_{\partial D}\log(1-|\varphi|)\,d\sigma > -\infty$$

then C_φ is not isolated in $\mathcal{C}(H^2)$.

PROOF   Let ω be a bounded outer function in D with $|\omega| = (1-|\varphi|)^{3/2}$ almost everywhere on ∂D. Consider $\varphi_t = \varphi + t\omega$ for $|t| < 1$. For almost all ζ in ∂D we have

$$|\varphi_t(\zeta)| \leq |\varphi(\zeta)| + |t||\omega(\zeta)| = |\varphi(\zeta)| + |t|(1-|\varphi(\zeta)|)^{3/2}$$
$$\leq |\varphi(\zeta)| + |t|(1-|\varphi(\zeta)|) < 1$$

so that $\varphi_t(D) \subset D$ for all t in $(-1,1)$, with $|\varphi_t| < 1$ almost everywhere on the unit circle. We use Lemma 9.12 to show that $t \mapsto C_{\varphi_t}$ is a continuous map of $[-\epsilon, \epsilon]$ into $\mathcal{C}(H^2)$ for ϵ chosen appropriately small. Notice that $1 - |\varphi_t| \geq 1 - |\varphi| - |t||\omega| = 1 - |\varphi| - |t|(1-|\varphi|)^{3/2} \geq \frac{1}{2}(1-|\varphi|)$ almost everywhere on ∂D when $|t| \leq \frac{1}{2}$ while

$$|1 - \overline{\varphi_t}\varphi_s| = |1 - (\overline{\varphi} + t\overline{\omega})(\varphi + s\omega)|$$
$$\geq 1 - |\varphi| - (|t|+|s|)(1-|\varphi|)^{3/2} - |st|(1-|\varphi|)^3$$
$$\geq \frac{1}{2}(1-|\varphi|)$$

provided $|s|, |t| \leq \frac{1}{8}$. Thus

$$\|C_{\varphi_t} - C_{\varphi_s}\|^2 \leq C\int_{\partial D}\frac{|s-t|^2|\omega|^2}{(1-|\varphi|)^2}\left(\frac{1}{1-|\varphi|}\right)\,d\sigma = C|s-t|^2$$

for some absolute constant C, provided $|s|, |t| \leq \frac{1}{8}$. Hence C_{φ_t} converges to C_{φ_s} as t approaches s.   ∎

This theorem can be generalized to $\mathcal{C}(\mathcal{D}_\alpha)$ for $\alpha > 1$ as outlined in Exercise 9.3.4.

The arc constructed in the proof of Theorem 9.14 has an additional property: If $|t|, |s| \leq \frac{1}{8}$ then $C_{\varphi_t} - C_{\varphi_s}$ is compact on $H^2(D)$. This is most easily verified by checking a stronger property, namely that $C_{\varphi_t} - C_{\varphi_s}$ is Hilbert–Schmidt. Recall that in general, a bounded operator T on $H^2(D)$ is a Hilbert–Schmidt operator if and only if

$$\|T\|_{HS}^2 \equiv \sum_{n=0}^{\infty} \|T(z^n)\|^2 < \infty$$

For a difference of two composition operators

$$\|(C_\varphi - C_\psi)(z^n)\|^2 = \|\varphi^n - \psi^n\|^2 = \int_{\partial D} |\varphi^n - \psi^n|^2 \, d\sigma$$

Inserting this in $\sum_0^\infty \|(C_\varphi - C_\psi)(z^n)\|^2$ and interchanging the sum and the integral gives

$$\|C_\varphi - C_\psi\|_{HS}^2 = \int_{\partial D} \frac{1}{1 - |\varphi|^2} + \frac{1}{1 - |\psi|^2} - 2\mathrm{Re}\, \frac{1}{1 - \overline{\varphi}\psi} \, d\sigma$$

If $|\varphi|$ and $|\psi|$ are almost everywhere less than 1 on ∂D (as they must be for this integral to have a chance to converge if $\varphi \neq \psi$) then the same algebra as in the proof of Lemma 9.11 shows that the integrand is equal almost everywhere on ∂D to

$$\frac{|\varphi - \psi|^2}{|1 - \overline{\varphi}\psi|^2} \left(\frac{1}{1 - |\varphi|^2} + \frac{1}{1 - |\psi|^2} - 1 \right)$$

Applying this to the maps $\varphi_t = \varphi + tw$ where φ is not extreme and $|w| = (1 - |\varphi|)^{3/2}$ a.e., and using the estimates from the proof of Theorem 9.14 gives

$$\|C_{\varphi_t} - C_{\varphi_s}\|_{HS}^2 \leq C \int_{\partial D} \frac{|t - s|^2 |w|^2}{(1 - |\varphi|)^2} \frac{1}{(1 - |\varphi|)} \, d\sigma = C|t - s|^2 < \infty$$

whenever $|t|, |s| \leq \frac{1}{8}$.

The next result uses Theorem 9.14.

PROPOSITION 9.15
Suppose $\varphi(D) \subset \Omega \subset D$ where $\overline{\Omega} \cap \partial D = \{\zeta\}$ and

$$\inf \left\{ \frac{1 - |z|}{|\zeta - z|^k} : z \in \Omega \right\} > 0$$

for some positive number k. Then C_φ is not isolated in $C(H^2)$.

PROOF By hypothesis there exists $k > 0$ and $\delta > 0$ so that for each $z \in D$

$$1 - |\varphi(z)| \geq \delta |\zeta - \varphi(z)|^k$$

or equivalently

$$\log(1 - |\varphi(z)|) \geq \log \delta + \log |\zeta - \varphi(z)|^k$$

This inequality persists at almost all points of the unit circle. Since $(\zeta - \varphi(z))^k$ is in $H^\infty(D)$, $\log|\zeta - \varphi(z)|^k$ is in $L^1(\partial D)$, and hence $\log(1 - |\varphi|)$ is also integrable. Thus Theorem 9.14 applies to conclude that C_φ is not isolated.   ∎

A generalization of this result is considered in Exercise 9.3.5.

The condition $\int_{\partial D} \log(1 - |\varphi|) = -\infty$ is *not* sufficient to guarantee that C_φ is isolated; an example that shows this is outlined in Exercise 9.3.6.

We next investigate the role of the angular derivative in studying the component structure of $\mathcal{C}(\mathcal{D}_\alpha)$, $\alpha \geq 1$. Recall that C_φ is compact on $\mathcal{D}_\alpha$, $\alpha > 1$ if and only if φ has no finite angular derivative at any point of ∂D, while for $\mathcal{D}_1 = H^2(D)$ this is a necessary but not sufficient condition for compactness (Corollary 3.14 and Theorem 3.22).

In the next result, we assume that φ has a finite angular derivative at some point and use kernel functions to estimate $\|C_\varphi - C_\psi\|_e^2$ on the spaces in question. Since the kernel functions have nicer expressions in the equivalent spaces $H^2(D)$ and $A_\gamma^2(D)$, we state the next theorem for these spaces.

THEOREM 9.16
Suppose $\varphi : D \to D$ has finite angular derivative $\varphi'(\zeta)$ at some point ζ in the unit circle. Let $\psi : D \to D$ and consider C_φ and C_ψ acting either on $H^2(D)$ or $A_\gamma^2(D)$ for $\gamma > -1$. Then unless both

(1) $\varphi(\zeta) = \psi(\zeta)$ (as radial limits) and
(2) $\varphi'(\zeta) = \psi'(\zeta)$

we have $\|C_\varphi - C_\psi\|_e^2 \geq |\varphi'(\zeta)|^{-\beta}$ where $\beta = 1$ for the space $H^2(D)$ and $\beta = \gamma + 2$ for the spaces $A_\gamma^2(D)$.

PROOF   Since the spaces under consideration are unitarily invariant, we may assume $\zeta = 1$, $\varphi(1) = 1$, and $\varphi'(1) = s < \infty$. We first estimate $\|C_\varphi - C_\psi\|$ by considering $(C_\varphi - C_\psi)^*$ acting on reproducing kernel functions K_z. Recall that on $H^2(D)$ we have $K_z(w) = (1 - \bar{z}w)^{-1}$ while on $A_\gamma^2(D)$, $K_z(w) = (\gamma + 1)(1 - \bar{z}w)^{-(\gamma+2)}$ (Exercise 2.1.5). Since

$$\|(C_\varphi - C_\psi)^* K_z\|^2 = \|K_{\varphi(z)} - K_{\psi(z)}\|^2$$
$$= \|K_{\varphi(z)}\|^2 + \|K_{\psi(z)}\|^2 - 2\operatorname{Re} K_{\varphi(z)}(\psi(z))$$

the formulas for the kernel functions K_z and their norms show that

$$\|C_\varphi - C_\psi\|^2 \geq \left(\frac{1 - |z|^2}{1 - |\varphi(z)|^2}\right)^\beta + \left(\frac{1 - |z|^2}{1 - |\psi(z)|^2}\right)^\beta - 2\operatorname{Re} \frac{K_{\varphi(z)}(\psi(z))}{\|K_z\|^2}$$

for all z in D, where $\beta = 1$ for the space $H^2(D)$ and $\beta = \gamma + 2$ for the spaces $A_\gamma^2(D)$. By the Julia–Carathéodory Theorem, $(1 - |z|^2)/(1 - |\varphi(z)|^2)$ has nontangential limit $1/s$ as z goes to 1. To deal with the term

$$\frac{2\text{Re}\,K_{\varphi(z)}(\psi(z))}{\|K_z\|^2} \tag{9.3.3}$$

we distinguish two cases:

First, if $\lim_{r\to 1}\psi(r) \neq 1$ find r_n increasing to 1 so that $\lim_{n\to\infty}\psi(r_n) = u \neq 1$. Setting $z = r_n$ in Expression (9.3.3) yields

$$2\text{Re}\left(\frac{1 - r_n^2}{1 - \overline{\varphi(r_n)}\psi(r_n)}\right)^{\beta}$$

which has limit 0 as n tends to infinity since $\lim \overline{\varphi(r_n)}\psi(r_n) = u \neq 1$. This combined with the above estimate on $(1 - r_n^2)/(1 - |\varphi(r_n)|^2)$ shows that $\|C_\varphi - C_\psi\|^2 \geq s^{-\beta}$.

On the other hand, if $\lim_{r\to 1}\psi(r) = 1$ but $\psi'(1) \neq \varphi'(1)$, either because ψ has finite angular derivative at 1 different from s or because $|\psi'(1)| = \infty$ then we first consider

$$\frac{1 - \overline{\varphi}(z)\psi(z)}{1 - |z|^2} = \frac{1 - \overline{\varphi}(z)\varphi(z) + \overline{\varphi}(z)\varphi(z) - \overline{\varphi}(z)\psi(z)}{1 - |z|^2}$$

$$= \frac{1 - |\varphi(z)|^2}{1 - |z|^2} + \overline{\varphi}(z)\frac{1 - z}{1 - |z|^2}\left(\frac{1 - \psi(z)}{1 - z} - \frac{1 - \varphi(z)}{1 - z}\right)$$

Let $\gamma_M = \{z \in D : \frac{|1-z|}{1-|z|^2} = M\}$, essentially the boundary of a nontangential approach region at 1. As z approaches 1 along γ_M, the Julia–Carathéodory Theorem shows that

- $(1 - |\varphi(z)|^2)/(1 - |z|^2) \to s$
- $(1 - \varphi(z))/(1 - z) \to s$
- $(1 - \psi(z))/(1 - z) \to \psi'(1)$ if $\psi'(1) < \infty$
  or $|(1 - \psi(z))/(1 - z)| \to \infty$ if $|\psi'(1)| = \infty$.

Thus given any positive number N we may, by choosing M sufficiently large, find a sequence z_n approaching 1 along γ_M so that for n sufficiently large

$$\left|\overline{\varphi}(z_n)\frac{1 - z_n}{1 - |z_n|^2}\left(\frac{1 - \psi(z_n)}{1 - z_n} - \frac{1 - \varphi(z_n)}{1 - z_n}\right)\right| > N$$

or equivalently, given $\delta > 0$ we may find a nontangential sequence $\{z_n\}$ approaching 1 so that for n sufficiently large

$$\left|\frac{2\text{Re}\,K_{\varphi(z_n)}(\psi(z_n))}{\|K_{z_n}\|^2}\right| \leq \frac{|2K_{\varphi(z_n)}(\psi(z_n))|}{\|K_{z_n}\|^2} < \delta$$

Our previous remark on the nontangential limit of $(1 - |z|^2)/(1 - |\varphi(z)|^2)$ and this inequality show that $\|C_\varphi - C_\psi\|^2 \geq s^{-\beta} - \delta$ for arbitrary positive δ.

To extend this estimate to $\|C_\varphi - C_\psi\|_e^2$ let Q be any compact operator. Then

$$\|C_\varphi - C_\psi - Q\| \geq \frac{\|(C_\varphi - C_\psi - Q)^* K_z\|}{\|K_z\|}$$

$$\geq \left\| (C_\varphi - C_\psi)^* \left(\frac{K_z}{\|K_z\|} \right) \right\| - \left\| Q^* \left(\frac{K_z}{\|K_z\|} \right) \right\|$$

Since Q is compact and the normalized reproducing kernel functions go to 0 weakly as z approaches the boundary of D we have

$$\|Q^*(K_z)\|/\|K_z\| \to 0$$

as $|z|$ approaches 1. Thus

$$\|C_\varphi - C_\psi\|_e^2 = \inf\{\|C_\varphi - C_\psi - Q\|^2 : Q \text{ compact}\} \geq s^{-\beta}$$

and we are done.   ∎

This theorem has several consequences, the first of which is an analogue of Berkson's isolation theorem.

COROLLARY 9.17

If φ has finite angular derivative on a set of positive measure, then C_φ is isolated in $\mathcal{C}(\mathcal{D}_\alpha)$, $\alpha \geq 1$. For any $\psi \neq \varphi$, $\|C_\varphi - C_\psi\|_e^2 \geq c(\alpha)s^{-\alpha}$ where s is the essential infimum of $\{|\varphi'(\zeta)| : |\varphi(\zeta)| = 1\}$ and $c(\alpha)$ is a constant depending only on α.

PROOF   Let $A(\varphi) = \{\zeta \in \partial D : |\varphi'(\zeta)| < \infty\}$ so that $A(\varphi) \subset E(\varphi) = \{\zeta : |\varphi(\zeta)| = 1\}$. Since $\psi \neq \varphi$, we have $|\{\zeta \in \partial D : \varphi(\zeta) = \psi(\zeta)\}| = 0$. Thus, for almost every ζ in $A(\varphi)$, Theorem 9.16 and the fact that $\mathcal{D}_1 = H^2(D)$ and $\mathcal{D}_\alpha = A_{\alpha-2}^2(D)$, with equivalent norms, shows that on $\mathcal{D}_\alpha$

$$\|C_\varphi - C_\psi\|_e^2 \geq c(\alpha)|\varphi'(\zeta)|^{-\alpha}$$

for some constant $c(\alpha)$ depending only on α (and reflecting the equivalence of norms) from which the result follows.   ∎

We will say that φ and ψ **have the same data** at a point ζ in ∂D if φ and ψ each have radial limit of modulus 1 at ζ with

(1) $\varphi(\zeta) = \psi(\zeta)$ and
(2) $|\varphi'(\zeta)| = |\psi'(\zeta)|$.

If (1) holds, then (2) actually implies that $\varphi'(\zeta) = \psi'(\zeta)$, provided $|\varphi'(\zeta)| < \infty$. Notice that Theorem 9.16 implies that if $C_\varphi - C_\psi$ is compact then φ and ψ have the same data at every point in $A(\varphi)$. The next result shows this also holds whenever C_φ and C_ψ lie in the same component of $\mathcal{C}(\mathcal{D}_\alpha)$.

THEOREM 9.18

Suppose C_ψ is in the component of $\mathcal{C}(\mathcal{D}_\alpha)$ containing C_φ. Then φ and ψ must have the same data at all points of the unit circle where φ has finite angular derivative.

PROOF Suppose $|\varphi'(\zeta)| < \infty$. Without loss of generality we may assume $\zeta = 1$, $\varphi(1) = 1$, and $\varphi'(1) = s < \infty$. If C_τ is in the component containing C_φ, denote by U_τ the open ball of radius $\frac{1}{2}\sqrt{c(\alpha)}s^{-\alpha/2}$ centered at C_τ, where $c(\alpha)$ is the constant from Corollary 9.17. From the collection $\{U_\tau\}$ extract a simple chain $\{U_{\tau_j}\}_{j=1}^n$ from C_φ to C_ψ; that is, C_φ in U_{τ_1}, C_ψ in U_{τ_n}, and $U_{\tau_i} \cap U_{\tau_j} \neq \emptyset$ if and only if $|i - j| \leq 1$ (see [HoY61, p. 108]). If C_{γ_1} lies in $U_{\tau_1} \cap U_{\tau_2}$, γ_1 must have the same data as φ at 1 by Theorem 9.16, since $\|C_{\gamma_1} - C_\varphi\| < \sqrt{c(\alpha)}s^{-\alpha/2}$. In particular $\gamma_1'(1) = s$. Similarly, γ_1 and τ_2 must have the same data at 1 and if C_{γ_2} is in $U_{\tau_2} \cap U_{\tau_3}$ so must γ_2 and τ_2. Continuing along the chain to C_ψ we conclude that ψ has the same data as φ at 1. ∎

COROLLARY 9.19

For $\alpha > 1$, the compact composition operators on $\mathcal{D}_\alpha$ form a path component in $\mathcal{C}(\mathcal{D}_\alpha)$.

PROOF We already know by Proposition 9.9 that the compact composition operator are in a path connected set in $\mathcal{C}(\mathcal{D}_\alpha)$. If C_ψ is not compact on $\mathcal{D}_\alpha$ for some $\alpha > 1$ then ψ has finite angular derivative at some point ζ in ∂D. But then C_ψ cannot be in the same component as C_φ for any compact C_φ, since ψ and φ cannot have the same data at ζ. ∎

It is an open question as to whether this last result holds for $\mathcal{D}_1 = H^2(D)$, although by Theorem 9.16 the component containing the compacts in $\mathcal{C}(H^2)$ cannot contain any C_φ where φ has finite angular derivative, nor can it contain any non-compact C_φ where φ has finite valency.

It is worth noting that we have not given a complete description of the isolated points in $\mathcal{C}(H^2)$ (or $\mathcal{C}(\mathcal{D}_\alpha)$). This question, as well as the more general one of a description of all components, remains open. The results presented here do suggest a conjecture: C_φ and C_ψ lie in the same component if and only if $C_\varphi - C_\psi$ is compact([ShS90a]). Whether or not this conjecture proves to be correct it would be of interest to have a characterization (perhaps analogous to Theorem 3.20 for single composition operators) of those pairs (φ, ψ) for which $C_\varphi - C_\psi$ is compact.

The questions addressed in this section have also been asked (and in some cases completely answered) for the space of composition operators on $H^2(D)$ in other topologies. Some of these results are indicated in the *Notes* at the end of the section.

Exercises

9.3.1 (a) If C_φ is isolated in $\mathcal{C}(\mathcal{D}_\alpha)$ then for every univalent ψ with $\psi(D) \supset \varphi(D)$, C_ψ is isolated also.

(b) If φ is univalent with C_φ non-isolated in $\mathcal{C}(\mathcal{D}_\alpha)$ and if ψ is any map with $\psi(D) \subset \varphi(D)$, then C_ψ is not isolated.

9.3.2 (This exercise extends Exercise 1.1.10 on the linear independence of composition operators to their independence in the Calkin algebra.) Suppose $\varphi_1, \varphi_2, \ldots, \varphi_n$ are distinct analytic maps of D into D and $a_1, a_2, \ldots,$ and a_n are complex constants. Recall that $E(\varphi) = \{\zeta \in \partial D : |\varphi(\zeta)| = 1\}$.

(a) Show that on $H^2(D)$

$$\left\| \sum_1^n a_j C_{\varphi_j} \right\|_e^2 \geq \sum_1^n |a_j|^2 |E(\varphi_j)|$$

(b) If ζ in ∂D is such that $\varphi_j(\zeta)$ exists and has modulus 1 for each $j = 1, 2, \ldots,$ n and if no pair $\{\varphi_i, \varphi_j\}, i \neq j$, has the same data at ζ then on $\mathcal{D}_\alpha$ for $\alpha \geq 1$

$$\left\| \sum_1^n a_j C_{\varphi_j} \right\|_e^2 \geq c(\alpha) \sum_1^n |a_j|^2 |\varphi_j'(\zeta)|^{-\alpha}$$

9.3.3 Let $\varphi(z) = 1 + \frac{1}{2}(z-1)$ and $\varphi_t(z) = 1 + \frac{1}{2}(z-1) + t(z-1)^2$.

(a) Show that $\varphi_t(D) \subset D$ for all t in $[0, 1]$.

(b) Is $C_\varphi - C_{\varphi_t}$ compact on $H^2(D)$ for any $t \neq 0$? Hint: Consider $(C_\varphi - C_{\varphi_t})^*$ acting on normalized reproducing kernel functions for a sequence of points approaching 1 along the circle $|1 - z|^2 = 1 - |z|^2$.

9.3.4 The goal of this problem is to extend Theorem 9.14 to $\mathcal{C}(\mathcal{D}_\alpha)$ for $\alpha > 1$.

(a) As a substitute for Lemma 9.12 first show that for f in $\mathcal{D}_\alpha$ and w in D

$$|f'(w)| \leq M_\alpha \frac{\|f\|_\alpha}{(1 - |w|^2)^{(\alpha+4)/2}}$$

where M_α is a constant depending only on α. From this show

$$\|C_\varphi f - C_\psi f\|_\alpha^2 \leq M_\alpha' \|f\|_\alpha^2 \int_D \frac{|\varphi(z) - \psi(z)|^2 (1 - |z|^2)^{\alpha-2}}{\left(\min(1 - |\varphi(z)|, 1 - |\psi(z)|)\right)^{\alpha+4}} \, dA(z)$$

(b) If $\log(1 - |\varphi|) \in L^1(\partial D)$ let ω be an outer function with $|\omega| = 1 - |\varphi|$ almost everywhere on the unit circle. Set $\upsilon = \omega^{(\alpha+4)/2}$ and $\varphi_t = \varphi + t\upsilon$ for sufficiently small t. Show that for such t, the parametrization $t \mapsto C_{\varphi_t}$ is a continuous arc in $\mathcal{C}(\mathcal{D}_\alpha)$. You will need the inequality $|\omega| + |\varphi| \leq 1$ on D. This can be obtained by considering the C^2-valued analytic function $F(z) = (\omega(z), \varphi(z))$.

9.3.5 For a positive integer k we say that $\varphi(D)$ has order of contact at most k with ∂D if for each ζ in ∂D there is a disk Δ centered at ζ so that

$$\inf\{\frac{1 - |z|}{|\zeta - z|^k}; z \in \varphi(D) \cap \Delta\} > 0$$

Show that if $\varphi(D)$ has order of contact at most k then C_φ is not isolated in $\mathcal{C}(H^2)$. By the previous exercise the same result holds in $\mathcal{C}(\mathcal{D}_\alpha)$.

9.3.6 The purpose of this exercise is to give an example of an extreme τ with C_τ not isolated. Let Ω be the region in D described by

$$\Omega = \{r\zeta : \zeta \in \partial D, 0 \leq r < 1 - \exp(-1/|1 - \zeta|)\}$$

bounded by the curve

$$\{re^{i\theta} : r = 1 - \exp(-1/(2|\sin\frac{\theta}{2}|))\}$$

so that $\overline{\Omega} \cap \partial D = \{1\}$, $\partial \Omega$ is a Jordan curve of class C^2, and Ω is symmetric about the real axis. There is a univalent map of D onto Ω which maps $(-1, 1)$ to the intersection of Ω with the real axis, the top half of D onto Ω^+, the top half of Ω, and the bottom half on D onto the bottom half of Ω. Furthermore, φ extends to a C^1 homeomorphism of $\overline{D}$ to $\overline{\Omega}$ with $\varphi(1) = 1$ and φ' never 0 on $\overline{D}$. ([Pom92, p. 48]) Let ψ be the self-map of D given by $\psi(z) = F^{-1}(\sqrt{F(z)})$ (principal value) where $F(z) = i(1 - z)/(1 + z)$, a linear fractional map of the disk onto the upper halfplane. Set $\tau = \psi \circ \varphi$.

(a) Show C_τ is compact on H^2.

(b) Obtain the estimates

- $1 - |F^{-1}(w)|^2 = 4\text{Im}\,w/(|i + w|^2) < 4\text{Im}\,w$ for w in the upper halfplane.
- $\text{Im}\,\sqrt{w} < \sqrt{\text{Im}\,w/2}$ for w in the upper halfplane with $\text{Re}\,w > 0$.
- $1 - |\psi(z)| < 4\sqrt{1 - |z|}$ for z in D with $\text{Re}\,z > 0$ and $\text{Im}\,z > 0$.
- $|1 - \varphi(\zeta)| \leq C|1 - \zeta|$ for ζ in $\tau^{-1}((\partial\Omega)^+)$ where $C = \sup\{|\varphi'(z)| : z \in D\}$.

(c) Use the above estimates and the description of Ω to show that $\log(1 - |\tau|)$ is not integrable.

Notes

E. Berkson's original isolation theorem appears in [Ber81] and is set in the context of $H^p(D)$, $1 \leq p < \infty$. The generalization of this result given in Proposition 9.10 is due to J. H. Shapiro and C. Sundberg [ShS90a]. Theorem 9.16 and its consequences are due to B. D. MacCluer [Mc89]. Most of the remaining results in this section, including those presented in Exercises 9.3.1, 9.3.5, and 9.3.6, are taken from [ShS90a] which we have followed quite closely. This paper also contains a number of results not presented here. In particular a generalization of Proposition 9.15 is given which shows non-isolation even when $\varphi(D)$ has limited exponential order contact. There is also in [ShS90a] a technically difficult result which is in the spirit of a partial converse to Theorem 9.14 and shows that a regular, univalent map of D onto a region making sufficiently close contact with ∂D will induce an isolated composition operator in $\mathcal{C}(H^2)$. In particular this allows the construction of the only known examples of mappings φ with $|\varphi| < 1$ a.e. and yet C_φ isolated in $\mathcal{C}(H^2)$.

Similar questions on the connectedness properties of $\mathcal{C}(H^2)$ in other topologies have also been studied. H. Hunziker, H. Jarchow and V. Mascioni [HuJM90] consider the topology generated by the open sets of the form

$$U(\varphi) = \{\psi : d(\psi, \varphi) < r\}$$

where $d(\cdot, \cdot)$ is the $[0, \infty]$ valued metric defined by

$$d(\psi, \varphi) = \|C_\varphi - C_\psi\|_{HS} = \left(\sum_0^\infty \|\varphi^n - \psi^n\|_{H^2}^2 \right)^{1/2}$$

They show that the isolated points of $\mathcal{C}(H^2)$ in this topology are precisely those composition operators induced by the extreme points of the unit ball in H^∞. Moreover, C_φ and C_ψ are in the same component (relative to this Hilbert–Schmidt topology) if and only if $C_\varphi - C_\psi$ is Hilbert–Schmidt.

9.4 Polynomial approximation

In this section we look at an application of the study of composition operators to a problem in polynomial approximation. The setting is as follows. Begin with a finite positive Borel measure μ with support in the complex plane and let $P^2(\mu)$ denote the closure in $L^2(\mu)$ of the analytic polynomials. It is an old problem, with many facets, to understand $P^2(\mu)$ for various μ. When $\mu = G\, dA/\pi$ with $G = G(|z|)$ a rotation invariant, positive, continuous weight function on the disk D we know that $P^2(\mu)$ is the Bergman space $A_G^2(D)$. If $\mu = d\theta$, Lebesgue measure on ∂D, then $P^2(\mu) = H^2(D)$, with norm multiplied by $\sqrt{2\pi}$. If $\mu = \chi_E\, d\theta$ where E is a Borel set in ∂D with $|\partial D \setminus E| > 0$ (where $|\cdot|$ denotes Lebesgue measure) then Szegö's theorem [Ga81, p. 144] shows that $P^2(\mu) = L^2(E)$. Motivated by these cases it is natural to consider the hybrid $\mu = \chi_E\, d\theta + G\, dA/\pi$ and to ask under what conditions on G and E the space $P^2(\mu)$ has a natural decomposition as a direct sum

$$P^2(\mu) = L^2(E) \oplus A_G^2(D) \tag{9.4.1}$$

The meaning of Equation (9.4.1) is that the natural isometry of $P^2(\mu)$ into $L^2(E) \oplus A_G^2(D)$, densely defined on the polynomials by $p \to \chi_{\partial D} p \oplus \chi_D p$, is *onto*. When Equation (9.4.1) holds for some $\mu = \chi_E\, d\theta + G\, dA/\pi$ we say that $P^2(\mu)$ *splits*. Asking whether $P^2(\mu)$ splits is really asking about the possibility of **simultaneous approximation**, a term which appears in the Soviet literature of the 1960's and 1970's where questions of this type were first systematically studied. It is the case that for every f in $L^2(E)$ and g in $A_G^2(D)$ there exist polynomials $\{p_n\}$ with p_n converging to f in $L^2(E)$ and p_n converging to g in $A_G^2(D)$ precisely when $P^2(\mu) = L^2(E) \oplus A_G^2(D)$. Conditions under which $P^2(\mu)$ splits, worked out by a number of authors, show a rather complicated interdependence on E and G. Our goal here is to discuss a result due to Hruščëv that gives a sufficient condition for splitting to occur when G decays sufficiently rapidly so that

$$\int_0^1 \exp\left(\frac{a}{1 - r}\right) G(r)\, dr < \infty$$

for all $a > 0$.

The proof of Hruščëv's theorem will require information about a composition operator induced by a special type of inner function. The relevant inner functions are constructed to have prescribed mapping properties and we turn next to a description of their construction.

Given a Borel set Q in ∂D with Lebesgue measure $|Q|$ satisfying $0 < |Q| < 2\pi$ set

$$\psi(z) = \psi_Q(z) = \exp\left(\frac{i}{2} \int_Q \frac{e^{i\theta} + z}{e^{i\theta} - z} \, d\theta\right)$$

for z in D. Note that $\psi(0) = e^{i|Q|/2}$.

Since

$$\arg \psi(z) = \mathrm{Im}\left(\frac{i}{2} \int_Q \frac{e^{i\theta} + z}{e^{i\theta} - z} \, d\theta\right)$$

$$= \frac{1}{2} \int_Q \mathrm{Re}\left(\frac{e^{i\theta} + z}{e^{i\theta} - z}\right) d\theta$$

$$= \frac{1}{2} \int_Q P_r(t - \theta) \, d\theta$$

where $z = re^{it}$ and

$$P_r(t - \theta) = \frac{1 - r^2}{1 - 2r\cos(t - \theta) + r^2} \geq 0$$

is the Poisson kernel, we have $0 < \arg \psi(z) < \pi$. Thus ψ maps D to the upper halfplane Π. Moreover $\arg \psi(z)$ has nontangential limit for almost all e^{it} in ∂D equal to $\pi \chi_Q(t)$. This means ψ has nontangential boundary values $\psi(e^{it}) < 0$ almost everywhere on Q and $\psi(e^{it}) > 0$ almost everywhere on $Q^c = \partial D \setminus Q$.

Composing ψ with the inverse Cayley transform $w \mapsto (w - i)/(w + i)$, or indeed with any linear fractional map of the upper halfplane onto D, produces an inner function in D. We will be interested the particular linear fractional map

$$\lambda(w) = \lambda_Q(w) = \frac{w - e^{i|Q|/2}}{w - e^{-i|Q|/2}}$$

which maps $(-\infty, 0)$ onto $V = \{e^{i\theta} : 0 < \theta < |Q|\}$ and $[0, \infty) \cup \{\infty\}$ onto $V^c = \partial D \setminus V$. The inner function $b \equiv \lambda \circ \psi$ has $b(0) = 0$ and its radial limit function maps Q to V and Q^c to V^c almost everywhere. If necessary for clarity we will write b_Q for $\lambda_Q \circ \psi_Q$ and refer to b_Q as the inner function associated with the set Q. To get some feeling for this definition the reader is invited in Exercises 9.4.2 and 9.4.4 to compute b_Q with several specific choices for Q. For example, if Q is the union of n disjoint arcs of equal length, symmetrically spaced around ∂D, then $b_Q = cz^n$ for some constant c of modulus 1. Since we always have $b(0) = 0$, Littlewood's Subordination Principle guarantees that C_b is bounded on *any* weighted Bergman space $A_G^2(D)$ given by a rotation invariant weight function G.

THEOREM 9.20

Let $\{Q_n\}_1^\infty$ be a sequence of Borel sets in ∂D with $0 < |Q_n| < 2\pi$. Denote by b_n the inner function associated with Q_n. Consider C_{b_n} acting on an arbitrary weighted Bergman space $A_G^2(D)$. If

$$\lim_{n \to \infty} \frac{1}{|Q_n|} \chi_{Q_n}(e^{i\theta})\, d\theta = \frac{d\theta}{2\pi}$$

in the weak* topology on the space of Borel measures on ∂D, then $\|C_{b_n} - P\|$ converges to 0 as n goes to infinity, where P is the projection of $A_G^2(D)$ onto the constant functions.

PROOF   Fix z in D. By hypothesis

$$\lim_{n \to \infty} \frac{1}{|Q_n|} \int_{Q_n} \frac{e^{i\theta} + z}{e^{i\theta} - z}\, d\theta = \int_{\partial D} \frac{e^{i\theta} + z}{e^{i\theta} - z} \frac{d\theta}{2\pi}$$

Since

$$\int_{\partial D} \frac{e^{i\theta} + z}{e^{i\theta} - z} \frac{d\theta}{2\pi} = 1$$

we have, for any given $\epsilon > 0$,

$$\left| \frac{i}{2} \int_{Q_n} \frac{e^{i\theta} + z}{e^{i\theta} - z}\, d\theta - \frac{i}{2}|Q_n| \right| < |Q_n|\epsilon$$

for sufficiently large n. If we set $\psi_n = \psi_{Q_n}$ and let

$$a_n = \frac{i}{2} \int_{Q_n} \frac{e^{i\theta} + z}{e^{i\theta} - z}\, d\theta$$

this says

$$|\psi_n(z) - e^{i|Q_n|/2}| = \left| \int_{i|Q_n|/2}^{a_n} e^z\, dz \right| \le \sup\{|e^z| : z \in D_n\}|Q_n|\epsilon$$

where D_n is the disk centered at $i|Q_n|/2$ with radius $|Q_n|\epsilon$. Assuming without loss of generality that $\epsilon < 1$ we certainly have $\sup\{|e^z| : z \in D_n\} < e^{2\pi}$, so that

$$|\psi(z) - e^{i|Q_n|/2}| \le e^{2\pi}|Q_n|\epsilon$$

By definition, $b_n = \lambda_n \circ \psi_n$ where

$$\lambda_n(w) = \frac{w - e^{i|Q_n|/2}}{w - e^{-i|Q_n|/2}}$$

Thus

$$|b_n(z)| = |b_n(z) - b_n(0)| = |\lambda_n(\psi_n(z)) - \lambda_n(e^{i|Q_n|/2})|$$

$$= \left| \int_L \lambda_n'(w)\, dw \right|$$

where L is the line segment from $e^{i|Q_n|/2}$ to $\psi_n(z)$. Since ψ_n takes values in the upper halfplane, a computation shows that

$$|\lambda'_n(w)| \le \frac{2}{\sin(|Q_n|/2)}$$

on L. Thus

$$|b_n(z)| \le \frac{|Q_n|/2}{\sin(|Q_n|/2)} 4e^{2\pi} \epsilon$$

If $|Q_n|$ is bounded away from 2π for all n then this shows that $|b_n(z)| \le c\epsilon$ for some absolute constant c. Since ϵ was arbitrary, we have $b_n(z)$ approaches 0 for each z in D. If there is a subsequence of $Q'_n s$ with $|Q_n|$ converging to 2π, consider the inner functions associated with Q_n^c, the complementary sets for this subsequence. The above argument, applied to these inner functions, show that

$$b_{Q_n^c}(z) \to 0$$

for each z in D. But by Exercise 9.4.1,

$$b_{Q_n} = e^{i|Q_n|} b_{Q_n^c}$$

so $b_n(z) = b_{Q_n}(z)$ goes to 0 in D. Theorem 5.12 now applies to give the conclusion. ∎

We are now ready to return to the question of when $P^2(\mu) \equiv P^2(\chi_E \, d\theta + G \, dA/\pi)$ has a direct sum decomposition as $L^2(E) \oplus A_G^2(D)$, where E is a Borel measurable subset of ∂D and $G(r)$ is a positive continuous function on $[0, 1)$. To set the stage for our work we state without proof two results which delineate the role played by convergence or divergence of the integral

$$\int_{1-\delta}^1 \log \log \frac{1}{G(r)} \, dr \tag{9.4.2}$$

for δ small. Since the cases of primary interest will be ones where $G(r)$ tends to 0 as r goes to 1, we also assume this and note that this guarantees that the integrand in Expression (9.4.2) makes sense for small δ. In the following two theorems G also must satisfy a mild regularity condition whose exact definition will not concern us here. The details can be found in [KrM90b].

THEOREM 9.21
If the integral in Expression (9.4.2) diverges then $P^2(\mu) = L^2(E) \oplus A_G^2(D)$ if and only if $|\partial D \setminus E| > 0$.

THEOREM 9.22
Suppose the integral in Expression (9.4.2) converges. If $P^2(\mu) = L^2(E) \oplus A_G^2(D)$ then $|I \setminus E| > 0$ for every open arc I in ∂D.

We will prove a result due to Hruščëv which gives a converse to Theorem 9.22 when the weight function G satisfies a certain growth condition.

THEOREM 9.23
Suppose that

$$\int_0^1 \exp\left(\frac{a}{1-r}\right) G(r) \, dr < \infty$$

for every $a > 0$. If $|I \setminus E| > 0$ for every open arc I in ∂D then $P^2(\mu) = L^2(E) \oplus A_G^2(D)$.

The proof of this theorem requires two lemmas, the first of which gives a simple condition equivalent to splitting.

LEMMA 9.24
For $\mu = \chi_E \, d\theta + G \, dA/\pi$ we have $P^2(\mu) = L^2(E) \oplus A_G^2(D)$ if and only if χ_D is in $P^2(\mu)$.

PROOF   The "only if" direction is obvious. For the converse suppose $P^2(\mu)$ contains χ_D. Since the polynomials are dense in $P^2(\mu)$, $P^2(\mu)$ is invariant under multiplication by z. Thus if p is a polynomial, $p\chi_D$ is in $P^2(\mu)$ and the map $p \mapsto p\chi_D$ extends to a bounded operator M on $P^2(\mu)$. Since $M^2 = M$ and $M^* = M$, M is an orthogonal projection. Density of the polynomials also shows that range $M = A_G^2(D)$ and range $(I - M) = L^2(E)$.   ∎

Thus, to prove Theorem 9.23 we will be interested in the distance from χ_D to $P^2(\mu)$; our goal is to show

$$\inf_{f \in P^2(\mu)} \int_{\overline{D}} |\chi_D - f|^2 \, d\mu = 0$$

when the hypotheses of Theorem 9.23 hold. It is in estimating this distance that the composition operators C_b will appear.

We will also have need of the function

$$\Theta(\epsilon) = \int_D \exp\left(\frac{-2\log\epsilon}{1-|z|}\right) G(|z|) \frac{dA(z)}{\pi}$$

defined for $0 < \epsilon < 1$. The hypothesis on G in Theorem 9.23 guarantees that $\Theta(\epsilon)$ is finite.

LEMMA 9.25
Let $Q \subset \partial D \setminus E$ be a Borel set of positive measure and let $b = b_Q$ be the associated inner function. For any $\epsilon, 0 < \epsilon < 1$

$$\inf_{f \in P^2(\mu)} \int_{\overline{D}} |\chi_D - f|^2 \, d\mu \leq 2\pi\epsilon + \|C_b - P\|^2 \Theta(\epsilon)$$

where $P(f) \equiv f(0)$ defines the orthogonal projection of $A_G^2(D)$ onto the constants and $\|C_b - P\|$ denotes the norm of $C_b - P$ as an operator on $A_G^2(D)$.

PROOF It is easy to see that $H^\infty(D) \subset P^2(\mu)$ (see Exercise 9.4.5), so it is enough to find g in $H^\infty(D)$ with

$$\int_{\overline{D}} |\chi_D - g|^2 \, d\mu \leq 2\pi\epsilon + \|C_b - P\|^2 \Theta(\epsilon)$$

Fix ϵ in $(0, 1)$ and set $V = \{e^{i\theta} : 0 < \theta < |Q|\}$. Let h be the outer function with boundary values

$$|h(e^{i\theta})|^2 = \begin{cases} \epsilon & \text{a.e. on } \partial D \setminus V \\ \epsilon^{1-(2\pi/|Q|)} & \text{a.e. on } V \end{cases}$$

Specifically

$$h(z) = \exp\left(\int_0^{2\pi} \frac{e^{i\theta} + z}{e^{i\theta} - z} \log |h(e^{i\theta})| \frac{d\theta}{2\pi} \right)$$

Setting $z = 0$ and using the boundary values of h we see that $h(0) = 1$. Now let $g = h \circ b$ so that $g \in H^\infty(D)$. By the mapping properties of b we see that $|g|^2 = \epsilon$ a.e. on $\partial D \setminus Q \supset E$. Moreover, $g - 1 = (C_b - P)h$ so

$$\int_{\overline{D}} |\chi_D - g|^2 \, d\mu = \int_E |g|^2 \, d\theta + \int_D |1 - g|^2 G \frac{dA}{\pi}$$

$$\leq 2\pi\epsilon + \int_D |(C_b - P)h|^2 G \frac{dA}{\pi}$$

$$\leq 2\pi\epsilon + \|C_b - P\|^2 \|h\|_G^2$$

Since

$$|h(z)|^2 = |h(re^{it})|^2 = \exp\left(2 \int_0^{2\pi} P_r(t - \theta) \log |h(e^{i\theta})| \frac{d\theta}{2\pi} \right)$$

$$\leq \exp\left(\int_0^{|Q|} P_r(t - \theta)(1 - \frac{2\pi}{|Q|}) \log \epsilon \frac{d\theta}{2\pi} \right)$$

$$\leq \exp\left(\int_0^{|Q|} \frac{2}{1 - |z|}(1 - \frac{2\pi}{|Q|}) \log \epsilon \frac{d\theta}{2\pi} \right)$$

$$\leq \exp\left(\frac{-2\log\epsilon}{1-|z|}\right)$$

the definition of $\Theta(\epsilon)$ shows that $\|h\|_G^2 \leq \Theta(\epsilon)$.   ∎

We can now assemble these lemmas into a proof of Hruščëv theorem.

PROOF (of Theorem 9.23)   We will first construct a sequence of Borel sets Q_n in $\partial D \setminus E$ each having positive measure and associate to each Q_n the inner function $b_{Q_n} = b_n$. For $n \geq 2$, Q_n is chosen as follows: Partition ∂D into n disjoint arcs $I_1, \dots, I_n$ each of length $2\pi/n$. By hypothesis, $|I_k \setminus E| > 0$ for $k = 1, \dots, n$ so we may find a Borel set $Q_n \subset \partial D \setminus E$ so that for each k, $|Q_n \cap I_k| > 0$ and independent of k. Thus we will have $|Q_n \cap I_k| = |Q_n|/n$ for $k = 1, \dots, n$. We claim that

$$\lim_{n\to\infty} \frac{1}{|Q_n|}\chi_{Q_n}(e^{i\theta})\,d\theta = \frac{d\theta}{2\pi} \tag{9.4.3}$$

weak*. Indeed, let f be continuous on ∂D and chose $\delta > 0$ arbitrary. Set

$$\nu_{1,k} = \frac{n}{|Q_n|}\chi_{Q_n\cap I_k}(e^{i\theta})\,d\theta$$

and

$$\nu_{2,k} = \frac{n}{2\pi}\chi_{I_k}(e^{i\theta})\,d\theta$$

so that $\nu_{1,k}$ and $\nu_{2,k}$ are probability measures for $k = 1, \dots, n$. When $n = n(\delta)$ is sufficiently large, uniform continuity of f guarantees that for ζ in I_k

$$\left|\int f\,d\nu_{i,k} - f(\zeta)\right| = \left|\int f\,d\nu_{i,k} - \int f(\zeta)\,d\nu_{i,k}\right| < \delta$$

for $i = 1, 2$ and $k = 1, \dots, n$. Thus for large n

$$\left|\frac{1}{|Q_n|}\int_{Q_n\cap I_k} f\,d\theta - \frac{1}{2\pi}\int_{I_k} f\,d\theta\right| < \frac{2\delta}{n}$$

for $k = 1, \dots, n$. By the triangle inequality

$$\left|\frac{1}{|Q_n|}\int_{Q_n} f\,d\theta - \frac{1}{2\pi}\int_{\partial D} f\,d\theta\right| < 2\delta$$

This verifies Equation (9.4.3).

By Theorem 9.20 the claim shows that $\lim_{n\to\infty} \|C_{b_n} - P\| = 0$, where b_n is the inner function associated with Q_n. Fix $\epsilon > 0$. For all n sufficiently large

$$\|C_{b_n} - P\|^2 < \frac{\epsilon}{\Theta(\epsilon)}$$

and, by Lemma 9.25, the square of the distance from χ_D to $P^2(\mu)$ is at most $2\pi\epsilon + \epsilon$. Since ϵ is arbitrary, χ_D belongs to $P^2(\mu)$ and we are done by Lemma 9.24.   ∎

Exercises

9.4.1 Let Q be a Borel subset of ∂D with $0 < |Q| < 2\pi$. Show directly from the definition that
$$b_{Q^c} = e^{-i|Q|}b_Q$$

9.4.2 (a) Let $Q = \{e^{i\theta} : r < \theta < s\}$ where $0 < s - r < 2\pi$. Show that the associated inner function is $b(z) = e^{-ir}z$.

(b) With Q as in (a) show that $\psi_Q(e^{i\theta})$ traverses $(-\infty, 0)$ once as $e^{i\theta}$ traverses Q and $\psi_Q(e^{i\theta})$ traverses $(0, \infty)$ once as $e^{i\theta}$ traverses Q^c.

(c) Suppose Q is a union of n non-degenerate arcs with pairwise disjoint closures. Show that $\psi_Q(e^{i\theta})$ traverses the entire real axis exactly n times (possibly with some backtracking) as $e^{i\theta}$ traverses ∂D once. Conclude that b_Q is a Blaschke product of degree n.

9.4.3 Suppose Q is a Borel set in ∂D with $0 < |Q| < 2\pi$ and $V = \{e^{i\theta} : 0 < \theta < |Q|\}$. Suppose B is an inner function with $B(0) = 0$ such that
- $B(e^{i\theta})$ is in V for almost all $e^{i\theta}$ in Q
- $B(e^{i\theta})$ is in V^c for almost all $e^{i\theta}$ in Q^c

Prove that $B = b_Q$. Hint: Set $\lambda(w) = (w - e^{i|Q|/2})/(w - e^{-i|Q|/2})$. Apply the Herglotz representation to $-i \log \psi(z)$ where $\psi \equiv \lambda^{-1} \circ B$.

9.4.4 Suppose Q is a union of n disjoint arcs of equal length symmetrically spaced around ∂D. Show $b_Q = cz^n$ for some c, $|c| = 1$.

9.4.5 If $\mu = \chi_E \, d\theta + G \, dA/\pi$, or more generally, $\mu = w \, d\theta + G \, dA/\pi$ where $w \geq 0$ is in $L^1(d\theta)$, then $H^\infty(D) \subset P^2(\mu)$.

9.4.6 Consider $\mu_1 = \chi_E \, d\theta + G_1 \, dA/\pi$ and $\mu_2 = \chi_E \, d\theta + G_2 \, dA/\pi$ where $G_1(r)$ and $G_2(r)$ are positive and continuous on $[0, 1)$. Suppose $G_1(r) \leq aG_2(r)$ for some $a > 0$ and $r \in (\rho, 1)$ for some $\rho < 1$. Show that if $P^2(\mu_2)$ splits then so does $P^2(\mu_1)$.

Note that this result is relevant to Theorem 9.22 where we have assumed that $G(r)$ tends to 0 as r tends to 1. If on the other hand $G_2(r)$ is bounded away from 0 as r tends to 1, we have $G_2(r) \geq 1 - r \equiv G_1(r)$ near 1 and if $P^2(\chi_E \, d\theta + G_2 \, dA/\pi)$ splits, so does $P^2(\chi_E \, d\theta + (1 - |z|) \, dA/\pi)$, where
$$\int_{1-\delta}^1 \log \log \frac{1}{1-r} \, dr < \infty.$$
Since a necessary condition for $P^2(\chi_E \, d\theta + (1 - |z|) \, dA/\pi)$ to split is $|I \setminus E| > 0$ for every open arc I in ∂D, so too is it necessary for $P^2(\chi_E \, d\theta + G_2 \, dA/\pi)$ to split.

9.4.7 Suppose $Q \subset \partial D$ with $0 < |Q| < 2\pi$. For $n \geq 2$ we say that Q is n-symmetric if there exist n pairwise disjoint arcs $I_1, \ldots, I_n$ in ∂D, each of length $2\pi/n$, with $|Q \cap I_k|$ independent of k. Fix a weighted Bergman space $A_G^2(D)$ and let δ_n denote the supremum of $\|C_b - P\|$ where b ranges over all inner functions b_Q with Q n-symmetric. Show δ_n tends to 0 as n goes to infinity.

9.4.8 Let $w \geq 0$ be in $L^1(d\theta)$ and set $\mu = w \, d\theta + G \, dA/\pi$ where G satisfies the hypothesis of Theorem 9.23. This exercise investigates a sufficient condition for
$$P^2(\mu) = L^2(w \, d\theta) \oplus A_G^2 \tag{9.4.4}$$
It is a fact that if the integral in Expression (9.4.2) converges and Equation (9.4.4) holds, then for every non-degenerate arc I in ∂D, $\log w$ is *not* in $L^1(I)$. It is not

known if the reverse implication is true. The point of this exercise is to show that splitting *does* occur under a hypothesis which is slightly stronger that requiring $\log w$ to not lie in *weak* $L^1(I)$ for all arcs I in ∂D. Recall that weak $L^1(I)$ is the space of measurable functions f on I for which

$$\lambda \left| \{ e^{i\theta} \in I : |f(e^{i\theta})| > \lambda \} \right|$$

is bounded as λ ranges over $(0, \infty)$. From this it follows that $\log w$ is not in weak $L^1(I)$ if and only if

$$\sup_{\epsilon > 0} \left| \{ e^{i\theta} \in I : w(e^{i\theta}) \le \epsilon \} \right| (-\log \epsilon) = \infty$$

For $0 < \delta < 2\pi$ and $\epsilon > 0$ define

$$\Omega(\delta, \epsilon) = \frac{1}{\delta} \min \left\{ \left| \{ e^{i\theta} \in I : w(e^{i\theta}) \le \epsilon \} \right| : |I| = \delta \right\}$$

where the minimum is taken over all arcs I in ∂D with $|I| = \delta$. Show that if there exists positive sequences δ_n and ϵ_n tending to 0 with

$$\lim_{n \to \infty} \Omega(\delta_n, \epsilon_n)(-\log \epsilon_n) = \infty$$

then $P^2(\mu) = L^2(w \, d\theta) \oplus A_G^2(D)$. Hints:

(i) Given $M > 0$ and an integer $N \ge 2$ show that there exists $n \ge N$, an n-symmetric set (see Exercise 9.4.7) Q, and $\epsilon > 0$ such that $w \le \epsilon$ on Q and $|Q| |\log \epsilon| = M$.

(ii) Approximate χ_D by $g = h \circ b$ where $b = b_Q$ and h is outer with $|h|^2 = \epsilon^{|Q|/(2\pi)}$ almost everywhere on Q^c and $|h|^2 = \epsilon^{|Q|/(2\pi)-1}$ almost everywhere on Q.

Notes

The functions $\psi_Q(z)$ were termed Cayley inner functions by M. Rosenblum and J. Rovnyak in [RoR78], where the basic properties of ψ_Q and the associated inner functions b_Q, as well as several applications, are presented.

The study of the structure of $P^2(\mu)$ for general μ has a rich history. Direct sum decompositions of the general type being considered here have been studied by various authors; the paper [KrM90b] contains a partial list of references. The importance of these questions about splitting is underscored by a recent theorem of J. Thomson [Th91] which asserts that a general $P^2(\mu)$ space is always an at most countable direct sum of analytic function spaces on disjoint simply connected domains and an L^2 space whose measure puts no mass on these domains. (Either type of summand may, of course, be absent.) As is the case here, identifying the summands requires special techniques suited to the μ in question.

Theorem 9.21 is due to T. L. Kriete [Kri79] and A. L. Vol'berg [Vol82]. See also [Koo88, p. 344]. Theorem 9.22 appears in [Kri79] and was generalized in [KrM90b]. Hruščëv's theorem (Theorem 9.23) appears in [Hr78] where it is established for a slightly larger class of weights G, including the exponential decay weights $G(r) = \exp(-c/(1-r)), c > 0$. The article [Hr78] contains versions of this theorem, and of Theorem 9.22 as well, appropriate for weights which decrease to 0 more slowly than these exponential decay weights. In these results the operative condition still takes the form $|I \setminus E| > 0$, but now I ranges over not just arcs in ∂D, but rather over a larger class of sets termed G-Carleson sets, defined by means of a Hayman–Carleson type condition related to G.

The techniques presented here for proving Theorem 9.23 are adapted from the work of T. L. Kriete and B. D. MacCluer [KrM90b] where measures of the more general form $\mu = w \, d\theta + G \, dA/\pi$, with $w \geq 0$ and integrable on ∂D are considered. This adaptation, which simplifies key ideas from [KrM90b] to highlight the role of composition operator norm estimates, was worked out by T. L. Kriete, and we are indebted to him for allowing us to present it here and for suggesting many of the exercises in this section.

The result of Exercise 9.4.8 appears in [KrM90b].

Bibliography

[Ab89a]   M. ABATE, Common fixed points of commuting holomorphic maps, *Math. Ann.* 283(1989), 645–655. (MR 90k#32074)

[Ab89b]   M. ABATE, *Iteration Theory of Holomorphic Maps on Taut Manifolds*, Research Notes in Mathematics, Mediterranean Press, Rende, Italy, 1989. (MR 92i#32032)

[Ab92]   M. ABATE, The infinitesimal generators of semigroups of holomorphic maps, *Ann. Mat. Pura Appl.* 161(1992), 167–180. (MR 93i#32029)

[AhC74]   P. R. AHERN AND D. N. CLARK, On inner functions with H^p derivative, *Michigan J. Math.* 21(1974), 115–127. (MR 49#9218)

[Alv82]   A. B. ALEKSANDROV, Existence of inner functions in the unit ball, *Math. USSR Sbornik* 46(1983), 143–159. (MR 83i#32002)

[Aln88]   A. ALEMAN, On the codimension of the range of a composition operator, *Rend. Sem. Mat. Univ. Politec. Torino* 46(1988), 323–326. (MR 92d#47042)

[Aln90]   A. ALEMAN, Compactness of resolvent operators generated by a class of composition semigroups on H^p, *J. Math. Anal. Appl.* 147(1990), 171–179. (MR 91b#47062)

[Aln93]   A. ALEMAN, Compact composition operators and iteration, *J. Math. Anal. Appl.* 173(1993), 550–556. (MR 94a#47051)

[Alp60]   L. ALPAR, Egyes hatványsorok absolút konvergenciája a konvergencia kör kerületén, *Matematikai Lapok* 11(1960), 312–322.

[ArF84]   J. ARAZY AND S. D. FISHER, Some aspects of the minimal, Mobius–invariant space of analytic functions on the unit disc, *Interpolation Spaces and Allied Topics in Analysis (Lund 1983)*, Springer–Verlag, Berlin, 1984, 24–44. (MR 86m#46024)

[ArF85]   J. ARAZY AND S. D. FISHER, The uniqueness of the Dirichlet space among Mobius–invariant Hilbert spaces, *Illinois J. Math.* 29(1985), 449–462. (MR 86j#30072)

[ArFP85]   J. ARAZY, S. D. FISHER, AND J. PEETRE, Mobius invariant function spaces, *J. reine angew. Math.* 363(1985), 110–145. (MR 87f#30104)

[AHHK84] W. ARVESON, D. W. HADWIN, T. B. HOOVER, AND E. E. KYMALA Circular operators, *Indiana Univ. Math. J.* 33(1984), 583–595. (MR 84b#47050)

[AtN72]   K. B. ATHREYA AND P. E. NEY, *Branching Processes*, Springer–Verlag, Berlin, 1972. (MR 51#9242)

[Att92]   K. R. M. ATTELE, Multipliers of the range of composition operators, *Tokyo J. Math.* 15(1992), 185–198. (MR 93j#47042)

[Ax88]   S. AXLER, Bergman spaces and their operators, *Surveys of Some Recent Results in Operator Theory, vol. I,* edited by J. Conway and B. Morrel, Longman Scientific and Technical, Harlow, 1988, 1–50. (MR 89b#47004)

[BaP79]   I. N. BAKER AND CH. POMMERENKE, On the iteration of analytic functions in a half plane II, *J. London Math. Soc. (2)* 20(1979), 255–258. (MR 83j#30024)

[Beh73]   D. F. BEHAN, Commuting analytic functions without fixed points, *Proc. Amer. Math. Soc.* 37(1973), 114–120. (MR 46#7492)

[Bel94]   S. R. BELL, Complexity of the classical kernel functions of potential theory, preprint, 1994.

[Ber80]   E. BERKSON, One parameter semigroups of isometries into H^p, *Pacific J. Math.* 86(1980), 403–413. (MR 82e#47051)

[Ber81]   E. BERKSON, Composition operators isolated in the uniform operator topology, *Proc. Amer. Math. Soc.* 81(1981), 230–232. (MR 82f#47039)

[BKP74]   E. BERKSON, R. KAUFMAN, AND H. PORTA, Mobius transformations of the disc and one-parameter groups of isometries on H^p, *Trans. Amer. Math. Soc.* 199(1974), 223–239. (MR 50#14365)

[BeP78]   E. BERKSON AND H. PORTA, Semigroups of analytic functions and composition operators, *Michigan J. Math.* 25(1978), 101–115. (MR 58#1112)

[BeP80]   E. BERKSON AND H. PORTA, The group of isometries on Hardy spaces of the *n*-ball and the polydisc, *Glasgow Math. J.* 21(1980), 199–204. (MR 81m#32006)

[Bern85]   B. BERNDTSSON, Interpolating sequences for H^∞ in the ball, *Math. Indag.* 47(1985), 1–10. (also *Proc. Kon. Nederl. Akad. Wetens.* 88A(1985), 1–10.) (MR 87a#32007)

[Bl92]   O. BLASCO, Operators on weighted Bergman spaces $(0 < p \leq 1)$
          and applications, *Duke Math. J.* 66(1992), 443–467.
          (MR 93h#47036)

[Boa87]   R. P. BOAS, *Invitation to Complex Analysis*, Random House,
          New York, 1987.

[BoH92]   A. BÖTTCHER AND H. HEIDLER, Algebraic composition opera-
          tors, *Integral Equations Operator Theory* 15(1992), 389–411.
          (MR 93b#47057)

[BoH93]   A. BÖTTCHER AND H. HEIDLER, Two papers on essentially alge-
          braic composition operators, Tech. Univ. Chemnitz Preprint, 1993.

[Bou87]   P. S. BOURDON, Density of polynomials in Bergman spaces, *Pa-
          cific J. Math.* 130(1987), 215–221. (MR 89a#46060)

[Bou90]   P. S. BOURDON, Fredholm multiplication and composition oper-
          ators on the Hardy space, *Integral Equations Operator Theory*
          13(1990), 607–610. (MR 91m#47038)

[BoS90]   P. S. BOURDON AND J. H. SHAPIRO, Cyclic composition oper-
          ators on H^2, *Proc. Symposia Pure Math.* 51(part 2)(1990), 43–53.
          (MR 91h#47028)

[BoS93]   P. S. BOURDON AND J. H. SHAPIRO, Cyclic phenomena for
          composition operators, preprint, 1993.

[Boy74]   D. M. BOYD, Composition operators on the Bergman space and
          analytic function spaces on the annulus, Thesis, University of North
          Carolina, 1974.

[Boy75]   D. M. BOYD, Composition operators on the Bergman space, *Colloq.
          Math.* 34(1975), 127–136. (MR 53#11416)

[Boy76]   D. M. BOYD, Composition operators on $H^p(A)$, *Pacific J. Math.*
          62(1976), 55–60. (MR 54#1002)

[dBr85]   L. DE BRANGES, A proof of the Bieberbach conjecture, *Acta Math.*
          154(1985), 137–152. (MR 86h#30026)

[Bu81]   R. B. BURCKEL, Iterating self maps of the discs, *Amer. Math.
          Monthly* 88(1981), 396–407. (MR 82g#30046)

[Cam72]   M. CAMBERN, The isometries of H_K^∞, *Proc. Amer. Math. Soc.*
          36(1972), 173–178. (MR 46#6042)

[CaJ89]   M. CAMBERN AND K. JAROSZ, The isometries of $H_{\mathcal{H}}^1$, *Proc.
          Amer. Math. Soc.* 107(1989), 205–214. (MR 90a#46083)

[Car89]   R. K. CAMPBELL–WRIGHT, On the Equivalence of Composition
          Operators, Thesis, Purdue University, 1989.

[Car91]   R. K. CAMPBELL–WRIGHT, Equivalent composition operators,
          Integral Equations Operator Theory 14(1991), 775–786.
          (MR 92h#47037)

[Car93]   R. K. CAMPBELL–WRIGHT, Similar compact composition opera-
            tors, *Acta Sci. Math. (Szeged)* 58(1993), 473–495.

[Car94]   R. K. CAMPBELL–WRIGHT, Unitarily equivalent compact com-
            position operators, *Houston J. Math.*, to appear.

[Cac60]   C. CARATHÉODORY, *Theory of Functions, Vol. II,* Chelsea, New
            York, 1960. (MR 16#346c)

[Cal62]   L. CARLESON, Interpolations by bounded analytic functions and the
            corona problem, *Annals Math.* 76(1962), 547–559. (MR 31#549)

[Caj85]   J. W. CARLSON, Weighted Composition Operators on ℓ^2, Thesis,
            Purdue University, 1985.

[Caj89]   J. W. CARLSON, Reducible weighted composition operators,
            preprint, 1989.

[Caj90a]   J. W. CARLSON, Hyponormal and quasinormal weighted composi-
            tion operators on ℓ^2, *Rocky Mountain J. Math.* 20(1990), 399–407.
            (MR 91f#47033)

[Caj90b]   J. W. CARLSON, The spectra and commutants of some weighted
            composition operators, *Trans. Amer. Math. Soc.* 317(1990), 631–
            654. (MR 90e#47019)

[CaC91]   T. CARROLL AND C. C. COWEN, Compact composition operators
            not in the Schatten classes, *J. Operator Theory* 26(1991), 109–120.
            (MR 94c#47045)

[Cau71]   J. G. CAUGHRAN, Polynomial approximation and spectral prop-
            erties of composition operators on H^2, *Indiana Univ. Math. J.*
            21(1971), 81–84. (MR 44#4213)

[CaS75]   J. G. CAUGHRAN AND H. J. SCHWARTZ, Spectra of compact
            composition operators, *Proc. Amer. Math. Soc.* 51(1975), 127–
            130. (MR 51#13750)

[ChS91]   K. C. CHAN AND J. H. SHAPIRO, The cyclic behavior of trans-
            lation operators on Hilbert spaces of entire functions, *Indiana Univ.
            Math. J.* 40(1991), 1421–1449. (MR 92m#47060)

[Che84]   G. CHEN, Iteration of holomorphic maps of the open unit ball and the
            generalized upper half plane of C^n, *J. Math. Anal. Appl.* 98(1984),
            305–313. (MR 85e#32001)

[ChK89]   J. S. CHOA AND H. O. KIM, Composition with a nonhomoge-
            neous bounded holomorphic function on the ball, *Canadian J. Math.*
            41(1989), 870–881. (MR 91a#32006)

[Cho89]   B. R. CHOE, Composition property of holomorphic functions on
            the ball, *Michigan J. Math.* 36(1989), 289–301. (MR 90g#32005)

[Cho92]   B. R. CHOE, The essential norms of composition operators, *Glas-
            gow Math. J.* 34(1992), 143–155. (MR 93h#47037)

[Cim77]   J. A. CIMA, A theorem on composition operators, *Banach Spaces of Analytic Functions,* Lecture Notes in Math., Vol. 604, Springer–Verlag, Berlin, 1977, 21–24. (MR 57#13562)

[CiH90]   J. A. CIMA AND L. J. HANSEN, Space–preserving composition operators, *Michigan J. Math.* 37(1990), 227–234. (MR 91m#47042)

[CiMa94] J. A. CIMA AND A. MATHESON, Completely continuous composition operators, *Trans. Amer. Math. Soc.* 344(1994), 849–856.

[CiMe94] J. A. CIMA AND P. R. MERCER, Composition operators between Bergman spaces on convex domains in C^n, *J. Operator Theory,* to appear.

[CiSW84] J. A. CIMA, C. S. STANTON, AND W. R. WOGEN, On boundedness of composition operators on $H^2(B_2)$, *Proc. Amer. Math. Soc.* 91(1984), 217–222. (MR 85j#47030)

[CiTW74] J. A. CIMA, J. THOMSON, AND W. R. WOGEN, On some properties of composition operators, *Indiana Univ. Math. J.* 24(1974), 215–220. (MR 50#2979)

[CiW74]   J. A. CIMA AND W. R. WOGEN, On algebras generated by composition operators, *Canadian J. Math.* 26(1974), 1234–1241. (MR 50#2978)

[CiW82]   J. A. CIMA AND W. R. WOGEN, A Carleson measure theorem for the Bergman space on the ball, *J. Operator Theory* 7(1982), 157–165. (MR 83f#46022)

[CiW87]   J. A. CIMA AND W. R. WOGEN, Unbounded composition operators on $H^2(B_2)$, *Proc. Amer. Math. Soc.* 99(1987), 477–483. (MR 88d#32009)

[Cir73]   E. M. ČIRKA, The Lindelöf and Fatou theorem in C^n, *Math. USSR Sb.* 21(1973), 619–641. (MR 49#3180)

[CoL66]   E. F. COLLINGWOOD AND A. J. LOHWATER, *The Theory of Cluster Sets,* Cambridge Univ. Press, Cambridge, 1966. (MR 38#325)

[Con90]   J. B. CONWAY, *A Course in Functional Analysis,* second edition, Springer–Verlag, New York, 1990. (MR 86h#46001)

[Con91]   J. B. CONWAY, *The Theory of Subnormal Operators,* American Math. Soc., Providence, 1991. (MR 92h#47026)

[Co78]   C. C. COWEN, The commutant of an analytic Toeplitz operator, *Trans. Amer. Math. Soc.* 239(1978), 1–31. (MR 58#2420)

[Co80a]   C. C. COWEN, The commutant of an analytic Toeplitz operator, II, *Indiana Univ. Math. J.* 29(1980), 1–12. (MR 82e#47038)

[Co80b]   C. C. COWEN, An analytic Toeplitz operator that commutes with a compact operator, *J. Functional Analysis* 36(1980), 169–184. (MR 81d#47020)

[Co81]   C. C. COWEN, Iteration and the solution of functional equations for functions analytic in the unit disk, *Trans. Amer. Math. Soc.* 265(1981), 69–95. (MR 82i#30036)

[Co82]   C. C. COWEN, Analytic solutions of Böttcher's functional equation in the unit disk, *Aequationes Mathematicae* 24(1982), 187–194. (MR 84h#30031)

[Co83]   C. C. COWEN, Composition operators on H^2, *J. Operator Theory* 9(1983), 77–106. (MR 84d#47038)

[Co84a]   C. C. COWEN, Commuting analytic functions, *Trans. Amer. Math. Soc.* 283(1984), 685–695. (MR 85i#30054)

[Co84b]   C. C. COWEN, Subnormality of the Cesàro operator and a semigroup of composition operators, *Indiana Univ. Math. J.* 33(1984), 305–318. (MR 86g#47034)

[Co88]   C. C. COWEN, Linear fractional composition operators on H^2, *Integral Equations Operator Theory* 11(1988), 151–160. (MR 89b#47044)

[Co90a]   C. C. COWEN, Composition operators on Hilbert spaces of analytic functions: A status report, *Proc. Symposia Pure Math.* 51(part 1)(1990), 131–145. (MR 91m#47043)

[Co90b]   C. C. COWEN, An application of Hadamard multiplication to operators on weighted Hardy spaces, *Linear Alg. Appl.* 133(1990), 21–32. (MR 92e#47046)

[Co92]   C. C. COWEN, Transferring subnormality of adjoint composition operators, *Integral Equations Operator Theory* 15(1992), 167–171. (MR 93g#47025)

[CoK88]   C. C. COWEN AND T. L. KRIETE, Subnormality and composition operators on H^2, *J. Functional Analysis* 81(1988), 298–319. (MR 90c#47055)

[CoL88]   C. C. COWEN AND S. LI, Hilbert space operators that are subnormal in the Krein space sense, *J. Operator Theory* 20(1988), 165–181. (MR 90b#47063)

[CoM94]   C. C. COWEN AND B. D. MACCLUER, Spectra of some composition operators, *J. Functional Analysis* 125(1994), 223–251.

[CoP82]   C. C. COWEN AND CH. POMMERENKE, Inequalities for the angular derivative of an analytic function in the unit disk, *J. London Math. Soc. (2)* 26(1982), 271–289. (MR 84a#30006)

[Ded72]   J. A. DEDDENS, Analytic Toeplitz and composition operators, *Canadian J. Math.* 24(1972), 859–865. (MR 46#9789)

[Den26]   A. DENJOY, Sur l'iteration des fonctions analytiques *C. R. Acad. Sci. Paris Sér. A.* 182(1926), 255–257.

[Di89]   S. DINEEN, *The Schwarz Lemma*, Oxford Mathematical Monographs, Clarendon Press, Oxford, 1989. (MR 91f#46064)

[Do72]   R. G. DOUGLAS, *Banach Algebra Techniques in Operator Theory*, Academic Press, New York, 1972. (MR 50#14335)

[Dug66]   J. DUGUNDJI, *Topology*, Allyn and Bacon, Boston, 1966. (MR 57#17581)

[DuS88]   N. DUNFORD AND J. SCHWARTZ, *Linear Operators, Part 1*, Wiley, New York, 1988. (MR 90g#47001a)

[Dur70]   P. L. DUREN, *Theory of H^p Spaces*, Academic Press, New York, 1970. (MR 42#3552)

[Dy72]   E. DYNKIN, Functions with given estimate for $\partial f/\partial \bar{z}$ and N. Levinson's theorem, *Math. USSR Sbornik* 18(1972), 181–189. (MR 48#4324)

[Em73]   M. R. EMBRY, A generalization of the Halmos–Bram criterion for subnormality, *Acta Sci. Math. (Szeged)* 35(1973), 61–64. (MR 48#6994)

[Er53]   A. ERDELYI, et. al., *Higher Transcendental Functions*, Vol. 1, McGraw–Hill, New York, 1953. (MR 15#419i)

[Er54]   A. ERDELYI, et. al., *Tables of Integral Transforms*, Vol. 1, McGraw–Hill, New York, 1954. (MR 15#868a)

[Es85]   M. ESSEN, D. F. SHEA, AND C. S. STANTON, A value–distribution criterion for the class $L \log L$ and some related questions, *Ann. Inst. Fourier (Grenoble)* 35, 4(1985), 127–150. (MR 87e#30041)

[Fig85a]   B. D. FIGURA, Composition operators on Hardy space in several complex variables, *J. Math. Anal. Appl.* 109(1985), 340–354. (MR 87c#47045)

[Fig85b]   B. D. FIGURA, Spectra of compact composition operators on Hardy spaces in several complex variables, *Bull. Polish Acad. Sci.* 33(1985), 299–304. (MR 87d#47042)

[Fig88]   B. D. FIGURA, The spectra of a certain class of composition operators, *Opuscula Math.* 4(1988), 31–43. (MR 90j#47030)

[Fis83a]   S. D. FISHER, Eigen-values and eigen-vectors of compact composition operators on $H^p(\Omega)$, *Indiana Univ. Math. J.* 32(1983), 843–847. (MR 84k#47028)

[Fis83b] S. D. FISHER, *Function Theory on Planar Domains*, John Wiley and Sons, New York, 1983. (MR 85d#30001)

[Fo64] F. FORELLI, The isometries of H^p, *Canadian J. Math.* 16(1964), 721–728. (MR 29#6336)

[Ga81] J. GARNETT, *Bounded Analytic Functions*, Academic Press, New York, 1981. (MR 83g#30037)

[Ge77] R. GELLAR, Circularly symmetric normal and subnormal operators, *J. D'analyse Math.* 32(1977), 93–117. (MR 58#12479)

[GLMR92] P. GORKIN, L. LAROCO, M. MARTINI, AND R. RUPP, Composition of inner functions, preprint, 1992.

[Gu89] J. GUYKER, On reducing subspaces of composition operators, *Acta Sci. Math. (Szeged)* 53(1989), 369–376. (MR 91a#47041)

[HaY91] K. T. HAHN AND E. H. YOUSSFI, Mobius invariant Besov p-spaces and Hankel operators in the Bergman space on the ball in C_n, *Complex Variables* 17(1991), 89–104. (MR 92m#47051)

[Haz67] G. HALÁSZ, On Taylor series absolutely convergent on the circumference of the circle of convergence, I, *Publ. Math., Debrecen* 14(1967), 63–68. (MR 36#3964)

[Hal50] P. R. HALMOS, Normal dilations and extensions of operators, *Summa Brasil. Math.* 2(1950), 125–134. (MR 13#359b)

[Hal74] P. R. HALMOS, *Measure Theory*, Springer–Verlag, New York, 1974, (c. 1950).

[Hal82] P. R. HALMOS, *A Hilbert Space Problem Book*, Springer–Verlag, New York, 1982. (MR 84e#47001)

[Hay20] G. H. HARDY, Note on a theorem of Hilbert, *Math. Zeit.* 6(1920), 314–317.

[Has63] T. E. HARRIS, *The Theory of Branching Processes*, Springer–Verlag, Berlin, 1963. (MR 29#664)

[Hast75] W. W. HASTINGS, A Carleson measure theorem for Bergman spaces, *Proc. Amer. Math. Soc.* 52(1975), 237–241. (MR 51#11082)

[Hat94] O. HATORI, Fredholm composition operators on spaces of holomorphic functions, *Integral Equations Operator Theory* 18(1994), 202–210.

[HaS71] T. L. HAYDEN AND T. J. SUFFRIDGE, Biholomorphic maps in Hilbert space have a fixed point, *Pacific J. Math.* 38(1971), 419–422. (MR 46#4288)

[He63] M. HERVÉ, Quelques propriétés des applications analytiques d'une boule à m dimensions dans elle–même, *J. Math. Pures Appl.* 42(1963), 117–147. (MR 28#3177)

[HiP57] E. HILLE AND R. S. PHILLIPS, *Functional Analysis and Semi-groups,* revised ed., American Math. Society, Providence, 1957. (MR 54#11077)

[HoY61] J. G. HOCKING AND G. S. YOUNG, *Topology,* Addison Wesley, Reading, 1961. (MR 23#A2857)

[Hof62] K. HOFFMAN, *Banach Spaces of Analytic Functions,* Prentice Hall, Englewood Cliffs, 1962. (MR 24#A2844)

[Hoo92] T. B. HOOVER, Isomorphic operator algebras and conjugate inner functions, *Michigan J. Math.* 39(1992), 229–237. (MR 93h#47056)

[Hor67] L. HORMANDER, L^p estimates for (pluri–)subharmonic functions, *Math. Scand.* 20(1967), 65–78. (MR 38#2323)

[Hr78] S. HRUŠČËV, The problem of simultaneous approximation and removal of singularities of Cauchy type integrals, *Proc. Steklov Inst. Math.* 130(1979), 133–203. (MR 80j#30055)

[Hu89] H. HUNZIKER, Kompositionsoperatoren auf klassischen Hardy-räumen, Thesis, University of Zürich, 1995.

[HuJ91] H. HUNZIKER, AND H. JARCHOW, Composition operators which improve integrability, *Math. Nachr.* 152(1991), 83–99. (MR 93d#47061)

[HuJM90] H. HUNZIKER, H. JARCHOW, AND V. MASCIONI, Some topologies on the space of analytic self–maps of the unit disk, *Geometry of Banach Spaces (Strobl, 1989),* Cambridge Univ. Press, Cambridge, 1990, 133–148. (MR 92k#47059)

[Hur94] P. R. HURST, A model for invertible composition operators on H^2, *Proc. Amer. Math. Soc.,* to appear.

[Hur95] P. R. HURST, Composition Operators on the Hardy and Bergman Spaces on the Disk, Thesis, Purdue University, 1995.

[Jaf90] F. JAFARI, On bounded and compact composition operators in poly-discs, *Canadian J. Math.* 42(1990), 869–889. (MR 91k#47065)

[Jaf92] F. JAFARI, Composition operators in Bergman spaces on bounded symmetric domains, *Contemp. Math.* 137(1992), 277–291. (MR 94c#47046)

[Jaf94] F. JAFARI, Composition operators in Bergman spaces on bounded symmetric domains, preprint, 1994. (revised version of [Jaf92].)

[Jar92] H. JARCHOW, Some factorization properties of composition operators, *Progress in Functional Analysis (Peniscola, 1990)* North Holland Math Studies, 170(1992), 405–413. (MR 93e#47037)

[Jar93] H. JARCHOW, Some functional analytic properties of composition operators, *Proc. Conf. at Kruger Park,* 1993. (MR 93e#47037)

[Jar94]   H. JARCHOW, Absolutely summing composition operators, *Functional Analysis: Proceedings of the Essen Conference* Marcel Dekker, 1994, 193–202.

[JaR93]   H. JARCHOW AND R. RIEDL, Factorization of composition operators through Bloch type spaces, *Illinois J. Math.*, to appear.

[JoM95]   M. JOVOVIC AND B. D. MACCLUER, Composition operators on Dirichlet spaces, preprint, 1995.

[Ju20]   G. JULIA, Extension nouvelle d'une lemme de Schwarz, *Acta Math.* 42(1920), 349–355.

[Kah85]   J. P. KAHANE, *Some random series of functions*, second edition, Cambridge University Press, Cambridge, 1985. (MR 87m#60119)

[Kam73]   H. KAMOWITZ, The spectra of endomorphisms of the disc algebra, *Pacific J. Math.* 46(1973), 433–440. (MR 49#5918)

[Kam75]   H. KAMOWITZ, The spectra of composition operators on H^p, *J. Functional Analysis* 18(1975), 132–150. (MR 53#11417) (Error in the proof of Theorem 3.1; see [Co83])

[Kam76]   H. KAMOWITZ, The spectra of endomorphisms of algebras of analytic functions, *Pacific J. Math.* 66(1976), 433–442. (MR 58#2426)

[Kam78]   H. KAMOWITZ, The spectra of a class of operators on the disc algebra, *Indiana Univ. Math. J.* 27(1978), 581–610. (MR 58#2427)

[Kam79]   H. KAMOWITZ, Compact operators of the form uC_φ, *Pacific J. Math.* 80(1979), 205–211. (MR 80f#47027)

[Ki85]   A. K. KITOVER, Weighted composition operators in spaces of analytic functions, *Zapiski Nauchnykh Seminarov Leningradskogo Otdeleniya Matematicheskogo Instituta imeni V. A. Steklova Akademii Nauk SSSR (LOMI)*, 41(1985), 154–161, 190–191. (MR 86j#47047)

[Koe84]   G. KOENIGS, Recherches sur les intégrales de certaines équations fonctionnelles, *Ann. Sci. École Norm. Sup. (Sér. 3)*, 1(1884), supplément, 3–41.

[Kol81]   C. J. KOLASKI, Isometries of Bergman spaces over bounded Runge domains, *Canadian J. Math.* 33(1981), 1157–1164. (MR 83b#32028)

[Kol82]   C. J. KOLASKI, Isometries of weighted Bergman spaces, *Canadian J. Math.* 34(1982), 910–915. (MR 84a#46054)

[Kön90]   W. KÖNIG, Semicocycles and weighted composition semigroups on H^p, *Michigan J. Math.* 37(1990), 469–476. (MR 91m#47057)

[Koo80]   P. KOOSIS, *Introduction to H_p Spaces*, Cambridge University Press, Cambridge, 1980. (MR 81c#30062)

[Koo88]   P. KOOSIS, *The Logarithmic Integral,* Cambridge University
           Press, Cambridge, 1988. (MR 90a#30097)

[Kra92]   S. G. KRANTZ, *Function Theory in Several Complex Variables,*
           2nd ed., Wadsworth & Brooks/Cole, Pacific Grove, 1992.
           (MR 93c#32001)

[Kri79]   T. L. KRIETE, On the structure of certain $H^2(\mu)$ spaces, *Indiana
           Univ. Math. J.* 28(1979), 757–773. (MR 80i#46045)

[Kri87]   T. L. KRIETE, Cosubnormal dilation semigroups on Bergman
           spaces, *J. Operator Theory* 17(1987), 191–200. (MR 88e#47041)

[Kri94]   T. L. KRIETE, Kernel function estimates and norms of composition
           operators, preprint, 1994.

[KrM90a]   T. L. KRIETE AND B. D. MACCLUER, Composition opera-
           tors and weighted polynomial approximation, *Proc. Symposia Pure
           Math.* 51(part 2)(1990), 175–182. (MR 91j#47034)

[KrM90b]   T. L. KRIETE AND B. D. MACCLUER, Mean–square approxi-
           mation by polynomials on the unit disk, *Trans. Amer. Math. Soc.*
           322(1990), 1–34. (MR 91b#30119)

[KrM92]   T. L. KRIETE AND B. D. MACCLUER, Composition operators on
           large weighted Bergman spaces, *Indiana Univ. Math. J.* 41(1992),
           755–788. (MR 93i#47031)

[KrM94]   T. L. KRIETE AND B. D. MACCLUER, A rigidity theorem for
           composition operators on certain Bergman spaces, *Mich. Math. J.,*
           to appear.

[KrR87]   T. L. KRIETE AND H. C. RHALY, Translation semigroups on
           reproducing kernel Hilbert spaces, *J. Operator Theory* 17(1987),
           33–83. (MR 88e#47080)

[KrT71]   T. L. KRIETE AND D. TRUTT, The Cesàro operator in ℓ^2 is sub-
           normal, *American J. Math.* 93(1971), 215–225. (MR 43#6744)

[KrT74]   T. L. KRIETE AND D. TRUTT, On the Cesàro operator, *Indiana
           Univ. Math. J.* 24(1974), 197–214. (MR 50#2981)

[Kub83]   Y. KUBOTA, Iteration of holomorphic maps of the unit ball into itself,
           Proc. Amer. Math. Soc. 88(1983), 476–485. (MR 85c#32047b)

[Kuc63]   M. KUCZMA, On the Schroeder equation, *Rozprawy Mat.* 34(1963).
           (MR 30#4082)

[La76]   A. LAMBERT, Subnormality and weighted shifts, *J. London Math.
           Soc. (2)* 14(1976), 476–480. (MR 55#8866)

[LaV29]   E. LANDAU AND G. VALIRON, A deduction from Schwarz's
           lemma, *J. London Math. Soc.* 4(1929), 162–163.

[Le70]   S. J. LEON, Composition operators on B^p, the containing Banach space of H^p, $0 < p < 1$, *Notices Amer. Math. Soc.* 17(1970), 784–784.

[Lik89]   K. Y. LI, Inequalities for fixed points of holomorphic functions, *Bull. London Math. Soc.* 22(1990), 446–452. (MR 92h#30002)

[Lis94]   S. Y. LI, Trace ideal criteria for composition operators on Bergman spaces, preprint, 1994.

[LiR95]   S. Y. LI AND B. RUSSO, On compactness of composition operators in Hardy spaces of several variables, *Proc. Amer. Math. Soc.* 123(1995), 161–171.

[Lin90]   P. K. LIN, The isometries of $H^\infty(E)$, *Pacific J. Math.* 143(1990), 69–77. (MR 91F#46075)

[Lin91]   P. K. LIN, The isometries of $H^p(K)$, *J. Austral. Math. Soc.* 50(1991), 23–33. (MR 92d#46091)

[Lit25]   J. E. LITTLEWOOD, On inequalities in the theory of functions, *Proc. London Math. Soc.* (2)23(1925), 481–519.

[LM]   B. A. LOTTO AND J. E. MCCARTHY, Composition preserves rigidity, *Bull. London Math. Soc.* 25(1993), 573–576.

[Løw82]   E. LØW, A construction of inner functions on the unit ball of C^p, *Invent. Math.* 67(1982), 223–229. (MR 84j#32008b)

[Lub75]   A. LUBIN, Isometries induced by composition operators and invariant subspaces, *Illinois J. Math.* 19(1975), 424–427. (MR 54#3477)

[Lue81]   D. H. LUECKING, Inequalities on Bergman spaces, *Illinois J. Math.* 25(1981), 1–11. (MR 82e#30072)

[Lue83]   D. H. LUECKING, A technique for characterizing Carleson measures on Bergman spaces, *Proc. Amer. Math. Soc.* 87(1983), 656–660. (MR 84e#32025)

[Lue87]   D. H. LUECKING, Trace ideal criteria for Toeplitz operators, *J. Functional Analysis* 73(1987), 345–368. (MR 88m#47046)

[LuZ92]   D. H. LUECKING AND K. ZHU, Composition operators belonging to the Schatten ideals, *American J. Math.* 114(1992), 1127–1145. (MR 93i#47032)

[Mc83]   B. D. MACCLUER, Iterates of holomorphic self–maps of the unit ball in C^n, *Michigan J. Math.* 30(1983), 97–106. (MR 85c#32047a)

[Mc84a]   B. D. MACCLUER, Spectra of compact composition operators on $H^p(B_N)$, *Analysis* 4(1984), 87–103. (MR 86e#47038)

[Mc84b]   B. D. MACCLUER, Spectra of automorphism–induced composition operators on $H^p(B_N)$, *J. London Math. Soc.* (2)30(1984), 95–104. (MR 86g#47036)

[Mc85]   B. D. MacCluer, Compact composition operators on $H^p(B_N)$, *Michigan J. Math.* 32(1985), 237–248. (MR 86g#47037)

[Mc87]   B. D. MacCluer, Composition operators on S^p, *Houston J. Math.* 13(1987), 245–254. (MR 88h#47044)

[Mc89]   B. D. MacCluer, Components in the space of composition operators, *Integral Equations Operator Theory* 12(1989), 725–738. (MR 91b#47070)

[McM93]   B. D. MacCluer and P. R. Mercer, Composition operators between Hardy and weighted Bergman spaces on convex domains in C^N, *Proc. Amer. Math. Soc.*, to appear.

[McS86]   B. D. MacCluer and J. H. Shapiro, Angular derivatives and compact composition operators on the Hardy and Bergman spaces, *Canadian J. Math.* 38(1986), 878–906. (MR 87h#47048)

[Mad93a]   K. M. Madigan, Composition Operators into Lipschitz Type Spaces, Thesis, SUNY Albany, 1993.

[Mad93b]   K. M. Madigan, Composition operators on analytic Lipschitz spaces, *Proc. Amer. Math. Soc.* 119(1993), 465–473. (MR 93k#47043)

[MaM93]   K. M. Madigan and A. Matheson, Compact composition operators on the Bloch space, *Proc. Amer. Math. Soc.*, to appear.

[MaS82]   D. E. Marshall and K. Stephenson, Inner divisors and composition operators, *J. Functional Analysis* 46(1982), 131–148. (MR 83h#46067)

[Mas85]   M. Masri, Compact Composition Operators on the Nevanlinna and Smirnov classes, Thesis, University of North Carolina, Chapel Hill, 1985

[Mat89]   V. Matache, Composition operators on H^p of the upper half-plane, *An. Univ. Timisoara Ser. Stiint. Mat.* 119(1989), 63–66. (MR 92k#47063)

[Mat93]   V. Matache, On the minimal invariant subspaces of the hyperbolic composition operator, *Proc. Amer. Math. Soc.* 27(1993), 837–841. (MR 93m#47038)

[MaY94]   Y. Matsugu and T. Yamada, On the isometries of $H_E^\infty(B)$, *Proc. Amer. Math. Soc.* 120(1994), 1107–1112. (MR 94f#46043)

[Mayd79]   D. H. Mayer, Spectral properties of certain composition operators arising in statistical mechanics, *Comm. Math. Phys.* 68(1979), 1–8. (MR 80h#47040)

[Mayd80]   D. H. Mayer, On composition operators on Banach spaces of holomorphic functions, *J. Functional Analysis* 35(1980), 191–206. (MR 81e#47025)

[Mayj79] J. MAYER, Isometries in Banach spaces of functions, holomorphic in the unit disc and smooth up to its boundary, *Moskow Univ. Math. Bull.* 34(1979), 38–43. (MR 80i#46023)

[Mer93] P. R. MERCER, Extremal disks and composition operators on convex domains in C^N, *Rocky Mtn. J. Math.*, to appear.

[Moc89] N. MOCHIZUKI, Algebras of holomorphic functions between H^p and N_*, *Proc. Amer. Math. Soc.* 105(1989), 898–902. (MR 90a#46137)

[Mor88] F. MORGAN, *Geometric Measure Theory,* Academic Press, Boston, 1988. (MR 89f#49036)

[Ne70] R. NEVANLINNA, *Analytic Functions,* translated by P. Emig, Springer–Verlag, Berlin, 1970. (MR 43#5003)

[No68] E. A. NORDGREN, Composition operators, *Canadian J. Math.* 20(1968), 442–449. (MR 36#6961)

[No78] E. A. NORDGREN, Composition operators on Hilbert spaces, *Hilbert Space Operators,* Lecture Notes in Math., Vol. 693, Springer–Verlag, Berlin, 1978, 37–63. (MR 80d#47046)

[NoRR84] E. A. NORDGREN, H. RADJAVI, AND P. ROSENTHAL, Composition operators and the invariant subspace problem, *C. R. Math. Rep. Acad. Sci. Canada* 6(1984), 279–282.

[NoRW87] E. A. NORDGREN, P. ROSENTHAL, AND F. S. WINTROBE, Invertible composition operators on H^p, *J. Functional Analysis* 73(1987), 324–344. (MR 89c#47044)

[Nos34] K. NOSHIRO, On the theory of schlicht functions, J. Fac. Sci. Hokaido U. (1) 2(1934–1935), 129–155.

[NoO85] W. P. NOVINGER AND D. M. OBERLIN, Linear isometries of some normed spaces of analytic functions, *Canadian J. Math.* 37(1985), 62–74. (MR 86f#46022)

[Pe90] M. M. PELOSO, Möbius invariant spaces on the unit ball, Thesis, Washington University, St. Louis, 1990.

[Pe92] M. M. PELOSO, Möbius invariant spaces on the unit ball, *Michigan J. Math.* 39(1992), 509–536. (MR 93k#46018)

[Poi07] H. POINCARÉ, Les fonctions analytiques de deux variables et la représentation conforme, *Rend. Circ. Matem. Palermo* 23(1907), 185–220.

[Pom79] CH. POMMERENKE, On the iteration of analytic functions in a half plane I, *J. London Math. Soc. (2)* 19(1979), 439–447. (MR 83j#30023)

[Pom92] CH. POMMERENKE, Boundary Behaviour of Conformal Maps, Springer–Verlag, Berlin, 1992.

[Pow85]   S. C. POWER, Hörmander's Carleson theorem for the ball, *Glasgow Math. J.* 26(1985), 13–17. (MR 86e#32007)

[Ri69]   W. C. RIDGE, Composition Operators, Thesis, Indiana University, 1969.

[Ri73]   W. C. RIDGE, Spectrum of a composition operator, *Proc. Amer. Math. Soc.* 37(1973), 121–127. (MR 46#5583)

[Ri74]   W. C. RIDGE, Characterization of abstract composition operators, *Proc. Amer. Math. Soc.* 45(1974), 393–396. (MR 49#11310)

[Rie94]   R. RIEDL, Composition operators and geometric properties of analytic functions, Inaugural–Dissertation, University of Zurich, 1994.

[RiN]   F. RIESZ AND B. SZ.–NAGY, *Functional Analysis,* Dover, New York, 1990. (MR 17#175i and 91g#00002)

[Roa76]   R. C. ROAN, Generators and composition operators, Thesis, University of Michigan, 1976.

[Roa78a] R. C. ROAN, Composition operators on H^p with dense range, *Indiana Univ. Math. J.* 27(1978), 159–162. (MR 57#13564)

[Roa78b] R. C. ROAN, Composition operators on the space of functions with H^p–derivative, *Houston J. Math.* 4(1978), 423–438. (MR 58#23735) (Error in Proposition 7; see [Mc87].)

[Roa80a] R. C. ROAN, Composition operators on a space of Lipschitz functions, *Rocky Mountain J. Math.* 10(1980), 371–379. (MR 81g#30046) (Error in proof of Theorem 2 and subsequent corollaries; see the *Notes* to Section 4.2 of this book.)

[Roa80b] R. C. ROAN, Weak–star generators for a class of subalgebras of ℓ^1, *J. Functional Analysis* 39(1980), 67–74. (MR 82a#46057)

[Rob92]   M. ROBBINS, Composition Operators between Hilbert Spaces of Analytic Functions, Thesis, University of Virginia, 1992.

[RoS76a] J. W. ROBERTS AND M. STOLL, Prime and principal ideals in the algebra N^+, *Arch. Math. (Basel)* 27(1976), 387–393. (MR 54#10625)

[RoS76b] J. W. ROBERTS AND M. STOLL, Composition operators on F^+, *Studia Math.* 57(1976), 217–228. (MR 55#8773)

[RoS78]   J. W. ROBERTS AND M. STOLL, Correction to the paper: "Prime and principal ideals in the algebra N^+" (Arch. Math. (Basel) **27**(1976), 387–393), *Arch. Math. (Basel)* **30**(1978), 672. (MR 58#11454)

[Roc93]   R. R. ROCHBERG, Projected composition operators on the Hardy space, preprint, 1993.

[RoR78]   M. ROSENBLUM AND J. ROVNYAK, Change of variable formulas with Cayley inner functions, *Topics in Functional Analysis,* Adv. in Math. Suppl. Stud., Vol. 3, Academic Press, New York, 1978, pp. 283–320. (MR 81d#30053)

[Ru55]   W. RUDIN, Analytic functions of class H_p, *Trans. Amer. Math. Soc.* 78(1955), 46–66. (MR 16#810)

[Ru56]   W. RUDIN, Boundary values of continuous analytic functions, *Proc. Amer. Math. Soc.* 7(1956), 808–811. (MR 18#472c)

[Ru78]   W. RUDIN, The fixed point sets of some holomorphic maps, *Bull. Maylaysian Math. Soc.* 1(1978), 25–28. (MR 80d#32002)

[Ru80]   W. RUDIN, *Function Theory in the Unit Ball of C^n*, Springer–Verlag, New York, 1980. (MR 82i#32002)

[Ru87]   W. RUDIN, *Real and Complex Analysis,* third edition, McGraw–Hill, New York, 1987. (MR 88k#00002)

[Ry66]   J. V. RYFF, Subordinate H^p functions, *Duke Math. J.* 33(1966), 347–354. (MR 33#289)

[Sad92]   H. SADRAOUI, Hyponormality of Toeplitz and composition operators, Thesis, Purdue University, 1992.

[Sar88]   D. SARASON, Angular derivatives via Hilbert space, *Complex Variables* 10(1988), 1–10. (MR 89f#30045)

[Sar90]   D. SARASON, Composition operators as integral operators, *Analysis and Partial Differential Equations,* Marcel Dekker, New York, 1990. (MR 92a#47040)

[Sar92]   D. SARASON, Weak compactness of holomorphic composition operators on H^1, *Functional Analysis and Operator Theory (New Delhi, 1990),* 75–79 Springer–Verlag, Berlin, 1992. (MR 93h#47041)

[Scr71]   E. SCHROEDER, Über itierte Funktionen, *Math. Ann.* 3(1871), 296–322.

[Scz69]   H. J. SCHWARTZ, Composition Operators on H^p, Thesis, University of Toledo, 1969.

[Shv86]   A. I. SHAKHBAZOV, Dimensions of the eigensubspaces of operators of weighted holomorphic substitution, *Akad. Nauk Azerbaidzhan. SSR Dokl.,* 42(1986), 3–5. (MR 88h#47046)

[Shv87]   A. I. SHAKHBAZOV, The spectrum of a compact weighted composition operator in some Banach spaces of holomorphic functions, *Teoriya Funkstii, Funktsionalnyi Anliz i ihk Prilozheniya,* 47(1987), 105–112. (MR 88m#47057)

[ShS61]   H. S. SHAPIRO AND A. L. SHIELDS, On some interpolation problems for analytic functions, *American J. Math.* 83(1961), 513–532. (MR 24#A3280)

[Sho87a]   J. H. SHAPIRO, The essential norm of a composition operator, *Annals Math.* 125(1987), 375–404. (MR 88c#47058)

[Sho87b]   J. H. SHAPIRO, Compact composition operators on spaces of boundary-regular holomorphic functions, *Proc. Amer. Math. Soc.* 100(1987), 49–57. (MR 88c#47059)

[Sho93]   J. H. SHAPIRO, Composition Operators and Classical Function Theory, Springer–Verlag, New York, 1993.

[ShSS92]   J. H. SHAPIRO, W. SMITH, AND D. A. STEGENGA, Geometric models and compactness of composition operators, preprint, 1992.

[ShS90a]   J. H. SHAPIRO AND C. SUNDBERG, Isolation amongst the composition operators, *Pacific J. Math.* 145(1990), 117–152. (MR 92g#47041)

[ShS90b]   J. H. SHAPIRO AND C. SUNDBERG, Compact composition operators on L^1, *Proc. Amer. Math. Soc.* 108(1990), 443–449. (MR 90d#47035)

[ShT73]   J. H. SHAPIRO AND P. D. TAYLOR, Compact, nuclear, and Hilbert–Schmidt composition operators on H^2, *Indiana Univ. Math. J.* 23(1973), 471–496. (MR 48#4816)

[Sha90]   S. D. SHARMA, Idempotent composition operators and several complex variables, *Bull. Calcutta Math. Soc.* 82(1990), 203–205. (MR 92d#47047)

[Sha83]   S. D. SHARMA, Compact and Hilbert–Schmidt composition operators on Hardy spaces of the upper half–plane, *Acta Sci. Math. (Szeged)* 46(1983), 197–202. (MR 85j#47032)

[ShaK91]   S. D. SHARMA AND R. KUMAR, Substitution operators on Hardy–Orlicz spaces, *Proc. Nat. Acad. Sci. India Sect. A* 61(1991), 535–541. (MR 93h#47042)

[Shi64]   A. L. SHIELDS, On fixed points of commuting analytic functions, *Proc. Amer. Math. Soc.* 15(1964), 703–706. (MR 29#2790)

[Shi74]   A. L. SHIELDS, Weighted shift operators and analytic function theory, *Topics in Operator Theory*, Math. Surveys, vol. 13, Amer. Math. Soc., Providence, 1974, 49–128. (MR 50#14341)

[ShW70]   A. L. SHIELDS AND L. J. WALLEN, The commutants of certain Hilbert Space operators, *Indiana Univ. Math. J.* 20(1970/71), 777–788. (MR 44#4558)

[Sie42]   C. L. SIEGEL, Iteration of analytic functions, *Annals Math.* 43(1942), 607–612. (MR 4#76c)

[Sin74]   R. K. SINGH, A relation between composition operators on $H^2(D)$ and $H^2(P^+)$, *Pure Appl. Math. Sci.* 1(1974), 1–5. (MR 56#12971)

[Sin80]   R. K. SINGH, Inner functions and composition operators on a Hardy space *Indian J. Pure Appl. Math.* 11(1980), 1297–1300. (MR 83e#47023)

[SiK85]   R. K. SINGH AND R. D. C. KUMAR, Weighted composition operators on functional Hilbert spaces, *Bull. Austral. Math. Soc.* 31(1985), 117–126. (MR 86d#47034)

[SiM93]   R. K. SINGH AND J. S. MANHAS, *Composition Operators on Function Spaces*, North Holland, New York, 1993.

[SiS79]   R. K. SINGH AND S. D. SHARMA, Composition operators on a functional Hilbert space, *Bull. Austral. Math. Soc.* 20(1979), 377–384. (MR 81c#47034)

[SiS80]   R. K. SINGH AND S. D. SHARMA, Non–compact composition operators, *Bull. Austral. Math. Soc.* 21(1980), 125–130. (MR 81m#47046)

[SiS92]   R. K. SINGH AND S. D. SHARMA, Compact composition operators on $H^2(D^n)$, *Indian J. Math* 34(1992), 73–79. (MR 94b#47044)

[SiS81]   R. K. SINGH AND S. D. SHARMA, Composition operators and several complex variables, *Bull. Austral. Math. Soc.* 23(1981), 237–247. (MR 82g#47023)

[Sis85]   A. G. SISKAKIS, Semigroups of composition operators and the Cesàro operator on $H^p(D)$, Thesis, University of Illinois, 1985.

[Sis86]   A. G. SISKAKIS, Weighted composition semigroups on Hardy spaces, *Linear Alg. Appl.* 84(1986), 359–371. (MR 88b#47058)

[Sis87a]   A. G. SISKAKIS, Composition semigroups and the Cesàro operator on H^p, *J. London Math. Soc. (2)* 36(1987), 153–164. (MR 89a#47048)

[Sis87b]   A. G. SISKAKIS, On a class of composition semigroups in Hardy spaces, *J. Math. Anal. Appl.* 127(1987), 122–129. (MR 89d#47093)

[Sis87c]   A. G. SISKAKIS, Semigroups of composition operators in Bergman spaces, *Bull. Austral. Math. Soc.* 35(1987), 397–406. (MR 88h#47047)

[Sis93]   A. G. SISKAKIS, The Koebe semigroup and a class of averaging operators on $H^p(D)$, *Trans. Amer. Math. Soc.* 339(1993), 337–350. (MR 93k#47044)

[Sis94a]   A. G. SISKAKIS, The Cesàro operator on Bergman spaces, preprint, 1994.

[Sis94b] A. G. SISKAKIS, Semigroups of composition operators on the Dirichlet space, preprint, 1994.

[Smr94]   R. C. SMITH, Local spectral theory for invertible composition operators on H^p, preprint, 1994.

[Smw94]   W. SMITH, Composition operators between Hardy and Bergman spaces, preprint, 1994.

[Sta86]   C. S. STANTON, Counting functions and majorization theorems for Jensen measures, *Pacific J. Math.* 125(1986), 459–468. (MR 88c#32002)

[Stg80]   D. A. STEGENGA, Multipliers of the Dirichlet space, *Illinois J. Math.* 24(1980), 113–139. (MR 81a#30027)

[Stp77]   K. STEPHENSON, Isometries of the Nevanlinna class, *Indiana Univ. Math. J.* 26(1977), 307–324. (MR 55#5885)

[Stp79]   K. STEPHENSON, Functions which follow inner functions, *Illinois J. Math.* 23(1979), 259–266. (MR 80h#30030)

[Sw76]   D. W. SWANTON, Compact composition operators on $B(D)$, *Proc. Amer. Math. Soc.* 56(1976), 152–156. (MR 53#11420)

[Su93]   S. L. SUN, Composition operators and the reducing subspaces of analytic Toeplitz operators, *Adv. in Math. (China)* 22(1993), 422–434. (MR 94m#47064)

[Ta93]   H. TAGAKI, Composition operators on spaces of analytic functions, *Research on Hardy spaces related to rings functions,* Surikaisekikenkyusho Kokyuroku, 825(1993), 95–100.

[Th91]   J. E. THOMSON, Approximation in the mean by polynomials, *Annals Math.* 133(1991), 477–507. (MR 93g#47026)

[Tom84]   B. TOMASZEWSKI, Interpolation and inner maps that preserve measure, *J. Functional Analysis* 55(1984), 63–67. (MR 85j#32007)

[Ton90]   Y. S. TONG, Composition operators on Hardy space, *J. Math. Res. Exposition* 10(1990), 59–64. (MR 91c#47058)

[Ts75]   M. TSUJI, *Potential Theory in Modern Function Theory,* Maruzen, Tokyo, 1975. (MR 54#2990)

[Tu58]   P. TURAN, A remark concerning the behaviour of a power–series on the periphery of its convergence circle, *Acad. Serbe Sci. Publ. Inst. Math.* 12(1958), 19–26. (MR 21#1381)

[Voa80]   C. VOAS, Toeplitz operators and univalent functions, Thesis, University of Virginia, 1980.

[Vol82]   A. L. VOL'BERG, The logarithm of an almost analytic function is summable, *Soviet Math. Dokl.* 26(1982), 238–243. (MR 84g#30035)

[VoJ87]   A. L. VOL'BERG AND B. JÖRICKE, Summability of the logarithm of an almost analytic function and generalizations of the Levinson–Cartwright theorem, *Math. USSR Sb.* 58(1987), 337–349. (MR 87k#30060)

[Wa35]   S. E. WARSHAWSKI, On the higher derivatives at the boundary in conformal mapping, *Trans. Amer. Math. Soc.* 38(1935), 310–340.

[Wog88]   W. R. WOGEN, The smooth mappings which preserve the Hardy space $H^2_{B_n}$, *Operator Theory: Advances Appl.* 35(1988), 249–267. (MR 91a#32006)

[Wog90]   W. R. WOGEN, Composition operators acting on spaces of holomorphic functions on domains in C^n, *Proc. Symposia Pure Math.* 51(part 2)(1991), 361–366. (MR 91k#47069)

[Wol26]   J. WOLFF, Sur l'iteration des fonctions, *C. R. Acad. Sci. Paris Sér. A.* 182(1926), 42–43, 200–201.

[Wol34]   J. WOLFF, L'intégrale d'une fonction holomorphe et à partie réele positive dans un demi plan est univalente, *C. R. Acad. Sci. Paris Sér. A.* 198(1934), 1209–1210.

[Xi94]   H. XIAN, Models of composition operators, Thesis, Washington University, St. Louis, 1994.

[Xu88]   X. M. XU, Composition operators on $H^p(\Omega)$, *Chinese Ann. Math.* 9(1988), 630–633. (MR 90h#47057)

[YaN78]   N. YANAGIHARA AND Y. NAKAMURA, Composition operators on N^+, *TRU Math.* 14(1978), 9–16. (MR 80f#30024)

[Zh90a]   K. ZHU, On certain unitary operators and composition operators, *Proc. Symposia Pure Math.* 51(part 2)(1990), 371–385. (MR 91m#47047)

[Zh90b]   K. ZHU, *Operator Theory in Function Spaces,* Marcel Dekker, New York, 1990. (MR 92c#47031)

[Zo87]   N. ZORBOSKA, Composition operators on weighted Hardy spaces, Thesis, University of Toronto, 1987.

[Zo89a]   N. ZORBOSKA, Composition operators induced by functions with supremum strictly smaller than 1, *Proc. Amer. Math. Soc.* 106(1989), 679–684. (MR 89k#47050)

[Zo89b]   N. ZORBOSKA, Compact composition operators on some weighted Hardy spaces, *J. Operator Theory* 22(1989), 233–241. (MR 91c#47062)

[Zo90a]   N. ZORBOSKA, Composition operators on S_a spaces, *Indiana Univ. Math. J.* 39(1990), 847–857. (MR 91k#47070)

[Zo91]   N. ZORBOSKA, Hyponormal composition operators on weighted Hardy spaces, *Acta Sci. Math. (Szeged)* 55(1991), 399–402. (MR 92m#47064)

[Zo94a]   N. ZORBOSKA, Angular derivative and compactness of composition operators on large weighted Hardy spaces, *Canad. Math. Bull.*, 37(1994), 428–432.

[Zo94b]   N. ZORBOSKA, Composition operators with closed range, *Trans. Amer. Math. Soc.* 344(1994), 791–801.

Symbol Index

Spaces

Sets

Operators

Functions

Φ	linear fractional map in model for iteration	62
Φ	Cayley transform	110
f^*	radial limit function	10
h_f	least harmonic majorant of $\lvert f \rvert^p$	10
K_x	kernel for evaluation at x	2
$k(z)$	generating function	16
$K_w^{(m)}(z)$	kernel for evaluation of mth derivative at w	19
$K_w^m(z)$	kernel for evaluation at w in $z^m H^2(D)$	290
M_f	nontangential maximal function	37
$N_\varphi(w)$	Nevanlinna counting function	33
$N_\varphi(w,r)$	partial counting function	137
$\Omega(z)$	a dilation ratio	167
φ_a	a special (involution) automorphism of B_N	98
$P_a(z)$	projection onto $[a]$	98
$Q_a(z)$	$(I - P_a)(z)$	98
σ	intertwining map in model for iteration	62
$\tau_\varphi(w)$	see page 155	155

Derivatives

$d(\zeta)$	angular derivative	49, 105
D_k	$\frac{\partial}{\partial z_k}$	96
$\psi'(z)$	derivative of an analytic map	96
$J_\psi(z)$	complex Jacobian of an analytic map	96
$\nabla f(z)$	gradient of f	168
$\mathcal{R}f(z)$	radial derivative of f	168

Measures

$dA(z)$	area measure in the disk	12
ν_N	normalized volume measure on the ball	23
σ_N	normalized surface measure on the sphere	22
τ_N	normalized Lebesgue measure on the torus	21

Other

$\|\cdot\|_e$	essential norm	15
$\|\cdot\|_{HS}$	Hilbert–Schmidt norm	145
$\langle f, g \rangle$	inner product on functional Hilbert spaces	10, 12, 13, 14, 23
$\langle z, w \rangle$	Euclidean inner product in C^N	21
$|z|$	Euclidean norm in C^N	21
z^α	monomial	21
$|\alpha|$	order of a multi-index	22
$\alpha!$	generalized factorial for multi-index	22
$\alpha < \beta$	ordering on multi-indices	25
$[\zeta]$	complex line through ζ	26
$\mathrm{Aut}(B_N)$	group of automorphisms of B_N	99
$\mathrm{Aut}(D)$	group of automorphisms of D	46
$d(z, w)$	quasi-metric on $\overline{B_N}$	172
γ_r	see page 135	135

Index

absorption property, 163, 173, 174
$\mathcal{AC}$, 180, 185
adjoint of a composition operator, *see*
    composition operator, adjoint
admissible limit, 103
affine set, 100
  preserved by automorphisms, 100
  preserved by Cayley transform, 115
Ahern, 142
Aleksandrov, 161
algebraically consistent,
    4–7, 18, 29, 150, 182
Alpar, 185
analytic map, 21
angular derivative, 50, 61
  in the ball, 104, 114
  infimum of, 54, 61, 205, 213
  of inner function, 142
Area Formula, 36, 135, 138, 203
area sub-mean-value property, 137
Athreya, 96
automorphism, 46, 63
  as symbol for C_φ, 119, 121, 172
  elliptic, 46, 59, 250
  fixed points of, 46, 115, 249, 256
  hyperbolic, 46, 60, 250, 253
  invariance, 117, 127, 179–181, 185,
    198, 210, 256, 267, 327
  of the ball, 97, 263
  of upper halfplane, 59
  parabolic, 46, 60, 250, 254, 255
Axler, 12, 29, 45

backward shift, 1, 6, 7
Baker, 95
ball, 21
Banach space of analytic functions,
  3
basis, 7, 11, 12, 14, 23, 272, 275,
    284, 288–290, 331
Behan, 62, 96
Bell, 29
Bergman space
  in the ball, 22
  in the disk, 12, 15
  weighted,
    15, 27, 41, 43, 133, 142, 197
Berkson, 332, 333, 342, 345
Beta function, 27
Bieberbach conjecture, x
biholomorphic, 21
binomial series formula, 27
Blaschke
  product,
    32, 33, 45, 161, 284, 307, 353
  sequence, 32, 255, 283
Bloch space, 194
  little, 194
Bochner, 316, 319
Böttcher's functional equation, 96
boundary function, 10
boundary-regular, 177
bounded below, 154
boundedness of C_φ, 6, 117, 118
  on $A_G^2(D)$, 202, 205, 206

381